Quantum field theory

QUANTUM FIELD THEORY

LEWIS H. RYDER

University of Kent at Canterbury

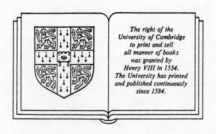

CAMBRIDGE UNIVERSITY PRESS

Cambridge

London New York New Rochelle

Melbourne Sydney

PHYSICS

5 2 5 7 3 1 6 3

Published by the Press Syndicate of the University of Cambridge
The Pitt Building, Trumpington Street, Cambridge CB2 1RP
32 East 57th Street, New York, NY 10022, USA
10 Stamford Road, Oakleigh, Melbourne 3166, Australia

First published 1985

Printed in Great Britain at the University Press, Cambridge

Library of Congress catalogue card number: 84-4183

British Library cataloguing in publication data
Ryder, Lewis H.
Quantum field theory
1.   Quantum field theory
I. Title
530.1'43   QC174.45

ISBN 0 521 23764 5

TM

FOR DANIEL

Yet nature is made better by no mean
But nature makes that mean: so, over that art
Which you say adds to nature, is an art
That nature makes.
William Shakespeare
(W. Tale, iv, 4, 89)

Omnia disce, videbis postea nihil
esse superfluum
(Learn everything, you will find
nothing superfluous.)
Hugh of St Victor

Contents

Preface

This book is designed for those students of elementary particle physics who have no previous knowledge of quantum field theory. It assumes a knowledge of quantum mechanics and special relativity, and so could be read by beginning graduate students, and even advanced third year undergraduates in theoretical physics.

I have tried to keep the treatment as simple as the subject allows, showing most calculations in explicit detail. Reflecting current trends and beliefs, functional methods are used almost throughout the book (though there is a chapter on canonical quantisation), and several chapters are devoted to the study of gauge theories, which at present play such a crucial role in our understanding of elementary particles. While I felt it important to make contact with particle physics, I have avoided straying into particle physics proper. The book is pedagogic rather than encyclopaedic, and many topics are not treated; for example current algebra and PCAC, discrete symmetries, and supersymmetry. Important as these topics are, I felt their omission to be justifiable in an introductory text.

I acknowledge my indebtedness to many people. Professors P.W. Higgs, FRS, and J.C. Taylor, FRS, offered me much valuable advice on early drafts of some chapters, and I have benefited (though doubtless insufficiently) from their deep understanding of field theory. I was lucky to have the opportunity of attending Professor J. Wess's lectures on field theory in 1974, and I thank him and the Deutscher Akademischer Austauschdienst for making that visit to Karlsruhe possible. I am also very grateful to Dr I.T. Drummond, Dr I.J.R. Aitchison, Professor G. Rickayzen and Dr W.A.B. Evans for reading various sections and making helpful suggestions. I wish to thank Miss Mary Watts for making a difficult and unattractive manuscript a handsome typescript, and this with constant good humour and cheerfulness. I also thank Mr Bernard Doolin for drawing the diagrams with such great care. I am grateful to the late Fr Eric Doyle, OFM, for unearthing the quotation by Hugh of St. Victor. I am grateful to the organisers of the first and second UK Theory Institutes in High Energy Physics for the opportunity they provided for stimulating discussions. I wish to record my special thanks to Dr Simon Mitton of Cambridge University Press for his constant encouragement and his indulgence over my failure to meet deadlines. Finally, and most of all, my thanks go to my wife, for her unfailing encouragement and support over a long period.

Canterbury, Kent   Lewis Ryder
August, 1984

1

Introduction: synopsis of particle physics

1.1 Quantum field theory

Quantum field theory has traditionally been a pursuit of particle physicists. In recent years, some condensed matter physicists have also succumbed to its charms, but the rationale adopted in this book is the traditional one: that the reason for studying field theory lies in the hope that it will shed light on the fundamental particles of matter and their interactions. Surely (the argument goes), a structure that incorporates quantum theory – which was so amazingly successful in resolving the many problems of atomic physics in the early part of this century – and field theory – the language in which was cast the equally amazing picture of reality uncovered by Faraday, Maxwell and Hertz – surely, a structure built on these twin foundations should provide some insight into the fundamental nature of matter.

And indeed it has done. Quantum electrodynamics, the first child of this marriage, predicted (to name only one of its successes) the anomalous magnetic moment of the electron correctly to six decimal places; what more could one want of a physical theory? Quantum electrodynamics was formulated in about 1950, many years after quantum mechanics. Planck's original quantum hypothesis (1901), however, was indeed that the electromagnetic *field* be quantised; the quanta we call photons. In the years leading up to 1925, the quantum idea was applied to the *mechanics* of atomic motion, and this resulted in particle–wave duality and the Schrödinger wave equation for electrons. It was only after this that a proper, systematic treatment of the quantised electromagnetic field was devised, thus coming, as it were, full circle back to Planck and completing the quantisation of a major area of classical physics.

Now, in a sense, quantisation blurs the distinction between particles and fields; 'point' particles become fuzzy and subject to a wave equation, and the

1

(electromagnetic) field, classically represented as a continuum, takes on a particlelike nature (the photon). It may then very well be asked: if we have charged particles (electrons, say) interacting with each other through the electromagnetic field, then in view of quantisation, which renders the particle and the field rather similar, is there an *essential* distinction between them? The answer to this question takes us into elementary particle physics. The salient point is that photons are the quanta of the field *which describes the interaction* between the particles of matter. The electrons 'happen to be there' and because they interact (if they did not we would not know they were there!) the electromagnetic field and, therefore, photons *become compulsory*! But this is not all. Muons and protons and all sorts of other charged particles also happen to exist, and to interact in the same way, through the electromagnetic field. The reason for the existence of all these particles is so far unknown, but we may summarise by saying that we have a *spectrum of particle states* $(e, \mu, p, \Sigma, \Omega,$ etc.) and a *field through which these particles interact* – an interaction, in short. This treatment and mode of comprehension of electrodynamics provides the paradigm, we believe, for a complete understanding of particle interactions. The idea is simply to apply the same methods and concepts to the other interactions known in nature. The only other interaction known in classical physics is the gravitational one, so let us first consider that.

1.2   Gravitation

The gravitational field is described by the general theory of relativity. It turns out, however, that the quantisation of this theory is beset by great problems. First, there are mathematical ones. Einstein's field equations are much more complicated than Maxwell's equations, and in fact are non-linear, so consistency with the superposition principle, the mathematical expression of wave–particle duality, which requires the existence of a linear vector space, would seem to be threatened. Second, there are conceptual problems. In Einstein's theory the gravitational field is manifested as a curvature of space–time. In electrodynamics, the field is, as it were, an actor on the space–time stage, whereas in gravity the actor becomes the space–time stage itself. In some sense, then, we are faced with quantisation of space–time; what is the meaning of this? Finally, there are practical problems. Maxwell's equations predict electromagnetic radiation, and this was first observed by Hertz. Quantisation of the field results in the possibility of observing individual photons; these were first seen in the photoelectric effect, in Einstein's classic analysis. Similarly, Einstein's equations for the gravitational field predict gravitational radiation, so there should, in principle, be a possibility of observing individual *gravitons*, quanta of the field. However, although some claims have been made that gravitational radiation has been observed, these are not unanimously accepted, and the observation of *individual gravitons*, a much more difficult enterprise, must be a next-generation problem! The basic reason for this is that gravity is so much weaker than the other forces in nature. In view of this, the particle physicist is justified in ignoring it; and, because of the difficulties mentioned above, is happy to! On the other hand, the methods that have been recently developed for the quantisation of non-abelian gauge fields, relevant for an understanding of the strong and weak nuclear forces, do seem to have relevance to gravity, and these will be briefly described in the book, where appropriate.

1.3   Strong interactions

After electromagnetism and gravity, the remaining interactions in nature are nuclear; the so-called strong and weak nuclear forces. The question is, can these forces be described by a *field*? Yukawa surmised that the strong force between protons and neutrons in a nucleus could, but that the field quantum had to have (unlike the photon) a *finite mass*; this is because the nuclear force has a finite range. Fig. 1.1 shows a Feynman diagram (explained in chapter 6) for the exchange of a virtual field quantum (π^+) between a proton and neutron. The uncertainty principle allows this to happen provided

$$\Delta E \Delta t = (m_\pi c^2)\Delta t \sim \hbar$$

where m_π is the mass of the field quantum (pion) and Δt is the time for which it exists. If the range of the force is r, then we may put $r = c(\Delta t)$, giving, with $r \approx 10^{-15}$ m,

$$m_\pi c^2 \approx \frac{\hbar c}{r} \approx 200 \text{ MeV}.$$

When the π^+ was discovered (in 1947) with a mass of $140 \text{ MeV}/c^2$ and possessing strong nuclear interactions, this was considered a triumph for Yukawa's theory. The view that the pion was the quantum of the strong field, however, began to run into difficulties:

Fig. 1.1. Exchange of a quantum of the strong field (pion) between the proton and neutron.

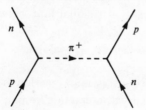

1. At high energies, the proton–neutron force was not well described by pion exchange.
2. The interaction between pions themselves could not (because of parity) be described by pion exchange.
3. With the discovery of strange particles (1950s and early 1960s), and the classification of particles by the group $SU(3)$, pions were found to be only three members of a supermultiplet of eight, the others being K and η – 'ordinary' elementary particles of 'matter'. If field quanta are in essence different from 'matter' quanta, they should surely not appear together in the same supermultiplet.
4. In the quark model (1964), pions are bound states of a quark–antiquark pair, just like all other mesons – their privileged role disappeared completely! (The photon, for example, is most definitely not made of quarks!)

In the sections below we review briefly the elementary particle spectrum and the quark model. But here we note that the quark model, while providing the death-blow to the pion as candidate for quantum of the strong field, also gives a clue to its successor; for the 'real' strong interaction is not that between nucleons, but that between quarks. What provides the interaction between quarks?

There is evidence (see §1.11 below) that the quark possesses a quantum number rather like electric charge except that (i) it has three degrees of freedom – three 'types' of charge, (ii) it is unobservable in the free state – this is to say that individual quarks are not observed, and the reason may be that systems carrying (non-zero values of) this quantum number are forbidden in the free state. The quantum number is known as colour, the degrees of freedom being chosen variously as red, white and blue, red, green and blue, etc. It is believed that colour, like electric charge, gives rise to a quantised field, massless and of spin one, like the photon. By means of this field, quarks interact. The quanta are known as gluons, and the dynamics of the quark–gluon system, quantum chromodynamics (QCD), to mirror quantum electrodynamics (QED). QCD is responsible, for example, for the binding of three quarks into protons and neutrons. No wonder the proton–neutron force is only approximately describable by pion exchange – it must in reality be a rather complicated force, a sort of 'residual' force between the quarks!

1.4   Weak interactions

To complete this preliminary account, we must mention the fourth known interaction in nature; the weak nuclear interaction, responsible for beta decay. In Fermi's original theory, this interaction was a *point* interaction between the four participating particles (in neutron decay $n \rightarrow p + e^- + \bar{v}_e$). In other words, there was no need for a field, because there is no effect to be propagated between one point and another. (It will be recalled that the notion of 'field' in physics was introduced partly to give a more satisfactory account of 'action-at-a-distance'; if there is no action at a distance, there is clearly no need for a field.) Fermi's theory worked extremely well; in fact, with a modification for parity violation, there was for several decades no direct evidence that it was wrong. It was, nevertheless, believed to be wrong, principally because it is non-renormalisable (see chapter 9). One of the recent triumphs in particle physics has been the appearance of a worthy successor to Fermi's theory, which was worked out by Glashow, Weinberg and Salam in the 1960s. In this theory, electromagnetism and the weak force are combined in a non-trivial way. The weak field acquires *quanta* – the W and Z bosons – which are more than 80 times as massive as the proton. In addition, *neutral current reactions* such as $v + p \rightarrow v + p +$ (neutral hadrons) are predicted, as well as a fourth (charmed) species of quark. Observation of all these things has established the Weinberg–Salam theory as the correct theory of weak interactions. Not the least important aspect of this theory is the fact that it is a *unified* theory – an 'electroweak' theory, in the new jargon. Weak and electromagnetic interactions are now unified. A putative 'grand unified theory' (GUT), which unifies the electroweak interaction with the strong interaction (QCD) would seem to be the next obvious thing to look for, but at the time of writing (1983) there is no hard evidence that the forces of nature are grand-unified.

These, then, are the ingredients of contemporary theories of fundamental particles and interactions. In the next sections of this chapter they will be described in greater detail, for the benefit of readers to whom the current scene in high energy physics is not familiar. The aim is to describe, in rather simple terms, the physical considerations which have led up to, and which follow from, the introduction of the quark model. This will provide a motivation for the study of quantum fields, as well as a chance to explain some of the concepts which will be used in the application of field theory to particle physics. It offers no pretence to completeness, but references to more detailed literature are given at the end of the chapter. Readers who know about particle physics are advised to skip to chapter 2. In the remainder of this chapter I shall use one or two concepts and techniques without adequate explanation. This applies particularly to Feynman diagrams. I ask the reader's indulgence to make the best he can of these sections until he meets the explanations later in the book.

Table 1.1. *Classification of fundamental particles*

Hadrons	Baryons	$p, n, \Lambda, \Sigma, \Omega, \Lambda_c, \ldots$, etc. (hundreds)
	Mesons	$\pi, K, \rho, \psi/J, \Upsilon, \ldots$, etc. (hundreds)
Leptons		$e^-, \nu_e, \mu^-, \nu_\mu, \tau^-, [\nu_\tau], \ldots$ (six only?)
Field quanta		Photon, γ
		Weak bosons, $W^\pm, Z^0$
		[Gluons]

Particles in square brackets have not yet been discovered

1.5   Leptonic quantum numbers

There is a basic classification of fundamental particles into those which experience the strong interaction – called *hadrons*, those which do not, called *leptons*, and, thirdly, the quanta of the interaction fields (see above). The hadrons are subdivided into *baryons* which have half-odd integral spin, and *mesons*, with integral spin (in units of $\hbar$ – spin is a purely quantum phenomenon). Table 1.1 gives a summary. Hundreds of baryons and mesons have been found. Five leptons have been found, and for aesthetic reasons it is suspected that the τ lepton, like e and μ, is accompanied by a neutrino ν_τ. Baryons and mesons come with all sorts of spins – as high as $\frac{7}{2}$ for baryons, and 3 for mesons, have been measured, but the leptons all have spin $\frac{1}{2}$. The field quanta all have spin 1. It is believed that the known hadrons are composite states of six types of quark (see below) so it would be nice if there were also six leptons.

In order for the division of spin $\frac{1}{2}$ particles into baryons and leptons to have any significance, transitions between these different types should not occur. For example, a proton should not decay into a positron (the antiparticle of the electron):

$$p \not\to e^+ + \gamma; \qquad (1.1)$$

and indeed this decay has not been observed. What could forbid it? Charge would be conserved, and so would mass-energy and angular momentum. There must be a *conserved quantum[‡] number* – call it *baryon number B* – such that baryons have $B = 1$ and other particles have $B = 0$. Conservation of baryon number then clearly forbids the decay (1.1).[‡‡] The reader might

---

[‡] It is common practice to call conserved quantities 'quantum numbers', but it is not at all obvious that baryon number has anything to do with the quantum theory. If it had, matter would become unstable by (1.1) as $h \to 0$.

[‡‡] It is characteristic of grand unified theories that baryon number is not absolutely conserved. They therefore predict proton instability. Observation of proton decay would count as important evidence that a GUT may be correct.

object that this is not so much an *explanation* as a restatement of the fact that the decay does not occur, in different language. That may well be true, but the language is at least more economical! Alternatively, instead of inventing baryon number, we could invent *lepton number L* – leptons have $L = 1$, everything else has $L = 0$, and conservation of L forbids proton decay. There is a further observation, however. Muons (which are just like electrons, but heavier) do not decay into electrons

$$\mu^- \not\to e^- + \gamma. \tag{1.2}$$

Why not? There must be a conserved quantity *muon number* L_μ; μ^- has $L_\mu = 1$, everything else has $L_\mu = 0$, so that (1.2) does not occur – in other words, the muon is *not* 'just like an electron, but heavier'. Alternatively, we could define 'electron number' L_e, in an obvious way. Now this has a very interesting consequence. The neutron pion decays into an electron and neutrino, or into a muon and neutrino:

$$\pi^0 \to e^- + \bar{\nu}_e$$
$$\pi^0 \to \mu^- + \bar{\nu}_\mu;$$

we label the neutrinos (actually, antineutrinos) with subscripts e and μ for ease of reference. If L_μ is conserved in these decays it follows that $\bar{\nu}_e$ and $\bar{\nu}_\mu$ are *distinct particles*; for since π^0 and e^- have $L_\mu = 0$, so does $\bar{\nu}_e$, but $\bar{\nu}_\mu$ must have $L_\mu = -1$. (This means that ν_μ has $L_\mu = 1$, which is why we call the particles emitted in π^0 decay antineutrinos.)

These simple considerations of stability lead to the conclusion that there are two distinct neutrinos in nature. It has been verified experimentally that ν_e and ν_μ are distinct. A similar argument may hold for the τ lepton, discovered in 1976 in e^+e^- colliding beam experiments. Events of the type

$$e^+e^- \to \mu^\pm e^\pm + \text{'missing energy'}$$

were eventually attributed to the process

$$e^+e^- \to \tau^+\tau^-$$
$$ \!\!\!\!\!\!\!\!\!\! \hookrightarrow \mu^- \bar{\nu}_\mu \nu_\tau$$
$$ \!\!\!\!\!\!\!\!\!\! \longrightarrow e^+ \nu_e \bar{\nu}_\tau$$

necessitating the introduction of a conserved quantity L_τ, and therefore of a third distinct neutrino ν_τ.

1.6   Hadronic quantum numbers

So much for the leptons. Now let us turn to the hadrons. Here also we find it convenient to introduce conserved quantum numbers, but the arguments are a bit more subtle. The original laboratory production of 'strange particles' in 1954 yielded, amongst other things,

$$\pi^- + p \rightarrow \Lambda + K^0 \tag{1.3}$$

with such a large cross section that this reaction must be due to the *strong interaction*. When the Λ and K^0 particles decay, however,

$$\Lambda \rightarrow p + \pi^-; \quad K^0 \rightarrow \pi^+ + \pi^-, \tag{1.4}$$

the lifetimes are so long ($\approx 10^{-10}$ s) that the decays must be due to the *weak interaction*, even though all the particles involved are hadrons. Why? Hadrons decaying into hadrons of a lower mass will surely do so by the strong interaction – unless violation of a conserved quantity is involved. We therefore introduce a quantum number S called 'strangeness', assign $S = 0$ to π and p, $S = -1$ to Λ and $S = +1$ to K^0, and invent the rule, *strong interactions conserve S, weak interactions change it*. Then it will be seen that in reaction (1.3), the algebraic sum of S on both sides of the reaction is zero, so S is conserved, and it is a strong interaction – as found. In the decays (1.4), however, S changes from -1 to 0 or $+1$ to 0, so the decays *cannot* proceed by the strong interaction, and so are weak.

A similar pattern occurs at higher energy. It was first observed in 1975 that neutrinos in a hydrogen bubble chamber reacted to give

$$\nu_\mu + p \rightarrow \mu^- + \Lambda + \pi^+ + \pi^+ + \pi^+ + \pi^-.$$

This cannot be a *single* weak interaction involving leptons, since they are all observed to obey the rule $\Delta S = \Delta Q$, where ΔS and ΔQ are the changes in strangeness and charge on the hadrons; the above reaction has $\Delta S = -\Delta Q$. It was concluded[8,9] that what is involved is a series of reactions:

$$\begin{aligned}
\nu_\mu + p &\rightarrow \mu^- + \Sigma_c^{++} \\
\Sigma_c^{++} &\rightarrow \Lambda_c^+ + \pi^+ \\
\Lambda_c^+ &\rightarrow \Lambda + \pi^+ + \pi^+ + \pi^-
\end{aligned} \tag{1.5}$$

The decay of Σ_c^{++} into Λ_c^+ is a strong one, but the decay of Λ_c^+ into Λ and pions is weak, and necessitates the introduction of a new quantum number 'charm' C; Λ_c^+ has $C = 1$, $S = 0$, and Λ has $C = 0$, $S = -1$, so the decay has $\Delta C = \Delta S = -1$. Σ_c^{++} also has $C = 1$, and strong interactions conserve C.

1.7   Resonances

During the 1950s and 1960s a 'population explosion' was observed amongst strongly interacting particles. The discovery of strange particles accounted for some of this, but a more significant contribution came from the discovery of *resonances*, of which hundreds are now known. To give only one example, the first one to be discovered (by Fermi, in 1952) was found as a large peak in the cross section for pion–nucleon scattering,

$$\pi^{\pm} + p \rightarrow \pi^{\pm} + p.$$

A sketch of the cross section is shown in Fig. 1.2. The sharp peak in σ is interpreted, following the usual Breit–Wigner theory, as a resonance, now called Δ, so that the reaction is effectively

$$\pi^{+} + p \rightarrow \Delta^{++} \rightarrow \pi^{+} + p.$$

The width $\Gamma \approx 110\,\text{MeV}$ may be measured directly from the curve, giving the lifetime from the uncertainty principle:

$$\tau \sim \hbar/\Gamma \approx 10^{-23}\,\text{s}.$$

This time is characteristic of *strong interactions*, by which resonances decay. Resonances are states with well-defined spin and parity, mass, charge, etc., and are to be treated on an equal footing with the other elementary particles. The 'stable' particles, i.e. those decaying by weak or electromagnetic interactions, are simply the lightest members of a set of particles with given values of isospin (see below), strangeness, charm, etc. In order for 'stable' particles to decay, these quantum numbers must change, therefore they decay only through the weak (or electromagnetic) interaction. Most particles, in fact, are resonances – the reader has no doubt seen lists of them in books on particle physics.

Fig. 1.2. Cross section for $\pi^{+}p$ elastic scattering, plotted against the pion laboratory energy. The peak, of width Γ, is interpreted as a resonance Δ.

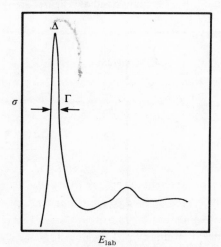

1.8   The quark model

To try to make sense of all these particles, perform the following exercise. Select a particular spin-parity J^P, and plot the charge Q and strangeness S of all particles (and resonances) with that value of J^P. As an example, the particles with $J^P = \frac{3}{2}^+$, are shown in Fig. 1.3. There are ten particles, and they form a regular pattern. This happens with *all* values of J^P (as long as all the particles have been found); there is always a regular pattern, though it more commonly contains eight particles rather than ten. Why is this?

It is because the particles are composite states of more fundamental entities known as 'quarks'. There are at present believed to be six quarks distinguished by Q, S, C and 'truth' and 'beauty' (or 'top' and 'bottom'). They are denoted u, d, s, c, t, b. The quantum numbers of u, d and s are given in Fig. 1.4. According to the quark model, baryons are bound states of three quarks (qqq), and mesons are quark–antiquark $(q\bar{q})$ states; quarks have spin $\frac{1}{2}$. We can now see how a supermultiplet of ten baryons may arise. Baryons are made of three identical fermions, so the possible states may be classified according to their symmetry under interchange of quark labels. Altogether there are $3^3 = 27$ states. One of these is totally antisymmetric:

$$uds + dsu + sud - usd - sdu - dus. \tag{1.6}$$

(Here of course the letters stand for the wave functions and we have not normalised the state.) There are ten totally symmetric states, which are

Fig. 1.3. Particles with $J^P = \frac{3}{2}^+$. The right-hand axis figures are the masses in MeV/c^2.

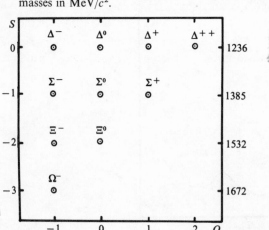

$$
\left.
\begin{array}{l}
uuu \\
ddd \\
sss \\
uud + udu + duu \\
uus + usu + suu \\
udd + ddu + dud \\
uss + ssu + sus \\
dds + dsd + sdd \\
dss + ssd + sds \\
uds + dsu + sud + usd + sdu + dus.
\end{array}
\right\} \qquad (1.7)
$$

This accounts for 11 of the 27 states. The remaining 16 split into two sets of 8, each with mixed symmetry properties – for example, symmetric on interchange of the first two labels, but antisymmetric on interchange of the last two. The 3-quark states will then manifest themselves as distinct supermultiplets of 10, 8, 8 and 1. The $\{10\}$ supermultiplet is identical with that in Fig. 1.3, so we identify $\Delta^{++} = (uuu)$, $\Sigma^0 = uds$, appropriately symmetrised, and so on. The proton and neutron (with $p = (uud)$, $n = (udd)$), are two states in an $\{8\}$ supermultiplet – for further details the reader is referred to the literature.[2–5] Similarly, there are $3^2 = 9$ meson states. One of these $(u\bar{u} + d\bar{d} + s\bar{s})$ is invariant under quark transformations, and the other 8 give the known pseudoscalar $(J^P = 0^-)$ and vector $(J^P = 1^-)$ mesons, with the quark spins 0 and 1, and no additional orbital angular momentum between the quarks. Addition of orbital angular momentum yields higher spin states – this is also true for baryons.

All the states predicted by the quark model (at least with the quarks u, d and s) have been found, and no others. When the fourth quark c is taken into account, many more states are expected. A good number of these have already been found, but not all of them, presumably because the c quark is heavier than u, d and s, so the particles are more difficult to produce.

Fig. 1.4. The u, d and s quarks.

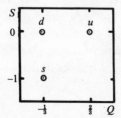

In summary, there is very good evidence that baryons and mesons are composite states of quarks. Almost all the known particles can be accounted for by postulating four quarks, but there is evidence[10],[11] for a fifth, and, if one believes that quarks come in pairs, for a sixth quark.

Particle decays are then explicable simply as quark decays. For example, neutron decay,

$$n \to p + e^- + \bar{v}_e,$$

is due to d decay,

$$d \to u + e^- + \bar{v}_e, \tag{1.8}$$

and Λ (leptonic) decay ($\Lambda = (uds)$),

$$\Lambda \to p + e^- + \bar{v}_e,$$

is due to s decay,

$$s \to u + e^- + \bar{v}_e. \tag{1.9}$$

From this we may deduce that u is the lightest quark, but since Λ is rather heavier than n, then presumably s is rather heavier than d. This explains previously observed selection rules. The rule $\Delta S = \Delta Q$ (Q is the charge on the hadron) explains the decay and non-decay of Σ^- and Σ^+:

$$\Sigma^- \to n + e^- + \bar{v}_e, \quad \Sigma^+ \not\to n + e^+ + v_e,$$

but these are much more simply explained by noting that $\Sigma^- = (dds)$, $\Sigma^+ = (uus)$, so decay of a *single quark* explains why Σ^- decays and Σ^+ does not (that is, does not decay into the particles above). Similarly, $\Xi^-(= (ssd))$ does not decay directly into a neutron:

$$\Xi^- \not\to n + e^- + \bar{v}_e,$$

and this indeed will not occur if baryon decays are due to single quark decays; the selection rule $|\Delta S| = 1$ was previously invoked to explain it.

1.9   SU(2), SU(3), SU(4), ...

The particles Δ (Fig. 1.3) have the quark content

$$\Delta^- = (ddd), \quad \Delta^0 = (udd), \quad \Delta^+ = (uud), \quad \Delta^{++} = (uuu)$$

(we ignore questions of symmetrisation and normalisation – they are not relevant here). The masses of these particles are very close: $m(\Delta^{++}) - m(\Delta^-) \approx 8 \text{ MeV}/c^2$; so, assuming the binding energy of each resonance is the same, we deduce that the u and d quarks are almost degenerate. There is a corresponding symmetry on mixing the u and d quarks. Putting

$$\psi = \begin{pmatrix} u \\ d \end{pmatrix}$$

if we let

$$\psi \to M\psi \tag{1.10}$$

where M is a 2×2 matrix, then in the limit of exact degeneracy this induces a mixing of the four Δ particles, which, in the absence of electromagnetism, is *unobservable*, and therefore a symmetry of the system. For mathematical simplicity the transformations (1.10) must form a group. This means that M must be either unitary or orthogonal (hermitian matrices do not obey the group law). Let us suppose it is unitary, then the group is $U(2)$, the group of unitary matrices in two (complex) dimensions. We then expect the hadron spectrum to show a $U(2)$ symmetry. Actually, it is conventional to consider simply $SU(2)$, which corresponds to M having unit determinant – this can always be done by taking out a phase factor, which corresponds to baryon number. A 2×2 unitary matrix with unit determinant can be written in the form

$$M = \exp\left(i\frac{\vec{\tau}}{2}\cdot\vec{\theta}\right) \tag{1.11}$$

where $\vec{\theta} = (\theta_1, \theta_2, \theta_3)$ are three arbitrary parameters and $\vec{\tau} = (\tau_1, \tau_2, \tau_3)$ are the traceless matrices

$$\tau_1 = \begin{pmatrix} 0 & 1 \\ 1 & 0 \end{pmatrix}, \quad \tau_2 = \begin{pmatrix} 0 & -i \\ i & 0 \end{pmatrix}, \quad \tau_3 = \begin{pmatrix} 1 & 0 \\ 0 & -1 \end{pmatrix} \tag{1.12}$$

known as the Pauli matrices. They obey the commutation relations

$$\frac{\tau_1}{2}\frac{\tau_2}{2} - \frac{\tau_2}{2}\frac{\tau_1}{2} \equiv \left[\frac{\tau_1}{2},\frac{\tau_2}{2}\right] = i\frac{\tau_3}{2} \tag{1.13}$$

and cyclic permutations. These are the same as the commutation relations for angular momentum $\vec{J} = \vec{r} \times \vec{p}$

$$[J_1, J_2] = iJ_3 \quad \text{and cyclic perms}$$

(where we have put $\hbar = 1$) showing that $SU(2)$ symmetry is a rotation symmetry, and the conserved quantity is a *vector* quantity like angular momentum – it is, in fact, *isospin*. This also shows that the group $SU(2)$ is the same (apart from topological properties) in the large as the rotation group $O(3)$, the group of orthogonal matrices in three (real) dimensions. This is a known mathematical equivalence. The angles $\vec{\theta}$ in (1.11) are the angles of rotation in 'isospin space'.

Quark degeneracy then leads to degenerate multiplets of hadrons, and, from the above reasoning, it follows that the multiplets act as bases for irreducible representations of the group $SU(2)$. This group is known to have 1, 2, 3, 4,... dimensional representations, and in fact the particles Ω, Ξ, Σ and Δ in Fig. 1.3 are bases for these first four representations.

These ideas, can, of course, be generalised to include more quarks. In the (hypothetical) case in which the s quark is degenerate with the u and d quarks, we expect invariance under

$$\psi \to M\psi; \quad \psi = \begin{pmatrix} u \\ d \\ s \end{pmatrix}$$

where M is now a 3×3 unitary matrix, and these matrices generate the group $SU(3)$. The dimensionalities of the irreducible representations of $SU(3)$ are given by[12-15]

$$N = \tfrac{1}{2}(p + 1)(q + 1)(p + q + 2) \tag{1.14}$$

where p and q are positive integers or zero; $p = 3$, $q = 0$ gives $N = 10$, which is the representation of Fig. 1.3; $p = q = 1$ gives $N = 8$, the famous 'octet' representation, and $p = 1$, $q = 0$ and $p = 0$, $q = 1$, give the 3-dimensional representation for quarks and antiquarks respectively. It will be recalled that the irreducible representation of Fig. 1.3 was obtained from considerations of symmetry, and it may be wondered what the connection is between this and the theory of group representations. The answer is that representations of $SU(n)$ can be obtained by writing tensors with p upper and q lower indices, symmetrised or antisymmetrised – there is thus a close connection between representations of $SU(n)$ and the symmetric group. The device used for realising the representations in this way is called a Young tableau.

A given representation of $SU(3)$ contains, in general, several representations of $SU(2)$; in other words, an $SU(3)$ supermultiplet contains several isospin multiplets, each with differing strangeness S. This is clear from Figs. 1.3 and 1.4. In Figs. 1.5 and 1.6 the $(p, q) = (1, 1)$, i.e. 8-dimensional representation of $SU(3)$ is drawn, showing the pseudoscalar mesons. The

Fig. 1.5. The $J^P = 0^-$ pseudoscalar octet of $SU(3)$. The masses of the (neutral) particles, in MeV/c^2, are $m(\pi) = 135$, $m(K) = 498$, $m(\eta) = 549$.

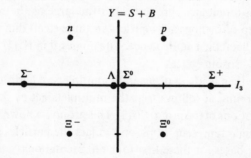

axes are hypercharge Y defined by

$$Y = B + S \tag{1.15}$$

and I_3, the third component of isospin, related to electric charge Q by

$$Q = I_3 + Y/2 \tag{1.16}$$

This is called the Gell-Mann–Nishijima relation.

$SU(3)$ symmetry is exact only in the limit in which all the particles in a supermultiplet are degenerate. From the masses in Figs. 1.3 and 1.5 it is seen that the symmetry is only very approximate.

Now let us include the fourth quark, c. With

$$\psi = \begin{pmatrix} u \\ d \\ s \\ c \end{pmatrix} \tag{1.17}$$

we are clearly now working with a – very approximate! – $SU(4)$ symmetry.

Fig. 1.6. The $J^P = \frac{1}{2}^+$ octet of baryons. The masses of the neutral particles, in MeV/c^2, are $m(N) = 939, m(\Sigma) = 1192, m(\Lambda) = 1115$, $m(\Xi) = 1315$.

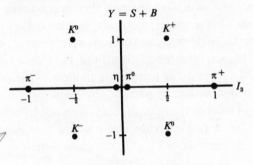

Fig. 1.7. The $\frac{1}{2}^+$ baryons in a 20-dimensional representation of $SU(4)$. The particles already found are indicated on the right. The $SU(3)$ content is a $\{3\}$, with $C = 2$, $\{6\}$ and $\{3\}$ with $C = 1$, and $\{8\}$ with $C = 0$, the uncharmed octet of Fig. 1.6.

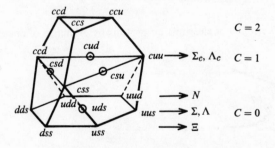

As before, 3-quark and quark–antiquark states fall into specific representations of $SU(4)$, each of which contains a number of representations of
$SU(3)$ with differing charm C. The Gell-Mann–Nishijima relation now
becomes

$$Q = I_3 + \frac{B+S+C}{2} \tag{1.18}$$

The 20-dimensional representation of $SU(4)$ appropriate to the spin $\frac{1}{2}$
baryons is shown in Fig. 1.7. The isosinglet (cud) state is identified with the
Λ_c^+ particle mentioned in (1.5) above. It has charm $C = 1$, as does the
particle Σ_c, with isospin 1, corresponding to the charge states
$\Sigma_c^0, \Sigma_c^+, \Sigma_c^{++}$. The other spin $\frac{1}{2}$ baryons with $C = 1$ have not yet been found,
and no baryons with $C = 2$ have been found. Λ_c has a mass of 2262 MeV/c^2
and Σ_c a mass of 2428 MeV/c^2. These are considerably heavier than the
$SU(3)$ octet with $C = 0$, indicating that the c quark must be much heavier
than u, d and s. $SU(4)$ symmetry is correspondingly a rather inexact
symmetry.

1.10   Dynamical evidence for quarks

From the classification of particle states, there is good evidence for
quarks. This is, however, not the only type of evidence there is. Experiments
on deep inelastic scattering of electrons and protons have revealed evidence
that the electrons scatter, not from the whole proton, but from a pointlike
constituent of it, called a *parton*. The temptation to identify partons with
quarks is one which is not resisted by particle physicists.

It will be recalled that measurements of *structure* are made by measuring
form factors. For the scattering of point electrons on protons, by the
one-photon-exchange mechanism of Fig. 1.8, the cross-section is

$$\frac{d\sigma}{d\Omega} = \left(\frac{d\sigma}{d\Omega}\right)_{\text{point}} \left(\frac{G_E^2 + \frac{q^2}{4M^2} G_M^2}{1 + \frac{q^2}{4M^2}} + \frac{q^2}{4M^2} 2G_M^2 \tan^2 \theta/2 \right) \tag{1.19}$$

Fig. 1.8. Scattering of electrons on protons by exchange of one
photon.

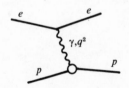

where $(d\sigma/d\Omega)_{point}$ is given by the Mott formula appropriate for the scattering of electrons off pointlike protons, $G_E(q^2)$ and $G_M(q^2)$ are the electric and magnetic form factors, q^2 is the square of the photon 4-momentum q (defined in chapter 2), θ is the angle through which the electron is scattered in the laboratory frame, and M is the proton mass. Measurement of $d\sigma/d\Omega$ and comparison with $(d\sigma/d\Omega)_{point}$ at varying θ allows a determination of the form factors $G_E(q^2)$ and $G_M(q^2)$, which turn out to have a dipole form

$$G(q^2) = \left(\frac{1}{1 + q^2/M_q^2}\right)^2 \tag{1.20}$$

with $M_q^2 = 0.71 \ (\text{GeV}/c^2)^2$. This is the Fourier transform of

$$\rho(r) = \int e^{-i\mathbf{q}\cdot\mathbf{r}} G(q^2) d^3\mathbf{q} \tag{1.21}$$

(a non-relativistic formula; $\mathbf{q}$ is a 3-momentum), which turns out to be an exponential charge distribution

$$\rho(r) \approx \exp(-M_q r). \tag{1.22}$$

This is, therefore, the structure of the proton revealed in this experiment. The interesting thing is that it does not become singular as $r \to 0$, indicating that there is *no hard core* to the proton. (This is not a trivial result: a simple pole instead of (1.20) would yield a Yukawa potential $\rho(r) \approx [\exp(-M_q r)]/r$ instead of (1.22), which *is* singular at $r = 0$.) So the proton is not like a plum, with a stone in the middle. It could be like a jelly, or it could be like a strawberry, with seeds scattered throughout, but *no accumulation* of them at the centre. To decide which, we turn to *inelastic* scattering, in which the final state is an arbitrary hadronic state:

$$e^- + p \to e^- + (\text{hadronic state}).$$

The Feynman diagram for this is shown in Fig. 1.9. M^* is the invariant mass of the final hadronic state (defined in chapter 2). One can show that

$$q^2 = M^2 - M^{*2} + 2Mv \tag{1.23}$$

Fig. 1.9. Inelastic electron–proton scattering. The initial and final electron energies are E_0 and E.

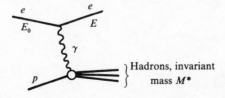

where q is the photon 4-momentum, v is the energy lost by the electron, $v = E_0 - E$, and M is the proton mass. Also we have

$$q^2 = 4EE_0 \sin^2(\theta/2).$$

In the elastic case, $M^* = M$ and $q^2 = 2Mv$; q^2 and v are therefore related – in experimental terms, θ and the energy loss $E_0 - E$ are related – and the form factors are functions of q^2 only. In the inelastic case, since it is only the final electron and not the final hadronic state which is observed, M^* is effectively an independent variable, so q^2 and v are independent, and the form factors $F(q^2)$ have to be replaced by so-called *structure factors* $W(q^2, v)$, functions of both q^2 and v. The differential scattering cross section is given by

$$\frac{d^2\sigma}{d\Omega\, dv} = \left(\frac{d\sigma}{d\Omega}\right)_{point} [W_2(v, q^2) + 2W_1(v, q^2)\tan^2(\theta/2)].$$

Measurements of this cross section then enable the structure functions $W_{1,2}$ to be found.

It is then possible to plot W_1 and vW_2 as functions of the dimensionless variable

$$\omega = \frac{2Mv}{q^2}.$$

What is found, as shown in Fig. 1.10, is that at high v and q^2, but ω finite, W_1 and vW_2 turn out to be functions of ω only, not of q^2 and v separately. This phenomenon is known as *Bjorken scaling*.

Scaling is interpreted as indicating that the scattering takes place off point-like constituents of the proton, called partons. For, if this is the case, the scattering of e^- from an individual parton of mass m must be elastic, and then $q^2 = 2mv$ from (1.23). It is then relatively straightforward to show that the structure factors are functions of ω only. Accepting this interpretation,

Fig. 1.10. vW_2 as a function of q^2 at $\omega = 4$ (after D.H. Perkins, *Introduction to High Energy Physics*, 2nd edition, Addison-Wesley, 1982).

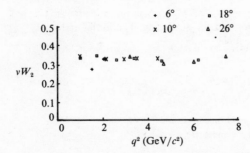

the parton spin may be found, for it may be shown that

$$\frac{(2v^2/q^2)W_2 - W_1}{2W_1} \xrightarrow[q^2,v \to \infty]{} 0 \text{ if parton spin} = \tfrac{1}{2},$$
$$\xrightarrow{\hspace{2cm}} \infty \text{ if parton spin} = 0.$$

Experimentally, this ratio tends to zero, so the partons have spin $\tfrac{1}{2}$; and are thus identified with quarks. These remarkable experiments are dynamical evidence for the existence of quarks.

1.11   Colour

There is a rather severe problem, which the reader may already have noticed, with quark statistics. Consider the hadrons belonging to the $\{10\}$ representation of $SU(3)$, in Fig. 1.3. It will be recalled that the three quarks are in a *symmetric* state, under interchange of the $SU(3)$ labels, Q and S. The complete wave function of one of these hadrons, say the Δ^{++}, is

$$\psi_{\Delta^{++}} = \psi_{\text{spin}}\psi_{SU(3)}\psi_{\text{space}};$$

it is the product of wave functions in the space, spin, and 'internal' $SU(3)$ variables. We have noted that the $SU(3)$ part is symmetric – in this case it is simply uuu. The three quarks are in an $S = \tfrac{3}{2}$, $L = 0$ state, so the three parallel spins clearly make ψ_{spin} symmetric and the symmetry of ψ_{space} is $(-1)^L$ which is symmetric when $L = 0$. The overall 3-quark wave function is therefore *symmetric* under exchange of all three labels; but this is forbidden by Fermi–Dirac statistics. So either quarks do not obey Fermi statistics, or the above argument is wrong – or, rather, incomplete. This is the problem of quark statistics.

Gell-Mann showed how to resolve it by introducing another degree of freedom for quarks, called *colour*; each quark carries, besides the 'labels' of spin and $Q, S, C, \ldots$, an *additional* label with three values R, W, B (red, white and blue), and the Δ wave function becomes

$$\psi_{\Delta} = \psi_{\text{spin}}\psi_{Q,S}\psi_{\text{space}}\psi_{\text{colour}} \tag{1.24}$$

and we simply stipulate that the first three functions are symmetric, but ψ_{colour} is antisymmetric under exchange of colour labels. We may say that the number of quarks is no longer 4 (or 5, 6, ...), but this number multiplied by 3. For example, in the 4-quark model, we now have four *flavours* (the generic name for $Q, S, C, \ldots$) and three colours, so the quarks may be written as

$$\begin{pmatrix} u_R & u_W & u_B \\ d_R & d_W & d_B \\ s_R & s_W & s_B \\ c_R & c_W & c_B \end{pmatrix}. \tag{1.25}$$

Just as the transformation group for flavours is $SU(4)_{\text{flavour}}$, we also postulate that the colour group is $SU(3)_{\text{colour}}$, so we have two types of symmetry, acting on the two types of quark label:

$$
\begin{array}{c}
\xrightarrow{\hspace{3cm}} SU(3)_{\text{colour}} \\[4pt]
\left.
\begin{array}{ccc}
u_R & u_W & u_B \\
d_R & d_W & d_B \\
s_R & s_W & s_B \\
c_R & c_W & c_B
\end{array}
\right\downarrow \\[4pt]
SU(4)_{\text{flavour}}
\end{array}
$$

Corresponding to the totally antisymmetric wave function (1.6) in three flavour variables, we may write the totally antisymmetric colour wave function to go in (1.24):

$$\psi_{\text{colour}} = RWB + WBR + BRW - RBW - BWR - WRB. \quad (1.26)$$

Since this totally antisymmetric function transforms as a *singlet* under $SU(3)_{\text{colour}}$, we have the interesting consequence that whereas quarks carry colour – transform according to a non-trivial representation of $SU(3)_{\text{colour}}$ – baryons do not transform under $SU(3)_{\text{colour}}$, so are colourless – as observed! Similar considerations follow for mesons, so we conclude that *hadrons are colourless.* This may have something to do with quark confinement – it may be that it is *colour* which is confined, so that particles with colour do not appear in the free state, but those with no colour do appear.

Actually, the introduction of colour in this way has a historical parallel, which the student may find reassuring. Consider the Bohr orbits of the two electrons in the helium atom. They are identical, and we may wonder if the exclusion principle, according to which the electrons must be in different states (or, equivalently, that the two-particle wave function must be antisymmetric), is violated. If no other variables are involved, then it is, but if we introduce *electron spin*, we save the Pauli principle if the two electrons are in an antisymmetric spin state, which corresponds to total spin zero. Electron spin in the helium problem plays the role of colour in the quark problem.

The introduction of colour, therefore, solves the problem of quark statistics. But is there any *independent* evidence for colour? In fact there are two pieces of evidence, which we now consider.

(i) $\pi^0 \to \gamma\gamma$ decay

Assuming that π^0 is made of quarks, $\pi^0 = (1/\sqrt{2})(\bar{u}u - \bar{d}d)$, the decay rate of π^0 into two photons is exactly calculable. The amplitude for the decay is determined by the diagram of Fig. 1.11. The amplitude is clearly additive for all the different types of quark (i.e. colours) which can contribute, and is therefore proportional to N_c, the number of colours. Calculation gives for the pion width

$$\Gamma(\pi^0 \to \gamma\gamma) = 7.87(N_c/3)^2 \text{ eV},$$

to be compared with the experimental value

$$\Gamma = 7.95 \pm 0.05 \text{ eV}.$$

We deduce that $N_c = 3$.

(ii) e^+e^- annihilation experiments

These spectacular experiments, using colliding beams, have been carried out at Stanford and elsewhere for a number of years. The electrons and positrons annihilate into virtual photons, which transform into e^+e^-, $\mu^+\mu^-, \tau^+\tau^-$, or into hadrons of all types. The production of hadrons must be due to

$$e^+e^- \to q\bar{q} \to \text{hadrons};\qquad\qquad(1.27)$$

quarks are produced first, they then find other quarks or antiquarks to combine with, so that the final products are hadrons. This is represented in Fig. 1.12. Actually, there is some evidence that the hadrons are produced in this way, for we should expect, on this mechanism, that they appear in *jets* in pairs of opposite directions; and there is substantial evidence for these jets.[20]

Fig. 1.11. Feynman diagram for $\pi^0 \to 2\gamma$ decay, through quarks.

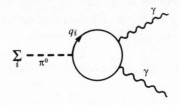

Fig. 1.12. e^+e^- annihilation into hadrons.

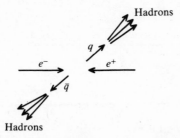

Now if quarks are point particles, just like e^- and μ^-, then because the process is electromagnetic, the amplitude only depends on the quark *charge*, so we expect for the ratio of cross sections

$$\frac{\sigma(e^+e^- \to q\bar{q})}{\sigma(e^+e^- \to \mu^+\mu^-)} = N_c Q_q^2$$

where Q_q is the quark charge and N_c the number of quark colours. Putting $N_c = 3$, the ratio of the cross section for producing hadrons to that for producing muons is then

$$R = \frac{\sigma(e^+e^- \to \text{hadrons})}{\sigma(e^+e^- \to \mu^+\mu^-)} = 3\sum Q_q^2$$
$$= 3(\tfrac{4}{9} + \tfrac{1}{9} + \tfrac{1}{9}) \quad \text{for 3 flavours}$$
$$= 2, \tag{1.28}$$

where we have assumed that the experiments are done at an energy low enough that only the first three flavours, u, d and s, are produced. As the energy increases, R will increase; when the c and b quarks with charges $\tfrac{2}{3}$ and $-\tfrac{1}{3}$ are produced, for example, R will increase to $3\tfrac{2}{3}$.

Fig. 1.13 shows a schematic plot of R as a function of centre of mass energy of e^+e^-. It is seen that R increases from about 2, at 2.5 GeV, to something over 4, at 4.7 GeV, the charm threshold. Above 20 GeV, R settles down at about $\tfrac{11}{3}$, indicating both the existence of u, d, s, c and b quarks, *and colour*. The peaks in Fig. 1.13 correspond to the spin 1 meson states $q\bar{q}$.

1.12   QCD

Finally, it is believed that colour is like electric charge, in that it gives rise to a dynamic field with a massless quantum analogous to the photon, called the *gluon*. So corresponding to electron–electron interaction

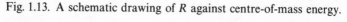

Fig. 1.13. A schematic drawing of R against centre-of-mass energy.

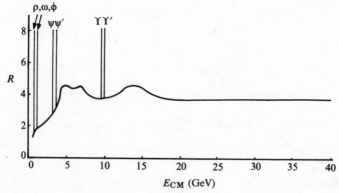

via photon exchange (Fig. 1.14) we have quark–quark interaction via gluon exchange. The difference is that photons are not themselves charged, but, because colour is '3-dimensional' quantity, and the interaction is an exchange interaction, the gluons themselves carry colour. There are eight of them, belonging to the {8} dimensional representation of $SU(3)_{colour}$.

Thus colour provides a mechanism for binding quarks together into hadrons, and the dynamics of this interaction is quantum chromodynamics (QCD). It will be shown in chapter 3 that electrodynamics is a gauge theory, therefore so is chromodynamics. The symmetry group of QED is $U(1)$, an abelian group, whereas that of QCD is $SU(3)$, which is non-abelian. The quantisation of non-Abelian gauge theories will be treated in chapter 7.

There is some evidence that gluons exist, for, just as accelerating electric charges radiate photons as bremsstrahlung, it is expected that quarks may radiate gluons in the same way, and therefore, at high energies, the quark produced in the e^+e^- experiments might emit gluons, as in Fig. 1.15. These gluons, will, like the quarks, somehow get converted into a jet of hadrons, so we will expect some *3-jet events*, as well as the 2-jet events discussed previously. There is some evidence[20] for these 3-jet events, which is interpreted as evidence for gluons.

Fig. 1.14. Electron–electron interaction due to virtual photon exchange, and quark–quark interaction due to virtual gluon exchange.

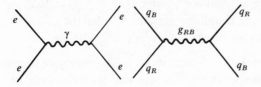

Fig. 1.15. Three jets of hadrons produced by quark–gluon bremsstrahlung in electron–positron annihilation experiments.

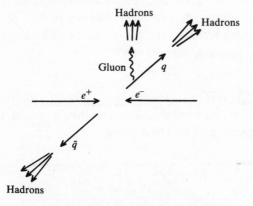

Why do gluons not appear in the free state? For that matter, why do quarks not appear? This, of course, is the famous and most intractable difficulty with the quark model, that quarks have never been found, despite intensive and extensive searches. It is believed there must exist a mechanism which confines quarks into (qqq) or $(q\bar{q})$ combinations – i.e. into hadrons; this is known as *quark confinement*. As mentioned above, it may be that what is really confined is *colour*; this would have the consequence that gluons, having colour, would also be confined, and so answer the question why they are not found. However, there is not yet in existence a completely satisfactory theory of colour confinement.

1.13   Weak interactions

This is not the best place to enlarge on weak interactions, which are the subject of chapter 8. Suffice it here to mention the Cabibbo hypothesis.[2-5] As was seen above, both the s and d quarks decay into the u quark (see (1.8) and (1.9)),

$$s \to u + e^- + \bar{v}_e; \quad d \to u + e^- + \bar{v}_e,$$

and this results in the strangeness-changing (with $|\Delta S| = 1$, $\Delta S = \Delta Q$) and strangeness-conserving ($\Delta S = 0$) decays of hadrons, for example

$$\pi^+ \to \pi^0 + e^+ + v_e, \quad K^+ \to \pi^0 + e^+ + v_e,$$

as well as n and Λ decay mentioned above. However, these decays are not of the same strength; the coupling constant, $G^{(1)}$, for the $|\Delta S| = 1$ decays is about 0.1 times that for the $\Delta S = 0$ decays, $G^{(0)}$. Putting

$$G^{(1)} = G \sin \theta_C$$
$$G^{(0)} = G \cos \theta_C$$

where θ_C, the *Cabibbo angle*, is found to be 0.247 ± 0.008, it turns out that G is equal in magnitude to the coupling constant for muon decay $\mu^- \to v_\mu + e^- + \bar{v}_\varepsilon$ – this is the non-trivial content of the Cabibbo hypothesis. A neat way of describing the leptonic decay mode of hadrons is, then, to say that the transition takes place between the u quark and the 'Cabibbo-rotated' d quark, denoted d_C;

$$d_C = \cos \theta_C d + \sin \theta_C s. \tag{1.29}$$

In the Weinberg–Salam theory, to be described in chapter 8, the particles u and d_C (actually their left-handed parts) are then assigned to a 2-dimensional representation of an $SU(2)$ group

$$\psi_L = \begin{pmatrix} u \\ d_C \end{pmatrix}_L \tag{1.30}$$

which generates so-called 'weak isospin'. It turns out that there are very interesting connections between this type of structure and the existence of the fourth, charmed, quark c. These matters will be dealt with in the proper place.

This concludes the introductory survey of particle physics. Its purpose was to introduce some of the more important concepts, to provide a motivation for the study of quantum fields. We begin this study, in the next chapter, with an examination of the wave equations for particles with spin 0 (the simplest case, though unphysical), $\frac{1}{2}$ (quarks and leptons) and 1 (photon, gluons and intermediate vector bosons).

Guide to further reading

An interesting, though simple, account of the role of fields in physics is given in
(1) A. Einstein & L. Infeld, *The Evolution of Physics*, Cambridge University Press, 1971.

Introductory accounts of particle physics appear in
(2) D.H. Perkins, *Introduction to High Energy Physics,* 2nd edn., Addison-Wesley, 1982.
(3) E. Segrè, *Nuclei and Particles*, 2nd edn., W.A. Benjamin, 1977.
(4) L.J. Tassie,*The Physics of Elementary Particles*, Longman, 1973.
(5) L.H. Ryder, *Elementary Particles and Symmetries*, Gordon & Breach, 1975.

The discovery of the τ lepton is described in
(6) Lectures of H.L. Lynch in M. Lévy *et al.* (eds.) *Quarks and Leptons* (Cargèse 1979), Plenum Press, 1980.
(7) Lectures of G. Wolf in M. Baldo Ceolin (ed.), *Weak Interactions* (Enrico Fermi School, Varenna, 1977), North-Holland Publishing Company, 1979.

The physics of charmed particles is reviewed in
(8) Lectures of F. Muller in M. Lévy *et al.* (eds.), ref. (6).
(9) Lectures of F. Muller in A. Zichichi (ed.), *The New Aspects of Subnuclear Physics* (Erice 1980) Plenum Press, 1981. See also ref. (7).

Recent reviews of the world of six quarks and six leptons are
(10) Lectures of M.K. Gallard & L. Maiani in M. Lévy *et al.* (eds.), ref. (6)
(11) Lectures of H. Fritzsch in M. Baldo Ceolin (ed.), ref. (7).

Accounts of group theory and its application to particle physics are
(12) R.E. Behrends *et al.*, *Reviews of Modern Physics*, **34**, 1 (1962).
(13) J.J. de Swart, *Reviews of Modern Physics*, **34**, 916 (1963).
(14) Lectures of S.M. Berman in A. Zichichi (ed.), *Symmetries in Elementary Particle Physics*, Academic Press, 1965.
(15) B.G. Wybourne, *Classical Groups for Physicists*, John Wiley & Sons, 1974.
(16) F.E. Close, *An Introduction to Quarks and Partons*, Academic Press, 1979.

Inelastic scattering experiments and the parton model are discussed by Close, ref. (16), and
(17) R.P. Feynman, *Photon–Hadron Interactions*, W.A. Benjamin, 1972.

Simplified accounts are given in refs. (2–5). Evidence for colour is given by Close, ref. (16), and

(18) M. Gell-Mann in P. Urban (ed.), *Elementary Particle Physics*, Springer-Verlag, 1972.

(19) S.L. Adler in S. Deser, M. Grisaru & H. Pendleton (eds.), *Lectures on Elementary Particles and Quantum Field Theory*, Vol. I (1970 Brandeis Summer Institute) MIT Press, 1970.

Jets are discussed in

(20) Lectures of H.L. Lynch, J. Ellis & C. Sachrajda, and H. Bøggild in M. Lévy *et al.* (eds.), ref. (6).

The following books contain good accounts of the subjects in this chapter:

(21) L.B. Okun, *Leptons and Quarks*, North-Holland, 1982.

(22) F. Halzen & A.D. Martin, *Quarks and Leptons*, Wiley, 1984.

2
Single-particle relativistic wave equations

2.1   Relativistic notation

Any theory of the fundamental nature of matter must, of course, be consistent with relativity, as well as with quantum theory. We therefore begin by establishing a notation for relativistic theories. We assume that the reader is familiar with special relativity.

Consider two events in space–time, (x, y, z, t) and $(x + dx, y + dy, z + dz, t + dt)$. We may generalise the notion of the distance between two points in space to the 'interval' between two points in space–time; call it ds. In order that ds be the same for all (inertial) observers, it must be invariant under Lorentz transformations and rotations, and so is given by

$$ds^2 = c^2 \, dt^2 - (dx^2 + dy^2 + dz^2). \tag{2.1}$$

Of course, we could have defined $ds^2 = dx^2 + dy^2 + dz^2 - c^2 \, dt^2$; we choose (2.1) for later convenience. With this definition, events which are separated by a *timelike* interval have $ds^2 > 0$; those separated by a *spacelike* interval $ds^2 < 0$; and those separated by a *null* or *lightlike* interval $ds^2 = 0$.

In 3-dimensional space (x, y, z) are regarded as the components of a 3-vector, and $dr^2 = dx^2 + dy^2 + dz^2$ is invariant under rotations. This quadratic form is the sum of squares, and so positive definite. To generalise to 4-dimensional space–time, we have the problem that the invariant interval is no longer positive definite. We therefore define

$$x^\mu = (x^0, x^1, x^2, x^3) = (ct, x, y, z)$$
$$x_\mu = (x_0, x_1, x_2, x_3) = (ct, -x, -y, -z) \tag{2.2}$$

and make the rule that the invariant is got by summing over *one upper and one lower index*:

$$ds^2 = \sum_{\mu=0}^{3} dx^\mu \, dx_\mu = c^2 \, dt^2 - dx^2 - dy^2 - dz^2. \tag{2.3}$$

A 4-vector like x^μ, with an upper index, is called a *contravariant vector* and one like x_μ, with a lower index, is called a *covariant vector*. The inner product of a contravariant and a covariant vector is an invariant (scalar). To simplify the notation, we adopt the *summation convention*: an index appearing once in an upper and once in a lower position is automatically summed from 0 to 3:

$$\sum_{\mu=0}^{3} V^\mu V_\mu \to V^\mu V_\mu. \tag{2.4}$$

The relation between x^μ and x_μ (or between any contravariant vector and its covariant counterpart) may be given by introducing a *metric tensor* $g_{\mu\nu}$:

$$
\begin{aligned}
x_\mu &= g_{\mu\nu} x^\nu \\
&= g_{\mu 0} x^0 + g_{\mu 1} x^1 + g_{\mu 2} x^2 + g_{\mu 3} x^3,
\end{aligned}
\tag{2.5}
$$

where the summation convention has been used. By inspection of (2.2), we have $x_0 = x^0$, $x_1 = -x^1$ etc., so from (2.5) it is clear that $g_{\mu\nu}$ may be written as a diagonal matrix

$$
g_{\mu\nu} = \begin{pmatrix}
1 & 0 & 0 & 0 \\
0 & -1 & 0 & 0 \\
0 & 0 & -1 & 0 \\
0 & 0 & 0 & -1
\end{pmatrix}, \tag{2.6}
$$

where rows and columns correspond to the 0, 1, 2 and 3 components. Since $g_{\mu\nu}$ has a non-zero determinant, its inverse exists, and is written

$$
g^{\mu\nu} = \begin{pmatrix}
1 & 0 & 0 & 0 \\
0 & -1 & 0 & 0 \\
0 & 0 & -1 & 0 \\
0 & 0 & 0 & -1
\end{pmatrix}; \tag{2.7}
$$

in fact, it has the same values as $g_{\mu\nu}$ in Minkowski space (in Cartesian co-ordinates), but this equality does not hold in general.

It is clear that $g_{\mu\nu}$ contains all the information about the geometry of the space – in this case, Minkowski space–time. In special relativity, however, the metric tensor only plays a passive role, and in fact need never be introduced. But in general relativity, it plays an active role since the geometry of the space is not fixed in advance, but depends on what matter is in it. The Einstein field equations, for example, are differential equations for $g_{\mu\nu}$.

It is common in particle physics to work in units where $c = 1$, so (2.3) becomes

$$ds^2 = dx^\mu dx_\mu = dt^2 - (dx^2 + dy^2 + dz^2). \tag{2.8}$$

Turning to differential operators, we define

$$\partial_\mu = \frac{\partial}{\partial x^\mu} = (\partial_0, \partial_1, \partial_2, \partial_3) = \left(\frac{1}{c} \frac{\partial}{\partial t}, \frac{\partial}{\partial x}, \frac{\partial}{\partial y}, \frac{\partial}{\partial z} \right)$$

$$= \left(\frac{1}{c} \frac{\partial}{\partial t}, \mathbf{\nabla} \right)$$

$$\partial^\mu = g^{\mu\nu} \partial_\nu = \left(\frac{1}{c} \frac{\partial}{\partial t}, -\mathbf{\nabla} \right), \tag{2.9}$$

giving the Lorentz invariant second order differential operator

$$\square = \partial^\mu \partial_\mu = \frac{1}{c^2} \frac{\partial^2}{\partial t^2} - \left(\frac{\partial^2}{\partial x^2} + \frac{\partial^2}{\partial y^2} + \frac{\partial^2}{\partial z^2} \right) = \frac{1}{c^2} \frac{\partial^2}{\partial t^2} - \nabla^2, \tag{2.10}$$

called the *d'Alembertian operator*.

The energy–momentum 4-vector of a particle is

$$p^\mu = \left(\frac{E}{c}, \mathbf{p} \right), \quad p_\mu = \left(\frac{E}{c}, -\mathbf{p} \right) \tag{2.11}$$

giving the invariant

$$p^2 = p^\mu p_\mu = \frac{E^2}{c^2} - \mathbf{p} \cdot \mathbf{p} = m^2 c^2 \tag{2.12}$$

or, when $c = 1$,

$$p^2 = E^2 - \mathbf{p}^2 = m^2. \tag{2.13}$$

We shall also use the notation $p \cdot x$ for $p_\mu x^\mu$:

$$p \cdot x = p_\mu x^\mu = Et - \mathbf{p} \cdot \mathbf{r}. \tag{2.14}$$

2.2 Klein–Gordon equation

We are now in a position to write down a wave equation for a particle with no spin – a scalar particle. Since it has no spin it has only one component, which we denote by ϕ. The wave equation is obtained from equation (2.12) by substituting differential operators for E and $\mathbf{p}$, in the fashion standard in quantum theory

$$E \to i\hbar \frac{\partial}{\partial t}, \quad \mathbf{p} \to -i\hbar \mathbf{\nabla}. \tag{2.15}$$

Equation (2.12) then gives

$$\left(\frac{1}{c^2} \frac{\partial^2}{\partial t^2} - \nabla^2 \right) \phi + \frac{m^2 c^2}{\hbar^2} \phi = 0,$$

which becomes, in units $\hbar = c = 1$ and using (2.10),

∎ $$(\square + m^2)\phi = 0. \tag{2.16}$$

This is known as the Klein–Gordon equation. Note that substituting (2.15) into the non-relativistic approximation to (2.12), $E = p^2/2m$ (here E is the *kinetic* energy only) yields the free-particle Schrödinger equation

$$\frac{\hbar^2}{2m}\nabla^2\phi = -i\hbar\frac{\partial\phi}{\partial t}. \tag{2.17}$$

It then follows that the ~~the~~ Schrödinger equation is the non-relativistic approximation to the Klein–Gordon equation.

The probability density for the Schrödinger equation is

$$\rho = \phi^*\phi \tag{2.18}$$

and the probability current is

$$\mathbf{j} = -\frac{i\hbar}{2m}(\phi^*\nabla\phi - \phi\nabla\phi^*). \tag{2.19}$$

They obey a continuity equation

$$\begin{aligned}
\frac{\partial\rho}{\partial t} + \nabla\cdot\mathbf{j} &= \frac{\partial}{\partial t}(\phi^*\phi) - \frac{i\hbar}{2m}(\phi^*\nabla^2\phi - \phi\nabla^2\phi^*) \\
&= \phi^*\left(\frac{\partial\phi}{\partial t} - \frac{i\hbar}{2m}\nabla^2\phi\right) + \phi\left(\frac{\partial\phi^*}{\partial t} + \frac{i\hbar}{2m}\nabla^2\phi^*\right) \\
&= 0,
\end{aligned}$$

where the Schrödinger equation and its complex conjugate have been used. What are the corresponding expressions for the Klein–Gordon equation? To be properly relativistic, ρ should not, as in (2.18), transform as a scalar, but as the time component of a 4-vector, whose space component is $\mathbf{j}$, given by (2.19). Then ρ is given by

$$\rho = \frac{i\hbar}{2m}\left(\phi^*\frac{\partial\phi}{\partial t} - \phi\frac{\partial\phi^*}{\partial t}\right) \tag{2.20}$$

and with

$$j^\mu = (\rho, \mathbf{j}) = \frac{i\hbar}{m}\phi^*(\overleftrightarrow{\partial_0}, -\overleftrightarrow{\nabla})\phi = \frac{i\hbar}{m}\phi^*\overleftrightarrow{\partial^\mu}\phi \tag{2.21}$$

where

$$A\overleftrightarrow{\partial^\mu}B \stackrel{\text{def}}{=} \tfrac{1}{2}[A\partial^\mu B - (\partial^\mu A)B], \tag{2.22}$$

and we have used (2.9), we have the continuity equation

$$\partial_\mu j^\mu = \frac{i\hbar}{2m}(\phi^*\Box\phi - \phi\Box\phi^*) = 0, \tag{2.23}$$

since ϕ^* also obeys the Klein–Gordon equation. Then ρ and $\mathbf{j}$ are the probability density and current we want. But this immediately presents a problem, because ρ, given by equation (2.20), unlike expression (2.18) for the Schrödinger equation, is *not positive definite*. Since the Klein–Gordon equation is second order, ϕ and $\partial\phi/\partial t$ can be fixed arbitrarily at a given time, so ρ may take on negative values, and its interpretation as a probability density has to be abandoned. The interpretation of the Klein–Gordon equation as a *single-particle equation*, with wave-function ϕ, therefore also has to be abandoned. In chapter 4 we shall re-interpret it as a *field equation*, and shall see that, on quantisation, we arrive, after all, at a successful particle interpretation.

It is worth remarking here that ϕ has been assumed to be complex. If ϕ is taken as real, then ρ in (2.20) vanishes, and so does **j**. It will be shown in the next chapter that the correct interpretation of complex ϕ is for the description of *charged* particles. Real ϕ corresponds to electrically neutral particles, and ρ and **j** are then the charge and current densities, rather than the probability and probability current densities.

There is another problem with the Klein–Gordon equation, and that is that the solution to (2.12), regarded as an equation for E, is

$$E = \pm (m^2c^4 + \mathbf{p}^2c^2)^{\frac{1}{2}}, \tag{2.24}$$

so solutions to the Klein–Gordon equation contain *negative energy* terms as well as positive energy ones. For a free particle, whose energy is thereby constant, this difficulty may be avoided, for we may choose the particle to have positive energy, and the negative energy states may be ignored. But an interacting particle may exchange energy with its environment, and there would then be nothing to stop it cascading down to infinite negative energy states, emitting an infinite amount of energy in the process. This, of course, does not happen, and so poses a problem for the single-particle Klein–Gordon equation. We shall see later that the interpretation of ϕ as a quantum field clears up this problem, as well as the previous one.

We now pass from scalar particles to particles with spin, starting with spin $\frac{1}{2}$ particles, which are described by the Dirac equation.

2.3   Dirac equation

The Dirac equation, unlike the Klein–Gordon equation, is of first order, and holds only for spin $\frac{1}{2}$ particles; since the Klein–Gordon equation expresses nothing more than the relativistic relation between energy, momentum, and mass, it must hold for particles of any spin. The Dirac equation, however (and the Maxwell and Proca equations, to be derived below), has an entirely different origin, and may be derived from the transformation properties of spinors under the Lorentz group. To avoid a straight plunge into this topic, we shall first briefly review the connection between the rotation group and $SU(2)$, and introduce the idea of spinors. This will then be extended to the Lorentz group.

$SU(2)$ *and the rotation group*

A general spatial rotation is of the form

$$\begin{pmatrix} x' \\ y' \\ z' \end{pmatrix} = (R) \begin{pmatrix} x \\ y \\ z \end{pmatrix} \quad \text{or} \quad r' = Rr; \tag{2.25}$$

R is the rotation matrix. Since rotations perserve distance from the origin, $x'^2 + y'^2 + z'^2 = x^2 + y^2 + z^2$, or $r'^T r' = r^T r$ (T = transpose), so

$$r^T R^T R r = r^T r$$
$$R^T R = 1, \tag{2.26}$$

and R is an orthogonal 3×3 matrix. These matrices form a group: if R_1 and R_2 are orthogonal, so is $R_1 R_2$:

$$(R_1 R_2)^T R_1 R_2 = R_2^T R_1^T R_1 R_2 = 1$$

This group is denoted $O(3)$; for matrices in n dimensions it is $O(n)$. Unitary matrices also form a group, denoted $U(n)$, but Hermitian matrices do not, unless they commute.

As an example of a rotation, consider a rotation of a vector $\mathbf{V}$ about the z axis as shown in Fig. 2.1. This rotation, considered as an *active rotation* (i.e. a rotation of the vector, leaving the co-ordinate axes fixed), is left-handed; considered as a *passive rotation* (i.e. rotating the axes, leaving the vector fixed) it is right-handed. We have

$$\begin{pmatrix} V'_x \\ V'_y \\ V'_z \end{pmatrix} = \begin{pmatrix} \cos\theta & \sin\theta & 0 \\ -\sin\theta & \cos\theta & 0 \\ 0 & 0 & 1 \end{pmatrix} \begin{pmatrix} V_x \\ V_y \\ V_z \end{pmatrix}, \tag{2.27}$$

Fig. 2.1.

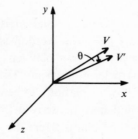

so may denote the rotation matrix by

$$R_z(\theta) = \begin{vmatrix} \cos\theta & \sin\theta & 0 \\ -\sin\theta & \cos\theta & 0 \\ 0 & 0 & 1 \end{vmatrix}. \qquad (2.28)$$

Similar matrices for rotations about the x and y axes are

$$R_x(\phi) = \begin{vmatrix} 1 & 0 & 0 \\ 0 & \cos\phi & \sin\phi \\ 0 & -\sin\phi & \cos\phi \end{vmatrix},$$

$$R_y(\psi) = \begin{vmatrix} \cos\psi & 0 & -\sin\psi \\ 0 & 1 & 0 \\ \sin\psi & 0 & \cos\psi \end{vmatrix}. \qquad (2.29)$$

Note that these matrices do not commute

$$R_x(\phi)R_z(\theta) \neq R_z(\theta)R_x(\phi); \qquad (2.30)$$

the rotation group, $O(3)$, is *non-Abelian*. It is a *Lie group*; that is, a continuous group, with an infinite number of elements, since the parameters of rotation, which are angles, take on a continuum of values. It is easy to see that a general rotation has three parameters; R has nine elements, and equation (2.26) gives six conditions on them. These parameters may, for example, be chosen to be the three Euler angles. Corresponding to three parameters are three *generators* defined by

$$J_z = \frac{1}{i}\frac{dR_z(\theta)}{d\theta}\bigg|_{\theta=0} = \begin{vmatrix} 0 & -i & 0 \\ i & 0 & 0 \\ 0 & 0 & 0 \end{vmatrix},$$

$$J_x = \frac{1}{i}\frac{dR_x(\phi)}{d\phi}\bigg|_{\phi=0} = \begin{vmatrix} 0 & 0 & 0 \\ 0 & 0 & -i \\ 0 & i & 0 \end{vmatrix},$$

$$J_y = \frac{1}{i}\frac{dR_y(\psi)}{d\psi}\bigg|_{\psi=0} = \begin{vmatrix} 0 & 0 & i \\ 0 & 0 & 0 \\ -i & 0 & 0 \end{vmatrix}. \qquad (2.31)$$

These generators are Hermitian, and infinitesimal rotations are given by, for example,

$$R_z(\delta\theta) = 1 + iJ_z\,\delta\theta, \quad R_x(\delta\phi) = 1 + iJ_x\,\delta\psi. \qquad (2.32)$$

The commutator $R_z(\delta\theta)R_x(\delta\theta)R_z^{-1}(\delta\theta)R_x^{-1}(\delta\theta)$ of these two rotations (compare (2.30)) may now be calculated using the easily verified commutation relations

$$J_xJ_y - J_yJ_x \equiv [J_x, J_y] = iJ_z \text{ and cyclic permutations.} \qquad (2.33)$$

To first order, it is found to be a rotation about the y axis. The relations (2.33), having a factor $\hbar$, will be recognised as the commutation relations for the components of angular momentum. So angular momentum operators are the generators of rotations.

It is now straightforward to write down the rotation matrix for finite rotations. The matrix corresponding to a rotation about the z axis through an angle $\theta = N \delta\theta$ ($N \to \infty$) is clearly

$$
\begin{aligned}
R_z(\theta) &= [R_z(\delta\theta)]^N \\
&= (1 + iJ_z \, \delta\theta)^N \\
&= \left(1 + iJ_z \frac{\theta}{N}\right)^N \\
&= e^{iJ_z\theta}.
\end{aligned}
\tag{2.34}
$$

We may check that this yields the required matrix (2.28). Defining the exponential by its power series expansion, we have

$$
e^{iJ_z\theta} = 1 + iJ_z\theta - J_z^2 \frac{\theta^2}{2!} - iJ_z^3 \frac{\theta^3}{3!} + \cdots
$$

$$
= \begin{vmatrix} 1 & 0 & 0 \\ 0 & 1 & 0 \\ 0 & 0 & 1 \end{vmatrix} + \theta \begin{vmatrix} 0 & 1 & 0 \\ -1 & 0 & 0 \\ 0 & 0 & 0 \end{vmatrix} + \frac{\theta^2}{2!} \begin{vmatrix} -1 & 0 & 0 \\ 0 & -1 & 0 \\ 0 & 0 & 0 \end{vmatrix}
$$

$$
+ \frac{\theta^3}{3!} \begin{vmatrix} 0 & -1 & 0 \\ 1 & 0 & 0 \\ 0 & 0 & 0 \end{vmatrix} + \cdots
$$

$$
= \begin{vmatrix} \cos\theta & \sin\theta & 0 \\ -\sin\theta & \cos\theta & 0 \\ 0 & 0 & 1 \end{vmatrix}
$$

which is (2.28). A finite rotation about an axis **n** through an angle θ is denoted

$$
R_n(\theta) = e^{iJ\cdot\theta} = e^{iJ\cdot n\theta}
\tag{2.35}
$$

where $\boldsymbol{\theta} = \mathbf{n}\theta$.

So much for the rotation group. Now consider the group $SU(2)$, consisting of 2×2 unitary matrices with unit determinant

$$
UU^+ = 1, \quad \det U = 1.
\tag{2.36}
$$

Putting

$$
U = \begin{pmatrix} a & b \\ c & d \end{pmatrix},
\tag{2.37}
$$

the unitarity condition reads $U^+ = U^{-1}$, which, since $\det U = 1$, becomes

$$\begin{pmatrix} a^* & c^* \\ b^* & d^* \end{pmatrix} = \begin{pmatrix} d & -b \\ -c & a \end{pmatrix}$$

and hence $a^* = d$, $b^* = -c$. Then $\det U = |a|^2 + |b|^2$, so we have

$$U = \begin{pmatrix} a & b \\ -b^* & a^* \end{pmatrix}, \quad |a|^2 + |b|^2 = 1. \tag{2.38}$$

This is regarded as the transformation matrix in a 2-dimensional complex space with basic spinor $\xi = \begin{pmatrix} \xi_1 \\ \xi_2 \end{pmatrix}$;

$$\xi \to U\xi, \quad \xi^+ \to \xi^+ U^+. \qquad \xi^+ = \begin{pmatrix} \xi_1^* & \xi_2^* \end{pmatrix} \tag{2.39}$$

It is clear that

$$\xi^+ \xi = |\xi_1|^2 + |\xi_2|^2$$

is invariant under (2.39). On the other hand, the outer product

$$\xi \xi^+ = \begin{pmatrix} |\xi_1|^2 & \xi_1 \xi_2^* \\ \xi_2 \xi_1^* & |\xi_2|^2 \end{pmatrix} \to U \xi \xi^+ U^+. \tag{2.40}$$

Observe that $\xi \xi^+$ is a Hermitian matrix.

We see from (2.39) that ξ and ξ^+ transform in different ways, but we may use the unitarity of U to show that $\begin{pmatrix} \xi_1 \\ \xi_2 \end{pmatrix}$ and $\begin{pmatrix} -\xi_2^* \\ \xi_1^* \end{pmatrix}$ transform in the same way under $SU(2)$. We have, comparing (2.38) and (2.39),

$$\left. \begin{array}{l} \xi_1' = a\xi_1 + b\xi_2, \\ \xi_2' = -b^*\xi_1 + a^*\xi_2, \end{array} \right\} \tag{2.41}$$

and hence

$$\left. \begin{array}{l} -\xi_2^{*\prime} = a(-\xi_2^*) + b\xi_1^*, \\ \xi_1^{*\prime} = -b^*(-\xi_2^*) + a^*\xi_1^*. \end{array} \right\} \tag{2.42}$$

Now

$$\begin{pmatrix} -\xi_2^* \\ \xi_1^* \end{pmatrix} = \begin{pmatrix} 0 & -1 \\ 1 & 0 \end{pmatrix} \begin{pmatrix} \xi_1^* \\ \xi_2^* \end{pmatrix} = \zeta \xi^* \tag{2.43}$$

where

$$\zeta = \begin{pmatrix} 0 & -1 \\ 1 & 0 \end{pmatrix}, \tag{2.44}$$

so we have shown that ξ and $\zeta \xi^*$ transform in the same way under $SU(2)$; in symbols

$$\xi \sim \zeta \xi^* \tag{2.45}$$

('$\sim$' means transforms like). Hence

$$\xi^+ \sim (\zeta\xi)^T = (-\xi_2, \xi_1), \tag{2.46}$$

and

$$\xi\xi^+ \sim \begin{pmatrix} \xi_1 \\ \xi_2 \end{pmatrix}(-\xi_2\,\xi_1) = \begin{pmatrix} -\xi_1\xi_2 & \xi_1^2 \\ -\xi_2^2 & \xi_1\xi_2 \end{pmatrix}. \tag{2.47}$$

Calling this matrix $-H$, we see from (2.40) that under an $SU(2)$ transformation

$$H \to UHU^+; \tag{2.48}$$

in addition, H is a traceless matrix.

We may now construct, from the position vector $\mathbf{r}$, a traceless 2×2 matrix transforming under $SU(2)$ like H. It is

$$h = \boldsymbol{\sigma}\cdot\mathbf{r} = \begin{pmatrix} z & x-iy \\ x+iy & -z \end{pmatrix} \tag{2.49}$$

where the matrices $\boldsymbol{\sigma}$ are the well-known Pauli matrices

$$\sigma_x = \begin{pmatrix} 0 & 1 \\ 1 & 0 \end{pmatrix}, \quad \sigma_y = \begin{pmatrix} 0 & -i \\ i & 0 \end{pmatrix}, \quad \sigma_z = \begin{pmatrix} 1 & 0 \\ 0 & -1 \end{pmatrix}; \tag{2.50}$$

h is Hermitian, and the transformation

$$h \to UhU^+ = h' \tag{2.51}$$

preserves the Hermiticity and tracelessness of h if U is unitary. In addition, if U belongs to $SU(2)$, and so has determinant 1, then $\det h' = \det h$, or

$$x'^2 + y'^2 + z'^2 = x^2 + y^2 + z^2; \tag{2.52}$$

the unitary transformation (2.51) on h induces a *rotation* of the position vector $\mathbf{r}$. Identifying H and h, we conclude finally that

■   An $SU(2)$ transformation on

$$\begin{pmatrix} \xi_1 \\ \xi_2 \end{pmatrix} \equiv O(3) \text{ transformation on } \begin{pmatrix} x \\ y \\ z \end{pmatrix} \text{ with}$$

$$x = \tfrac{1}{2}(\xi_2^2 - \xi_1^2), \quad y = \frac{1}{2i}(\xi_1^2 + \xi_2^2), \quad z = \xi_1\xi_2. \tag{2.53}$$

The parameters of an $SU(2)$ transformation are a, b, both complex, with one condition: $|a|^2 + |b|^2 = 1$. There are thus three real parameters, just like there are for a rotation. We shall now find an explicit relation between the two sets of parameters. Taking the square and the product of the two relations (2.41), and using the expressions for x, y and z given above (equation (2.53)), we have, under $SU(2)$,

$$x' = \tfrac{1}{2}(a^2 + a^{*2} - b^2 - b^{*2})x - \frac{i}{2}(a^2 - a^{*2} + b^2 - b^{*2})y$$

$$- (a^*b^* + ab)z,$$

$$y' = \frac{i}{2}(a^2 - a^{*2} - b^2 + b^{*2})x + \tfrac{1}{2}(a^2 + a^{*2} + b^2 + b^{*2})y$$

$$- i(ab - a^*b^*)z,$$

$$z' = (ab^* + ba^*)x + (ba^* - ab^*)iy + (|a|^2 - |b|^2)z. \tag{2.54}$$

Now we put $a = e^{i\alpha/2}$, $b = 0$ (which obeys $|a|^2 + |b|^2 = 1$). Equation (2.54) gives

$$x' = x\cos\alpha + y\sin\alpha,$$

$$y' = -x\sin\alpha + y\cos\alpha,$$

$$z' = z,$$

which is a rotation about the z axis through an angle α (cf. (2.27)). Hence the correspondence between the $SU(2)$ matrix (2.38) and the $O(3)$ matrix (2.28) is

$$U = \begin{pmatrix} e^{i\alpha/2} & 0 \\ 0 & e^{-i\alpha/2} \end{pmatrix} \leftrightarrow R = \begin{pmatrix} \cos\alpha & \sin\alpha & 0 \\ -\sin\alpha & \cos\alpha & 0 \\ 0 & 0 & 1 \end{pmatrix}. \tag{2.55}$$

In terms of the generators J_z (equation (2.31)) and σ_z (equation (2.50)) we may write

$$U = e^{i\sigma_z\alpha/2}, \quad R = e^{iJ_z\alpha} \tag{2.56}$$

where, as usual, the exponential expression for U is defined by its power series expansion.

In a similar way, putting $a = \cos\beta/2$, $b = \sin\beta/2$, we have the correspondence

$$U = \begin{pmatrix} \cos\beta/2 & \sin\beta/2 \\ -\sin\beta/2 & \cos\beta/2 \end{pmatrix} \leftrightarrow R = \begin{pmatrix} \cos\beta & 0 & -\sin\beta \\ 0 & 1 & 0 \\ \sin\beta & 0 & \cos\beta \end{pmatrix} \tag{2.57}$$

which may be written

$$U = e^{i\sigma_y\beta/2}, \quad R = e^{iJ_y\beta}; \tag{2.58}$$

and, lastly, putting $a = \cos\gamma/2$, $b = i\sin\gamma/2$, gives

$$U = \begin{pmatrix} \cos\gamma/2 & i\sin\gamma/2 \\ i\sin\gamma/2 & \cos\gamma/2 \end{pmatrix} \leftrightarrow R = \begin{pmatrix} 1 & 0 & 0 \\ 0 & \cos\gamma & \sin\gamma \\ 0 & -\sin\gamma & \cos\gamma \end{pmatrix} \tag{2.59}$$

with

$$U = e^{i\sigma_x \gamma/2}, \quad R = e^{iJ_x \gamma}. \tag{2.60}$$

In general, then, the correspondence between an $SU(2)$ transformation in the spinor space $\begin{pmatrix} \xi_1 \\ \xi_2 \end{pmatrix}$ and an $O(3)$ transformation in the space $\begin{pmatrix} x \\ y \\ z \end{pmatrix}$ is

$$U = e^{i\sigma \cdot \theta/2} = \cos \theta/2 + i\sigma \cdot \mathbf{n} \sin \theta/2 \leftrightarrow R = e^{iJ \cdot \theta}. \tag{2.61}$$

This correspondence between $SU(2)$ and $O(3)$ implies that the groups must have a similar structure, and therefore that their generators obey the same commutation relations. In fact, it may easily be checked that the Pauli matrices obey

$$\left[\frac{\sigma_x}{2}, \frac{\sigma_y}{2} \right] = i\frac{\sigma_z}{2} \text{ and cyclic perms.} \tag{2.62}$$

These are the same relations as (2.33) for **J**, and we also see that the factors $\frac{1}{2}$ in (2.62) are the ones in (2.61), showing that the spinor rotates through half the angle that the vector rotates through. This is responsible for a global topological distinction between $SU(2)$ and $O(3)$, for, as may be seen from equations (2.55)–(2.61), increasing the angle α (say) by 2π gives $U \rightarrow -U, R \rightarrow R$; so the elements U and $-U$ in $SU(2)$ both correspond to the rotation R in $O(3)$: *there is a 2 to 1 mapping of the elements of SU(2) onto those of O(3).* In particular

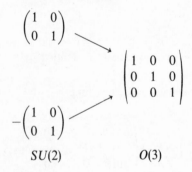

$$SU(2) \qquad\qquad O(3)$$

SL(2, C) and the Lorentz group

Analogous to the correspondence between $SU(2)$ and the rotation group, there is a correspondence between $SL(2, C)$ and the Lorentz group, which we now explore. We shall then employ this correspondence to derive the Dirac equation.

Pure 'boost' Lorentz transformations are those connecting two inertial frames, moving with relative speed v. If the relative motion is along the common x axis, the equations are

$$x' = \frac{x + vt}{(1 - v^2/c^2)^{\frac{1}{2}}}, \quad y' = y, \quad z' = z, \quad t' = \frac{t + vx/c^2}{(1 - v^2/c^2)^{\frac{1}{2}}}.$$

Putting $\gamma = (1 - v^2/c^2)^{-\frac{1}{2}}, \beta = v/c, x^0 = ct, x^1 = x$ etc., these are expressed as

$$x^{0'} = \gamma(x^0 + \beta x^1), \quad x^{1'} = \gamma(\beta x^0 + x^1), \quad x^{2'} = x^2, \quad x^{3'} = x^3.$$

Observing that $\gamma^2 - \beta^2\gamma^2 = 1$, we may put

$$\gamma = \cosh \phi, \quad \gamma\beta = \sinh \phi, \tag{2.63}$$

thus parameterising the transformation in terms of the variable ϕ, with $\tanh \phi = v/c$, and we have

$$\begin{pmatrix} x^{0'} \\ x^{1'} \\ x^{2'} \\ x^{3'} \end{pmatrix} = \begin{pmatrix} \cosh \phi & \sinh \phi & 0 & 0 \\ \sinh \phi & \cosh \phi & 0 & 0 \\ 0 & 0 & 1 & 0 \\ 0 & 0 & 0 & 1 \end{pmatrix} \begin{pmatrix} x^0 \\ x^1 \\ x^2 \\ x^3 \end{pmatrix}. \tag{2.64}$$

Let us call the above matrix the boost matrix B. The generator K_x of this boost transformation along the x axis is defined by analogy with (2.31):

$$K_x = \frac{1}{i}\frac{\partial B}{\partial \phi}\bigg|_{\phi=0} = -i \begin{pmatrix} 0 & 1 & 0 & 0 \\ 1 & 0 & 0 & 0 \\ 0 & 0 & 0 & 0 \\ 0 & 0 & 0 & 0 \end{pmatrix}. \tag{2.65}$$

Similarly, the other boost generators are

$$K_y = -i \begin{pmatrix} 0 & 0 & 1 & 0 \\ 0 & 0 & 0 & 0 \\ 1 & 0 & 0 & 0 \\ 0 & 0 & 0 & 0 \end{pmatrix}, \quad K_z = -i \begin{pmatrix} 0 & 0 & 0 & 1 \\ 0 & 0 & 0 & 0 \\ 0 & 0 & 0 & 0 \\ 1 & 0 & 0 & 0 \end{pmatrix}. \tag{2.66}$$

In this 4×4 matrix notation, the rotation generators (2.31) may be written

$$J_x = -i \begin{pmatrix} 0 & 0 & 0 & 0 \\ 0 & 0 & 0 & 0 \\ 0 & 0 & 0 & 1 \\ 0 & 0 & -1 & 0 \end{pmatrix}, \quad J_y = -i \begin{pmatrix} 0 & 0 & 0 & 0 \\ 0 & 0 & 0 & -1 \\ 0 & 0 & 0 & 0 \\ 0 & 1 & 0 & 0 \end{pmatrix},$$

$$J_z = -i \begin{pmatrix} 0 & 0 & 0 & 0 \\ 0 & 0 & 1 & 0 \\ 0 & -1 & 0 & 0 \\ 0 & 0 & 0 & 0 \end{pmatrix}. \tag{2.67}$$

The most general Lorentz transformation is composed of boosts in three directions, and rotations about three axes, and the six generators are those above. Their commutation relations may be calculated explicitly, and we find

$$[K_x, K_y] = -iJ_z \text{ and cyclic perms,}$$
$$[J_x, K_x] = 0 \text{ etc.,} \qquad (2.68)$$
$$[J_x, K_y] = iK_z \text{ and cyclic perms,}$$

together with (2.33), involving Js only. An interesting consequence of these relations is that *pure Lorentz transformations do not form a group*, since the generators $\mathbf{K}$ do not form a closed algebra under commutation. So the commutator of two infinitesimal boosts in different directions

$$e^{iK_x\delta\phi}e^{iK_y\delta\psi}e^{-iK_x\delta\phi}e^{-iK_y\delta\psi}$$
$$= 1 - [K_x, K_y]\delta\phi\,\delta\psi + K_x^2(\delta\phi)^2 K_y^2(\delta\psi)^2 + \cdots$$

contains, by virtue of the first of equations (2.68), a *rotation* about the z axis. This is the origin of Thomas precession.[4]

How do Pauli spinors transform under Lorentz transformations? We may make a guess, because the above commutation relations are satisfied by

$$\mathbf{K} = \pm i\frac{\boldsymbol{\sigma}}{2}; \qquad (2.69)$$

so there should be two types of spinor, corresponding to the two possible signs of $\mathbf{K}$. This may be made clearer by elucidating the nature of the Lorentz group, with the six generators $\mathbf{J}$ and $\mathbf{K}$. Let us define the *non-Hermitian* generators

$$\mathbf{A} = \tfrac{1}{2}(\mathbf{J} + i\mathbf{K}),$$
$$\mathbf{B} = \tfrac{1}{2}(\mathbf{J} - i\mathbf{K}). \qquad (2.70)$$

The commutation relations (2.68) and (2.33) then become

$$[A_x, A_y] = iA_z \text{ and cyclic perms,}$$
$$[B_x, B_y] = iB_z \text{ and cyclic perms,} \qquad (2.71)$$
$$[A_i, B_j] = 0 \quad (i, j = x, y, z).$$

This shows that $\mathbf{A}$ and $\mathbf{B}$ each generate a group $SU(2)$, and the two groups commute. The Lorentz group is then essentially $SU(2) \otimes SU(2)$, and states transforming in a well-defined way will be labelled by two angular momenta (j, j'), the first one corresponding to A, and the second to B. As special cases, one or the other will correspond to spin zero:

$$(j, 0) \to \mathbf{J}^{(j)} = i\mathbf{K}^j \quad (\mathbf{B} = 0),$$
$$(0, j) \to \mathbf{J}^{(j)} = -i\mathbf{K}^{(j)} \quad (\mathbf{A} = 0), \qquad (2.72)$$

and this in fact corresponds to the two possibilities in (2.69). We may now

$$\vec{J} = i\vec{K}$$

define two types of spinor:

Type I: $(\frac{1}{2}, 0)$: $\mathbf{J}^{(\pm)} = \boldsymbol{\sigma}/2,$ $\mathbf{K}^{(\pm)} = -\mathrm{i}\boldsymbol{\sigma}/2.$

We denote the spinor ξ. If $(\boldsymbol{\theta}, \boldsymbol{\phi})$ are the parameters of a rotation and pure Lorentz transformation, ξ transforms as

$$\xi \to \exp\left(\mathrm{i}\frac{\boldsymbol{\sigma}}{2} \cdot \boldsymbol{\theta} + \frac{\boldsymbol{\sigma}}{2} \cdot \boldsymbol{\phi} \right)\xi$$

$$= \exp\left[\mathrm{i}\frac{\boldsymbol{\sigma}}{2} \cdot (\boldsymbol{\theta} - \mathrm{i}\boldsymbol{\phi}) \right]\xi \equiv M\xi. \tag{2.73}$$

Type II: $(0, \frac{1}{2})$: $\mathbf{J}^{(\pm)} = \boldsymbol{\sigma}/2,$ $\mathbf{K}^{(\pm)} = \mathrm{i}\boldsymbol{\sigma}/2.$

This spinor is denoted η and transforms like

$$\eta \to \exp\left[\mathrm{i}\frac{\boldsymbol{\sigma}}{2} \cdot (\boldsymbol{\theta} + \mathrm{i}\boldsymbol{\phi}) \right]\eta \equiv N\eta. \tag{2.74}$$

It is important to note that these are *inequivalent* representations of the Lorentz group, i.e. there is no matrix S such that $N = SMS^{-1}$. They are in fact related by

$$N = \zeta M^* \zeta^{-1} \quad \text{with}\ \zeta = -\mathrm{i}\sigma_2, \tag{2.75}$$

as defined in (2.44) above. This follows by observing that

$$\sigma_2 \boldsymbol{\sigma}^* \sigma_2 = -\sigma_2^2 \boldsymbol{\sigma} = -\boldsymbol{\sigma},$$

for then

$$\zeta M^* \zeta^{-1} = \sigma_2 \exp\left[-\frac{\mathrm{i}}{2}\boldsymbol{\sigma}^* \cdot (\boldsymbol{\theta} + \mathrm{i}\boldsymbol{\phi}) \right]\sigma_2$$

$$= \sigma_2^2 \exp\left[\frac{\mathrm{i}}{2}\boldsymbol{\sigma} \cdot (\boldsymbol{\theta} + \mathrm{i}\boldsymbol{\phi}) \right]$$

$$= N.$$

We note that $\det M = \det N = 1$, so M and N are complex 2×2 matrices with unit determinant. Such matrices form a group, $SL(2, C)$. It clearly has six parameters, for the matrices are of the form

$$M = \begin{pmatrix} a & b \\ c & d \end{pmatrix}, \quad ad - bc = 1;$$

they consist of four complex numbers, with two conditions. These six parameters are related to the three angles and three velocities of the general Lorentz transformations. Let us pause to summarise what we have found. Besides 3-vectors, there are 2-component Pauli spinors, which have a well-defined transformation (2.61) under rotations. Under general Lorentz transformations, however, there are two *different* types of 2-component spinor, transforming respectively by (2.73) and (2.74). In the older literature, these are known as 'dotted' and 'undotted' spinors[1-4]. They correspond to the representations $(\frac{1}{2}, 0)$ and $(0, \frac{1}{2})$ of the Lorentz group. In essence, the Dirac equation is a relation between these spinors.

Let us introduce the parity operation, under which the velocity in the Lorentz boost changes sign: $\mathbf{v} \to -\mathbf{v}$. Hence the generators $\mathbf{K}$ change sign, $\mathbf{K} \to -\mathbf{K}$, like the components of a *vector*, whereas $\mathbf{J}$ does not change sign, $\mathbf{J} \to +\mathbf{J}$, behaving like an *axial vector* or *pseudovector*, which indeed is how angular momentum transforms under parity. If follows that the $(j, 0)$ and $(0, j)$ representations become interchanged,

$$(j, 0) \leftrightarrow (0, j), \quad \text{under parity} \tag{2.76}$$

and hence

$$\xi \leftrightarrow \eta.$$

If we consider parity, then, it is no longer sufficient to consider the 2-spinors ξ and η separately, but the 4-spinor

$$\psi = \begin{pmatrix} \xi \\ \eta \end{pmatrix}. \tag{2.77}$$

Under Lorentz transformations ψ transforms as follows:

$$\begin{pmatrix} \xi \\ \eta \end{pmatrix} \to \begin{pmatrix} e^{\frac{1}{2}\sigma \cdot (\theta - i\phi)} & 0 \\ 0 & e^{\frac{1}{2}\sigma \cdot (\theta + i\phi)} \end{pmatrix} \begin{pmatrix} \xi \\ \eta \end{pmatrix}$$

$$= \begin{pmatrix} D(\Lambda) & 0 \\ 0 & \bar{D}(\Lambda) \end{pmatrix} \begin{pmatrix} \xi \\ \eta \end{pmatrix}, \tag{2.78}$$

with

$$\bar{D}(\Lambda) = \zeta D^*(\Lambda) \zeta^{-1}, \tag{2.79}$$

and Λ denotes the Lorentz transformation (2.64), which we may write as

$$x'^{\mu} = \Lambda^{\mu}_{\nu} x^{\nu}. \tag{2.80}$$

Under parity, ψ transforms as

$$\begin{pmatrix} \xi \\ \eta \end{pmatrix} \to \begin{pmatrix} 0 & 1 \\ 1 & 0 \end{pmatrix} \begin{pmatrix} \xi \\ \eta \end{pmatrix}. \tag{2.81}$$

The 4-spinor ψ is an *irreducible* representation of the Lorentz group *extended by parity*. Note, however, that the representation (2.78) is *not unitary*; this is because the generators $\mathbf{A}$ and $\mathbf{B}$ are not Hermitian. In general, in quantum mechanics, one is only interested in *unitary* representations of a symmetry group[8], since it is only these that preserve the transition probability between two states, as measured in different frames of reference. From a fundamental point of view, then, there is something less than satisfactory about our present pursuit. It is related to the fact that the Lorentz group, unlike the rotation group, is *non-compact*. This corresponds roughly to the observation that velocities, which are the parameters of Lorentz boosts, take on values along an open line, from $v/c = 0$ to $v/c = 1$, whereas angles of rotation extend from $\theta = 0$ to $\theta = 2\pi$, and these points are *identified*, so the line becomes closed into a circle. The group space of the rotation group is finite, but that of the Lorentz group is infinite, so the Lorentz group is non-compact. There is, moreover, a theorem that unitary representations of non-compact groups are infinite-dimensional. What we have above is a negative illustration of this – a finite-dimensional. and non-unitary representation of the Lorentz group. Actually, it was realised by Wigner many years ago that the fundamental group for particle physics is not the (homogeneous) Lorentz group considered above, but the *inhomogeneous* Lorentz group, commonly called the *Poincaré group*, consisting of Lorentz boosts and rotations, and also translations in space and time. An analysis of this group gives a true understanding of, and some surprising insights into, the nature of spin. Some more will be said about the Poincaré group below.

Now let us specialise the transformations (2.78) to the case of a Lorentz boost ($\theta = 0$), and at the same time relabel the 2-spinors ξ and η:

$$\xi \to \phi_R, \quad \eta \to \phi_L, \tag{2.82}$$

R and L standing for right and left. We have

$$\phi_R \to e^{\frac{1}{2}\sigma \cdot \phi}\phi_R$$
$$= [\cosh(\phi/2) + \sigma \cdot \mathbf{n} \sinh(\phi/2)]\phi_R \tag{2.83}$$

where $\mathbf{n}$ is a unit vector in the direction of the Lorentz boost. Suppose that the original spinor refers to a particle at rest, $\phi_R(0)$, and the transformed one to a particle with momentum $\mathbf{p}$, $\phi_R(\mathbf{p})$. From (2.63) we have $\cosh(\phi/2) = [(\gamma + 1)/2]^{\frac{1}{2}}$, $\sinh(\phi/2) = [(\gamma - 1)/2]^{\frac{1}{2}}$, so (2.83) becomes

$$\phi_R(\mathbf{p}) = \left[\left(\frac{\gamma+1}{2}\right)^{\frac{1}{2}} + \sigma \cdot \hat{\mathbf{p}}\left(\frac{\gamma-1}{2}\right)^{\frac{1}{2}}\right]\phi_R(0). \tag{2.84}$$

Since, for a particle with (total) energy E, mass m and momentum $\mathbf{p}$, $\gamma = E/m$ ($c = 1$), (2.84) then becomes

$$\phi_R(\mathbf{p}) = \frac{E + m + \sigma \cdot \mathbf{p}}{[2m(E+m)]^{\frac{1}{2}}}\phi_R(0). \tag{2.85}$$

In a similar way, we find

$$\phi_L(\mathbf{p}) = \frac{E + m - \sigma \cdot \mathbf{p}}{[2m(E+m)]^{\frac{1}{2}}}\phi_L(0). \tag{2.86}$$

Now when a particle is at rest, one cannot define its spin as either left- or right-handed, so $\phi_R(0) = \phi_L(0)$. It then follows, with a bit of algebra, from (2.85) and (2.86) that

$$\phi_R(\mathbf{p}) = \frac{E + \boldsymbol{\sigma}\cdot\mathbf{p}}{m}\phi_L(\mathbf{p}) \tag{2.87}$$

and hence

$$\phi_L(\mathbf{p}) = \frac{E - \boldsymbol{\sigma}\cdot\mathbf{p}}{m}\phi_R(\mathbf{p}). \tag{2.88}$$

We may rewrite these equations as

$$\left.\begin{array}{l} -m\phi_R(\mathbf{p}) + (p_0 + \boldsymbol{\sigma}\cdot\mathbf{p})\phi_L(\mathbf{p}) = 0, \\ (p_0 - \boldsymbol{\sigma}\cdot\mathbf{p})\phi_R(\mathbf{p}) - m\phi_L(\mathbf{p}) = 0, \end{array}\right\} \tag{2.89}$$

or, in matrix form,

$$\begin{pmatrix} -m & p_0 + \boldsymbol{\sigma}\cdot\mathbf{p} \\ p_0 - \boldsymbol{\sigma}\cdot\mathbf{p} & -m \end{pmatrix}\begin{pmatrix} \phi_R(\mathbf{p}) \\ \phi_L(\mathbf{p}) \end{pmatrix} = 0. \tag{2.90}$$

Defining the 4-spinor

$$\psi(p) = \begin{pmatrix} \phi_R(\mathbf{p}) \\ \phi_L(\mathbf{p}) \end{pmatrix} \tag{2.91}$$

and the 4×4 matrices

$$\gamma^0 = \begin{pmatrix} 0 & 1 \\ 1 & 0 \end{pmatrix}, \quad \gamma^i = \begin{pmatrix} 0 & -\sigma^i \\ \sigma^i & 0 \end{pmatrix}, \tag{2.92}$$

equation (2.90) becomes

$$(\gamma^0 p_0 + \gamma^i p_i - m)\psi(p) = 0 \tag{2.93}$$

(note that $p_\mu = (E, -\mathbf{p})$ (see (2.11)), so $\gamma^0 p_0 + \gamma^i p_i = \gamma^0 p_0 - \boldsymbol{\gamma}\cdot\mathbf{p}$), or

$$\blacksquare \qquad (\gamma^\mu p_\mu - m)\psi(p) = 0. \tag{2.94}$$

This is the Dirac equation for massive spin $\frac{1}{2}$ particles. In the case of massless particles, it is clear, for example from (2.89), that the equation decouples into two equations, each for a 2-component spinor

$$\begin{array}{l} (p_0 + \boldsymbol{\sigma}\cdot\mathbf{p})\phi_L(\mathbf{p}) = 0 \\ (p_0 - \boldsymbol{\sigma}\cdot\mathbf{p})\phi_R(\mathbf{p}) = 0 \end{array} \tag{2.95}$$

These are known as the *Weyl equations*, and ϕ_L and ϕ_R are Weyl spinors. Since, for a massless particle, $p_0 = |\mathbf{p}|$, these equations read

$$\boldsymbol{\sigma}\cdot\hat{\mathbf{p}}\phi_L = -\phi_L, \quad \boldsymbol{\sigma}\cdot\hat{\mathbf{p}}\phi_R = \phi_R.$$

The operator $\boldsymbol{\sigma}\cdot\hat{\mathbf{p}}$ measures the component of the spin in the direction of momentum, and this quantity is called *helicity*. Thus Weyl spinors are eigenstates of helicity, the left-handed (right-handed) spinor having negative (positive) helicity.

Traditionally it has been supposed that neutrinos are massless particles, and therefore are described by (one of) the Weyl equations, but there is recent speculation that neutrinos may have a mass.[‡] The question is at present unresolved.

The derivation of the Dirac equation given above differs from Dirac's original one. Dirac's aim was to find an equation which did not share the problems of the Klein–Gordon equation, discussed above. We now consider these problems.

2.4   Prediction of antiparticles

We saw above that the Klein–Gordon equation suffers from two defects: the probability density is not positive definite, and negative energy states occur. For these reasons the Klein–Gordon equation was discarded and Dirac looked for an equation to replace it, which was, unlike the Klein–Gordon equation, a *first order* equation – in those days that was regarded as a desirable asset. He then discovered equation (2.94), and deduced that the γ^μ matrices must be 4×4 matrices. It is instructive to reproduce this reasoning. First let us write the Dirac equation in co-ordinate space. Substituting $i\partial_\mu$ for p_μ in (2.94) gives

■ $$(i\gamma^\mu \partial_\mu - m)\psi = 0. \tag{2.96}$$

This is a first order differential equation. Now applying the operator $i\gamma^\mu \partial_\mu$ again to this equation:

$$[-(\gamma^\mu \partial_\mu)(\gamma^\nu \partial_\nu) - i(\gamma^\mu \partial_\mu)m]\psi = 0,$$
$$(\gamma^\mu \gamma^\nu \partial_\mu \partial_\nu + m^2)\psi = 0.$$

Now $\partial_\mu \partial_\nu = \partial_\nu \partial_\mu$, so $\gamma^\mu \gamma^\nu$ may be replaced by the symmetric combination

$$\tfrac{1}{2}(\gamma^\mu \gamma^\nu + \gamma^\nu \gamma^\mu) \equiv \tfrac{1}{2}\{\gamma^\mu, \gamma^\nu\} \tag{2.97}$$

to give

$$\tfrac{1}{2}\{\gamma^\mu, \gamma^\nu\}\partial_\mu \partial_\nu \psi + m^2 \psi = 0.$$

On the other hand, relativity requires that the energy–momentum–mass relation be satisfied, and therefore that each component of ψ satisfies the Klein–Gordon equation

$$(\Box + m^2)\psi(x) = 0. \tag{2.98}$$

[‡] For a short review, see, for example, W.J. Marciano, *Comments on Nuclear and Particle Physics*, **9**, 169 (1981).

It therefore follows that the coefficient of $\partial_\mu \partial_\nu$ is $g^{\mu\nu}$, so

$$\{\gamma^\mu, \gamma^\nu\} = 2g^{\mu\nu}. \tag{2.99}$$

This is the general relation which the coefficients γ^μ must satisfy. Taking in turn $\mu = \nu = 0$, $\mu = \nu = i$ and $\mu \neq \nu$, we have

$$(\gamma^0)^2 = 1, \quad (\gamma^i)^2 = -1, \quad \gamma^\mu\gamma^\nu = -\gamma^\nu\gamma^\mu \quad (\nu \neq \mu). \tag{2.100}$$

It is an easy matter to check that the matrices (2.92) do satisfy these relations. (As an exercise, the student may also convince himself that it is impossible to satisfy (2.100) with 2×2 matrices.) It is also clear that if four γ^μs satisfy (2.99), so will

$$\gamma'^\mu = S\gamma^\mu S^{-1} \tag{2.101}$$

where S is a unitary 4×4 matrix, and the Dirac equation will then be satisfied by

$$\psi' = S\psi. \tag{2.102}$$

Let us now construct a probability current j^μ, analogous to (2.21) for the Klein–Gordon equation, and check to see whether the density is positive. We write the Dirac equation in the form (2.96)

$$(i\gamma^\mu\partial_\mu - m)\psi = 0, \tag{2.96}$$

where $\gamma^\mu\partial_\mu = \gamma^0\partial_0 + \gamma^i\partial_i$. Now take the Hermitian conjugate of this equation, noting from (2.92) or (2.100) that $\gamma^{0\dagger} = \gamma^0$, $\gamma^{i\dagger} = -\gamma^i$. This gives

$$\psi^\dagger(-i\gamma^0\overleftarrow{\partial}_0 + i\gamma^i\overleftarrow{\partial}_i - m) = 0.$$

(Here $\psi^\dagger$ is a row vector, and $\overleftarrow{\partial}_0$ and $\overleftarrow{\partial}_i$ operate on it to the left.) This does not look very appealing, so we multiply on the right by γ^0, and use, from (2.100), $\gamma^i\gamma^0 = -\gamma^0\gamma^i$ to give

$$\bar{\psi}(i\gamma^\mu\overleftarrow{\partial}_\mu + m) = 0 \tag{2.103}$$

where

$$\bar{\psi} = \psi^+\gamma^0 \tag{2.104}$$

is called the *adjoint spinor* to ψ. Equations (2.96) and (2.103) may now be used to show that the current

$$j^\mu = \bar{\psi}\gamma^\mu\psi \tag{2.105}$$

is conserved:

$$\begin{aligned}\partial_\mu j^\mu &= (\partial_\mu\bar{\psi})\gamma^\mu\psi + \bar{\psi}\gamma^\mu(\partial_\mu\psi) \\ &= (im\bar{\psi})\psi + \bar{\psi}(-im\psi) = 0.\end{aligned} \tag{2.106}$$

The density j^0 is therefore

$$j^0 = \bar{\psi}\gamma^0\psi = \psi^+\psi = |\psi_1|^2 + |\psi_2|^2 + |\psi_3|^2 + |\psi_4|^2$$

and is *positive*; j^0 is therefore fit to serve as the probability density for the particle in question, and the Dirac equation is seen to have cleared this hurdle, which the Klein–Gordon equation failed to clear.

We now come to the other difficulty faced by the Klein–Gordon equation, that of negative energy states. Here, we appear to be not so lucky. In fact, it is easy to see from (2.94) that a Dirac particle at rest obeys

$$\left.\begin{array}{l} \gamma^0 p_0 \psi = m\psi, \\ p_0 \psi = m\gamma^0 \psi. \end{array}\right\} \tag{2.107}$$

The eigenvalues of γ^0 are clearly $+1$ (twice) and -1 (twice), so there are two positive energy solutions $(+m)$ and two negative energy solutions $(-m)$. In fact, it is straightforward to show, by writing out all four components of (2.94), that the eigenvalues of E are

$$E = +(m^2 + p^2)^{\frac{1}{2}} \text{ twice,}$$
$$E = -(m^2 + p^2)^{\frac{1}{2}} \text{ twice.}$$

For each value of p, there are two positive energy solutions, corresponding to the states of a spin $\frac{1}{2}$ particle, and two additional negative energy solutions. This potential catastrophe was turned by Dirac into a triumph.

The energy spectrum of solutions is shown in Fig. 2.2, and the problem it poses is the same as that met in the Klein–Gordon equation. An electron in a positive energy state may[‡] jump to a negative energy state, and then cascade down to $E = -\infty$, emitting in the process an infinite amount of energy (say, electromagnetic radiation). Dirac's solution to this problem relies on the fact that electrons have spin $\frac{1}{2}$ and therefore obey Pauli's *exclusion principle*. Dirac supposed that the negative energy states are already *completely filled*, and the exclusion principle prevents any more electrons being able to enter the 'sea' of negative energy states.

This 'Dirac sea' is the *vacuum*; so, on Dirac's theory, the vacuum is not

Fig. 2.2. The energy spectrum of solutions to the Dirac equation.

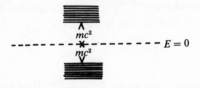

[‡] If the electron is in *interaction* with another particle or field – say, the electromagnetic field. The Dirac equation above holds only for *free* electrons, but the problem of negative energy states remains after interactions have been included.

'nothing', but an infinite sea of negative energy electrons, protons, neutrinos, neutrons and all other spin $\frac{1}{2}$ particles!

Now this ingenious theory makes an important prediction, for suppose there occurs one vacancy in the electron sea – a 'hole' with energy $-|E|$. An electron with energy E may fill this hole, emitting energy $2E$, and leaving the vacuum:

$$e^- + \text{hole} \rightarrow \text{energy}, \tag{2.108}$$

so the 'hole' effectively has charge $+e$ and positive energy, and is called a *positron*, the *antiparticle* of the electron. This theory of Dirac's predicted the existence of antiparticles for all spin $\frac{1}{2}$ particles, and in time the $e^+, \bar{p}, \bar{n}, \bar{\nu}$ etc. were all found. It turns out that bosons also have antiparticles, but to see this requires treating ϕ, the Klein–Gordon 'wave function', as a quantised field, which we shall do in chapter 4.

The prediction and discovery of antiparticles is one of the outstanding episodes in the history of particle physics, and inspires considerable confidence in the Dirac equation. In fact the Dirac equation has been outstandingly successful in its predictions and applications[11,12] but in this book we shall content ourselves with studying, firstly, the rather formal construction and properties of Dirac spinors, and, secondly, the successful prediction of the electron magnetic moment.

We conclude this account of antiparticles by noting that, despite the successful resolution of the problem of negative energies, the Dirac equation is *no longer a single-particle equation*, since it describes both particles and antiparticles. The only consistent philosophy is to regard the spinor ψ as a *field*, such that $|\psi|^2$ gives a measure of the number of particles at a particular point. This field is naturally a quantum field; we shall see in chapter 4 how this reproduces the particle and antiparticle interpretation, as well as the proper connection between spin and statistics.

2.5   Construction of Dirac spinors: algebra of γ matrices

First, it will be useful to know the Lorentz-transformation properties of bilinear expressions like $\bar{\psi}\psi$, $\bar{\psi}\gamma^\mu\psi$, etc. We begin by showing that $\bar{\psi}\psi$ is a scalar quantity. We work in the basis (2.91)

$$\psi = \begin{pmatrix} \phi_R \\ \phi_L \end{pmatrix}, \tag{2.109}$$

and we know that under a Lorentz transformation (including rotation) (θ, ϕ) we have, from (2.78),

$$\phi_R \rightarrow \exp\left[\frac{i}{2}\sigma\cdot(\theta - i\phi)\right]\phi_R, \quad \phi_L \rightarrow \exp\left[\frac{i}{2}\sigma\cdot(\theta + i\phi)\right]\phi_L, \tag{2.110}$$

hence

$$\phi_R^\dagger \to \phi_R^\dagger \exp\left[\frac{-i}{2}\boldsymbol{\sigma}\cdot(\boldsymbol{\theta}+i\boldsymbol{\phi})\right], \quad \phi_L^\dagger \to \phi_L^\dagger \exp\left[\frac{-i}{2}\boldsymbol{\sigma}\cdot(\boldsymbol{\theta}-i\boldsymbol{\phi})\right],$$

(2.111)

and it is immediately evident that

$$\psi^\dagger\psi = \phi_R^\dagger\phi_R + \phi_L^\dagger\phi_L \tag{2.112}$$

is *not* invariant. However, the adjoint spinor $\bar{\psi}$ defined in (2.104) has components

$$\bar{\psi} = \psi^\dagger\gamma^0 = (\phi_R^\dagger\phi_L^\dagger)\begin{pmatrix}0 & 1 \\ 1 & 0\end{pmatrix} = (\phi_L^\dagger\phi_R^\dagger), \tag{2.113}$$

so it is easy to see that

$$\bar{\psi}\psi = \phi_L^\dagger\phi_R + \phi_R^\dagger\phi_L \tag{2.114}$$

is invariant (i.e. a scalar) under Lorentz transformations. Moreover, under parity

$$\phi_R \leftrightarrow \phi_L, \tag{2.115}$$

so $\bar{\psi}\psi \to \bar{\psi}\psi$, and is a true scalar, i.e. does not change sign under space reflection.

Now define the 4×4 matrix

$$\gamma^5 = \begin{pmatrix}1 & 0 \\ 0 & -1\end{pmatrix}. \tag{2.116}$$

(Remember that each entry here is a 2×2 matrix.) This defines γ^5 in the basis (2.109). In an arbitrary basis it is defined by

$$\gamma^5 = i\gamma^0\gamma^1\gamma^2\gamma^3 = \gamma_5. \tag{2.117}$$

We then see that

$$\begin{aligned}\bar{\psi}\gamma^5\psi &= (\phi_L^\dagger\phi_R^\dagger)\begin{pmatrix}1 & 0 \\ 0 & -1\end{pmatrix}\begin{pmatrix}\phi_R \\ \phi_L\end{pmatrix} \\ &= \phi_L^\dagger\phi_R - \phi_R^\dagger\phi_L.\end{aligned} \tag{2.118}$$

It is clear from (2.110), (2.111) and (2.115) that this is invariant under Lorentz transformations, but changes sign under parity. For that reason, it is called a *pseudoscalar* quantity.

Now consider the quantity $\bar{\psi}\gamma^\mu\psi$, which we suspect of transforming like a 4-vector under Lorentz transformations. Its time and space components are

$$\bar{\psi}\gamma^0\psi = \phi_R^\dagger\phi_R + \phi_L^\dagger\phi_L, \tag{2.119}$$

$$\bar{\psi}\gamma\psi = (\phi_R^\dagger \phi_L^\dagger)\begin{pmatrix} 0 & -\boldsymbol{\sigma} \\ \boldsymbol{\sigma} & 0 \end{pmatrix}\begin{pmatrix} \phi_R \\ \phi_L \end{pmatrix}$$

$$= -\phi_R^\dagger \boldsymbol{\sigma}\phi_L + \phi_L^\dagger \boldsymbol{\sigma}\phi_R. \tag{2.120}$$

Under spatial rotations ($\boldsymbol{\theta} \neq 0, \boldsymbol{\phi} = 0$) we have

$$\bar{\psi}\gamma^0\psi \to \bar{\psi}\gamma^0\psi \tag{2.121}$$

and, if $\boldsymbol{\theta}$ is infinitesimal,

$$\begin{aligned}\bar{\psi}\gamma\psi \to &- \phi_R^\dagger e^{-(i/2)\boldsymbol{\sigma}\cdot\boldsymbol{\theta}}\boldsymbol{\sigma} e^{(i/2)\boldsymbol{\sigma}\cdot\boldsymbol{\theta}}\phi_L \\ &+ \phi_L^\dagger e^{-(i/2)\boldsymbol{\sigma}\cdot\boldsymbol{\theta}}\boldsymbol{\sigma} e^{(i/2)\boldsymbol{\sigma}\cdot\boldsymbol{\theta}}\phi_R \\ = &- \phi_R^\dagger\left(1-\frac{i}{2}\boldsymbol{\sigma}\cdot\boldsymbol{\theta}\right)\boldsymbol{\sigma}\left(1+\frac{i}{2}\boldsymbol{\sigma}\cdot\boldsymbol{\theta}\right)\phi_L \\ &+ \phi_L^\dagger\left(1-\frac{i}{2}\boldsymbol{\sigma}\cdot\boldsymbol{\theta}\right)\boldsymbol{\sigma}\left(1+\frac{i}{2}\boldsymbol{\sigma}\cdot\boldsymbol{\theta}\right)\phi_R \\ = &- \phi_R^\dagger(\boldsymbol{\sigma}-\boldsymbol{\theta}\times\boldsymbol{\sigma})\phi_L + \phi_L^\dagger(\boldsymbol{\sigma}-\boldsymbol{\theta}\times\boldsymbol{\sigma})\phi_R \\ = &\,\bar{\psi}\gamma\psi - \boldsymbol{\theta}\times(\bar{\psi}\gamma\psi).\end{aligned} \tag{2.122}$$

(The student may find it a useful exercise to check this, using the commutation relations (2.62), expressed in the form

$$[\sigma_i, \sigma_j] = 2i\varepsilon_{ijk}\sigma_k, \tag{2.123}$$

where

$$\varepsilon_{ijk} = \begin{cases} +1 \text{ if } (ijk) \text{ is an even permutation of (123),} \\ -1 \text{ if } (ijk) \text{ is an odd permutation of (123),} \\ 0 \text{ otherwise,} \end{cases} \tag{2.124}$$

and putting $\boldsymbol{\sigma}\cdot\boldsymbol{\theta} = \sigma_i\theta_i$ (summation convention implied!).) Equation (2.122) describes the behaviour of a vector under rotations: for infinitesimal θ, equation (2.27) reads $V'_x = V_x + \theta V_y, V'_y = V_y - \theta V_x, V'_z = V_z$, which is the z-component of

$$\mathbf{V}' = \mathbf{V} - \boldsymbol{\theta}\times\mathbf{V}.$$

In addition, the time component, by (2.121), is invariant under rotations, so $\bar{\psi}\gamma^\mu\psi$ indeed behaves like a 4-vector under rotations. The student is invited to check that it also behaves like one under Lorentz boost transformations, i.e. like x^μ in (2.64) for boosts along the x axis. Moreover, under parity it is easy to see that

$$\bar{\psi}\gamma^0\psi \to \bar{\psi}\gamma^0\psi, \quad \bar{\psi}\gamma\psi \to -\bar{\psi}\gamma\psi \tag{2.125}$$

like a polar vector does. This is summarised by saying that $\bar{\psi}\gamma^\mu\psi$ transforms as a *vector*. In like manner, it may be shown that $\bar{\psi}\gamma^\mu\gamma^5\psi$ behaves like an *axial vector* (or *pseudovector*) – it behaves like a 4-vector under Lorentz transformations, including rotations, but has the opposite behaviour to (2.125) under parity, so that the 'space' part transforms like an antisymmetric rank 2 tensor – the corresponding symmetric tensor is, by virtue of (2.99), invariant. This completes all the possibilities, for in four dimensions, a rank 3 tensor transforms like a pseudovector, and a rank 4 tensor like a pseudoscalar; in three dimensions, this reduction happens sooner – a rank 2 tensor, like $\mathbf{r} \times \mathbf{p}$ (or $\mathbf{r} \wedge \mathbf{p}$), an axial vector, and a rank 3 tensor like the volume element $dxdydz$ (properly written $dx \wedge dy \wedge dz$) is a pseudoscalar[‡].

These results are summarised thus:

$$
\begin{aligned}
&\bar{\psi}\psi && \text{scalar,} \\
&\bar{\psi}\gamma_5\psi && \text{pseudoscalar,} \\
&\bar{\psi}\gamma^\mu\psi && \text{vector,} \\
&\bar{\psi}\gamma^\mu\gamma^5\psi && \text{axial vector,} \\
&\bar{\psi}(\gamma^\mu\gamma^\nu - \gamma^\nu\gamma^\mu)\psi && \text{antisymmetric tensor.}
\end{aligned}
\tag{2.126}
$$

Let us now construct spinors corresponding to an arbitrary state of motion of a Dirac particle. We do this by a method essentially the same as that used to obtain (2.85), (2.86) and (2.91), except that we want explicit expressions for the spinors, and shall not work in the representation (2.91), which we may call the *chiral* representation (since ϕ_R and ϕ_L are eigenstates of chirality, whose operator is γ^5) but in the so-called *standard* representation, in which γ^0 is diagonal. From (2.107), this is clearly the representation appropriate for describing particles at rest. The plane wave solutions of the Dirac equation for a particle at rest are clearly

$$
\left.
\begin{aligned}
\psi(x) &= u(0)\,e^{-imt} && \text{positive energy,} \\
\psi(x) &= v(0)\,e^{imt} && \text{negative energy,}
\end{aligned}
\right\}
\tag{2.127}
$$

with the two positive and two negative energy spinors

$$
u^{(1)}(0) = \begin{pmatrix} 1 \\ 0 \\ 0 \\ 0 \end{pmatrix}, \quad
u^{(2)}(0) = \begin{pmatrix} 0 \\ 1 \\ 0 \\ 0 \end{pmatrix}, \quad
v^{(1)}(0) = \begin{pmatrix} 0 \\ 0 \\ 1 \\ 0 \end{pmatrix}, \quad
v^{(2)}(0) = \begin{pmatrix} 0 \\ 0 \\ 0 \\ 1 \end{pmatrix}.
\tag{2.128}
$$

Here γ^0 is, in the standard representation,

$$
\gamma^0 = \begin{pmatrix} 1 & 0 & 0 & 0 \\ 0 & 1 & 0 & 0 \\ 0 & 0 & -1 & 0 \\ 0 & 0 & 0 & -1 \end{pmatrix}
\tag{2.129}
$$

[‡] For a nice account of these matters, see, for example, D. Hestenes, *Space–Time Algebra*, Gordon & Breach, 1966.

which we write as usual in the condensed form:

$$\gamma^0 = \begin{pmatrix} 1 & 0 \\ 0 & -1 \end{pmatrix}. \tag{2.130}$$

This is obtained from the chiral representation (2.92)

$$\gamma^0_{SR} = S\gamma^0_{CR}S^{-1},$$

$$S = \frac{1}{\sqrt{2}}\begin{pmatrix} 1 & 1 \\ 1 & -1 \end{pmatrix} \tag{2.131}$$

so that, in the standard representation,

$$\psi = S\begin{pmatrix} \phi_R \\ \phi_L \end{pmatrix} = \frac{1}{\sqrt{2}}\begin{pmatrix} \phi_R + \phi_L \\ \phi_R - \phi_L \end{pmatrix}. \tag{2.132}$$

For a Lorentz boost to a moving frame, we have, from (2.78) with $\boldsymbol{\theta} = 0$,

$$\begin{pmatrix} \phi_R \\ \phi_L \end{pmatrix} \to \begin{pmatrix} \phi'_R \\ \phi'_L \end{pmatrix} = \begin{pmatrix} e^{\frac{1}{2}\boldsymbol{\sigma}\cdot\boldsymbol{\phi}} & 0 \\ 0 & e^{-\frac{1}{2}\boldsymbol{\sigma}\cdot\boldsymbol{\phi}} \end{pmatrix}\begin{pmatrix} \phi_R \\ \phi_L \end{pmatrix} = M\begin{pmatrix} \phi_R \\ \phi_L \end{pmatrix}, \tag{2.133}$$

so in the standard representation the boost matrix is

$$M_{SR} = SM_{CR}S^{-1} = \begin{pmatrix} \cosh(\phi/2) & \boldsymbol{\sigma}\cdot\mathbf{n}\sinh(\phi/2) \\ \boldsymbol{\sigma}\cdot\mathbf{n}\sinh(\phi/2) & \cosh(\phi/2) \end{pmatrix} \tag{2.134}$$

and since

$$\cosh(\phi/2) = \left(\frac{E+m}{2m}\right)^{\frac{1}{2}}, \quad \sinh(\phi/2) = \left(\frac{E-m}{2m}\right)^{\frac{1}{2}},$$

$$\tanh(\phi/2) = \frac{p}{E+m}$$

with $p = (E^2 - m^2)^{\frac{1}{2}}$, we obtain

$$M_{SR} = \left(\frac{E+m}{2m}\right)^{\frac{1}{2}}\begin{pmatrix} 1 & 0 & \dfrac{p_z}{E+m} & \dfrac{p_x - ip_y}{E+m} \\ 0 & 1 & \dfrac{p_x + ip_y}{E+m} & \dfrac{-p_z}{E+m} \\ \dfrac{p_z}{E+m} & \dfrac{p_x - ip_y}{E+m} & 1 & 0 \\ \dfrac{p_x + ip_y}{E+m} & \dfrac{-p_z}{E+m} & 0 & 1 \end{pmatrix} \tag{2.135}$$

and the corresponding plane wave spinors are

$$\psi^{(\alpha)}(x) = u^{(\alpha)}(p)e^{-ip\cdot x} \quad \text{positive energy,}$$
$$\psi^{(\alpha)}(x) = v^{(\alpha)}(p)e^{ip\cdot x} \quad \text{negative energy,} \tag{2.136}$$

where $\alpha = 1, 2$, and $u^{(\alpha)}(p)$ and $v^{(\alpha)}(p)$ are obtained by multiplying M_{SR} by the rest spinors (2.128), giving

$$u^{(1)} = \sqrt{\frac{E+m}{2m}} \begin{pmatrix} 1 \\ 0 \\ \dfrac{p_z}{E+m} \\ \dfrac{p_+}{E+m} \end{pmatrix}, \quad u^{(2)} = \sqrt{\frac{E+m}{2m}} \begin{pmatrix} 0 \\ 1 \\ \dfrac{p_-}{E+m} \\ \dfrac{-p_z}{E+m} \end{pmatrix} \tag{2.137}$$

$$v^{(1)} = \sqrt{\frac{E+m}{2m}} \begin{pmatrix} \dfrac{p_z}{E+m} \\ \dfrac{p_+}{E+m} \\ 1 \\ 0 \end{pmatrix}, \quad v^{(2)} = \sqrt{\frac{E+m}{2m}} \begin{pmatrix} \dfrac{p_-}{E+m} \\ \dfrac{-p_z}{E+m} \\ 0 \\ 1 \end{pmatrix} \tag{2.138}$$

where $p_\pm = p_x \pm ip_y$. The normalisation of the u spinors is

$$\bar{u}^{(1)}u^{(1)} = \left(\frac{E+m}{2m}\right)\begin{pmatrix} 1 & 0 & \dfrac{p_z}{E+m} & \dfrac{p_-}{E+m} \end{pmatrix}\begin{pmatrix} 1 & & & \\ & 1 & & \\ & & -1 & \\ & & & -1 \end{pmatrix}\begin{pmatrix} 1 \\ 0 \\ \dfrac{p_z}{E+m} \\ \dfrac{p_+}{E+m} \end{pmatrix} = 1,$$

and similarly for $u^{(2)}$. The final results are

$$\bar{u}^{(\alpha)}(p)u^{(\alpha')}(p) = \delta_{\alpha\alpha'}$$
$$\bar{v}^{(\alpha)}(p)v^{(\alpha')}(p) = -\delta_{\alpha\alpha'}$$
$$\bar{u}^{(\alpha)}(p)v^{(\alpha')}(p) = 0$$
$$u^{(\alpha)+}(p)u^{(\alpha')}(p) = v^{(\alpha)+}(p)v^{(\alpha')}(p) = \frac{E}{m}\delta_{\alpha\alpha'}. \tag{2.139}$$

In addition, from (2.95) and (2.136), u and v satisfy

$$\left.\begin{array}{c} (\gamma \cdot p - m)u(p) = 0, \\ (\gamma \cdot p + m)v(p) = 0, \end{array}\right\} \tag{2.140}$$

and it follows that the adjoint spinors obey

$$\left.\begin{array}{c} \bar{u}(p)(\gamma \cdot p - m) = 0, \\ \bar{v}(p)(\gamma \cdot p + m) = 0. \end{array}\right\} \tag{2.141}$$

The operator

$$P_+ = \sum_\alpha u^{(\alpha)}(p)\bar{u}^{(\alpha)}(p) \tag{2.142}$$

is important in many applications. It is a projection operator, since, in view of (2.139)

$$P_+^2 = \sum_{\alpha,\beta} u^{(\alpha)}(p)\bar{u}^{(\alpha)}(p)u^{(\beta)}(p)\bar{u}^{(\beta)} = \sum_{\alpha} u^{(\alpha)}(p)\bar{u}^{(\alpha)}(p) = P_+. \qquad (2.143)$$

It clearly projects out positive energy states. We shall now find an expression for P_+. Because of (2.140),

$$(\gamma \cdot p - m)P_+ = 0,$$

therefore

$$\frac{\gamma \cdot p}{m} P_+ = P_+. \qquad (2.144)$$

Now we assume that P_+ is of the form $a + b\gamma \cdot p$. Inserting (2.144) into this gives $a = mb$. Then using $P_+^2 = P_+$ gives $b = 1/2m$, so finally we have

$$P_+ = \sum_{\alpha} u^{(\alpha)}(p)\bar{u}^{(\alpha)}(p) = \frac{\gamma \cdot p + m}{2m}. \qquad (2.145)$$

Similarly the projection operator for negative energy states is

$$P_- = -\sum_{\alpha} v^{(\alpha)}(p)\bar{v}^{(\alpha)}(p) = \frac{-\gamma \cdot p + m}{2m}. \qquad (2.146)$$

As expected, $P_+ + P_- = 1$.

It is appropriate here to note some trace formulae which will be useful later. They all involve the γ^μ matrices. Since these are 4×4 matrices, we have

$$\text{Tr } 1 = 4$$

and because of the cyclicity of the trace, we have

$$\begin{aligned}
\text{Tr}(\gamma \cdot a)(\gamma \cdot b) &= \text{Tr}(\gamma \cdot b)(\gamma \cdot a) \\
&= \tfrac{1}{2}\text{Tr } a_\mu b_\nu (\gamma^\mu \gamma^\nu) \\
&= a \cdot b \, \text{Tr } 1 = 4a \cdot b.
\end{aligned} \qquad (2.147)$$

Next, we prove that the trace of an odd number of γ matrices is zero. To do this, we use the fact that γ^5, defined by (2.117), has the properties

$$(\gamma^5)^2 = 1, \quad \{\gamma^5, \gamma^\mu\} = 0. \qquad (2.148)$$

Let us, for convenience, also adopt the 'slash' notation

$$a_\mu \gamma^\mu = a \cdot \gamma \equiv \rlap{/}a. \qquad (2.149)$$

Then we have

$$\begin{aligned}
\text{Tr } \rlap{/}a_1 \ldots \rlap{/}a_n &= \text{Tr } \rlap{/}a_1 \ldots \rlap{/}a_n \gamma_5 \gamma_5 \\
&= \text{Tr } \gamma_5 \rlap{/}a_1 \ldots \rlap{/}a_n \gamma_5,
\end{aligned}$$

using the cyclicity condition. We now move the γ_5 on the left past each $\not{a}$ in turn, with a consequent change of sign each time, obtaining eventually

$$\mathrm{Tr}\,\not{a}_1 \ldots \not{a}_n = (-1)^n\,\mathrm{Tr}\,\not{a}_1 \ldots \not{a}_n \gamma_5 \gamma_5$$

so that

$$\mathrm{Tr}\,\not{a}_1 \ldots \not{a}_n = 0, \quad n \text{ odd}. \tag{2.150}$$

Finally, substituting

$$\gamma^\mu \gamma^\nu = -\gamma^\nu \gamma^\mu + 2g^{\mu\nu}$$

into the first two factors of $\mathrm{Tr}(\gamma \cdot a)(\gamma \cdot b)(\gamma \cdot c)(\gamma \cdot d)$, we obtain

$$\mathrm{Tr}(\gamma \cdot a)(\gamma \cdot b)(\gamma \cdot c)(\gamma \cdot d) = -\mathrm{Tr}(\gamma \cdot b)(\gamma \cdot a)(\gamma \cdot c)(\gamma \cdot d)$$
$$+ 2a \cdot b\,\mathrm{Tr}(\gamma \cdot c)(\gamma \cdot d). \tag{2.151}$$

These formulae will be referred to when we come to calculate scattering cross sections.

2.6 Non-relativistic limit and the electron magnetic moment

Particles with spin also possess an 'intrinsic' magnetic moment. Now a charge e circulating in a closed orbit of angular momentum l interacts with a magnetic field and so possesses an effective magnetic moment[‡]

$$\mu = \frac{e}{2m}\mathbf{l}.^\ddagger \tag{2.152}$$

If nature were simple, the proportionality constant between the electron spin $\mathbf{S} = \frac{1}{2}\hbar\boldsymbol{\sigma}$ and its magnetic moment would be the same, $e/2m$, so the intrinsic magnetic moment would be $(e/2m)|\mathbf{S}| = e\hbar/4m$. The consequent shift in the frequencies of spectral lines would then be those of the 'normal' Zeeman effect. Experiments revealed, however, an 'anomalous' Zeeman effect, explicable if the proportionality constant for spin is *twice* that for orbital motion so the electron magnetic moment is $-\boldsymbol{\mu}$ where

$$\mu = 2\frac{e}{2m}\mathbf{S} = \frac{e}{m}\mathbf{S} = \frac{e\hbar}{2m}\boldsymbol{\sigma}. \tag{2.153}$$

(The charge on the electron is here taken as $-e$). The factor of 2 is sometimes called the Landé g factor, $g_s = 2$. One of the successes of Dirac's theory of the electron is that it gives the correct value of g_s. To derive this we must consider the equation, not for a free electron, but for an electron in the presence of an electromagnetic field. There is a prescription for doing this, known as the 'minimal' prescription, which we shall consider more properly

---

[‡] This formula is correct in the SI system of units. It also holds in the c.g.s. system if $c = 1$.

in the next chapter. It consists in replacing the momentum p^μ by

$$p^\mu \to p^\mu - eA^\mu \tag{2.154}$$

or, with $p^\mu = (E, \mathbf{p})$, $A^\mu = (\phi, \mathbf{A})$,

$$E \to E - e\phi, \quad \mathbf{p} \to \mathbf{p} - e\mathbf{A}. \tag{2.155}$$

The Dirac equation (2.94) then becomes

$$\gamma^0(E - e\phi)\psi - \boldsymbol{\gamma} \cdot (\mathbf{p} - e\mathbf{A})\psi = m\psi. \tag{2.156}$$

In the standard representation of the γ matrices (see (2.130) and (2.131))

$$\gamma^0 = \begin{pmatrix} 1 & 0 \\ 0 & -1 \end{pmatrix}, \quad \boldsymbol{\gamma} = \begin{pmatrix} 0 & \boldsymbol{\sigma} \\ -\boldsymbol{\sigma} & 0 \end{pmatrix}, \quad \psi = \begin{pmatrix} u \\ v \end{pmatrix} \tag{2.157}$$

this becomes

$$(E - e\phi)\begin{pmatrix} u \\ -v \end{pmatrix} - (\mathbf{p} - e\mathbf{A}) \cdot \begin{pmatrix} 0 & \boldsymbol{\sigma} \\ -\boldsymbol{\sigma} & 0 \end{pmatrix}\begin{pmatrix} u \\ v \end{pmatrix} = m\begin{pmatrix} u \\ v \end{pmatrix}$$

which we write explicitly as two equations

$$(E - e\phi)u - \boldsymbol{\sigma} \cdot (\mathbf{p} - e\mathbf{A})v = mu, \tag{2.158}$$

$$-(E - e\phi)v + \boldsymbol{\sigma} \cdot (\mathbf{p} - e\mathbf{A})u = mv. \tag{2.159}$$

The second equation gives

$$v = \frac{\boldsymbol{\sigma} \cdot (\mathbf{p} - e\mathbf{A})}{E + m - e\phi} u.$$

In the non-relativistic limit $E \approx m$, $p \approx mv$, so (ignoring the fact that $(E + m - e\phi)$ and $\boldsymbol{\sigma} \cdot (\mathbf{p} - e\mathbf{A})$ do not commute)

$$v \approx \frac{1}{2m} \boldsymbol{\sigma} \cdot (\mathbf{p} - e\mathbf{A})u = 0\left(\frac{v}{c}\right)u \tag{2.160}$$

and we see that the bottom two components of ψ are much smaller than the top two. Inserting (2.160) into (2.158) gives

$$Eu = \frac{\boldsymbol{\sigma} \cdot \boldsymbol{\pi}\, \boldsymbol{\sigma} \cdot \boldsymbol{\pi}}{2m} u + mu + e\phi u$$

where $\boldsymbol{\pi} = \mathbf{p} - e\mathbf{A}$, and, with $E = m + W$, we have

$$Wu = \left[\frac{1}{2m}(\boldsymbol{\sigma} \cdot \boldsymbol{\pi})(\boldsymbol{\sigma} \cdot \boldsymbol{\pi}) + e\phi\right]u. \tag{2.161}$$

Now using $\sigma_i\sigma_j = \delta_{ij} + i\varepsilon_{ijk}\sigma_k$ it follows that

$$(\boldsymbol{\sigma} \cdot \mathbf{A})(\boldsymbol{\sigma} \cdot \mathbf{B}) = \mathbf{A} \cdot \mathbf{B} + i\boldsymbol{\sigma} \cdot (\mathbf{A} \times \mathbf{B}) \tag{2.162}$$

so

$$(\boldsymbol{\sigma} \cdot \boldsymbol{\pi})^2 = \boldsymbol{\pi} \cdot \boldsymbol{\pi} + i\boldsymbol{\sigma} \cdot (\boldsymbol{\pi} \times \boldsymbol{\pi})$$

$$= (\mathbf{p} - e\mathbf{A})^2 + i\boldsymbol{\sigma} \cdot (\mathbf{p} - e\mathbf{A}) \times (\mathbf{p} - e\mathbf{A}). \tag{2.163}$$

The only non-zero part of the cross product in the last term is

$$\mathbf{p} \times \mathbf{A} + \mathbf{A} \times \mathbf{p}. \tag{2.164}$$

Using the operator equation

$$[p_i, A_j] = -i\hbar\partial_i A_j$$

we have, taking the difference with the same equation with $i \leftrightarrow j$,

$$(p_i A_j - p_j A_i) + (A_i p_j - A_j p_i) = -i\hbar(\partial_i A_j - \partial_j A_i).$$

Multiplying both sides by ε_{ijk} and summing over i and j gives the k component of

$$\mathbf{p} \times \mathbf{A} + \mathbf{A} \times \mathbf{p} = -i\hbar\mathbf{\nabla} \times \mathbf{A} = -i\hbar\mathbf{B},$$

so we have a value for (2.164). Substituting finally in (2.161) gives $Wu = Hu$ where

$$H = \frac{1}{2m}(\mathbf{p} - e\mathbf{A})^2 + e\phi - \frac{e\hbar}{2m}\boldsymbol{\sigma}\cdot\mathbf{B}. \tag{2.165}$$

The first two terms give the classical Hamiltonian, and the last term is the interaction energy of a magnetic moment (2.153) in a magnetic field. Hence the Dirac equation predicts the correct electron magnetic moment, with $g_s = 2$. The other terms, which we neglected (see the remark above (2.160)), give a spin–orbit interaction with the correct Thomas precession factor of 2.

2.7 The relevance of the Poincaré group: spin operators and the zero mass limit

We have stated repeatedly that the Dirac equation holds for particles with spin $\frac{1}{2}$; and the way we derived it would seem to guarantee this, for we started off with spinor representations of $SU(2)$. But to complete the picture we need to find a *spin operator* S_i $(i = 1, 2, 3)$ with the right commutation relations

$$[S_i, S_j] = i\varepsilon_{ijk}S_k \tag{2.166}$$

and its square must be an invariant of the group, i.e. it must commute with all the generators:

$$\mathbf{S}\cdot\mathbf{S} = S^2 = s(s + 1) \tag{2.167}$$

where s is the spin of the particle. In addition, the fact that there are two solutions to the equations $\gamma\cdot pu = mu$ means that $\mathbf{S}$ must commute with $\gamma\cdot p$

$$[\mathbf{S}, \gamma\cdot p] \Rightarrow 0. \tag{2.168}$$

The search for $\mathbf{S}$ turns out to be a particularly difficult one, and we shall not pursue it to the end, but we shall find out enough to provide some valuable insights into the nature of spin.

As a first guess we may try

$$S = \tfrac{1}{2}\Sigma \equiv \tfrac{1}{2}\begin{pmatrix} \sigma & 0 \\ 0 & \sigma \end{pmatrix}. \tag{2.169}$$

This clearly has the right eigenvalues $\pm\tfrac{1}{2}$ for both positive and negative energy solutions, and obeys the commutation relations (2.166), but it does not obey (2.168) because

$$[\Sigma, \gamma^\mu] \neq 0.$$

This may be seen directly by working out one or two commutators, or by observing that, if we define

$$\sigma^{\mu\nu} = \frac{i}{2}[\gamma^\mu, \gamma^\nu], \tag{2.170}$$

then, in the standard representation (2.157),

$$\sigma^{ij} = \varepsilon_{ijk}\begin{pmatrix} \sigma^k & 0 \\ 0 & \sigma^k \end{pmatrix} = \varepsilon_{ijk}\Sigma_k, \tag{2.171}$$

and it then follows that

$$[\Sigma_i, \gamma_\mu] = i\varepsilon_{ijk}(\gamma_j g_{k\mu} - \gamma_k g_{j\mu}).$$

The relativistic spin operator, then, is not $\tfrac{1}{2}\Sigma$. (For a particle at rest, however, $\gamma \cdot p = E\gamma^0$, and then $\tfrac{1}{2}\Sigma$ is a good spin operator.) Σ is, of course, the matrix representative of $\mathbf{J}$, so we conclude that the relativistic spin operator is not $\tfrac{1}{2}\mathbf{J}$. This is confirmed by the fact that $\mathbf{J}\cdot\mathbf{J} = J^2$ (cf. (2.167)) does not commute with all the generators of the Lorentz group. For example,

$$[J^2, K_1] = [J_1^2, K_1] + [J_2^2, K_1] + [J_3^2, K_1]$$

and it is easy to see that (see (2.68))

$$[J_1^2, K_1] = 0,$$
$$[J_2^2, K_1] = -i(J_2 K_3 + K_3 J_2),$$
$$[J_3^2, K_1] = i(J_3 K_2 + K_2 J_3),$$

so that

$$[J^2, K_1] \neq 0,$$

and hence

$$[J^2, K_i] \neq 0. \tag{2.172}$$

Spin is a 'kinematic' property of elementary particles. The other obvious kinematic property they possess is mass. Both of these should be describable by quantities invariant under relativistic transformation. Now mass M is given by

$$M^2 = P_\mu P^\mu \tag{2.173}$$

and P_μ, the momentum operator, does not appear in the analysis of the (homogeneous) Lorentz group, outlined above. This is because, as we shall show, P_μ is the generator of *space–time translations*

$$x^\mu \to x'^\mu = x^\mu + a^\mu, \tag{2.174}$$

which we never considered. What we must do is to adjoin these translations to those of the Lorentz group. This gives the *inhomogeneous Lorentz group*, commonly called the *Poincaré group*. It was Wigner who performed the first analysis of this group, in a paper[13] which has become a classic. What he found was that mass and spin are indeed the two properties which characterise systems invariant under the Poincaré group, and that spin also corresponds to a rotation group symmetry $SU(2)$, *but only if* $M^2 > 0$, i.e. if the momentum is 'timelike'. In the case $M^2 = 0$, spin is not longer described by $SU(2)$, and this, in fact, is why the polarisation states of a massless particle with spin J are $J_z = \pm J$ only; for example, physical photons do not exist in a $J_z = 0$ state, whereas massive spin 1 particles do. In the case of spacelike momenta, $M^2 < 0$, 'spin' is different again, and may indeed correspond to a continuous parameter.

It is not my intention to explain this theory in any detail, but merely to give the reader enough of an idea of it to see how the notion of spin emerges. Let us start by finding the structure of the Poincaré group. If we perform a translation through a distance a^μ, $e^{iP\cdot a}$, then a boost to a frame moving with velocity $\mathbf{v} = \tanh \boldsymbol{\phi}$, $e^{i\mathbf{K}\cdot\boldsymbol{\phi}}$, then translate back through $-a^\mu$, and finally boost back by applying a velocity $-\mathbf{v}$, then what is the final result? The extent to which it differs from the starting point is, of course, the measure of the *structure* of the group. The standard practice is to consider infinitesimal transformations (from which finite ones may be generated) and so

$$\begin{aligned}
e^{-i\mathbf{K}\cdot\boldsymbol{\phi}}e^{-iP\cdot a}e^{i\mathbf{K}\cdot\boldsymbol{\phi}}e^{iP\cdot a} &= (1 - i\mathbf{K}\cdot\boldsymbol{\phi})(1 - iP\cdot a)(1 + i\mathbf{K}\cdot\boldsymbol{\phi})(1 + iP\cdot a) \\
&= 1 + [P_\mu, P_\nu]a^\mu a^\nu + 2[P_\mu, K_i]a^\mu \phi_i \\
&\quad + [K_i, K_j]\phi_i\phi_j.
\end{aligned}$$

Hence the structure of the group is known when the commutation relations between the generators are known. The inhomogeneous Lorentz group has ten generators: three Js for rotations, three Ks for boosts and four Ps for translations. The commutation relations between the Js and Ks have already been found (equation (2.68)), but they were derived using matrix forms for the Js and Ks. It will be instructive now to derive expressions for $\mathbf{J}$ and $\mathbf{K}$ as *differential operators* rather than matrices. They will, of course, obey the same commutation relations (2.68). It will then be straightforward to derive an expression for P_μ, and hence to calculate the complete set of commutation relations among the generators of the Poincaré group. (The commutation relations of the generators of a group are called the *Lie algebra* of the group – hence (2.68) is the Lie algebra of the Lorentz group.)

Let us begin by deriving an expression for J_z. It generates rotations about the z axis:

$$x' = x \cos \theta + y \sin \theta,$$
$$y' = - x \sin \theta + y \cos \theta,$$
$$z' = z.$$

The generator J_z is defined through its operation on a function $f(x, y, z)$ as

$$J_z f(x, y, z) = i \lim_{\theta \to 0} \left[\frac{f(x', y', z) - f(x, y, z)}{\theta} \right]$$

$$= i \lim_{\theta \to 0} \left[\frac{f(x + y\theta, y - x\theta, z) - f(x, y, z)}{\theta} \right]$$

$$= i \left(y \frac{\partial f}{\partial x} - x \frac{\partial f}{\partial y} \right) \tag{2.175}$$

or

$$J_z = - i \left(x \frac{\partial}{\partial y} - y \frac{\partial}{\partial x} \right). \tag{2.176}$$

Similarly

$$J_x = - i \left(y \frac{\partial}{\partial z} - z \frac{\partial}{\partial y} \right), \quad J_y = - i \left(z \frac{\partial}{\partial x} - x \frac{\partial}{\partial z} \right), \tag{2.177}$$

and we can easily prove that

$$[J_x, J_y] = i J_z \quad \text{and cyclic perms.} \tag{2.178}$$

Expressions (2.176) and (2.177) are, of course, the quantum mechanical expressions for angular momentum operators (having the factor $\hbar$), hence the observation that angular momentum operators generate rotations. The formula (2.175) may be rewritten according to the following definition. The generator corresponding to a parameter a^α is defined to be

$$X_\alpha = i \left(\frac{\partial x'}{\partial a^\alpha} \bigg|_{a=0} \frac{\partial}{\partial x} + \frac{\partial y'}{\partial a^\alpha} \bigg|_{a=0} \frac{\partial}{\partial y} + \frac{\partial z'}{\partial a^\alpha} \bigg|_{a=0} \frac{\partial}{\partial z} + \frac{\partial t'}{\partial a^\alpha} \bigg|_{a=0} \frac{\partial}{\partial t} \right)$$

$$= i \frac{\partial x'^\mu}{\partial a^\alpha} \frac{\partial}{\partial x^\mu} \quad (\alpha = 1, \ldots, r). \tag{2.179}$$

This refers to an r-parameter group of transformations. The reader may check that with $a^\alpha = \theta$, this definition gives J_z. Now we apply the formula (2.179) to pure Lorentz transformations:

$$x' = \gamma(x + vt), \quad y' = y, \quad z' = z, \quad t' = \gamma(t + vx), \quad \gamma = (1 - v^2)^{-\frac{1}{2}}.$$

We find that the generator K_x is given by

$$K_x = i\left(t\frac{\partial}{\partial x} + x\frac{\partial}{\partial t} \right) \tag{2.180}$$

and similarly

$$K_y = i\left(t\frac{\partial}{\partial y} + y\frac{\partial}{\partial t} \right), \quad K_z = i\left(t\frac{\partial}{\partial z} + z\frac{\partial}{\partial t} \right). \tag{2.181}$$

This gives

$$\left.\begin{array}{ll} [K_x, K_y] = -iJ_z & \text{and cyclic perms,} \\ [K_x, J_y] = iK_z & \text{and cyclic perms,} \\ [K_x, J_x] = 0, & \text{etc.,} \end{array}\right\} \tag{2.182}$$

exactly as in (2.68). We see again that pure boosts do not form a subgroup of the Lorentz group, but rotations do. Relations (2.178) and (2.182) form the Lie algebra of the Lorentz group, which may be written in a unified way by defining

$$J_{\mu\nu}(\mu, \nu = 0, \dots, 3)\begin{cases} J_{ij} = -J_{ji} = \varepsilon_{ijk}J_k \\ J_{i0} = -J_{0i} = K_i \end{cases} (i, j, k = 1, 2, 3). \tag{2.183}$$

We then have

$$[J_{\mu\nu}, J_{\rho\sigma}] = i(g_{\nu\rho}J_{\mu\sigma} - g_{\mu\rho}J_{\nu\sigma} + g_{\mu\sigma}J_{\nu\rho} - g_{\nu\sigma}J_{\mu\rho}). \tag{2.184}$$

To get the Lie algebra of the Poincaré group we have to adjoin the generators of translations

$$x'^\mu = x^\mu + a^\mu$$

which are, by (2.179),

$$P_x = i\frac{\partial}{\partial x}, \quad \text{etc.,}$$

or

$$P_\mu = i\frac{\partial}{\partial x^\mu}, \tag{2.185}$$

thus justifying the symbol P_μ – up to a factor $\hbar$, the generators of translations are the energy–momentum operators. It is now straightforward to prove the commutation relations

$$[P_\mu, P_\nu] = 0, \tag{2.186}$$

$$[P_\mu, J_{\rho\sigma}] = i(g_{\mu\rho}P_\sigma - g_{\mu\sigma}P_\rho). \tag{2.187}$$

Relation (2.186) shows that translations in different directions commute (which is intuitively obvious), and (2.186) and (2.187) show that both **J** and

P commute with the Hamiltonian $P_0 = H$, but **K** does not, so does not give a conserved quantity. Relations (2.184), (2.186) and (2.187) show that the Lie algebra involving the ten generators is indeed *closed* – the operators on the right-hand sides are contained within the set – and hence that the transformations do generate a group. This is, indeed, easy to verify directly, and for completeness we note the transformation laws. A general inhomogeneous Lorentz transformation (that is, including boosts, rotations and translations) is

$$x'^{\mu} = \Lambda^{\mu}{}_{\nu}x^{\nu} + a^{\mu}. \qquad (2.188)$$

The matrix Λ (a generalisation of (2.64) to include rotations) must preserve the 'length' of x: $x'^{\mu}x'_{\mu} = x^{\mu}x_{\mu}$, hence

$$\Lambda^{\mu}{}_{\rho}\Lambda^{\nu}{}_{\sigma}g_{\mu\nu} = g_{\rho\sigma} \qquad (2.189)$$

or

$$\Lambda^{\mu}{}_{\rho}\Lambda^{\sigma\rightarrow}_{\leftarrow\mu} = \delta^{\sigma}_{\rho} \qquad (2.190)$$

so the inverse of Λ^{μ}_{ν} is

$$(\Lambda^{-1})^{\mu}{}_{\nu} = \Lambda_{\nu}{}^{\mu} \qquad (2.191)$$

Now we perform a second transformation $\bar{\Lambda}$ on x'^{μ} above

$$\begin{aligned} x''^{\mu} &= \bar{\Lambda}^{\mu}{}_{\nu}(x'^{\nu}) + \bar{a}^{\mu} \\ &= \bar{\Lambda}^{\mu}{}_{\nu}\Lambda^{\nu}{}_{\kappa}x^{\kappa} + \bar{\Lambda}^{\mu}{}_{\nu}a^{\nu} + \bar{a}^{\mu}. \end{aligned} \qquad (2.192)$$

This is of the form (2.188), and we may express the group law as

$$\{\bar{\Lambda}, \bar{a}\}\{\Lambda, a\} = \{\bar{\Lambda}\Lambda, \bar{\Lambda}a + \bar{a}\}. \qquad (2.193)$$

The unit element is of course $\{1, 0\}$.

We come now to the Wigner method. It relies on the fact that for a state with momentum p^{μ}, the effect of a Lorentz transformation is to change p^{μ}, but to leave $p^{\mu}p_{\mu}$ unchanged. Indeed, a state $|p\rangle$ with

$$P^{\mu}|p\rangle = p^{\mu}|p\rangle \qquad (2.194)$$

is converted, by a transformation (Λ, a), into

$$U(\Lambda, a)|p\rangle = |\Lambda p\rangle \qquad (2.195)$$

with

$$P^{\mu}|\Lambda p\rangle = (\Lambda p)^{\mu}|\Lambda p\rangle \qquad (2.196)$$

but because of (2.190)

$$(\Lambda p)^2 = (\Lambda p)^{\mu}(\Lambda p)_{\mu} = P^{\mu}P_{\mu} = P^2. \qquad (2.197)$$

Hence a Lorentz transformation leaves $P^{\mu}P_{\mu}$ invariant. Actually, this is because $P^{\mu}P_{\mu}$ commutes with all the generators of the group (as the reader should check, following the method used in (2.172)), so is an invariant, called the *first Casimir invariant* C_1,

$$C_1 = P^\mu P_\mu. \tag{2.198}$$

As a consequence, all the states obtained by Lorentz transformations from an initial state have the same value of p^2. Moreover, since the sign of p^0 is unchanged by a Lorentz transformation, the complete catalogue of states forming bases for representations of the group falls into *six distinct classes*:

$$\left.\begin{array}{ll}
\text{(i)} & p^2 = m^2 > 0, \quad p^0 > 0, \\
\text{(ii)} & p^2 = m^2 > 0, \quad p^0 < 0, \\
\text{(iii)} & p^2 = 0, \quad p^0 > 0, \\
\text{(iv)} & p^2 = 0, \quad p^0 < 0, \\
\text{(v)} & p^\mu \equiv 0, \\
\text{(vi)} & p^2 < 0.
\end{array}\right\} \tag{2.199}$$

The first and third classes correspond to physical massive and massless particles, the fifth class is the vacuum, and the sixth should correspond to virtual particles (which often have spacelike momentum). The other classes are probably unphysical.

Having chosen a particular p^μ, belonging to a particular class $\{p^\mu\}$, an important observation, which we shall prove below, is that the subgroup of the Poincaré group that leaves p^μ invariant – which is called the *little group* of p^μ – has the same structure for all momenta in $\{p^\mu\}$. Now consider class (i), $p^2 = m^2$. A particular p^μ is the particle rest-frame – let us denote it k^μ:

$$k^\mu = (m, 0, 0, 0). \tag{2.200}$$

What is its little group? It is clearly the rotation group, since this will have no effect on k^μ:

The little group for k^μ is the rotation group $SU(2)$. $\qquad$ (2.201)

Hence, for a timelike momentum, to know the effect of an arbitrary Lorentz transformation requires only a knowledge of representations of the rotation group – which we all have! This is the conclusion of Wigner's work, and to understand it more properly we shall take a closer look.

Consider an arbitrary timelike p^μ. There clearly exists a Lorentz transformation which transforms k^μ (the rest-frame momentum above) into p^μ. Call it $L(p)$:

$$p^\mu = L^\mu{}_\nu(p)k^\nu \tag{2.202}$$

(In general, L will be a product $R^{-1}BR$ where R is a rotation which carries $\hat{p}$ into the z axis and B is a boost of the form (2.64)). We denote the states in Hilbert space $|p, \sigma\rangle$ and $|k, \sigma\rangle$: σ is a 'spin' index. Corresponding to (2.202) in space–time, we have the Hilbert space relation

$$|p, \sigma\rangle = U(L(p))|k, \sigma\rangle. \tag{2.203}$$

Here $U(L(p))$ is a unitary operator (matrix) which represents $L(p)$. Now consider an arbitrary Lorentz transformation Λ,

$$p^\mu \rightarrow p'^\mu = \Lambda^\mu_{\ \nu} p^\nu, \tag{2.204}$$

and the corresponding unitary transformation

$$|p, \sigma\rangle \rightarrow U(\Lambda)|p, \sigma\rangle. \tag{2.205}$$

We need to find $U(\Lambda)|p, \sigma\rangle$, and proceed as follows. First we use (2.203):

$$U(\Lambda)|p, \sigma\rangle = U(\Lambda)U(L(p))|k, \sigma\rangle$$

then multiply by what is clearly the identity

$$= U(L(\Lambda p))U^{-1}(L(\Lambda p))U(\Lambda)U(L(p))|k, \sigma\rangle$$

then use the group law $U^{-1}(A) = U(A^{-1})$

$$= U(L(\Lambda p))U(L^{-1}(\Lambda p))U(\Lambda)U(L(p))|k, \sigma\rangle$$

and the group law again in the form $U(A)U(B)U(C) = U(ABC)$

$$= U(L(\Lambda p))U(L^{-1}(\Lambda p)\Lambda L(p))|k, \sigma\rangle. \tag{2.206}$$

Now $L^{-1}(\Lambda p)\Lambda L(p)$ is a matrix which, when operating on k^μ, gives k^μ again, since $L(p)$ changes k into p (equation (2.202)), Λ change p into Λp, and $L^{-1}(\Lambda p)$ changes Λp back into k. $L^{-1}(\Lambda p)\Lambda L(p)$ is therefore a *rotation* (see (2.201)), and $U(L^{-1}(\Lambda p)\Lambda L(p))$ is therefore a matrix of the form $\exp i\mathbf{J}\cdot\boldsymbol{\theta}$, whose elements we denote $D_{\sigma'\sigma}(R)$, with $R = L^{-1}(\Lambda p)\Lambda L(p)$, so

$$U(\Lambda)|p, \sigma\rangle = U(L(\Lambda p))\sum_{\sigma'} D_{\sigma'\sigma}(R)|k, \sigma'\rangle$$

$$= \sum_{\sigma'} D_{\sigma'\sigma}(R)U(L(\Lambda p))|k, \sigma'\rangle$$

$$= \sum_{\sigma'} D_{\sigma'\sigma}(R)|\Lambda p, \sigma'\rangle, \tag{2.207}$$

using (2.203) in the last line. We conclude that, to know the representations of the Lorentz group for a timelike state, we need only know the representations of the rotation group! Hence spin, defined as whatever other label states may carry and which is affected by Lorentz transformations, is given by the rotation group, and we have proved that (2.201) applies to *all* timelike momenta. The reader should appreciate what an amazing result this is. Ever since we were undergraduates we have assumed that spin was a sort of angular momentum and therefore (though we might not have thought in this language!) given by a representation of the rotation group, but it is only as a result of this work of Wigner, that we understand why this is true! On the other hand, when we come to treat class (iii), states with lightlike momenta, we find that it is not true; spin is no longer given by the rotation group.

We have still to ask: if mass corresponds to the Casimir operator (2.198), what invariant operator corresponds to spin? We saw in (2.172) that it is not J^2. First we define the *Pauli–Lubanski pseudovector* W_μ:

$$W_\mu = -\tfrac{1}{2}\varepsilon_{\mu\nu\rho\sigma}J^{\nu\rho}P^\sigma. \tag{2.208}$$

Here $\varepsilon_{\mu\nu\rho\sigma}$ is the totally antisymmetric symbol in four dimensions. It is clear that W_μ is orthogonal to P^μ:

$$W_\mu P^\mu = 0, \tag{2.209}$$

so that in the particle rest-frame W is spacelike, $W_\mu = (0, \mathbf{W})$ with

$$\begin{aligned} W_i &= -\tfrac{1}{2}\varepsilon_{i\nu\rho\sigma}J^{\nu\rho}P^\sigma \\ &= -\frac{m}{2}\varepsilon_{ijk0}J^{jk} \\ &= -m\Sigma_i, \end{aligned}$$

showing that W_i reduces essentially to Σ_i in the rest-frame. It may be shown that the second Casimir invariant is

$$C_2 = W_\mu W^\mu = -m^2 s(s+1) \tag{2.210}$$

where s is the spin of the particle. The Poincaré group is of rank 2, so there are only two Casimir invariants, and we have now found them. What we still have not found, however, is the spin operators themselves. They cannot be $W_i (i = k, 2, 3)$ because these are only three components of a 4-vector, and, in any case, W_i do not have the required $SU(2)$ commutation relations (2.166). The correct spin operators are rather complicated in form, and the interested reader is referred to the literature.[15]

Let us briefly, and finally, consider the case of lightlike particles, $p^2 = 0$. Let us choose, corresponding to the rest frame in the timelike case,

$$k^\mu = (k, 0, 0, k) \tag{2.211}$$

which describes a massless particle moving along the z axis. We rely on Wigner's little group method again, so need to know what is the most general Lorentz transformation which leaves k^μ invariant. It turns out that it is a particular combination of boosts (with parameters u and v) and rotations (parameters θ and (believe it or not!) u, v), and, in place of $U(L^{-1}(\Lambda p)\Lambda L(p))$ in (2.206) we get

$$\begin{aligned} U &= 1 + i\theta J_3 + iu(K_1 - J_2) + iv(K_2 + J_1) \\ &\equiv 1 + i\theta J_3 + iuL_1 + ivL_2. \end{aligned} \tag{2.212}$$

The generators L_1, L_2 and J_3 form a Lie algebra which is easily seen to be

$$[L_1, L_2] = 0,$$
$$[J_3, L_1] = iL_2, \qquad (2.213)$$
$$[L_2, J_3] = iL_1.$$

This is clearly not the Lie algebra of the rotation group $SU(2)$, because of the zero in the first relation. In fact it corresponds to the group of rotations (generated by J_3) and translations (generated by L_1, L_2) in the plane – the so-called Euclidean group $E(2)$. The physical significance of this is obscure, but we can see that 'spin' for massless particles is not what it is for massive ones. We can in fact learn something about it by noting that, since $m^2 = 0$, and from (2.210), we have

$$W \cdot W |k\rangle = 0, \quad P \cdot P |k\rangle = 0 \qquad (2.214)$$

and from (2.209)

$$W \cdot P |k\rangle = 0,$$

so W^μ and P^μ are orthogonal and both lightlike. This means they must be proportional,

$$(W^\mu - \lambda P^\mu)|k\rangle = 0, \qquad (2.215)$$

and we have the result that the state of a massless particle is characterised by *one number* λ, which is the ratio of W^μ and P^μ and so has the dimensions of angular momentum. It is called *helicity*. If parity is included, the helicity takes on two values, λ and $-\lambda$. What seems (at least to me) to be a mystery is why λ is integral or half-integral.

This reproduces what we know about the neutrino and the photon (assuming the neutrino, or one of them, is massless). A left-handed massless neutrino obeys the Weyl equation and has $\lambda = -\frac{1}{2}$. Photons come in both right- and left-circularly polarised states, with $\lambda = \pm 1$ – but not $\lambda = 0$, which would appear if the photon were massive.

2.8   Maxwell and Proca equations

We turn now to particles of spin 1. Photons have no mass and are described by Maxwell's equations, and massive spin 1 particles (for example, the intermediate bosons $W^\pm$ of weak interactions) are described by the Proca equation.

Maxwell's equations are, of course, well known. Our only concern here is to show how they are cast in a *manifestly* Lorentz covariant form. It is clear that they are, in fact, Lorentz covariant: it was Einstein's observation that Maxwell's equations were Lorentz covariant that gave birth to the theory of relativity. We seek only a notation which expresses the covariance as neatly as possible.

Maxwell's equations (in Heaviside–Lorentz rationalised units, so that $e^2/4\pi hc = \alpha = 1/137$) are

(a) $\operatorname{div} \mathbf{B} = 0$,   (b) $\operatorname{curl} \mathbf{E} + \dfrac{\partial \mathbf{B}}{\partial t} = 0$,

(c) $\operatorname{div} \mathbf{E} = \rho$,   (d) $\operatorname{curl} \mathbf{B} - \dfrac{\partial \mathbf{E}}{\partial t} = \mathbf{j}$.   (2.216)

(a) tells us there are no magnetic charges. (b) is Faraday's law; a changing magnetic field produces an electric field. (c) is Gauss's law; the total charge inside a closed surface may be obtained by integrating the normal component of $\mathbf{E}$ over the surface. (d) is Ampère's law, $\operatorname{curl} \mathbf{B} = \mathbf{j}$, with Maxwell's additional term $\partial \mathbf{E}/\partial t$, stating that changing electric fields produce magnetic fields. The equations (a) and (b) are known as the homogeneous equations, (c) and (d) as the inhomogeneous ones.

Introducing the 4-vector potential

$$A^\mu = (\phi, \mathbf{A}) \tag{2.217}$$

with

$$\mathbf{B} = \operatorname{curl} \mathbf{A}, \quad \mathbf{E} = -\frac{\partial \mathbf{A}}{\partial t} - \nabla \phi, \tag{2.218}$$

equations (a) and (b) are *automatically satisfied*, since div curl $\equiv 0$ and curl grad $\equiv 0$. Now observe that the right-hand sides of equations (2.218) are the components of a 4-dimensional curl, defined by

$$F^{\mu\nu} = -F^{\nu\mu} = \partial^\mu A^\nu - \partial^\nu A^\mu. \tag{2.219}$$

It has components (recall that $\partial^i = -\partial_i$)

$$\begin{aligned}
F^{0i} &= \partial^0 A^i - \partial^i A^0 \\
&= \left(\frac{\partial \mathbf{A}}{\partial t} + \nabla \phi\right)_i \\
&= -E^i
\end{aligned} \tag{2.220}$$

and

$$\begin{aligned}
F^{ij} &= \partial^i A^j - \partial^j A^i \\
&= -\varepsilon^{ijk} B^k,
\end{aligned} \tag{2.221}$$

where $\varepsilon^{ijk} = \varepsilon_{ijk}$ is the totally antisymmetric Levi–Civita symbol (2.124). Equations (2.219) and (2.220) may be displayed in matrix form, with the rows and columns corresponding to the numbers $0, 1, 2, 3$:

$$F^{\mu\nu} = \begin{pmatrix} 0 & -E^1 & -E^2 & -E^3 \\ E^1 & 0 & -B^3 & B^2 \\ E^2 & B^3 & 0 & -B^1 \\ E^3 & -B^2 & B^1 & 0 \end{pmatrix}. \tag{2.222}$$

$F^{\mu\nu}$ is called the *electromagnetic field tensor*. It transforms, under Lorentz transformations, like an antisymmetric second rank tensor:

$$F^{\mu\nu} \to \Lambda^\mu{}_\alpha \Lambda^\nu{}_\beta F^{\alpha\beta}.$$

To summarise so far: if we write the electric and magnetic fields in terms of the tensor $F^{\mu\nu}$, then the statement that $F^{\mu\nu}$ is a 4-dimensional curl means that the first two (homogeneous) Maxwell equations are automatically satisfied.

Now consider the inhomogeneous equations. It is straightforward to verify that they are both contained in the covariant equation

$$\partial_\mu F^{\mu\nu} = j^\nu \tag{2.223}$$

with

$$j^\nu = (\rho, \mathbf{j}). \tag{2.224}$$

For putting $\nu = 0$ gives

$$\partial_1 F^{10} + \partial_2 F^{20} + \partial_3 F^{30} = \rho,$$

$$\text{div } \mathbf{E} = \rho,$$

which is (c), and putting $\nu = 1$ gives

$$\partial_0 F^{01} + \partial_2 F^{21} + \partial_3 F^{31} = j^1,$$

$$-\frac{\partial E^1}{\partial t} + \frac{\partial}{\partial x_2} B^3 - \frac{\partial}{\partial x_3} B^2 = j^1,$$

which is the '1' component of (d).

It is perhaps useful here to insert a remark about gauge transformations. Although (2.217) specifies the electric and magnetic fields in terms of $\mathbf{A}$ and ϕ, it does not do so *uniquely*, for under a *gauge transformation*

$$\mathbf{A} \to \mathbf{A} - \nabla\chi, \quad \phi \to \phi + \frac{\partial\chi}{\partial t} \tag{2.225}$$

which has the covariant form

$$A^\mu \to A^\mu + \partial^\mu\chi, \tag{2.226}$$

where χ is an arbitrary scalar function, and $\mathbf{E}$ and $\mathbf{B}$ remain unchanged; equivalently $F^{\mu\nu}$ is unchanged:

$$F^{\mu\nu} \to F^{\mu\nu} + (\partial^\mu \partial^\nu - \partial^\nu \partial^\mu)\chi = F^{\mu\nu}. \tag{2.227}$$

Substituting (2.219) into (2.223) we see that A^μ satisfies

$$\Box A^\nu - \partial^\nu(\partial_\mu A^\mu) = j^\nu. \tag{2.228}$$

We may now make use of the freedom (2.226) and choose a particular χ so that the transformed A^μ satisfies the *Lorentz gauge condition*:

$$\partial_\mu A^\mu = \frac{\partial\phi}{\partial t} + \nabla\cdot\mathbf{A} = 0. \tag{2.229}$$

In this 'choice of gauge' (2.228) becomes

$$\Box A^\mu = j^\mu \tag{2.230}$$

which of course stands for the well-known equations

$$\frac{\partial^2 \phi}{\partial t^2} - \nabla^2 \phi = \rho, \quad \frac{\partial^2 \mathbf{A}}{\partial t^2} - \nabla^2 \mathbf{A} = \mathbf{j} \tag{2.231}$$

whose solutions give the Liénard–Wiechert potentials. *In vacuo*, equation (2.230) becomes

$$\Box A^\mu = 0, \tag{2.232}$$

which means that the electromagnetic field, when its quantum nature is fully exploited, will be seen to correspond to massless particles (which therefore travel at the speed of light; hence relativity, which was where we came in).

We have now cast Maxwell's equations into a manifestly covariant form. The homogeneous equations (*a*) and (*b*) are summarised in equation (2.219). The inhomogeneous equations (*c*) and (*d*) are summarised in (2.223). We shall now show that there is a neat way of combining equations (2.218) and (2.219), so that no explicit reference to the vector potential A^μ need be made. From (2.219) it follows that

$$\partial^\lambda F^{\mu\nu} + \partial^\mu F^{\nu\lambda} + \partial^\nu F^{\lambda\mu} = 0 \tag{2.233}$$

as may trivially be checked. Now we define the *dual tensor* $\tilde{F}^{\mu\nu}$ by

$$\tilde{F}^{\mu\nu} = \tfrac{1}{2}\varepsilon^{\mu\nu\rho\sigma} F_{\rho\sigma} \tag{2.234}$$

where $\varepsilon^{\mu\nu\rho\sigma}$ is the Levi–Civita symbol in four dimensions (with $\varepsilon^{0123} = 1$). Its elements are easily seen to be

$$\tilde{F}^{\mu\nu} = \begin{pmatrix} 0 & -B^1 & -B^2 & -B^3 \\ B^1 & 0 & E^3 & -E^2 \\ B^2 & -E^3 & 0 & E^1 \\ B^3 & E^2 & -E^1 & 0 \end{pmatrix}. \tag{2.235}$$

Because of the antisymmetry of $\varepsilon^{\mu\nu\rho\sigma}$, it follows that the equation

$$\partial_\mu \tilde{F}^{\mu\nu} = 0 \tag{2.236}$$

yields (2.233); alternatively, by looking at (2.235) the reader will soon satisfy himself that it gives Maxwell's equations (*a*) and (*b*). In conclusion, then, Maxwell's equations may be written in the compact form

■ $$\partial_\mu F^{\mu\nu} = j^\nu, \quad \partial_\mu \tilde{F}^{\mu\nu} = 0. \tag{2.237}$$

Massive spin 1 particles obey equations which generalise Maxwell's equations. They are known as the *Proca equations*:

■ $$F^{\mu\nu} = \partial^\mu A^\nu - \partial^\nu A^\mu; \quad \partial_\mu F^{\mu\nu} + m^2 A^\nu = 0. \tag{2.238}$$

Taking the divergence of this we have

$$m^2 \partial_\nu A^\nu = 0, \tag{2.239}$$

and since $m^2 \neq 0$, we find $\partial_\nu A^\nu = 0$; the Lorentz condition, as it were, always holds, and we have lost the freedom of gauge transformations which Maxwell's equations had. In fact, since $F^{\mu\nu}$ is gauge invariant, it follows directly from (2.238) that *the equations for massive spin 1 particles are not gauge invariant*. Substituting (2.239) into (2.238) gives

$$(\Box + m^2)A^\mu = 0 \tag{2.240}$$

as well as

$$\partial_\mu A^\mu = 0. \tag{2.241}$$

Equation (2.240) shows, as expected, that we have particles of mass m. Equation (2.241) is one condition imposed on the four components of A^μ, so there are only three independent components. This is indeed appropriate for a massive spin 1 particle.

Let us conclude this section with a general remark on wave equations, which, for convenience, are reproduced together.

(Klein–Gordon)	$(\Box + m^2)\phi = 0,$	(2.16)
(Dirac)	$(i\gamma^\mu \partial_\mu - m)\psi = 0,$	(2.96)
	$(\Box + m^2)\psi = 0,$	(2.98)
(Maxwell)	$\partial_\mu F^{\mu\nu} = 0,$	(2.223)
	$\Box A^\mu = 0,$	(2.232)
(Proca)	$\partial_\mu F^{\mu\nu} + m^2 A^\nu = 0,$	(2.238)
	$(\Box + m^2)A^\mu = 0.$	(2.240)

In the case of spin $\frac{1}{2}$ and spin 1 fields, there are *first order* wave equations, but there is no first order wave equation for scalar fields. On the other hand, every component of the spin $\frac{1}{2}$ and spin 1 field satisfies a Klein–Gordon equation (with $m = 0$ for the photon), which is, after all, only a requirement of relativity $(E^2 - p^2 = m^2)$ and quantum theory $(E \to i(\partial/\partial t),\ \mathbf{p} \to -i\nabla)$. Thus the Dirac, Maxwell and Proca equations are of a different type from the Klein–Gordon equation. We saw above that the Dirac equation could be *derived* by considering the transformation of spinors under the Lorentz group, and it may be shown[18], though we did not do so here, that the Maxwell and Proca equations can be obtained in the same way. So these equations, for fields with non-zero spin, are simply a relation between the spin components; in Weinberg's words, they are a confession that we have too many spin components. The Klein–Gordon equation is not of this nature, since we only have one component.

One final observation: in our derivation of the Dirac equation we relied crucially on the assumption that the components of the spin $\frac{1}{2}$ field form a *linear vector space*, suitable as a basis for constructing a representation of the Lorentz group. This assumption, while it may look mathematically innocuous, is physically highly non-trivial, for it corresponds to a *principle of superposition* and therefore to wave–particle duality and the quantum theory. In other words, the fields we have found are *already* quantum fields. The statement, often found in the literature, that we must now subject these fields to 'second quantisation' is, in this light, misleading. It is better to say that we shall explore further the implication that these fields are quantum fields, by (for example) writing down the commutation relations which must hold between them; this we do in chapter 4.

2.9 Maxwell's equations and differential geometry

Maxwell's equations (2.237) relate antisymmetric tensors and vectors, but, as indicated by the indices, do so component by component. Compared with equations such as $\mathbf{V} \cdot \mathbf{B} = 0$, this may be considered a retrograde step; $\nabla \cdot \mathbf{B}$ is a more economical notation than $\nabla^i B_i$, to which it is equivalent. So we are led to ask, is there a way of writing Maxwell's equations in terms of the tensor F and current j, without making explicit reference to the components? Because of the development of differential geometry, there is indeed a way to do this, and Maxwell's equations take on the elegant from $dF = 0$, $d*F = J$; the antisymmetry of the field tensor F is automatically included! In this section we shall explain this notation. A common reaction of physicists to this type of mathematical development is one of impatience. After all, they point out, the equation $d*F = J$ has to be translated into the form $\partial_\mu F^{\mu\nu} = j^\nu$ before it can be dealt with in a particular co-ordinate system. This may be true, but in the opinion of a growing number of physicists the development of notation actually corresponds to a deepening in our understanding. The point is that, like many developments in contemporary mathematics, this one has evolved by introducing new concepts, and making distinctions that have not in the past been made. In the case of Maxwell's equations, light is shed on their geometrical interpretation. As will be discussed in the next chapter, electromagnetism is a gauge theory, having a $U(1)$ invariance group. Gauge theories with non-Abelian invariance groups ($SU(2) \times U(1)$, $SU(3)$) play a central role in contemporary particle physics, and the geometrical interpretation of them, along the lines mentioned, may indeed play no minor part in the understanding of their ultimate significance.

To begin, consider the meaning of ordinary line and surface integrals:

$$\left.\begin{aligned}
I_1 &= \int_C F_x\,dx + F_y\,dy + F_z\,dz = \int_C \mathbf{F}\cdot d\mathbf{r}, \\
I_2 &= \int_S (G_x\,dy \wedge dz + G_y\,dz \wedge dx + G_z\,dx \wedge dy) \\
&= \int_S \mathbf{G}\cdot d\mathbf{S}.
\end{aligned}\right\} \qquad (2.242)$$

I_1 and I_2 are *numbers*. I_1 is the integral of something over a line C, and I_2 the integral of something else over a surface S. In some sense, then, the 'something' is *dual* to the 'line', since when they are 'combined' (by the integral) the result is a pure number. Similarly, for I_2, the 'something else' is dual to the 'surface'. We systematise this by coining new words; the line and the surface are called 'chains', and the objects integrated over the chains are called 'differential forms' or simply 'forms'. Thus forms are dual to chains.

We shall call a line a 1-chain, since it has one dimension, a surface, a 2-chain, etc., and denote the generic chain C_n, with n dimensions. So we have

$$\left.\begin{aligned}
C_0 \quad &\text{0-chain = point,} \\
C_1 \quad &\text{1-chain = line,} \\
C_2 \quad &\text{2-chain = area,} \\
C_3 \quad &\text{3-chain = volume,} \\
C_n \quad &\text{n-chain.}
\end{aligned}\right\} \qquad (2.243)$$

Now the *boundary* of an n-chain is an $(n-1)$-chain. The boundary of an area is a line, and that of a line two points. We define a *boundary operator* ∂ which maps C_n into C_{n-1}

$$C_n \xrightarrow{\partial} C_{n-1} \quad \text{or} \quad \partial C_n = C_{n-1}. \qquad (2.244)$$

Some chains have no boundaries: the surface of a sphere is a 2-chain (area) with no boundary, and a closed line like a circle is a 1-chain with no boundary. Such *closed chains* are called *cycles* and denoted Z_n. Since they have no boundary, it is clear that

$$\partial Z_n = 0. \qquad (2.245)$$

(Z_n is actually the kernel of the mapping (2.244).) On the other hand, there are some chains which *themselves are boundaries* of higher dimensional chains, and these are denoted B_n:

$$B_n = \partial C_{n+1}. \qquad (2.246)$$

(B_{n-1} is actually the image of the map (2.244).) For example, a closed surface B_2 is the boundary of a volume, and a closed line B_1 is the boundary of an area. It is clear that the B_ns themselves have no boundary (are closed):

$$\partial B_n = 0. \tag{2.247}$$

Combining these last two equations gives

■ $$\partial^2 = 0. \tag{2.248}$$

In words, 'the boundary of a boundary is zero', or a chain which is a boundary is also closed. An interesting consideration is whether the converse holds: is a closed chain necessarily the boundary of another chain? In Euclidean spaces the answer is yes, so that $Z_n = B_n$. In general, however, there are closed chains which are not boundaries, so $Z_n \supset B_n$. For example, on a torus, a closed curve like C_1 in Fig. 2.3 is not the boundary of any part of the surface of the torus, whereas C_2 obviously is. Similarly, on the space of a circle S^1, the circle itself is not the boundary of any part of the space; it may not be thought of as the boundary of the area enclosed, because the area, which is 2-dimensional, is not part of S^1, which is 1-dimensional. This completes what we need to say about chains.

We now turn to *forms*. As mentioned above, the integral of a form over a chain is a number. We write

$$\int_{C_n} \omega_n \equiv \int_{C_n} f_{i_1 \ldots i_n} \, dx_{i_1} \wedge dx_{i_2} \wedge \cdots \wedge dx_{i_n} = \text{number.} \tag{2.249}$$

The wedge product, $\wedge$, appears above because the *orientation* of a curve or surface, etc., is important. The existence of integrals implies a duality between forms and chains, which we shall see manifested. A 1-form ω_1 is something to be integrated over a line (1-chain), so in 3-dimensional space is of the form $A\,dx + B\,dy + C\,dz$. Other forms follow the same pattern, so we have (in 3-dimensional space)

$$\left.\begin{array}{lll} \omega_0 & \text{0-form} & \text{function,} \\ \omega_1 & \text{1-form} & A\,dx + B\,dy + C\,dz, \\ \omega_2 & \text{2-form} & f\,dx \wedge dy + g\,dy \wedge dz + h\,dz \wedge dx, \\ \omega_3 & \text{3-form} & F\,dx \wedge dy \wedge dz, \end{array}\right\} \tag{2.250}$$

where

$$dx \wedge dy = -dy \wedge dx, \quad dx \wedge dx = 0, \quad \text{etc.} \tag{2.251}$$

Fig. 2.3. Closed curves on a torus; C_1 is not the boundary of a surface belonging to the torus, but C_2 is.

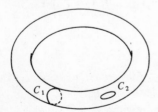

Because of (2.251), it is clear that in an n-dimensional space there are n-forms, but not $(n+1)$- or higher degree forms. Clearly, if we differentiate an n-form we will get something like an $(n+1)$-from: we shall get *precisely* an $(n+1)$-from if we build into the differentiation the antisymmetrisation above. Hence we define a so-called *exterior derivative operator* d:

$$d\omega_n = \omega_{n+1}.$$

(2.252)

Its action on a 1-form (in 3-space) is

$$d(A\,dx + B\,dy + C\,dz)$$

$$= \frac{\partial A}{\partial y}dy \wedge dx + \frac{\partial A}{\partial z}dz \wedge dx + \frac{\partial B}{\partial x}dx \wedge dy$$

$$+ \frac{\partial B}{\partial z}dz \wedge dy + \frac{\partial C}{\partial x}dx \wedge dz + \frac{\partial C}{\partial y}dy \wedge dz$$

$$= \left(\frac{\partial B}{\partial x} - \frac{\partial A}{\partial y}\right)dx \wedge dy + \left(\frac{\partial C}{\partial y} - \frac{\partial B}{\partial z}\right)dy \wedge dz$$

$$+ \left(\frac{\partial A}{\partial z} - \frac{\partial C}{\partial x}\right)dz \wedge dx.$$

(2.253)

By studying this example, the student should learn how d operates on any form, and so convince himself that the 2-form

$$\omega_2 = f\,dx \wedge dy + g\,dy \wedge dz + h\,dz \wedge dx$$

has exterior derivative

$$d\omega_2 = \left(\frac{\partial f}{\partial z} + \frac{\partial g}{\partial x} + \frac{\partial h}{\partial y}\right)dx \wedge dy \wedge dz.$$

(2.254)

In the first example, (2.253), the quantities

$$\frac{\partial C}{\partial y} - \frac{\partial B}{\partial z}, \quad \frac{\partial A}{\partial z} - \frac{\partial C}{\partial x}, \quad \frac{\partial B}{\partial x} - \frac{\partial A}{\partial y}$$

are the three components of curl $\mathbf{F}$ where $\mathbf{F} = \mathbf{i}A + \mathbf{j}B + \mathbf{k}C$. In the second example, putting $(g, h, f) = \mathbf{W}$, a vector, the quantity $\partial g/\partial x + \partial h/\partial y + \partial f/\partial z$ is div $\mathbf{W}$.

Now note that, in view of (2.254), if we calculate the exterior derivative of (2.253), we get identically zero; in other words

$$d[d(A\,dx + B\,dy + C\,dz)] = d^2(A\,dx + B\,dy + C\,dz) = 0$$

or, in general,

■ $$d^2 = 0.$$

(2.255)

In component language, this reads

$$\text{div curl} = 0;$$

d is sometimes called the *coboundary operator*, to emphasize the fact that $d^2 = 0$ is the dual of $\partial^2 = 0$, equation (2.248); $d^2 = 0$ is known as the *Poincaré lemma*.

An n-form ω_n is called *closed* if $d\omega_n = 0$.
An n-form ω_n is called *exact* if it is the derivative of an $(n-1)$-form,
$$\omega_n = d\omega_{n-1}.$$

The Poincaré lemma tells us that all exact forms are closed, since $d(d\omega_{n-1}) = d^2\omega_{n-1} = 0$, but it is not in general true that all closed forms are exact, though this is true in Euclidean spaces. Again, this is because of the duality between chains and forms: in Euclidean spaces, all closed chains are boundaries.

Well-known results follow from *Stokes' formula*, which states that if ω is a p-form, and c a $(p+1)$-chain, then

$$\int_{\partial c} \omega = \int_c d\omega. \tag{2.256}$$

As an example, put $p = 2$. Let ω then be the 2-form (in 3-dimensional space),

$$\omega_2 = A_x \, dy \wedge dz + A_y \, dz \wedge dx + A_z \, dx \wedge dy,$$

and let C_3 be a domain V, with boundary ∂V. Then Stokes' formula gives

$$\int_{\partial V} A_x \, dy \wedge dz + A_y \, dz \wedge dx + A_z \, dx \wedge dy$$

$$= \int_V \left(\frac{\partial A_x}{\partial x} + \frac{\partial A_y}{\partial y} + \frac{\partial A_z}{\partial z} \right) dx \wedge dy \wedge dz$$

where we have used (2.254); and this is

$$\oint_{\partial V} \mathbf{A} \cdot d\mathbf{S} = \int_V \operatorname{div} \mathbf{A} \, dV \tag{2.257}$$

which is the divergence theorem (or Gauss's theorem). As an exercise, the reader may show that with $p = 1$ we recover Stokes' theorem

$$\oint_{\partial A} \mathbf{A} \cdot d\mathbf{l} = \int_A \operatorname{curl} \mathbf{A} \cdot d\mathbf{S}. \tag{2.258}$$

We have now seen the relation between the exterior derivative operator d and the usual differential operators of grad, div and curl. However, as seen in the derivation of (2.257) above, if the components of $\mathbf{A}$ are the coefficients of a 2-form, div $\mathbf{A}$ is the coefficient of the 3-form obtained by operating d on the 2-form. In the ordinary language of vectors, the operator ∇ changes a scalar into a vector, and a vector either into a scalar (div) or into an axial vector (curl). There is, however, an operator which does not change the vectorial character; this is the *Laplacian* ∇^2 (the d'Alembertian $\square$ in four dimensions). $\nabla^2 \phi$ is a scalar, $\nabla^2 \mathbf{A}$ is a vector, etc. How is this represented in the language of forms? d changes a p-form into a $(p + 1)$-form, so we need to combine it with another operator (δ) which changes a p-form into a $(p - 1)$-form. We shall now show how this is done.

For definiteness, let us work in 3-dimensional space. The space of 1-forms is obviously 3-dimensional, with basis dx, dy and dz. The space of 2-forms is also 3-dimensional; in fact the bases may be written as follows:

$$\left.\begin{array}{ll} \text{Basis of } \omega_p \text{ for } n = 3 \\ \omega_0: & 1, \\ \omega_1: & dx, dy, dz, \\ \omega_2: & dx \wedge dy, dy \wedge dz, dz \wedge dx. \\ \omega_3: & dx \wedge dy \wedge dz. \end{array}\right\} \tag{2.259}$$

Obviously, there are no 4-forms in 3-dimensional space. It is clear that the dimensionality of the space of p-forms is the same as that of the space of $(n - p)$-forms, so we may define an operator which converts one into the other. It is known as the *Hodge* operator* or *duality transformation*. In a Euclidean (flat) space it is defined by

$$*(dx^{i_1} \wedge dx^{i_2} \wedge \cdots \wedge dx^{i_p})$$
$$= \frac{1}{(n - p)!} \varepsilon_{i_1 i_2 \ldots i_p i_{p+1} i_n} dx^{i_{p+1}} \wedge dx^{i_{p+2}} \wedge \cdots \wedge dx^{i_n}. \tag{2.260}$$

Thus in the case $n = 3$ we have the following basis for $*\omega$:

$$\left.\begin{array}{ll} \text{Basis of } *\omega_p \text{ for } n = 3 \\ *\omega_0: & dx \wedge dy \wedge dz, \\ *\omega_1: & dy \wedge dz, dz \wedge dx, dx \wedge dy, \\ *\omega_2: & dz, dx, dy, \\ *\omega_3: & 1. \end{array}\right\} \tag{2.261}$$

Repeating the operator $*$ on a p-form ω_p gives

$$**\omega_p = (-1)^{p(n-p)} \omega_p. \tag{2.262}$$

(To convince oneself that sign changes occur, simply consider the case of $p = 1$, $n = 2$.) It is clear that whereas $d\omega_p \sim \omega_{p+1}$, $d(*\omega_p) \sim *\omega_{p-1}$, so we define the operator δ:

$$\delta = (-1)^{np+n+1} *d*, \tag{2.263}$$

where p is the degree of the form ω_p on which δ is applied and n is the dimension of the space; δ is called the *adjoint exterior derivative operator*, and it should be clear that $\delta\omega$ is of degree $(p-1)$.

As an example, let us show that δ changes the 1-form $\mathbf{v}\cdot\mathbf{ds}$ into a 0-form:

$$
\begin{aligned}
\delta(\mathbf{v}\cdot\mathbf{ds}) &= \delta(v_x\,dx + v_y\,dy + v_z\,dz) \\
&= -\,{}^*d{}^*(v_x\,dx + v_y\,dy + v_z\,dz) \\
&= -\,{}^*d(v_x\,dy \wedge dz + v_y\,dz \wedge dx + v_z\,dx \wedge dy) \\
&= -\,{}^*\!\left(\frac{\partial v_x}{\partial x} + \frac{\partial v_y}{\partial y} + \frac{\partial v_z}{\partial z}\right)dx \wedge dy \wedge dz \\
&= -\operatorname{div}\mathbf{v}.
\end{aligned}
\tag{2.264}
$$

It is easy to see that, like d, δ has a square of zero:

$$
\begin{aligned}
\delta\delta &= (-1)^{np+n+1}(-1)^{n(p-1)+n+1}{}^*d{}^*{}^*d{}^* \\
&= (-1)^{pn-p^2+n}{}^*d^2{}^*,
\end{aligned}
$$

and in view of (2.255) we have

$$
\delta^2 = 0. \tag{2.265}
$$

Finally, the *Laplacian* Δ changes p-forms into p-forms and is defined by

$$
\Delta = (d + \delta)^2 = d\delta + \delta d. \tag{2.266}
$$

After this long preamble, it is now an easy matter to show how Maxwell's equations may be put into a geometric (or 'intrinsic') form. The space we work in is of course 4-dimensional Minkowski space–time. Lowering the indices on the field tensor $F^{\mu\nu}$ in (2.222) gives ($E_x \equiv E_1$, etc.)

$$
\left.
\begin{aligned}
F_{01} &= E_x, & F_{02} &= E_y, & F_{03} &= E_z, \\
F_{12} &= -B_z, & F_{31} &= -B_y, & F_{23} &= -B_x.
\end{aligned}
\right\}
\tag{2.267}
$$

We then define the *Faraday 2-form* F by

$$
\begin{aligned}
F &= -\tfrac{1}{2}F_{\mu\nu}\,dx^\mu \wedge dx^\nu \\
&= (E_x\,dx + E_y\,dy + E_z\,dz) \wedge dt \\
&\quad + B_z\,dx \wedge dy + B_x\,dy \wedge dz + B_y\,dz \wedge dx.
\end{aligned}
\tag{2.268}
$$

The dual form (also a 2-form) ${}^*F$ is (see (2.260))

$$
\begin{aligned}
{}^*F &= -\tfrac{1}{2}F_{\mu\nu}{}^*(dx^\mu \wedge dx^\nu) \\
&= -E_x\,dy \wedge dz - E_y\,dz \wedge dx - E_z\,dx \wedge dy \\
&\quad + (B_x\,dx + B_y\,dy + B_z\,dz) \wedge dt.
\end{aligned}
\tag{2.269}
$$

It is seen that the components of ${}^*F$ in the basis $dx^\mu \wedge dx^\nu$ are $-\tfrac{1}{2}\tilde{F}_{\mu\nu}$ as shown in (2.235)

$$
{}^*F = -\tfrac{1}{2}\tilde{F}_{\mu\nu}\,dx^\mu \wedge dx^\nu. \tag{2.270}
$$

Finally, we define the current density 3-form J:

$$J = (j_x\, dy \wedge dz + j_y\, dz \wedge dx + j_z\, dx \wedge dy) \wedge dt$$
$$- \rho\, dx \wedge dy \wedge dz. \tag{2.271}$$

A simple exercise shows that Maxwell's equations are

■ $dF = 0, \quad d*F = J. \tag{2.272}$

This may be seen either by putting $F = -\frac{1}{2}F_{\mu\nu}\, dx^{\mu} \wedge dx^{\nu}$ and then observing that $dF = 0$ implies equation (2.233), which is equivalent to the two homogeneous Maxwell's equations. Alternatively, using equation (2.268), we find explicitly

$$dF = \frac{\partial E_x}{\partial y}\, dy \wedge dx \wedge dt + \frac{\partial E_x}{\partial z}\, dz \wedge dx \wedge dt$$

$$+ \frac{\partial E_y}{\partial x}\, dx \wedge dy \wedge dt + \cdots$$

$$+ \frac{\partial B_z}{\partial z}\, dz \wedge dx \wedge dy + \frac{\partial B_z}{\partial t}\, dt \wedge dx \wedge dy + \cdots$$

$$= \left(\frac{\partial E_y}{\partial x} - \frac{\partial E_x}{\partial y} + \frac{\partial B_z}{\partial t} \right) dx \wedge dy \wedge dt + \cdots$$

$$+ (\text{div } \mathbf{B}) dx \wedge dy \wedge dz. \tag{2.273}$$

$dF = 0$ then implies the two equations

$$\text{curl } \mathbf{E} + \frac{\partial \mathbf{B}}{\partial t} = 0; \quad \text{div } \mathbf{B} = 0,$$

which are the homogeneous Maxwell equations. Similar manipulations show that $d*F = J$ yields the inhomogeneous equations.

In Euclidean space (which, for our purposes, may be extended to Minkowski space), the converse of the Poincaré lemma holds: all closed forms are exact, so if $dF = 0$, then there is a 1-form A such that

$$F = dA. \tag{2.274}$$

The 1-form A will be, in a co-ordinate basis,

$$A = A_{\mu}\, dx^{\mu} \tag{2.275}$$

and it follows immediately that equation (2.274) is equivalent to $F_{\mu\nu} = \partial_{\mu}A_{\nu} - \partial_{\nu}A_{\mu}$, as in (2.219). The 1-form A has a further geometric significance; it is the *connection form* which is used to define covariant derivatives – this is treated in the next chapter. If, however, we start from the point of view that the presence of electromagnetism gives a covariant derivative, and therefore a connection 1-form A, then $F = dA$ is known as the 'curvature' 2-form, and the identity $dF = 0$ is known as the *Bianchi identity*.

Summary[‡]

[2]Scalar particles are described by the Klein–Gordon equation, but the difficulties of interpretation are that the probability density is not positive definite, and negative energy states exist. It is concluded that the Klein–Gordon equation is not suitable as a single-particle equation. [3]The Dirac equation describes spin $\frac{1}{2}$ particles, and may be derived from the group $SL(2, C)$ which is shown to embody Lorentz transformations, extended by parity. The Dirac equation is found to have positive probability density, but has negative, as well as positive energy states. Massless spin $\frac{1}{2}$ particles are shown to obey the Weyl equation. [4]The hypothesis that the negative energy states are filled leads to the successful prediction of antiparticles. [5]It is shown how to construct spinors which satisfy the Dirac equation, and the transformation properties of bilinear forms $\bar{\psi}O\psi$ are studied, as well as various algebraic identities involving spinors and γ-matrices. [6]The Dirac equation gives the correct gyromagnetic ratio for the electron. [7]It is observed that the 'Casimir' invariant operator for spin, $s(s + 1)$, is not constructed from generators of the Lorentz group, but from those of the inhomogeneous Lorentz (= Poincaré) group, whose other Casimir invariant is (mass)^2. Particles with $m^2 > 0$ have a 'little group' $SU(2)$, which is interpreted as spin, whereas those with $m^2 = 0$ and $m^2 < 0$ have non-compact little groups, so their spin is not described by a rotation group. [8]Maxwell's equations are exhibited in manifestly covariant form, and the Proca equations, for massive spin 1 particles, are written down. [9]Chains and differential forms are explained, and Maxwell's equations exhibited in terms of differential forms.

Guide to further reading

Good accounts of spinors and the relation between $O(3)$ and $SU(2)$ are to be found in

(1) B.L. van der Waerden, *Group Theory and Quantum Mechanics*, Springer-Verlag, 1974.
(2) L.D. Landau & E.M. Lifshitz, *Quantum Mechanics*, §§55, 56, Pergamon Press, 1977.
(3) C.W. Misner, K.S. Thorne & J.A. Wheeler, *Gravitation*, W.H. Freeman & Co., 1973.
(4) R.U. Sexl & H.K. Urbantke, *Relativität, Gruppen, Teilchen*, Springer-Verlag, 1976.
(5) J.M. Normand, *A Lie group: Rotations in Quantum Mechanics*, North-Holland Publishing Company, 1980.

[‡] Superscripts refer to section numbers.

The topological distinction between $SU(2)$ and $O(3)$ is explained in ref. (4) §7.6, and

(6) D. Speiser, in F. Gürsey (ed.), *Group Theoretical Methods and Concepts in Elementary Particle Physics*, Gordon & Breach, 1964.

(7) F. Gürsey, in B. DeWitt & C. DeWitt (eds.), *Relativity, Groups and Topology*, Blackie & Sons Ltd, (London) and Gordon & Breach (New York), 1964.

An excellent account (in French) of the Lorentz group and the connection with $SL(2, C)$ appears in

(8) A.S. Wightman, in C. DeWitt & R. Omnès (eds.), *Relations de Dispersion et Particules Élémentaires*, Hermann & John Wiley, 1960.

See also refs. (1), (3), (4) and

(9) R. Omnès & M. Froissart in C. DeWitt & M. Jacob (eds.), *High Energy Physics*, Gordon & Breach, 1965.

(10) V.B. Berestetskii, E.M. Lifshitz & L.P. Pitaevskii, *Relativistic Quantum Theory, Part 1*, §§17, 18, Pergamon Press, 1971.

For full accounts of the Dirac equation, see, for example,

(11) J.D. Bjorken & S.D. Drell, *Relativistic Quantum Mechanics*, McGraw-Hill, 1964.

(12) C. Itzykson & J.B. Zuber, *Quantum Field Theory*, McGraw-Hill, 1980.

Wigner's analysis of the Poincaré group appears in

(13) E.P. Wigner, *Annals of Mathematics*, **40**, 149 (1939).

See also

(14) E.P. Wigner in *Theoretical Physics*, International Atomic Energy Agency, Vienna, 1963.

The relativistic treatment of spin is considered in refs. (8) and (9) and by

(15) F. Gürsey in C. DeWitt & M. Jacob (eds.), ref. (9).

An early, and rather brief, account is given by

(16) W. Pauli, *Ergebnisse der exakten Naturwissenschaften*, **37**, 85 (1965).

The connection between the Poincaré group and wave equations goes back to

(17) V. Bargmann & E.P. Wigner, *Proceedings of the National Academy of Sciences (USA)*, **34**, 211 (1948).

and is explored further by

(18) S. Weinberg, *Physical Review*, **133B**, 1318 (1964); **134B**, 882 (1964), and in K. Johnson, D.B. Lichtenberg, J. Schwinger & S. Weinberg, *Lectures on Particles and Field Theory* (Brandeis Summer Institute in Theoretical Physics, 1964, Vol. 2), Prentice-Hall, Inc., 1965.

Introductions to differential forms, suitable for the physicist, are

(19) H. Flanders, *Differential Forms with Applications to the Physical Sciences*, Academic Press, 1963.

(20) Y. Choquet-Bruhat, C. Morette-DeWitt & M. Bleik-Dillard, *Analysis, Manifolds and Physics*, North-Holland Publishing Company, 1977.

(21) Y. Choquet-Bruhat, *Géometrie Différentielle et Systèmes Extérieurs*, Dunod, 1968.

(22) B. Schutz, *Geometrical Methods of Mathematical Physics*, Cambridge University Press, 1980.

(23) T. Eguchi, P.B. Gilkey & A.J. Hanson, *Physics Reports*, **66**, 213 (1980).

(24) C. von Westenholz, *Differential Forms in Mathematical Physics*, North-Holland Publishing Company, 1978.

Less complete, but very readable accounts appear in ref. (3), chapters 3, 4, 8, 9 and

(25) C.W. Misner & J.A. Wheeler, *Annals of Physics*, **2**, 525 (1957).

3

Lagrangian formulation, symmetries and gauge fields

We have, up until now in this book, considered only the fields of the *free* spin $\frac{1}{2}$ particle, like the electron, and the spin 1 particle, like the photon. It is almost tautological, however, to state that it is only by virtue of the fact that particles *interact*, that we may observe them. It is, therefore, of great importance to enquire how we may describe particle interactions in field theory. In the old days – the 1950s and 1960s – there was a degree of arbitrariness in the choice of interaction. For example, the pion–nucleon interaction could be represented as $\bar{\psi}_N \gamma_5 \psi_N \phi_\pi$ or as $\bar{\psi}_N \gamma^\mu \gamma_5 \psi_N \partial_\mu \phi_\pi$. Present day theories, however, allow no such freedom. They state that interactions between *fundamental* fields (like electrons, quarks, weak vector bosons, and so on) are dictated by a gauge principle. This principle arises from the requirement that quantities which are conserved are conserved *locally* and not merely globally. Globally, conservation of electric charge would be satisfied if 10 C of charge suddenly disappeared from the Earth, and simultaneously appeared on Mars. But by 'conservation of charge' we usually mean something more than this; in fact this process would not be allowed, since ordinary notions of continuity demand that the disappearance of charge at one point be accompanied by a current, which makes possible its appearance at another point.

These matters are the concern of Noether's theorem, which connects symmetries and conservation laws, using a Lagrangian formulation of field theory. At the same time it emerges that in order to have a local symmetry, we need a spin 1 massless gauge field, whose interaction with the 'matter' fields is dictated uniquely. This gives us the electromagnetic, weak and strong interactions, the respective gauge fields being the photon, the weak boson and the gluons.

We begin, then, with a discussion of the Lagrangian formulation of field theory, which plays such a central role in the contemporary understanding

81

of interactions and symmetries. In the context of classical physics, by 'Lagrangian formulation' we mean 'the principle of least action'. This gives us a dynamical principle on which we base our formulation of the problem. When, in later chapters, we come to consider the quantum theory of fields, we shall also employ a Lagrangian formulation, in the shape of Feynman's path integral formalism, which is closely related to the least action formulation of classical field theory.

The simplest classical field to consider is the scalar field, but in order to familiarise ourselves with Lagrangian methods, let us first consider the classical mechanics of point particles, and derive Newton's law of motion from the principle of least action.

3.1 Lagrangian formulation of particle mechanics

In classical mechanics, a particle is idealised as a point of mass m, say. Let its position at time t be $x(t)$. If it moves in a region where the potential energy is $V(x)$, then Newton's second law tells us that

$$m\frac{d^2x}{dt^2} = F = -\frac{dV}{dx} \tag{3.1}$$

where F is the force acting on the particle. The principle of least action is a way of deriving this. The Lagrangian L is defined by

$$L = T - V = \tfrac{1}{2}m\left(\frac{dx}{dt}\right)^2 - V(x) \tag{3.2}$$

where T and V are the kinetic and potential energy; and the action S by

$$S = \int L\,dt \tag{3.3}$$

where the integral is taken over an entire path, from t_1 to t_2, as in Fig. 3.1;

$$S = \int_{t_1}^{t_2} L(x, \dot{x})\,dt. \tag{3.4}$$

Here, and in what follows, we put $dx/dt = \dot{x}$, $d^2x/dt^2 = \ddot{x}$. As the particle

Fig. 3.1. A path in space–time.

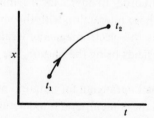

moves from $x(t_1)$ to $x(t_2)$, there is an infinite number of possible paths it may take, some of which are shown in Fig. 3.2. Which one does the particle actually take? The answer is provided by the principle of least action, which states that the actual motion is the one for which S is a minimum. Let us now show that this yields Newton's law.

Consider a variation in the path

$$x(t) \rightarrow x'(t) = x(t) + a(t), \quad a \ll x.$$

The particle is constrained to be at $x(t_1)$ at time t_1 and at $x(t_2)$ at time t_2, so the end-points of the motion are fixed, and consequently

$$a(t_1) = a(t_2) = 0. \tag{3.5}$$

On the substitution $x \rightarrow x'$, the action S becomes

$$\begin{aligned}
S \rightarrow S' &= \int_{t_1}^{t_2} \left[\frac{m}{2}(\dot{x} + \dot{a})^2 - V(x + a) \right] dt \\
&= \int_{t_1}^{t_2} \{ \tfrac{1}{2}m\dot{x}^2 + m\dot{x}\dot{a} - [V(x) + aV'(x)] \} \, dt + O(a^2) \\
&= S + \int_{t_1}^{t_2} [m\dot{x}\dot{a} - aV'(x)] dt \\
&\equiv S + \delta S
\end{aligned}$$

where

$$\delta S = \int_{t_1}^{t_2} (m\dot{x}\dot{a} - aV'(x)) dt \tag{3.6}$$

and $V'(x) = dV(x)/dx$. If S is a minimum under the variation in $x, \delta S = 0$. The first term in δS may be integrated by parts, giving

$$\int_{t_1}^{t_2} \dot{x}\dot{a} \, dt = \dot{x}a|_{t_1}^{t_2} - \int_{t_1}^{t_2} a\ddot{x} \, dt = - \int_{t_1}^{t_2} a\ddot{x} \, dt$$

where the last step follows from (3.5). We then have

$$\delta S = - \int_{t_1}^{t_2} [ma\ddot{x} + aV'(x)] dt = 0;$$

Fig. 3.2. Some of the infinite number of paths between two points.

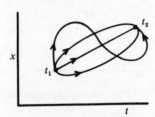

and this is fulfilled if

$$m\ddot{x} = - V'(x) \tag{3.7}$$

which is Newton's law, (3.1).

3.2   The real scalar field; variational principle and Noether's theorem

The passage from a point particle at position $x(t)$ to a field $\phi(x^{\mu}) = \phi(x, y, z, t)$ may be visualised as the 'replacement' of x by ϕ, and of t by x^{μ}. The scalar field ϕ obeys the Klein–Gordon equation (2.16)

$$(\Box + m^2)\phi = 0, \tag{3.8}$$

but it should be clear that in this treatment ϕ is interpreted merely as a *scalar field*, not as a single-particle wave function, as in chapter 2. In this case, it may well be asked, what is m? The answer to this is only forthcoming in chapter 4, when we see that on *quantisation* of the classical scalar field, we arrive at a particle interpretation, and the particles have mass m.

We now show how to derive (3.8) from a variational principle applied to an action

$$S = \int \mathscr{L}(\phi, \partial_{\mu}\phi)\mathrm{d}^4x \tag{3.9}$$

where $\partial_{\mu}\phi \equiv \partial\phi/\partial x^{\mu}$. Equation (3.9) is to be compared with equations (3.3) and (3.4). The function $\mathscr{L}$ is properly called the Lagrangian density of the theory, since $L = \int \mathscr{L} \, \mathrm{d}^3x$ is evidently the Lagrangian. We shall, however, simply refer to $\mathscr{L}$ as the Lagrangian. It is usual to assume that $\mathscr{L}$ depends

Fig. 3.3. A field ϕ traces out a region R of space–time.

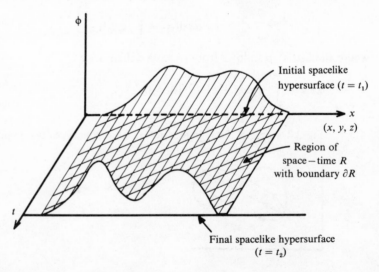

Initial spacelike
hypersurface ($t = t_1$)

x
(x, y, z)

Region of
space–time R
with boundary ∂R

Final spacelike hypersurface
($t = t_2$)

only on ϕ and its first derivatives; this is not necessary, in the sense that the deductions to be made below also follow if $\mathscr{L}$ depends on second and higher derivatives of ϕ; but we assume it for simplicity. We shall see that the Klein–Gordon equation can be derived from the Lagrangian

$$\mathscr{L} = \tfrac{1}{2}(\partial_\mu \phi)(\partial^\mu \phi) - \frac{m^2}{2}\phi^2$$

$$= \tfrac{1}{2}[(\partial_0 \phi)^2 - (\vec{\nabla}\phi)^2 - m^2\phi^2]. \tag{3.10}$$

The field ϕ traces out a 4-dimensional region R of space–time, as sketched in Fig. 3.3. The initial and final spacelike hypersurfaces may be taken to be the time-slices $t = t_1$, and $t = t_2$, which form part of the boundary ∂R of the region R. We now subject both the field variable ϕ and the co-ordinates x to a variation, which vanishes on the boundary ∂R:

$$\left. \begin{aligned} x^\mu \to x'^\mu &= x^\mu + \delta x^\mu, \\ \phi(x) \to \phi'(x) &= \phi(x) + \delta\phi(x). \end{aligned} \right\} \tag{3.11}$$

It is convenient to consider the case where $\mathscr{L}$ depends explicitly on x^μ

$$\mathscr{L} = \mathscr{L}(\phi, \partial_\mu \phi, x^\mu); \tag{3.12}$$

this happens if ϕ interacts with an external source, and so does not describe a closed system.

It is important to note that $\delta\phi$ as defined in (3.11) is merely the *functional* variation in ϕ; ϕ' is compared with ϕ at the *same point* in space–time x^μ. We may in addition define the *total* variation in ϕ, $\Delta\phi$, by

$$\phi'(x') = \phi(x) + \Delta\phi(x) \tag{3.13}$$

and it then follows, to first order in δx, that

$$\begin{aligned} \Delta\phi &= \phi'(x') - \phi(x') + \phi(x') - \phi(x) \\ &= \delta\phi + (\partial_\mu \phi)\delta x^\mu. \end{aligned} \tag{3.14}$$

The variation in the action is

$$\delta S = \int \mathscr{L}(\phi', \partial_\mu \phi', x'^\mu)\,\mathrm{d}^4 x' - \int \mathscr{L}(\phi, \partial_\mu \phi, x^\mu)\,\mathrm{d}^4 x,$$

where $\mathrm{d}^4 x' = J(x'/x)\mathrm{d}^4 x$, $J(x'/x)$ being the Jacobian of the transformation $x \to x'$. From (3.11),

$$\frac{\partial x'^\mu}{\partial x^\lambda} = \delta^\mu_\lambda + \partial_\lambda \delta x^\mu$$

so

$$J\left(\frac{x'}{x}\right) = \det\left(\frac{\partial x'^\mu}{\partial x^\lambda}\right) = 1 + \partial_\mu(\delta x^\mu);$$

and hence

$$\delta S = \int (\delta\mathcal{L} + \mathcal{L}\partial_\mu \delta x^\mu)\mathrm{d}^4 x \qquad (3.15)$$

where

$$\delta\mathcal{L} = \frac{\partial\mathcal{L}}{\partial\phi}\delta\phi + \frac{\partial\mathcal{L}}{\partial(\partial_\mu\phi)}\delta(\partial_\mu\phi) + \frac{\partial\mathcal{L}}{\partial x^\mu}\delta x^\mu. \qquad (3.16)$$

From (3.11) we see that $\delta(\partial_\mu\phi) = \partial_\mu\delta\phi$, so (3.15) and (3.16) give

$$\delta S = \int_R \left[\frac{\partial\mathcal{L}}{\partial\phi}\delta\phi + \frac{\partial\mathcal{L}}{\partial(\partial_\mu\phi)}\partial_\mu(\delta\phi) + \partial_\mu(\mathcal{L}\,\delta x^\mu)\right]\mathrm{d}^4 x \qquad (3.17)$$

where, as indicated, the integral is taken over the region R of space–time. The third term in this equation is a total divergence. The second term may be rewritten so as to introduce a total divergence:

$$\frac{\partial\mathcal{L}}{\partial(\partial_\mu\phi)}\partial_\mu(\delta\phi) = \partial_\mu\left[\frac{\partial\mathcal{L}}{\partial(\partial_\mu\phi)}\delta\phi\right] - \partial_\mu\left[\frac{\partial\mathcal{L}}{\partial(\partial_\mu\phi)}\right]\delta\phi,$$

and the resulting integrals of total divergences over R may be written as surface integrals over ∂R, using a 4-dimensional generalisation of Gauss's theorem (2.257). This gives

$$\delta S = \int_R \left\{\frac{\partial\mathcal{L}}{\partial\phi} - \partial_\mu\left[\frac{\partial\mathcal{L}}{\partial(\partial_\mu\phi)}\right]\right\}\delta\phi\,\mathrm{d}^4 x$$
$$+ \int_{\partial R}\left[\frac{\partial\mathcal{L}}{\partial(\partial_\mu\phi)}\delta\phi + \mathcal{L}\,\delta x^\mu\right]\mathrm{d}\sigma_\mu. \qquad (3.18)$$

Now we have already agreed that the variation in ϕ and x^μ on the boundary of R vanishes:

$$\delta\phi = 0, \quad \delta x^\mu = 0 \quad \text{on } \partial R,$$

so the second term in equation (3.18) vanishes, and the condition for a stationary action is

$$\blacksquare \qquad \frac{\partial\mathcal{L}}{\partial\phi} - \frac{\partial}{\partial x^\mu}\left[\frac{\partial\mathcal{L}}{\partial(\partial_\mu\phi)}\right] = 0. \qquad (3.19)$$

This is known as the *Euler–Lagrange equation* for ϕ. It is the equation of motion for the field ϕ, analogous to Newton's equation of motion for point masses. As we know already, the required equation of motion is the Klein–Gordon equation, and it is now an easy matter to confirm, as promised, that the Lagrangian given by (3.10), when substituted into the Euler–Lagrange equation (3.19), yields the Klein–Gordon equation. In fact, writing (3.10) as

$$\mathcal{L} = \tfrac{1}{2}g^{\kappa\lambda}(\partial_\kappa\phi)(\partial_\lambda\phi) - \frac{m^2}{2}\phi^2$$

(remembering that κ and λ are dummy suffixes) gives

$$\frac{\partial \mathcal{L}}{\partial \phi} = -m^2 \phi, \quad \frac{\partial \mathcal{L}}{\partial(\partial_\mu \phi)} = g^{\mu\nu}(\partial_\nu \phi) = \partial^\mu \phi,$$

so the Euler–Lagrange equation gives

$$\partial_\mu \partial^\mu \phi + m^2 \phi \equiv \Box \phi + m^2 \phi = 0,$$

which is the Klein–Gordon equation. In case it is suspected that there is anything miraculous going on, let me say there is not: the Lagrangian is written as in (3.10) *deliberately* so as to give the Klein–Gordon equation.

We now explore another consequence of the use of a variational principle. It is that if the action is unchanged by a re-parameterisation of x^μ and ϕ, i.e. is invariant under some group of transformations on x^μ and ϕ, then there exist one or more conserved quantities, that is, combinations of fields and their derivatives which are invariant under the transformations. This crucial result is known as Noether's theorem and it plays an important role in field theory and particle physics. It accounts for conservation of energy, momentum, angular momentum, and any other 'quantum'[‡] numbers which particles happen to carry, like charge, isospin, colour, etc. (There is, however, another type of conserved quantity which is 'topological' in nature, and whose conservation has nothing to do with Noether's theorem. These matters will be considered in chapter 10.) We shall first give a rather general account of Noether's theorem and then apply it to conservation of energy, momentum and angular momentum.

To start, let us return to (3.18) for the variation in the action, and rewrite the surface term as follows

$$\delta S = \int_R \left\{ \frac{\partial \mathcal{L}}{\partial \phi} - \partial_\mu \left[\frac{\partial \mathcal{L}}{\partial(\partial_\mu \phi)} \right] \right\} \delta \phi \, d^4 x$$

$$+ \int_{\partial R} \left\{ \frac{\partial \mathcal{L}}{\partial(\partial_\mu \phi)} [\delta \phi + (\partial_\nu \phi) \delta x^\nu] \right.$$

$$\left. - \left[\frac{\partial \mathcal{L}}{\partial(\partial_\mu \phi)} \partial_\nu \phi - \delta^\mu_\nu \mathcal{L} \right] \delta x^\nu \right\} d\sigma_\mu;$$

we have simply added and subtracted a term. The first square bracket in the surface integral is the 'total' variation $\Delta \phi$ defined in (3.14). The second

---

[‡] It is a loose but common habit to refer to isospin, strangeness and so on, as 'quantum' numbers. Unlike the n, l and m quantum numbers in the hydrogen atom, it is not at all clear that they have anything to do with quantum theory, i.e. depend on h. Electric charge certainly does not, so is not a *quantum* number.

square bracket defines the *energy–momentum tensor* θ^{μ}_{ν}:

$$\theta^{\mu}_{\nu} = \frac{\partial \mathscr{L}}{\partial(\partial_{\mu}\phi)} \partial_{\nu}\phi - \delta^{\mu}_{\nu}\mathscr{L}; \tag{3.20}$$

this name will be justified below. We have, then,

$$\delta S = \int_{R} \left\{ \frac{\partial \mathscr{L}}{\partial \phi} - \partial_{\mu}\left[\frac{\partial \mathscr{L}}{\partial(\partial_{\mu}\phi)} \right] \right\} \delta\phi \, \mathrm{d}^4 x$$

$$+ \int_{\partial R} \left[\frac{\partial \mathscr{L}}{\partial(\partial_{\mu}\phi)} \Delta\phi - \theta^{\mu}_{\nu}\delta x^{\nu} \right] \mathrm{d}\sigma_{\mu}. \tag{3.21}$$

Now let us suppose that the action S is invariant under a group of transformations on x^{μ} and ϕ, which for infinitesimal transformations, take the form

$$\Delta x^{\mu} = X^{\mu}_{\nu}\delta\omega^{\nu}, \quad \Delta\phi = \Phi_{\mu}\delta\omega^{\mu}, \tag{3.22}$$

characterised by the infinitesimal parameter $\delta\omega^{\nu}$. Here X^{μ}_{ν} is a matrix and Φ_{μ} is a set of numbers. If ν is a single index, then this is a 4-parameter group of transformations, but we are not restricted to this case; ν could be a double index, as it needs to be to generate Lorentz transformations, so the number of parameters is not necessarily four, despite the form of (3.22). In addition, we may want to consider the more general case in which we have a multiplet of scalar fields ϕ_i. In that case the transformation would change ϕ_i by

$$\Delta\phi_i = \Phi_{ij}\delta\omega_j \tag{3.23}$$

where Φ is a matrix. Taking the form of (3.22) at face value, however, then if we further assume that the transformed ϕ obeys (3.19), the requirement that $\delta S = 0$ gives, from (3.21) and (3.22),

$$\int_{\partial R} \left[\frac{\partial \mathscr{L}}{\partial(\partial_{\mu}\phi)} \Phi_{\nu} - \theta^{\mu}_{\kappa}X^{\kappa}_{\nu} \right] \delta\omega^{\nu}\mathrm{d}\sigma_{\mu} = 0$$

or, since $\delta\omega^{\nu}$ is arbitrary,

$$\int_{\partial R} J^{\mu}_{\nu}\mathrm{d}\sigma_{\mu} = 0 \tag{3.24}$$

where

$$J^{\mu}_{\nu} = \frac{\partial \mathscr{L}}{\partial(\partial_{\mu}\phi)} \Phi_{\nu} - \theta^{\mu}_{\kappa}X^{\kappa}_{\nu}. \tag{3.25}$$

It follows from Gauss's theorem that $\int_{R}\partial_{\mu}J^{\mu}_{\nu}\mathrm{d}^4 x = 0$, and thence, since R is arbitrary, that

$$\partial_{\mu}J^{\mu}_{\nu} = 0. \tag{3.26}$$

We therefore have a *conserved (divergenceless) current* J^{μ}_{ν} whose existence follows from the invariance of the action under the transformations (3.22). This gives rise to a *conserved (time independent) charge* Q_v defined by

$$Q_v = \int_\sigma J_v^\mu d\sigma_\mu$$

where the integral is taken over a spacelike hypersurface σ_μ. If this is chosen to be $t = \text{const}$, then

$$Q_v = \int_V J_v^0 d^3x \qquad (3.27)$$

where the integral is taken over the 3-volume V. Conservation of Q_v follows by integrating (3.26) over V:

$$\int_V \partial_0 J_v^0 d^3x + \int_V \partial_i J_v^i d^3x = 0.$$

The second term is transformed into a surface integral by (the 3-dimensional) Gauss's theorem, and vanishes, as may be seen by taking the surface far enough away, leaving

$$\frac{d}{dt}\int J_v^0 d^3x = \frac{dQ_v}{dt} = 0. \qquad (3.28)$$

This is *Noether's theorem*. Let us apply it to the case where the transformation (3.22) is simply a translation of the origin of space and time. In that case

$$\Delta x^\mu = \varepsilon^\mu, \quad \Delta\phi = 0 \qquad (3.29)$$

or

$$X_v^\mu = \delta_v^\mu, \quad \Phi_\mu = 0. \qquad (3.30)$$

This requirement is very fundamental: it is the condition that (for a closed system) the laws of physics are translationally invariant, i.e. the same in London, New York and Andromeda, and also invariant under time displacement, i.e. the same yesterday, today and tomorrow. If it were not for these conditions, it is obvious that science itself would be impossible. From equations (3.30) and (3.25) we see that the conserved current in this case is

$$J_v^\mu = -\theta_v^\mu \qquad (3.31)$$

and the corresponding conservation law is

$$\frac{d}{dt}\int \theta_v^0 d^3x = 0. \qquad (3.32)$$

It will now be shown that $P_v = \int \theta_v^0 d^3x$ is the 4-momentum, or energy–momentum, of the field ϕ; this justifies calling θ_v^μ the energy–momentum tensor. Referring to (3.20), we have

$$\int \theta_0^0 \mathrm{d}^3 x = \int \left\{ \frac{\partial \mathscr{L}}{\partial(\partial_0 \phi)} \partial_0 \phi - \mathscr{L} \right\} \mathrm{d}^3 x$$

$$= \int \left\{ \frac{\partial \mathscr{L}}{\partial \dot{\phi}} \dot{\phi} - \mathscr{L} \right\} \mathrm{d}^3 x \tag{3.33}$$

where $\dot{\phi} = \partial\phi/\mathrm{d}t$. We recall that in point particle mechanics, the relation between the Hamiltonian and Lagrangian functions is (see, for example, Goldstein, ref. (1) chapter 7)

$$H = \sum_i p_i \dot{q}_i - L$$

where the momentum p_i is defined as

$$p_i = \frac{\partial L}{\partial \dot{q}_i}.$$

On transcribing these relations for the case of fields we see that the right-hand side of (3.33) is the energy of the field. The proof that $\int \theta_1^0 \, \mathrm{d}^3 x$ is the momentum follows immediately on the observation that $\partial\phi/\partial x^\mu$ is a 4-vector under Lorentz transformations, just as energy–momentum is.

Conservation of energy and momentum, then, holds for any system whose Lagrangian (and therefore action) does not depend on x^μ. This is in line with the observation above, following (3.12), that such a system will not be able to exchange energy or momentum with the outside. Substituting the Lagrangian (3.10) into the energy–momentum tensor $\theta^{\mu\nu}$ (3.20) gives

$$\theta^{\mu\nu} = (\partial^\mu \phi)(\partial^\nu \phi) - g^{\mu\nu} \mathscr{L}. \tag{3.34}$$

This is clearly symmetric in μ and ν. Hence, for a scalar field, the energy–momentum tensor is symmetric. This is not always true, however; it is clear from the definition (3.20) that $\theta^{\mu\nu}$ is not, in general, symmetric. On the other hand, neither is it unique, for we may add a term $\partial_\lambda f^{\lambda\mu\nu}$ where $f^{\lambda\mu\nu} = -f^{\mu\lambda\nu}$, so that

$$\partial_\mu \partial_\lambda f^{\lambda\mu\nu} \equiv 0. \tag{3.35}$$

Hence, if we define the quantity

$$T^{\mu\nu} = \theta^{\mu\nu} + \partial_\lambda f^{\lambda\mu\nu} \tag{3.36}$$

which we call the *canonical* energy–momentum tensor, we have

$$\partial_\mu T^{\mu\nu} = \partial_\mu \theta^{\mu\nu} = 0; \tag{3.37}$$

and $f^{\lambda\mu\nu}$ may be so chosen as to make $T^{\mu\nu}$ symmetric. The total 4-

momentum of the system is the same, since $(i = 1, 2, 3)$

$$\int \partial_\lambda f^{\lambda 0 v} \, d^3x = \int \partial_i f^{i0v} \, d^3x$$

$$= \int_\sigma f^{i0v} \, d\sigma_i$$

$$= 0$$

where Gauss's theorem has been used. The last equality follows since the surface σ is at infinity, where no fields are present. Hence although the energy–momentum tensor is not unique, the energy and momentum in the field are.

There are two reasons for wanting the canonical energy–momentum tensor to be symmetric. One is that, according to the theory of general relativity, it is this tensor which determines the curvature of space, according to Einstein's field equations

$$R_{\mu\nu} - \tfrac{1}{2} g_{\mu\nu} R = -\frac{8\pi G}{c^2} T_{\mu\nu}. \tag{3.38}$$

The Ricci tensor $R_{\mu\nu}$ and the metric tensor are both symmetric, so $T_{\mu\nu}$ must be also.

The second reason for wanting a symmetric $T_{\mu\nu}$ becomes apparent when we consider angular momentum. We demand that the action be invariant under spatial rotations

$$\delta x^i = \varepsilon^{ij} x^j, \quad \varepsilon^{ij} = -\varepsilon^{ji} \quad (i, j = 1, 2, 3), \tag{3.39}$$

$$\delta \phi = 0. \tag{3.40}$$

Since the rotation group is a subgroup of the Lorentz group, we may generalise (3.39) to

$$\delta x^\mu = \varepsilon^\mu_\nu x^\nu, \quad \varepsilon^{\mu\nu} = -\varepsilon^{\nu\mu}. \tag{3.41}$$

This may be written

$$\delta x^\mu = X^\mu_{\rho\sigma} \varepsilon^{\rho\sigma}, \quad X^\mu_{\rho\sigma} = \delta^\mu_\rho x_\sigma \tag{3.42}$$

to bring it into the same form as (3.22); the index ν in that equation has now become a double index. The conserved Noether current is, from (3.25) (with $\Phi = 0$, and substituting $T^{\mu\nu}$ for $\theta^{\mu\nu}$),

$$J^{\mu\rho\sigma} = -T^\mu_\kappa X^{\kappa\rho\sigma}$$

$$= -T^{\mu\rho} x^\sigma. \tag{3.43}$$

Since, however, $\varepsilon^{\rho\sigma}$ in (3.42) is antisymmetric in ρ and σ, only that part of X which is antisymmetric in its bottom indices will contribute in (3.42), so we may make the replacement

$$X^\mu_{\rho\sigma} \to \tfrac{1}{2}(X^\mu_{\rho\sigma} - X^\mu_{\sigma\rho})$$

which has the effect of replacing (3.43) by

$$J^{\mu\rho\sigma} = -\tfrac{1}{2}(T^{\mu\rho}x^\sigma - T^{\mu\sigma}x^\rho). \tag{3.44}$$

The $\mu = 0$ component of this current is (apart from a numerical factor) the angular momentum density of the field. The angular momentum is given by

$$M^{\mu\nu} = \int (T^{0\mu}x^\nu - T^{0\nu}x^\mu)\,\mathrm{d}^3x \tag{3.45}$$

or

$$M^{\mu\nu} = \int \mathscr{M}^{0\mu\nu}\,\mathrm{d}^3x \tag{3.46}$$

where

$$\mathscr{M}^{0\mu\nu} = T^{0\mu}x^\nu - T^{0\nu}x^\mu. \tag{3.47}$$

$M^{\mu\nu}$ is a conserved quantity:

$$\frac{\mathrm{d}}{\mathrm{d}t}M^{\mu\nu} = 0,$$

and the tensor $\mathscr{M}^{\rho\mu\nu}$ is divergenceless:

$$\partial_\rho \mathscr{M}^{\rho\mu\nu} = 0. \tag{3.48}$$

Substituting $\mathscr{M}^{\rho\mu\nu} = T^{\rho\mu}x^\nu - T^{\rho\nu}x^\mu$ in this equation, and noting that $\partial_\rho T^{\rho\alpha} = 0$ (equation (3.37)) and $\partial_\rho x^\alpha = \delta^\alpha_\rho$ gives

$$T^{\mu\nu} = T^{\nu\mu}; \tag{3.49}$$

so conservation of angular momentum requires the energy–momentum tensor to be symmetric. This is our second reason for requiring a symmetric $T^{\mu\nu}$.

To be precise we should remark that the three components of the angular momentum of the system are the three space–space components of $M^{\mu\nu}$ in (3.45); M^{12}, M^{23}, M^{31}. The three time–space components M^{01}, M^{02}, and M^{03} are related to the centre of mass of the system and are conserved by virtue of invariance under pure Lorentz ('boost') transformations – cf. equation (3.41).

This completes our account of the origin of the conservation laws for energy, momentum and angular momentum; they all follow from space–time symmetries via Noether's theorem, and so involve, in the machinations above, a non-zero X^μ_ν in (3.25).

Let us now turn our attention to electric charge, which we also know to be conserved. If this is due to a symmetry of the action, what is the symmetry? It clearly may not involve an X_ν^μ in (3.25), since these symmetries are completely accounted for. The maximum symmetry we may have in Minkowski space is symmetry under translations, time displacements, rotations and Lorentz transformations; and we have considered all of these already. Any additional symmetry must therefore involve Φ_μ; in other words, the scalar field must have *more than one component*. (We dismiss the case of a spinor or vector field; in these cases the different components are related by *space–time* transformations, with the consequence that the conserved angular momentum tensor contains additional terms, interpreted as the intrinsic spin, which is nothing to do with charge! The Dirac field will be considered below in this context.) The simplest possibility is for ϕ to have two components; and a field with two real components is mathematically equivalent to a complex field. We now consider this.

3.3 Complex scalar fields and the electromagnetic field

If the scalar field now has two real components ϕ_1 and ϕ_2, we may put

$$\phi = (\phi_1 + i\phi_2)/\sqrt{2}, \tag{3.50}$$
$$\phi^* = (\phi_1 - i\phi_2)/\sqrt{2}, \tag{3.51}$$

and since the action is real, we put

$$\mathscr{L} = (\partial_\mu \phi)(\partial^\mu \phi^*) - m^2 \phi^* \phi. \tag{3.52}$$

Regarding ϕ and ϕ^* as independent fields, the Euler–Lagrange equations (3.19) give the two Klein–Gordon equations

$$(\Box + m^2)\phi = 0, \tag{3.53}$$
$$(\Box + m^2)\phi^* = 0. \tag{3.54}$$

The Lagrangian is clearly invariant under the transformation

$$\phi \to e^{-i\Lambda}\phi, \quad \phi^* \to e^{i\Lambda}\phi^* \tag{3.55}$$

where Λ is a real constant. This is known as a *gauge transformation of the first kind*. Its infinitesimal form is

$$\delta\phi = -i\Lambda\phi, \quad \delta\phi^* = i\Lambda\phi^* \tag{3.56}$$

and so

$$\delta(\partial_\mu\phi) = -i\Lambda\partial_\mu\phi, \quad \delta(\partial_\mu\phi^*) = i\Lambda\partial_\mu\phi^* \tag{3.57}$$

Since the transformation (3.55) does not involve space–time (it is purely 'internal'), in the notation of (3.22) we have

$$\Phi = -i\phi, \quad \Phi^* = i\phi^*, \quad X = 0 \tag{3.58}$$

Noether's theorem then gives a conserved current, which is, from equation (3.25),

$$J^\mu = \frac{\partial \mathscr{L}}{\partial(\partial_\mu \phi)}(-i\phi) + \frac{\partial \mathscr{L}}{\partial(\partial_\mu \phi^*)}(i\phi^*). \tag{3.59}$$

(Here, the 'internal indices' of ϕ have effectively been summed over, giving separate contributions from ϕ and ϕ^*.) Substituting (3.52) into (3.59) then gives

$$J^\mu = i(\phi^* \partial^\mu \phi - \phi \partial^\mu \phi^*). \tag{3.60}$$

It follows immediately from (3.53) and (3.54) that this current has a vanishing 4-divergence, as of course it should have:

$$\partial_\mu J^\mu = 0, \tag{3.61}$$

and the corresponding conserved quantity is, from (3.27),

$$\begin{aligned}
Q &= \int J^0 \, dV \\
&= \frac{i}{c} \int \left(\phi^* \frac{\partial \phi}{\partial t} - \phi \frac{\partial \phi^*}{\partial t} \right) dV.
\end{aligned} \tag{3.62}$$

This (real) quantity we should like to identify with electric charge. Let us therefore note the following things:

(1) It is a conserved quantity: $dQ/dt = 0$.
(2) It contains no mention of e, the charge on the proton.
(3) It is a *classical* quantity, not a quantum one; it contains no h.
(4) It is not 'integral', or 'quantised'. In other words, it does not account for the fact that real electric charges all seem to be multiples of a basic quantity.
(5) When ϕ is real, $\phi = \phi^*$, $Q = 0$, and there is no conserved quantity. This is the situation we dealt with above.

We may express the gauge transformation (3.55) in a geometric form. To do this, first we substitute equations (3.50) and (3.51) into (3.52), to give

$$\mathscr{L} = (\partial_\mu \phi_1)(\partial^\mu \phi_1) + (\partial_\mu \phi_2)(\partial^\mu \phi_2) - m^2(\phi_1^2 + \phi_2^2) \tag{3.63}$$

and then write

$$\boldsymbol{\phi} = \mathbf{i}\phi_1 + \mathbf{j}\phi_2 \tag{3.64}$$

as a vector in a 2-dimensional space with orthonormal basis vectors $\mathbf{i}$ and $\mathbf{j}$. The Lagrangian is now

$$\mathscr{L} = (\partial_\mu \boldsymbol{\phi}) \cdot (\partial^\mu \boldsymbol{\phi}) - m^2 \boldsymbol{\phi} \cdot \boldsymbol{\phi}. \tag{3.65}$$

The gauge transformation (3.55) may be written

$$\phi_1' + i\phi_2' = e^{-i\Lambda}(\phi_1 + i\phi_2),$$
$$\phi_1' - i\phi_2' = e^{i\Lambda}(\phi_1 - i\phi_2),$$

which is equivalent to

$$\left. \begin{array}{l} \phi_1' = \phi_1 \cos\Lambda + \phi_2 \sin\Lambda, \\ \phi_2' = -\phi_1 \sin\Lambda + \phi_2 \cos\Lambda. \end{array} \right\} \tag{3.66}$$

This is clearly a rotation in the (1, 2) plane of the vector ϕ through an angle Λ, as shown in Fig. 3.4. Rotations in two dimensions form the group $O(2)$. On the other hand, since the transformation was equivalently represented as $e^{i\Lambda}$, which is a unitary 'matrix' in one dimension

$$e^{i\Lambda}(e^{i\Lambda})^+ = 1,$$

the group concerned is also $U(1)$. We are thus concerned with gauge transformations which generate the group $O(2) \approx U(1)$. (It is easy to see that these groups are the same: each element of $O(2)$ is given uniquely by an angle ϕ, the angle of rotation in the plane. The group space is then the space of the values of ϕ. We must identify ϕ with $\phi + 2\pi$, $\phi + 4\pi$, etc., for these correspond to the same rotation. The group space is then a *circle* (see Fig. 3.5). The group $U(1)$, on the other hand, is the group of all real numbers of the form $e^{i\alpha} = \cos\alpha + i\sin\alpha$, and since $\cos^2\alpha + \sin^2\alpha = 1$, the space of these is also a circle.)

Returning to the main theme, we have identified a conserved quantity Q, as a result of invariance of the action under the gauge transformation (3.55), or (3.66). Since Λ is a constant, however, this gauge transformation must be the same at all points in space–time – it is a 'global' gauge transformation. So

Fig. 3.4. A rotation of the field ϕ in the internal space.

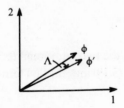

Fig. 3.5. The group space of $O(2)$ is a circle.

when we perform a rotation in the internal space of ϕ at one point, through an angle Λ, we must perform the same rotation *at all other points at the same time*. If we take this physical interpretation seriously, we see that it is impossible to fulfil, since it contradicts the letter and spirit of relativity, according to which there must be a minimum time delay equal to the time of light travel. To get round this problem we simply abandon the requirement that Λ is a constant, and write it as an arbitrary function of space–time, $\Lambda(x^\mu)$. This is called a 'local' gauge transformation, since it clearly differs from point to point. It is also called a *gauge transformation of the second kind*.

We now see that, for $\Lambda \ll 1$,

$$\phi \rightarrow \phi - i\Lambda\phi$$

therefore

$$\delta\phi = -i\Lambda\phi, \tag{3.67}$$

$$\partial_\mu\phi \rightarrow \partial_\mu\phi - i(\partial_\mu\Lambda)\phi - i\Lambda(\partial_\mu\phi)$$

therefore

$$\delta(\partial_\mu\phi) = -i\Lambda(\partial_\mu\phi) - i(\partial_\mu\Lambda)\phi \tag{3.68}$$

and, similarly,

$$\delta\phi^* = i\Lambda\phi^*, \tag{3.69}$$

$$\delta(\partial_\mu\phi^*) = i\Lambda(\partial_\mu\phi^*) + i(\partial_\mu\Lambda)\phi^*. \tag{3.70}$$

Comparing these equations with (3.56) and (3.57), we note the extra term in $(\partial_\mu\Lambda)$ occurring in the transformation of the derivatives of the fields. Because of this extra term, for example comparing (3.67) and (3.68), we say that $\partial_\mu\phi$ does not transform *covariantly*, i.e. not the same way as ϕ itself. Moreover, we shall now see that these extra terms cause the action itself to be *no longer invariant*. In fact, the change in the Lagrangian is

$$\delta\mathscr{L} = \frac{\partial\mathscr{L}}{\partial\phi}\delta\phi + \frac{\partial\mathscr{L}}{\partial(\partial_\mu\phi)}\delta(\partial_\mu\phi) + (\phi \rightarrow \phi^*), \tag{3.71}$$

where '$(\phi \rightarrow \phi^*)$' is a shorthand way of writing the two additional terms in ϕ^*. Substituting the Euler–Lagrange equation (3.19) into the first term, and also using (3.67) and (3.68) gives

$$\partial\mathscr{L} = \partial_\mu\left[\frac{\partial\mathscr{L}}{\partial(\partial_\mu\phi)}\right](-i\Lambda\phi) + \frac{\partial\mathscr{L}}{\partial(\partial_\mu\phi)}(-i\Lambda\partial_\mu\phi - i\phi\partial_\mu\Lambda)$$

$$+ (\phi \rightarrow \phi^*)$$

$$= -i\Lambda\partial_\mu\left[\frac{\partial\mathscr{L}}{\partial(\partial_\mu\phi)}\phi\right] - i\frac{\partial\mathscr{L}}{\partial(\partial_\mu\phi)}(\partial_\mu\Lambda)\phi + (\phi \rightarrow \phi^*).$$

The first term above is a total divergence, so the corresponding change in the action is zero, and we ignore it. Using the explicit form (3.52) for the Lagrangian then gives

$$\delta\mathscr{L} = i\partial_\mu\Lambda(\phi^*\partial^\mu\phi - \phi\partial^\mu\phi^*)$$
$$= J^\mu\partial_\mu\Lambda \tag{3.72}$$

where the current J^μ is given by equation (3.60). The action, as it stands, is therefore not invariant under gauge transformation of the second kind. To make it invariant, we introduce a new 4-vector A_μ which couples directly to the current J^μ giving an *extra term in* $\mathscr{L}$:

$$\mathscr{L}_1 = -eJ^\mu A_\mu$$
$$= -ie(\phi^*\partial^\mu\phi^* - \phi\partial_\mu\phi^*)A_\mu. \tag{3.73}$$

The coupling constant e is a number such that eA_μ has the same units as $\partial/\partial x^\mu$. We also demand that, under gauge transformations of the second kind,

$$A_\mu \to A_\mu + \frac{1}{e}\partial_\mu\Lambda, \tag{3.74}$$

so that

$$\delta\mathscr{L}_1 = -e(\delta J^\mu)A_\mu - eJ^\mu(\delta A_\mu)$$
$$= -e(\delta J^\mu)A_\mu - J^\mu\partial_\mu\Lambda. \tag{3.75}$$

The last term above cancels $\delta\mathscr{L}$ in equation (3.72), but we now need to cancel the first term! We have

$$\delta J^\mu = i\delta(\phi^*\partial^\mu\phi - \phi\partial^\mu\phi^*)$$
$$= 2\phi^*\phi\partial^\mu\Lambda \tag{3.76}$$

so that

$$\delta\mathscr{L} + \delta\mathscr{L}_1 = -2eA_\mu(\partial^\mu\Lambda)\phi^*\phi. \tag{3.77}$$

We therefore add another term to $\mathscr{L}$:

$$\mathscr{L}_2 = e^2 A_\mu A^\mu\phi^*\phi. \tag{3.78}$$

By virtue of (3.74), we have

$$\delta\mathscr{L}_2 = 2e^2 A_\mu\delta A^\mu\phi^*\phi$$
$$= 2eA_\mu(\partial^\mu\Lambda)\phi^*\phi \tag{3.79}$$

and hence

$$\delta\mathscr{L} + \delta\mathscr{L}_1 + \delta\mathscr{L}_2 = 0. \tag{3.80}$$

The total Lagrangian $\mathscr{L} + \mathscr{L}_1 + \mathscr{L}_2$ is now invariant, by virtue of our having introduced a field A_μ which couples to the current J_μ of the complex ϕ field. The field A_μ, however, must presumably contribute by itself to the

Lagrangian. Since $\mathscr{L} + \mathscr{L}_1 + \mathscr{L}_2$ is invariant, we need an $\mathscr{L}_3$ which is also gauge invariant, and to construct it we define the 4-dimensional curl of A_μ

$$F_{\mu\nu} = \partial_\mu A_\nu - \partial_\nu A_\mu. \tag{3.81}$$

It is clear that, under the gauge transformation (3.74), $F_{\mu\nu}$ is itself invariant. The scalar Lagrangian $\mathscr{L}_3$ is then

$$\mathscr{L}_3 = -\tfrac{1}{4} F^{\mu\nu} F_{\mu\nu}. \tag{3.82}$$

Collecting the formulae together, we have

$$\begin{aligned}
\mathscr{L}_{\text{tot}} &= \mathscr{L} + \mathscr{L}_1 + \mathscr{L}_2 + \mathscr{L}_3 \\
&= (\partial_\mu \phi)(\partial^\mu \phi^*) - ie(\phi^* \partial^\mu \phi - \phi \partial^\mu \phi^*) A_\mu \\
&\quad + e^2 A_\mu A^\mu \phi^* \phi - m^2 \phi^* \phi - \tfrac{1}{4} F^{\mu\nu} F_{\mu\nu},
\end{aligned}$$

$$\blacksquare \quad \mathscr{L}_{\text{tot}} = (\partial_\mu \phi + ieA_\mu \phi)(\partial^\mu \phi^* - ieA^\mu \phi^*) - m^2 \phi^* \phi - \tfrac{1}{4} F^{\mu\nu} F_{\mu\nu}. \tag{3.83}$$

It will be recognised that $F_{\mu\nu}$ defined in (3.81) is the *electromagnetic field tensor* (2.219), whose six components are three electric and three magnetic field components. What we have done, therefore, is to show how the electromagnetic field arises *naturally* by demanding invariance of the action under gauge transformation of the second kind, i.e., under *local* (x-dependent) rotations in the internal space of the complex ϕ field. The gauge potential A_μ couples to the current J_μ with coupling strength e, which is the charge of the field ϕ. To spell these conclusions out in some more detail, it is convenient to divide our observations into four points:

1. Comparing the Lagrangian (3.83) with (3.52), we note that $\partial_\mu \phi$ is replaced by $(\partial_\mu + ieA_\mu)\phi$. This is known as the 'covariant derivative' $D_\mu \phi$:

$$\blacksquare \quad D_\mu \phi = (\partial_\mu + ieA_\mu)\phi \tag{3.84}$$

since, unlike $\partial_\mu \phi$, it transforms covariantly under gauge transformation, i.e. like ϕ itself. Indeed, from equations (3.67), (3.68) and (3.74) it follows immediately that

$$\begin{aligned}
\delta(D_\mu \phi) &= \delta(\partial_\mu \phi) + ie(\delta A_\mu)\phi + ieA_\mu \delta\phi \\
&= -i\Lambda(\partial_\mu \phi + ieA_\mu \phi) \\
&= -i\Lambda(D_\mu \phi), \tag{3.85}
\end{aligned}$$

which is the rule for covariant transformation. The rule that ∂_μ becomes replaced by $\partial_\mu + ieA_\mu$ in an electromagnetic field is equivalent to a result already familiar from classical physics. Since, with $c = 1$,

$$\partial_\mu = (\partial_0, \nabla), \quad A_\mu = (\phi, -\mathbf{A}),$$

then the 'space' part of the substitution rule is

$$\nabla \to \nabla - ie\mathbf{A}.$$

Then putting $\mathbf{p} = -i\hbar\mathbf{V}$ gives, with $\hbar = 1$,

$$\mathbf{p} \to \mathbf{p} - e\mathbf{A}. \tag{3.86}$$

In the presence of electromagnetism, if the generalised momentum $\mathbf{P}$ is defined as $\partial L / \partial \mathbf{v}$, where L is the Lagrangian, then for a particle with charge e it is given by

$$\mathbf{P} = \mathbf{p} + e\mathbf{A} = \gamma m\mathbf{v} + e\mathbf{A} \tag{3.87}$$

and it can be shown that the Hamiltonian for a particle with charge e interacting with an electromagnetic field is

$$H = \frac{1}{2m}(\mathbf{P} - e\mathbf{A})^2 + e\phi.$$

This corresponds to the substitution (3.86). (For further details see ref. (8), section 16, and ref. (2), volume 3, section 21–3.)

It is now clear that ϕ describes a field with charge e, and that ϕ^*, whose covariant derivative is

$$\blacksquare \qquad D_\mu\phi^* = (\partial_\mu - ieA_\mu)\phi^* \tag{3.88}$$

describes a field with charge $-e$. The fields ϕ_1 and ϕ_2 in (3.50) and (3.51) are therefore not eigenstates of charge.

2. Because of the identity of equations (3.81) and (2.219), we are assured that the compensating vector potential introduced in (3.73) is actually the electromagnetic vector potential, and therefore that the homogeneous Maxwell equations hold for $F_{\mu\nu}$. In addition, the gauge transformation (3.74) where $\Lambda = e\chi$, is identical with (2.226). So we arrive at a new interpretation of the electromagnetic field: *it is the gauge field which has to be introduced to guarantee invariance under local U(1) gauge transformations.*

It is worth noting that the inhomogeneous Maxwell equations are obtained from (3.83) by varying A_μ. The Euler–Lagrange equation

$$\frac{\partial \mathscr{L}}{\partial A_\mu} - \partial_\nu \left[\frac{\partial \mathscr{L}}{\partial(\partial_\nu A_\mu)} \right] = 0$$

gives

$$\begin{aligned} \partial_\nu F^{\mu\nu} &= -ie(\phi^*\partial^\mu\phi - \phi\partial^\mu\phi^*) + 2e^2 A^\mu|\phi|^2 \\ &= -ie(\phi^* D^\mu\phi - \phi D^\mu\phi^*) \\ &\equiv -e\mathscr{J}^\mu \end{aligned} \tag{3.89}$$

where

$$\mathscr{J}^\mu = i(\phi^* D^\mu\phi - \phi D^\mu\phi^*) \tag{3.90}$$

is the covariant version of J^μ in (3.60). Because of the antisymmetry of $F^{\mu\nu}$,

(3.89) implies immediately that

$$\partial_\mu \mathscr{J}^\mu = 0; \tag{3.91}$$

it is the 'covariant' current $\mathscr{J}^\mu$ which is conserved when the electromagnetic field is present, not the current J^μ.

3. From the work in the last chapter, we know already that the electromagnetic field is *massless*, but it is worth re-emphasising this point. A mass term in the Lagrangian would be of the form

$$\mathscr{L}_M = M^2 A_\mu A^\mu \tag{3.92}$$

and it is clear that this term is not invariant under the gauge transformation (3.74). Hence *gauge invariance requires that the gauge field is massless*. In the usual approach to the electromagnetic field, it is relativity which requires that the field is massless, and therefore travels with the speed of light.

4. The charge e appears as a *coupling constant*, via the covariant derivatives (3.84) and (3.88). Indeed, from the Lagrangian (3.83), we see that the ϕ field couples to the electromagnetic field with strength e. This brings to light the *dual role of electric charge*; it is both a conserved quantity, as we originally stated (in eqn. (3.62), which now must be modified by replacing J^0 with $\mathscr{J}^0$ – see above), and it also measures the strength with which a particle interacts with electric and magnetic fields. This dynamical aspect of charge is a consequence of the 'gauge principle', and it is this which has come to play such an important role in contemporary particle physics. When, for example, we discover conserved quantities (like isospin, strangeness, etc.) we may ask if they also have associated massless gauge fields which couple to the particles carrying those quantum numbers with a fixed strength. We shall consider this question, in §3.5, in which the present considerations will be generalised to the cases where the Lagrangian is invariant under the group $SU(2)$ rather than $U(1)$. Before doing that, however, it will be instructive to discuss a very interesting and beautiful consequence of the gauge invariance of electrodynamics, namely the Bohm–Aharonov effect.

Fig. 3.6. The 2-slit interference experiment with electrons.

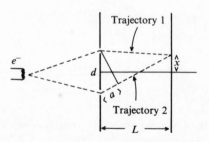

3.4   Topology and the vacuum: the Bohm–Aharonov effect

In classical physics, the force on a particle with charge e in an electromagnetic field is given by the Lorentz formula

$$\mathbf{F} = e\mathbf{E} + e\mathbf{v} \times \mathbf{B}. \tag{3.93}$$

$\mathbf{E}$ and $\mathbf{B}$ are the electric field and magnetic induction, related to the vector potential $\mathbf{A}$ and scalar potential ϕ by (2.218),

$$\mathbf{E} = -\frac{\partial \mathbf{A}}{\partial t} - \nabla \phi, \quad \mathbf{B} = \operatorname{curl} \mathbf{A}. \tag{2.218}$$

As we saw in the last chapter, $\mathbf{A}$ and ϕ are not unique; they may be changed by the gauge transformation (2.225) (equivalently (3.74)) under which $\mathbf{E}$ and $\mathbf{B}$ are invariant. In this context, it is usual to make the assertion that the only physical effect of an electromagnetic field on a charge is the Lorentz force, (3.93), and this only exists in regions where $\mathbf{E}$ and/or $\mathbf{B}$ are non-zero. The Bohm–Aharonov effect demonstrates that this is not so in quantum mechanics; physical effects occur in regions where $\mathbf{E}$ and $\mathbf{B}$ are both zero, but A_μ is not. Hence A_μ has more physical significance than was formerly thought!

The effect concerns the prototype quantum mechanical experiment, the 2-slit experiment with electrons, shown in Fig. 3.6. Because of the wave nature of the electrons, as long as it is not detected which slit they pass through, they produce a characteristic interference pattern. If the electron wavelength is λ, the phase difference between waves from the two slits is

$$\delta = 2\pi \frac{a}{\lambda} = \frac{a}{\lambdabar}.$$

If $x \ll L$, then $a = (x/L)d$, so

$$\delta = \frac{x}{L} \frac{d}{\lambdabar}; \quad x = \frac{L\lambdabar}{d} \delta. \tag{3.94}$$

Fig. 3.7. The Bohm–Aharonov effect; a solenoid is placed between the slits.

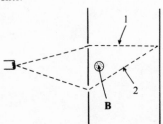

Maxima occur at $\delta = 2n\pi$ and minima at $(2n+1)\pi$, so this formula tells us the interference pattern.

The idea of Bohm and Aharonov[10] was to introduce a small solenoid behind the wall between the slits, as shown in Fig. 3.7. There are lines of magnetic induction **B** *inside* the solenoid, but not outside, so, as long as the solenoid is small enough, the electrons always move in a field-free region. It is an easy matter to write down the form of **A** which gives a solenoidal magnetic field. It is shown schematically in Fig. 3.8. In cylindrical polar coordinates, **A** has only a ϕ component, given by

$$\textit{Inside} \quad \left. \begin{array}{l} A_r = A_z = 0, \\[2mm] A_\phi = \dfrac{Br}{2}, \end{array} \right\} \tag{3.95}$$

$$\textit{Outside} \quad \left. \begin{array}{l} A_r = A_z = 0, \\[2mm] A_\phi = \dfrac{BR^2}{2r}, \end{array} \right\} \tag{3.96}$$

where R is the radius of the solenoid. Since $\mathbf{B} = \mathbf{V} \times \mathbf{A}$, we have, in cylindrical polars,

$$B_z = \frac{1}{r}\left[\frac{\partial(rA_\phi)}{\partial r} - \frac{\partial A_r}{\partial \phi} \right]$$

and similar formulae for B_r and B_ϕ, giving

$$\textit{Inside} \quad \left. \begin{array}{l} B_r = B_\phi = 0, \\[1mm] B_z = B, \end{array} \right\} \tag{3.97}$$

$$\textit{Outside} \quad \mathbf{B} = 0, \tag{3.98}$$

as required.

Fig. 3.8. The forms of **A** and **B** in a solenoid.

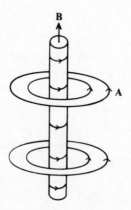

So much for the field of a solenoid. We now ask, how does this field affect an electron? The wave function of the electron in free space is

$$\psi = |\psi| \exp \frac{i}{\hbar} \mathbf{p} \cdot \mathbf{r} \equiv |\psi| \exp i\alpha \tag{3.99}$$

and the effect of an electromagnetic field is to change $\mathbf{p}$, as in (3.86):

$$\mathbf{p} \rightarrow \mathbf{p} - e\mathbf{A}$$

where e is the (negative) charge on the electron. The phase α of the wave function then changes according to

$$\alpha \rightarrow \alpha - \frac{e}{\hbar} \mathbf{A} \cdot \mathbf{r} \tag{3.100}$$

and the change in phase over an entire trajectory is

$$\Delta\alpha = -\frac{e}{\hbar} \int_{\text{trajectory}} \mathbf{A} \cdot d\mathbf{r} \tag{3.101}$$

Over trajectories 1 and 2 (see Figs. 3.6 and 3.7), then, we have

$$\Delta\alpha_1 = -\frac{e}{\hbar} \int_1 \mathbf{A} \cdot d\mathbf{r}, \quad \Delta\alpha_2 = -\frac{e}{\hbar} \int_2 \mathbf{A} \cdot d\mathbf{r}$$

and the change in the phase difference δ is then

$$\Delta\delta = \Delta\alpha_1 - \Delta\alpha_2$$

$$= \frac{e}{\hbar} \oint_{2-1} \mathbf{A} \cdot d\mathbf{r}$$

$$= \frac{e}{\hbar} \int_{2-1} \text{curl}\,\mathbf{A} \cdot d\mathbf{S}$$

$$= \frac{e}{\hbar} \int \mathbf{B} \cdot d\mathbf{S}$$

$$= \frac{e}{\hbar} \Phi \tag{3.102}$$

where Φ is the flux through the solenoid. The interference pattern therefore moves upwards by an amount

$$\Delta x = \frac{L\lambdabar}{d} \Delta\delta = \frac{L\lambdabar}{d} \frac{e}{\hbar} \Phi. \tag{3.103}$$

The net effect, then, is that *the presence of the solenoid causes a shift in the interference pattern, even though the electrons only ever move through regions of no magnetic field.*

The experiment is not an easy one to perform, because the solenoid has to be so small. A single magnetised iron whisker was used in the original experiment by Chambers,[11] in which the effect was observed, and found to be in agreement with the theoretical prediction. Since then (1960), it has been confirmed in several other experiments.

The significance of this effect is that, in quantum theory, an electron is influenced by the vector potential **A**, even though it travels entirely in regions where **B** = 0. On the other hand, from (3.103) and (3.102), the physical effects depend only on curl **A**, so we deduce that an electron is influenced by fields which are only non-zero in regions inaccessible to it. More formally, this amounts to a *non-locality* in the integral $\oint \mathbf{A} \cdot d\mathbf{r}$.

We shall now show that the Bohm–Aharonov effect owes its existence to the non-trivial topology of the vacuum, and the fact that electrodynamics is a gauge theory. In fact, it has recently been realised that the vacuum, in gauge theories, has a rich mathematical structure, with associated physical consequences, which will be discussed more fully in later chapters. The Bohm–Aharonov effect is the simplest illustration of the importance of topology in this branch of physics.

Outside the solenoid, **E** = 0 and **B** = 0 so the energy density of the electromagnetic field $U = 0$ and we have a vacuum. On the other hand, **A** ≠ 0 so the vacuum has a 'structure'.

Since curl **A** = 0, we may write **A** = **∇**χ for some function χ which may be found by noting that, from (3.95),

$$A_\phi = \frac{1}{r}\frac{\partial \chi}{\partial \phi} = \frac{BR^2}{2r}$$

giving

$$\chi = \frac{BR^2}{2}\phi \tag{3.104}$$

where we have ignored an arbitrary constant of integration. Now χ is not a single-valued function, since it increases by $\pi R^2 B$ when $\phi \to \phi + 2\pi$. Indeed, from (3.102),

$$\Delta\delta = \Delta\alpha = \frac{e}{\hbar}\int \mathbf{A} \cdot d\mathbf{r}$$

$$= \frac{e}{\hbar}\int \mathbf{\nabla}\chi \cdot d\mathbf{r}$$

$$= \frac{e}{\hbar}[\chi]_{\phi=0}^{\phi=2\pi}$$

$$= \frac{e}{\hbar}\pi R^2 B$$

$$= \frac{e}{\hbar}\Phi. \tag{3.105}$$

Regular non-single-valued functions, however, may only exist in *non-simply connected spaces*. A simply connected space is one in which all closed curves may be shrunk continuously to a point. A non-simply connected space is one in which not all curves may be continuously shrunk to a point. The relevant space in this problem is the space of the vacuum, i.e. the space outside the solenoid, and that is not simply connected. This is clear from Fig. 3.9. The curve c_1 can be shrunk to a point, but c_2 cannot; moreover, a curve going round the solenoid n times cannot be shrunk to one going round $m(\neq n)$ times.

The group function χ, then, is a many-valued function, and this is possible because it is defined in a space which is not simply connected. Equivalently and conversely, we may say that if χ *were* single-valued, then $\mathbf{B} = \operatorname{curl} \mathbf{A} = \operatorname{curl} \operatorname{grad} \chi \equiv 0$ everywhere, so there could be no magnetic flux at all.

It is thus an essential condition for the Bohm–Aharonov effect to occur that the configuration space of the vacuum is not simply connected. In (3.104) the function χ is a function in the group space of the gauge group $U(1)$:

$$A_\mu = \partial_\mu \chi. \tag{3.106}$$

A_μ, in fact, is a gauge transform of the 'true' vacuum $A_\mu = 0$. The corresponding transformation on a charged particle is $e^{-ie\chi}$, which is an element of $U(1)$. As mentioned above (see the discussion below (3.66)) the group space of $U(1)$ is the circle, which we denote by S^1. This group space is not simply connected, because a path which goes twice round a circle cannot be continuously deformed (while staying only on the circle) into one which goes round once. In fact, there is an infinite number of inequivalent paths in the group space. We now define this notion more formally.

A path a in a space X is defined as a continuous function $a(s)$ of a real parameter s, so that each value of s in the interval $0 \leqslant s \leqslant 1$ corresponds to a point $a(s)$ in the space X. If a path a connects the points P and Q we have $a(0) = P$, $a(1) = Q$. If $a(0) = a(1) = P$ we have a *closed path* (or loop) at P. Consider two closed paths $a(s)$ and $b(s)$ and suppose there is a function $L(t, s)$ such that $L(0, s) = a(s)$, $L(1, s) = b(s)$. Then a and b are *homotopic*, and we

Fig. 3.9. The field-free space in the Bohm–Aharonov effect is not simply connected.

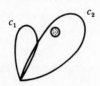

write $a \sim b$. The *inverse* of a path a is written a^{-1} and defined by

$$a^{-1}(s) = a(1 - s) \tag{3.107}$$

so that it corresponds to the same path traversed in the opposite direction. If the end-point of path a coincides with the starting-point of path b, so that $a(1) = b(0)$, we may define the *product path* $c = ab$ by

$$\left.\begin{array}{ll} c(s) = a(2s) & \text{for } 0 \leqslant s \leqslant \tfrac{1}{2}, \\ c(s) = b(2s - 1) & \text{for } \tfrac{1}{2} \leqslant s \leqslant 1. \end{array}\right\} \tag{3.108}$$

A *null path* consists of a single point. If $a \sim b$, then ab^{-1} is homotopic to a null path.

We may now construct a group by introducing the *class* of paths homotopic to a, denoted by $[a]$. They must, of course, have the same end-points. These homotopy classes may be multiplied, by defining the multiplication law

$$[a][b] = [ab]. \tag{3.109}$$

It is easy to see that this multiplication law defines a group, called the *fundamental group* or *first homotopy group* of the space X, and denoted $\pi_1(X)$. It was introduced by Poincaré in 1895, and is called the first homotopy group of X in anticipation of the fact that there are higher homotopy groups of X, introduced by Hurewicz in 1935. We need only check that the group laws are satisfied:

1. Closure: if $[a] \in \pi_1(X)$ and $[b] \in \pi_1(X)$, then in view of (3.109) $[a][b] = [ab] \in \pi_1(X)$.
2. Associativity: since $(ab)c \sim a(bc)$ we have $([a][b])[c] = [a]([b][c])$.
3. Identity element: is the class of the null path $[1]$, since $[a][1] = [a]$.
4. Inverse: $[a^{-1}][a] = [1]$, so $[a]^{-1} = [a^{-1}]$.

It now follows that the first homotopy group of $U(1)$ is isomorphic to the group of integers under addition (the integer n corresponding to a closed path that goes n times round the circle S^1 in group space). This group is denoted $\mathbb{Z}$. It is an infinite Abelian group. We have

$$\pi_1(U_1) = \mathbb{Z}; \tag{3.110}$$

in other words, it is because the gauge group of electromagnetism, U_1, is not simply connected that the Bohm–Aharonov effect is possible. (A *simply connected* space X is one in which every closed path is homotopic to the null path, so $\pi_1(X) = 1$.)

The configuration space of the Bohm–Aharonov experiment is the plane

$\mathbb{R}^2$ (symbolising the set of pairs of real numbers) with a hole in, and this is, topologically, the direct product of the line $\mathbb{R}^1$ and the circle S^1; $\mathbb{R}^1 \times S^1$. The line, for example, may be parameterised by r, and the circle by ϕ. The gauge function χ is a mapping from the group space G onto configuration space X:

$$\chi: G \to X. \tag{3.111}$$

In our case, G is S^1 and X is $S^1 \times \mathbb{R}$; however, functions mapping S^1 onto $\mathbb{R}$ are all deformable to a constant, so the non-trivial part of χ is given by

$$\chi: S^1 \to S^1. \tag{3.112}$$

(Actually, in our problem, this is the full form of χ, since it has no r dependence.) This, however, is the way to define the homotopy groups. To be exact, let $[X, Y]$ be the set of all homotopy classes of continuous maps from X onto Y. They generally form a group. Then

$$[S^1, Y] = \pi_1(Y) \tag{3.113}$$

is the first homotopy group of Y, Poincaré's 'fundamental group'. Similarly

$$[S^n, Y] = \pi_n(Y) \tag{3.114}$$

is the nth homotopy group of Y, which is an Abelian group for $n \geqslant 2$. (The first homotopy group is not always Abelian, but in all cases of interest to physics, it is Abelian.) It follows then from (3.112) that the gauge functions χ are divided into disjoint classes, generating the group

$$[S^1, S^1] = \pi_1(S^1) = \mathbb{Z}. \tag{3.115}$$

After this long mathematical digression, we may summarise. In the Bohm–Aharonov experiment, electrons are diffracted either side of an infinite solenoid, through a *null field*. There is, nevertheless, a positive effect on the interference fringes. The mathematical reason for this is that the configuration space of the null field (vacuum) is the plane with a hole in; which is $S^1 \times \mathbb{R}$. The vector potential A_μ is derived from a gauge function χ which maps the gauge space onto the configuration space. These mappings fall into disjoint classes, and since

$$[S^1, U_1] = \pi_1(U_1) = \mathbb{Z}$$

they are not all deformable to a constant gauge function $\chi = $ const (which, of course, would give $A_\mu = 0$, and no Bohm–Aharonov effect). It follows that if the gauge group is a simply connected group like $SU(2)$,

$$\pi_1(SU(2)) = 1, \tag{3.116}$$

then all mappings are deformable to the constant map, and no Bohm–Aharonov effect exists.

3.5   The Yang–Mills field

We now wish to generalise the results of §3.3 to the case where the Lagrangian has a higher symmetry than $O(2)$ or $U(1)$. The simplest generalisation is to $SU(2)$. This group, as well as the more complicated ones which are considered in physics, is non-Abelian, so what we are studying is the subject of *non-Abelian gauge fields*.

In §3.3 the scalar field we considered had two components which were related by rotations in the plane. An obvious way to generalise this is to consider the case where ϕ has three components, $\phi = (\phi_1, \phi_2, \phi_3)$, in an 'internal' space and the gauge transformations (of the first kind) are rotations in this space. This will give a 'vector' quantity which is conserved, instead of charge. It will be like isospin.

The hypothesis that isospin symmetry is a local symmetry was first advanced by Yang and Mills[17] in 1954. For many years their work was regarded as a blind alley – though an interesting one – since, as discussed below, there is no evidence that isospin is associated with a gauge symmetry. In recent years, however, the idea of Yang and Mills has been resurrected to apply to the strong interactions between quarks (colour gauge symmetry) and the unified weak and electromagnetic interactions (weak isospin and hypercharge gauge symmetry). In what follows we shall nevertheless refer to the conserved vector quantity as isospin, and the internal symmetry space as isospin space or isospace. The reader will remember that these words are intended to be understood generically.

Refering for a minute to Fig. 3.4, let us regard rotations in the $(1–2)$ plane as rotations about the 3 axis, in other words, rotations through an angle Λ_3, as shown in Fig. 3.10. We have, then, from (3.66),

$$\left.\begin{aligned}
\phi_1' &= \cos\Lambda_3\phi_1 + \sin\Lambda_3\phi_2, \\
\phi_2' &= -\sin\Lambda_3\phi_1 + \cos\Lambda_3\phi_2, \\
\phi_3' &= \phi_3.
\end{aligned}\right\} \tag{3.117}$$

Fig. 3.10. Rotation about the 3 axis in the internal symmetry space.

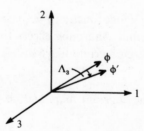

If Λ_3 is infinitesimal, then

$$\left.\begin{array}{l} \phi_1' = \phi_1 + \Lambda_3\phi_2, \\ \phi_2' = -\Lambda_3\phi_1 + \phi_2, \\ \phi_3' = \phi_3. \end{array}\right\} \tag{3.118}$$

which is the '3' component of the equation

$$\phi \to \phi' = \phi - \Lambda \times \phi, \tag{3.119}$$

which therefore corresponds to a general rotation through an angle Λ. The meaning of this is that $|\Lambda|$ is the angle of rotation, and $\Lambda/|\Lambda|$ is the axis of rotation. We have, then,

$$\delta\phi = -\Lambda \times \phi. \tag{3.120}$$

This is a gauge transformation of the first kind, and is to be compared with the two equations (3.67) and (3.69) – (3.120) is, of course, effectively three equations.

Our strategy now is to proceed as far as possible in analogy with the case of electromagnetism. We shall find that the present case is more complicated, however, and this is directly traceable to the fact that in the present case the rotations form the group $O(3)$ which is non-Abelian. Its non-Abelian nature is responsible for that fact that $\mathbf{a} \times \mathbf{b} = -\mathbf{b} \times \mathbf{a}$ – the vector product is not commutative. It will be seen below how this complicates matters. These complications have direct physical consequences.

First note that (3.120) is an instruction to perform a rotation in the internal space of ϕ through the *same* angle Λ at all points in space–time. We modify this to the more reasonable demand that Λ depends on x^μ. We then have

$$\partial_\mu\phi \to \partial_\mu\phi' = \partial_\mu\phi - \partial_\mu\Lambda \times \phi - \Lambda \times \partial_\mu\phi$$

or

$$\delta(\partial_\mu\phi) = -\Lambda \times \partial_\mu\phi - \partial_\mu\Lambda \times \phi. \tag{3.121}$$

Expressed in words, $\partial_\mu\phi$ does not transform covariantly, like ϕ does. We must construct a 'covariant derivative', analogous to that of (3.84). This will involve introducing a gauge potential analogous to A_μ. We then write the covariant derivative as

$$\blacksquare \qquad D_\mu\phi = \partial_\mu\phi + g\mathbf{W}_\mu \times \phi. \tag{3.122}$$

$\mathbf{W}_\mu$ is the gauge potential analogous to A_μ; note that it is a vector in the internal space, whereas A_μ only had one component; g is a coupling constant, analogous to electric charge e. By comparison with (3.120) we then require that

$$\delta(D_\mu\phi) = -\Lambda \times (D_\mu\phi). \tag{3.123}$$

How should $\mathbf{W}_\mu$ transform to satisfy this requirement? We may make a guess: first, $\mathbf{W}_\mu$ is an (internal) vector, so, as in (3.120), there will be a term $\mathbf{\Lambda} \times \mathbf{W}_\mu$; secondly, since A_μ picks up a term in $(1/e)\partial_\mu \Lambda$ (see (3.74)), we might expect $\mathbf{W}_\mu$ to pick up $(1/g)\partial_\mu \mathbf{\Lambda}$. So we write

$$\mathbf{W}_\mu \to \mathbf{W}_\mu - \mathbf{\Lambda} \times \mathbf{W}_\mu + \frac{1}{g}\partial_\mu \mathbf{\Lambda}$$

or

$$\delta \mathbf{W}_\mu = -\mathbf{\Lambda} \times \mathbf{W}_\mu + \frac{1}{g}\partial_\mu \mathbf{\Lambda}. \tag{3.124}$$

This would give, using (3.122), (3.121) and (3.120),

$$\begin{aligned}
\delta(D_\mu\boldsymbol{\phi}) &= \delta(\partial_\mu\boldsymbol{\phi}) + g(\delta\mathbf{W}_\mu) \times \boldsymbol{\phi} + g\mathbf{W}_\mu \times (\delta\boldsymbol{\phi}) \\
&= -\mathbf{\Lambda} \times \partial_\mu\boldsymbol{\phi} - \partial_\mu\mathbf{\Lambda} \times \boldsymbol{\phi} - g(\mathbf{\Lambda} \times \mathbf{W}_\mu) \times \boldsymbol{\phi} + \partial_\mu\mathbf{\Lambda} \times \boldsymbol{\phi} \\
&\quad - g\mathbf{W}_\mu \times (\mathbf{\Lambda} \times \boldsymbol{\phi}) \\
&= -\mathbf{\Lambda} \times \partial_\mu\boldsymbol{\phi} - g[(\mathbf{\Lambda} \times \mathbf{W}_\mu) \times \boldsymbol{\phi} + \mathbf{W}_\mu \times (\mathbf{\Lambda} \times \boldsymbol{\phi})].
\end{aligned} \tag{3.125}$$

Now the vector identity

$$(\mathbf{A} \times \mathbf{B}) \times \mathbf{C} + (\mathbf{B} \times \mathbf{C}) \times \mathbf{A} + (\mathbf{C} \times \mathbf{A}) \times \mathbf{B} = 0$$

gives, on rearrangement,

$$(\mathbf{A} \times \mathbf{B}) \times \mathbf{C} + \mathbf{B} \times (\mathbf{A} \times \mathbf{C}) = \mathbf{A} \times (\mathbf{B} \times \mathbf{C}). \tag{3.126}$$

Using this to rewrite the terms in the square brackets in (3.125), we have

$$\delta(D_\mu\boldsymbol{\phi}) = -\mathbf{\Lambda} \times (\partial_\mu\boldsymbol{\phi} + g\mathbf{W}_\mu \times \boldsymbol{\phi}) = -\mathbf{\Lambda} \times D_\mu\boldsymbol{\phi} \tag{3.127}$$

as required – this is the rule for covariant transformation. Our guess was right; introducing a gauge potential $\mathbf{W}_\mu$ which transforms as in (3.124), the covariant derivative of the vector $\boldsymbol{\phi}$ is given by (3.122).

$\mathbf{W}_\mu$ is the analogue of A_μ; what is the analogue of the field strength $F_{\mu\nu} = \partial_\mu A_\nu - \partial_\nu A_\mu$? Let us call it $\mathbf{W}_{\mu\nu}$. Unlike $F_{\mu\nu}$, which is a scalar under $O(2)$, $\mathbf{W}_{\mu\nu}$ will be a vector under $O(3)$, and so will transform like $\boldsymbol{\phi}$ itself:

$$\delta(\mathbf{W}_{\mu\nu}) = -\mathbf{\Lambda} \times \mathbf{W}_{\mu\nu}. \tag{3.128}$$

$\mathbf{W}_{\mu\nu} = \partial_\mu\mathbf{W}_\nu - \partial_\nu\mathbf{W}_\mu$ will not do, because

$$\begin{aligned}
\delta(\partial_\mu\mathbf{W}_\nu - \partial_\nu\mathbf{W}_\mu) &= \partial_\mu\left(-\mathbf{\Lambda} \times \mathbf{W}_\nu + \frac{1}{g}\partial_\nu\mathbf{\Lambda}\right) \\
&\quad - \partial_\nu\left(-\mathbf{\Lambda} \times \mathbf{W}_\mu + \frac{1}{g}\partial_\mu\mathbf{\Lambda}\right) \\
&= -\mathbf{\Lambda} \times (\partial_\mu\mathbf{W}_\nu - \partial_\nu\mathbf{W}_\mu) \\
&\quad - (\partial_\mu\mathbf{\Lambda} \times \mathbf{W}_\nu - \partial_\nu\mathbf{\Lambda} \times \mathbf{W}_\mu).
\end{aligned} \tag{3.129}$$

The last term is unwanted, and so must be cancelled. We now observe that

$$\delta(g\mathbf{W}_\mu \times \mathbf{W}_\nu) = g\left(-\mathbf{\Lambda} \times \mathbf{W}_\mu + \frac{1}{g}\partial_\mu\mathbf{\Lambda}\right) \times \mathbf{W}_\nu$$
$$+ g\mathbf{W}^\mu \times \left(-\mathbf{\Lambda} \times \mathbf{W}_\nu + \frac{1}{g}\partial_\nu\mathbf{\Lambda}\right).$$

The first and third terms may be combined according to (3.126), giving

$$\delta(g\mathbf{W}_\mu \times \mathbf{W}_\nu) = -g\mathbf{\Lambda} \times (\mathbf{W}_\mu \times \mathbf{W}_\nu)$$
$$+ (\partial_\mu\mathbf{\Lambda} \times \mathbf{W}_\nu - \partial_\nu\mathbf{\Lambda} \times \mathbf{W}_\mu). \tag{3.130}$$

The last term in the above equation is the same as in (3.129), so if we define

$$\mathbf{W}_{\mu\nu} = \partial_\mu\mathbf{W}_\nu - \partial_\nu\mathbf{W}_\mu + g\mathbf{W}_\mu \times \mathbf{W}_\nu \tag{3.131}$$

then $\mathbf{W}_{\mu\nu}$ transforms in the required way, by (3.128). The field strength $\mathbf{W}_{\mu\nu}$ is a vector, so $\mathbf{W}_{\mu\nu}\cdot\mathbf{W}^{\mu\nu}$ is a scalar and will appear in the Lagrangian, which is, therefore,

■ $$\mathcal{L} = (D_\mu\boldsymbol{\phi})\cdot(D^\mu\boldsymbol{\phi}) - m^2\boldsymbol{\phi}\cdot\boldsymbol{\phi} - \tfrac{1}{4}\mathbf{W}_{\mu\nu}\cdot\mathbf{W}^{\mu\nu}. \tag{3.132}$$

The equations of motion are obtained by functional variation of this Lagrangian in the usual way, so that the Euler–Lagrange equation

$$\frac{\partial\mathcal{L}}{\partial(W^i_\mu)} = \partial_\nu\left[\frac{\partial\mathcal{L}}{\partial(\partial_\nu W^i_\mu)}\right],$$

where i is an internal index, gives, after some algebra,

$$\partial^\nu\mathbf{W}_{\mu\nu} + g\mathbf{W}^\nu \times \mathbf{W}_{\mu\nu} = g[(\partial_\mu\boldsymbol{\phi}) \times \boldsymbol{\phi} + g(\mathbf{W}_\mu \times \boldsymbol{\phi}) \times \boldsymbol{\phi}] \tag{3.133}$$

or, by virtue of (3.122),

$$D^\nu\mathbf{W}_{\mu\nu} = g(D_\mu\boldsymbol{\phi}) \times \boldsymbol{\phi} \equiv g\mathbf{J}_\mu. \tag{3.134}$$

This equation is analogous to Maxwell's equation (3.89), so that $\mathbf{W}_{\mu\nu}$ is the 'isospin' gauge field, $\mathbf{J}_\mu$ is the source or 'matter' term, and instead of ordinary derivatives there are covariant ones. The equation (3.134) thus relates the covariant divergence of the gauge field to the 'matter' current. Whereas Maxwell's equations are linear in A_μ, however, this equation is non-linear in $\mathbf{W}_\mu$. The consequence of this is that, whereas in the absence of matter ($\phi \to 0$) Maxwell's equation (3.89) becomes

$$\partial^\nu F_{\mu\nu} = 0 \to \mathbf{\nabla}\cdot\mathbf{E} = 0, \quad \frac{\partial\mathbf{E}}{\partial t} - \text{curl}\,\mathbf{B} = 0, \tag{3.135}$$

indicating that there is no source term for the electromagnetic field, the non-Abelian gauge-field equation (3.133) becomes, in the absence of matter,

$$D^\nu\mathbf{W}_{\mu\nu} = 0 \to \partial^\nu\mathbf{W}_{\mu\nu} = -g\mathbf{W}^\nu \times \mathbf{W}_{\mu\nu}, \tag{3.136}$$

indicating that the field $\mathbf{W}_{\mu\nu}$ acts as a *source* for itself. This corresponds to the fact that the electromagnetic field $F_{\mu\nu}$ carries no charge, and is therefore not its own source, but the 'isospin' field $\mathbf{W}_{\mu\nu}$ carries 'isospin' (it has $I = 1$) so *does* act as a source for itself, since it is the field which is generated by the existence of any particle with 'isospin'. It will be recalled that this is a direct consequence of the fact that the symmetry group $O(3)$ is non-Abelian. We shall find in chapter 7, however, that there are also consequences in the quantum theory of gauge fields. In fact, substituting $D_{\mu}\phi$ from (3.122) and $\mathbf{W}_{\mu\nu}$ from (3.131) into the Lagrangian (3.132) we find that there are several terms of order 3 and 4 in the fields ϕ and $\mathbf{W}_{\mu}$. For example, looking at terms of order 3, whereas the Lagrangian of electrodynamics (3.83) contains a $(\phi\phi^{*}A)$ coupling which gives rise to

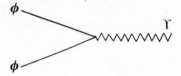

the present Lagrangian contains $(\phi\phi\mathbf{W})$ and $(\mathbf{WWW})$ terms giving the vertices

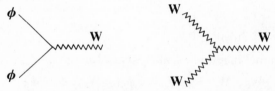

Of course, this is no more than what we expect from the field equations; as discussed above, in electrodynamics, only charged fields, like ϕ, may emit photons, but in non-Abelian gauge theories, the field $\mathbf{W}$ itself, as well as the field ϕ, may emit gauge particles. The detailed consequences of this are left until chapter 7. It is, however, worth remarking here that a similar situation holds in general relativity. There the gravitational field itself carries energy, which is equivalent to mass, and therefore is a source of gravitation. In terms of the field equation (3.38)

$$G_{\mu\nu} \equiv R_{\mu\nu} - \tfrac{1}{2}g_{\mu\nu}R = -\frac{8\pi G}{c^2}T_{\mu\nu},$$

both sides of this equation are tensors whose *covariant* divergence vanishes. The covariant derivative concerned is analogous to the one considered here, but has a strictly geometric origin, related to the fact that space–time itself is no longer Euclidean in the presence of gravitating objects. Therefore, even in the absence of matter ($T_{\mu\nu} = 0$), the 'real' divergence of the Einstein tensor $G_{\mu\nu}$ is not zero, and this is the mathematical expression of the fact that the gravitational field couples to itself. It is clear that the non-Abelian nature of the gauge group is responsible for a much richer structure than exists in electrodynamics, and to illustrate this once more, consider the equations analogous to the *homogeneous* Maxwell equations (2.233):

$$\partial_\lambda F_{\mu\nu} + \partial_\mu F_{\nu\lambda} + \partial_\nu F_{\lambda\mu} = 0. \tag{2.233}$$

These equations imply, as we saw in chapter 2, that

$$\text{div } \mathbf{B} = 0, \quad \frac{\partial B}{\partial t} + \text{curl } \mathbf{E} = 0,$$

the first of which says there are no magnetic charges as sources of magnetic field. The non-Abelian generalisation of (2.233) is, as will be proved in the next section,

$$D_\lambda \mathbf{W}_{\mu\nu} + D_\mu \mathbf{W}_{\nu\lambda} + D_\nu \mathbf{W}_{\lambda\mu} = 0 \tag{3.137}$$

and, because it is the covariant and not the ordinary derivative that appears, it follows that

$$\text{div } \vec{\mathbf{B}} \neq 0 \tag{3.138}$$

where $\vec{\mathbf{B}}$ is the 'magnetic induction isovector' – it is a vector in 'isospace' as well as in ordinary space. For these next few lines $\rightarrow$ indicates a vector in configuration space, and bold type a vector in isospin space. Since this isovector magnetic field has non-zero divergence, there exist isospin 1 magnetic monopoles. Or, to press the point home, if we introduce a scalar isovector field ϕ, and *define*

$$\vec{B}_{HP} = \boldsymbol{\phi} \cdot \vec{\mathbf{B}} \tag{3.139}$$

to be the 'ordinary' magnetic field, then

$$\text{div } \vec{B}_{HP} \neq 0 \tag{3.140}$$

and this theory has ordinary (isoscalar) magnetic monopoles. The subscript HP here stands for 't Hooft and Polyakov, who were the first to realise that non-Abelian gauge fields offer the possibility of magnetic monopoles. Their theory will be explained in more detail in chapter 10.

Not everything is different when we go to the non-Abelian case, however. One thing that is the same is that *the 'isospin' field* $\mathbf{W}_{\mu\nu}$ *must be massless*, just like the electromagnetic field. The reason is the same – to account for a field with mass m, an extra term

$$\mathscr{L}_m = m^2 \mathbf{W}_\mu \cdot \mathbf{W}^\mu \tag{3.141}$$

must be added to the Lagrangian – this is analogous to (3.92). The equation of motion (3.134) is then changed to

$$D^\nu \mathbf{W}_{\mu\nu} = g \mathbf{J}_\mu + m^2 \mathbf{W}_\mu. \tag{3.142}$$

The term (3.141), however, is clearly not gauge invariant, so, as before, we see that gauge invariance implies zero mass for the gauge field.

It is appropriate now to ask the question: *is isospin described by a gauge-field theory?* The concept of isospin was outlined in chapter 1, where it was seen to be a conserved 'vector' quantum number. Is it therefore analogous to electric charge? If it is, its conservation follows from a *local* symmetry and there must exist a massless isospin 1, spin 1 gauge field. What is more, our field ϕ above describes an *arbitrary* isovector field. Its covariant derivative $D_\mu\phi$, equation (3.122), contains a coupling constant g, which, through the Lagrangian (3.132) describes the strength of the coupling of ϕ to the gauge field. This same g, however, occurs in the definition of the gauge field $\mathbf{W}_{\mu\nu}$ itself, equation (3.131), and so is not to be tampered with! It follows that *all* isovector fields couple with the *same* strength to the gauge field. This *universality* of coupling is yet another consequence of the non-Abelian symmetry. It is not true in electromagnetism, where particles may have an *arbitrary* charge; here, all particles have the same 'charge'. This, incidentally, is one reason for believing that non-Abelian gauge theories may be relevant for physics – they would explain charge quantisation. So we return to the question: is isospin described by a gauge theory?

The answer is no. In the 1960s, however, when a lot of work was done on the newly discovered vector (spin 1) mesons, a number of people, particularly Sakurai,[18] believed that the ρ meson was the isospin gauge field, and, further, that the ω or ϕ meson (or some combination) was a gauge field for baryon number conservation, and that the $K(890)$ meson was the gauge field coupled to strangeness. An exposition of these ideas would be totally out of place in this book, but suffice it to say that the type of theory advocated, known as 'vector meson dominance', did receive a fair amount of experimental support. Nowadays, however, we believe that the ρ meson cannot be the isospin gauge field because it is massive, and this destroys gauge invariance. Actually, this problem can be overcome by 'spontaneous symmetry breakdown' (see chapter 8), and this is believed to explain the fact that the W-bosons are massive. A more telling observation is that, according to the quark model, ρ is simply

$$(\rho^+, \rho^0, \rho^-) = \left(u\bar{d}, \frac{u\bar{u} - d\bar{d}}{\sqrt{2}}, d\bar{u} \right);$$

it is made of quarks, just like all other hadrons. If it were a gauge field, it would not be made of quarks. Similar remarks apply to ω, ϕ and $K(890)$ as candidates for gauge fields.

A summary of this position is given in Table 3.1. Conservation of charge is described by local symmetry, and has a dynamical manifestation, the

Table 3.1. *A comparison of various conservation laws*

Conserved quantity	Associated dynamical (gauge) field	Conservation law
Charge, Q	E.m. γ	Local
Isospin, $\mathbf{I}$	—	Global
Strangeness, S	—	Global
Baryon number, B	—	Global
Lepton number, L	—	Global

electromagnetic field, whose quanta are photons. Conservation of isospin, strangeness, baryon number and lepton number are all described by global symmetries, and have no dynamical manifestations, that is, no corresponding fields. We might be tempted, then, to abandon the idea that gauge theories (save for electromagnetism) have anything to do with particle physics. But, in point of fact, they are now seen as the central feature of it. We will discuss the reasons for this as they arise out of our theoretical discussion of the gauge idea.

In this section, we introduced the idea of a non-Abelian gauge symmetry by treating the case of an isovector field ϕ, transforming under the symmetry group $O(3)$, or $SU(2)$. The calculations involved were rather cumbersome. What we need to do is to tidy up our whole approach to the subject by making it more abstract and general. We will then be able to see, for example, how to write the covariant derivative of an isospinor field, and how to generalise to groups other than $O(3)$. The way to achieve this is to adopt a geometrical point of view which is reminiscent of general relativity (but requires no knowledge of it!).

3.6   The geometry of gauge fields

The gauge transformations that we have been considering have deliberately been chosen to be different at different points in space, so it is natural that we should try to cast our work into an explicitly geometrical language, which we shall do in this section.

To begin with, let us rewrite the formula for rotation of an isovector through an angle Λ in isospace, equation (3.119):

$$\phi \to \phi' = \phi - \Lambda \times \phi. \tag{3.119}$$

This is the infinitesimal form of

$$\phi \to \phi' = \exp(i\mathbf{I}\cdot\Lambda)\phi \tag{3.143}$$

where the matrix generators $\mathbf{I}$ are given by

$$I_1 = \begin{pmatrix} 0 & 0 & 0 \\ 0 & 0 & -i \\ 0 & i & 0 \end{pmatrix}, \quad I_2 = \begin{pmatrix} 0 & 0 & i \\ 0 & 0 & 0 \\ -i & 0 & 0 \end{pmatrix}, \quad I_3 = \begin{pmatrix} 0 & -i & 0 \\ i & 0 & 0 \\ 0 & 0 & 0 \end{pmatrix},$$

(3.144)

and, as may easily be seen, have the matrix elements

$$(I_i)_{mn} = -i\varepsilon_{imn} \tag{3.145}$$

where ε_{imn} is the usual Levi–Città symbol. Expanding equation (3.143) to order Λ, gives (with summation over repeated indices)

$$\begin{aligned} \phi'_m &= (1 + iI_i\Lambda_i)_{mn}\phi_n \\ &= (\delta_{mn} + \varepsilon_{imn}\Lambda_i)\phi_n \\ &= \phi_m - \varepsilon_{min}\Lambda_i\phi_n \\ &= (\boldsymbol{\phi} - \boldsymbol{\Lambda} \times \boldsymbol{\phi})_m, \end{aligned}$$

which is (3.119). Now let Λ depend on x^μ, and write (3.143) as

$$\boldsymbol{\phi} \to \boldsymbol{\phi}' = \exp[i\mathbf{I} \cdot \boldsymbol{\Lambda}(x)]\boldsymbol{\phi} = S(x)\boldsymbol{\phi}. \tag{3.146}$$

The matrices $\mathbf{I}$ are representations of the generators of $O(3)$ (or $SU(2)$), and hence obey

$$[I_i, I_j] = i\varepsilon_{ijk}I_k = C_{ijk}I_k. \tag{3.147}$$

This equation identifies $i\varepsilon_{ijk}$ as the *structure constants* C_{ijk} of the group $SU(2)$. Since the generators M_i of any group obey the Jacobi identity

$$[[M_i, M_j], M_k] + [[M_j, M_k], M_i] + [[M_k, M_i], M_j] = 0,$$

the structure constants C_{ijk}, which are totally antisymmetric in i, j, k, obey the condition

$$C_{lim}C_{mjk} + C_{ljm}C_{mki} + C_{lkm}C_{mij} = 0. \tag{3.148}$$

Returning to $SU(2)$, the explicit representation (3.144) with matrix elements

$$(I_i)_{mn} = C_{ijk}$$

is called the *adjoint representation* of the group. From our treatment of spin, in the previous chapter, we know that an *isospinor* ψ transforms like

$$\psi \to \psi' = \exp\left[\frac{i}{2}\boldsymbol{\tau} \cdot \boldsymbol{\Lambda}(x)\right]\psi = S(x)\psi \tag{3.149}$$

where $S(x)$ is, here, a 2×2 matrix, and the Pauli matrices τ_i, or, more precisely, $\tau_i/2$, obey the relations (3.147), as required for a representation of the group. Let us now write the general, n-dimensional, case as

$$\psi(x) \to \psi'(x) = \exp[iM^a\Lambda^a(x)]\psi(x) = S(x)\psi(x) \tag{3.150}$$

where the index a is summed from 1 to 3, ψ is an n-component vector and

M^a are three $n \times n$ matrices representing the generators, and having the commutation relation (3.147). It is clear that $\partial_\mu \psi$ does not transform covariantly:

$$\partial_\mu \psi' = S(\partial_\mu \psi) + (\partial_\mu S)\psi. \tag{3.151}$$

The problem is that we are performing a different 'isorotation' at each point in space, which we may express by saying that the 'axes' in isospace are oriented differently at each point. The reason $\partial \psi / \partial x^\mu$ is not covariant is that $\psi(x)$ and $\psi(x + \mathrm{d}x) = \psi(x) + \mathrm{d}\psi$ are measured in different co-ordinate systems. This is sketched in Fig. 3.11; ψ is a field, and so has different values at different points, but $\psi(x)$ and $\psi(x + \mathrm{d}x) = \psi(x) + \mathrm{d}\psi$ are measured with respect to different axes. The quantity $\mathrm{d}\psi$, then, carries information about the variation of the field ψ itself with distance, but also about the rotation of the axes in isospace on moving from x to $x + \mathrm{d}x$. To form a properly covariant derivative, we should compare $\psi(x + \mathrm{d}x)$ not with $\psi(x)$ but with the value $\psi(x)$ would have if it were 'carried' from x to $x + \mathrm{d}x$ keeping the axes in isospace fixed – this we may call *parallel transport in isospace*, and it is illustrated in Fig. 3.12. The resulting vector is denoted $\psi + \delta\psi$. Note that $\delta\psi$ is not zero, because $\psi + \delta\psi$ is the vector which, when measured in the local isoco-ordinate system at $x + \mathrm{d}x$, is equal ('parallel') to the vector ψ, measured in the local isoco-ordinate system at x. These co-ordinate systems are not the same, so neither are the vectors. What is $\delta\psi$? It is sensible to assume that it is proportional to ψ itself, and also to $\mathrm{d}x^\mu$, the distance over which the vector is carried, so we put

$$\delta\psi = \mathrm{i}gM^a A_\mu^a \,\mathrm{d}x^\mu \psi \tag{3.152}$$

where g is a number put in to get the dimension right and A_μ^a is an additional

Fig. 3.11. $\mathrm{d}\psi$ carries information about the variation in ψ, as well as the change in co-ordinate axes between x and $x + \mathrm{d}x$.

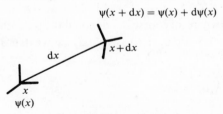

Fig. 3.12. $\delta\psi$ is defined by parallel transport – see text.

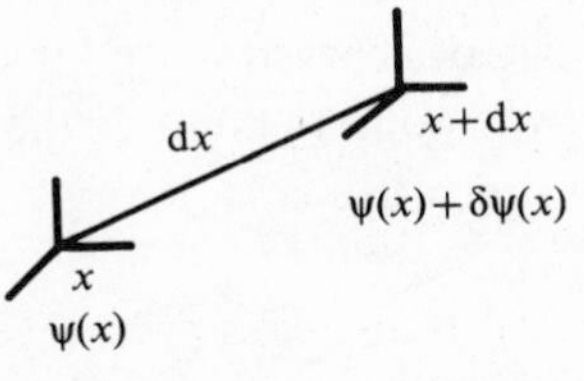

field or potential – Feynman calls it a 'universal influence' – which tells us to what extent the axes in isospace differ from point to point.

We now have two vectors at the point $x + dx$; $\psi + d\psi$ and $\psi + \delta\psi$. The 'true' derivative of ψ is given by taking the difference between these vectors:

$$D\psi = (\psi + d\psi) - (\psi + \delta\psi)$$

$$= d\psi - \delta\psi$$

$$= d\psi - igM^a A_\mu^a \, dx^\mu \psi,$$

■ $$\frac{D\psi}{dx^\mu} = D_\mu \psi = \partial_\mu \psi - igM^a A_\mu^a \psi. \tag{3.153}$$

The above equation defines the covariant derivative of an arbitrary field ψ transforming under an arbitrary group where generators are represented by the matrices M^a appropriate to the representation of ψ. Let us check that it gives the same covariant derivatives as those we have already found. To begin with, we consider the group $U(1)$. Comparing (3.55) and (3.150), we see that we require $M \to -1$. In addition, substituting $g \to e$ gives

$$U(1): \quad D_\mu = \partial_\mu + ieA_\mu$$

in agreement with (3.84). Next, for the group $SU(2)$, in the vector representation we have, from (3.145),

$$(M^a)_{mn} = -i\varepsilon_{amn} \tag{3.154}$$

where the indices take on the values 1, 2, 3 (for internal indices we make no distinction between upper and lower position). So taking the m component of (3.153) gives

$$D_\mu\phi_m = \partial_\mu\phi_m - ig(M^a)_{mn}A_\mu^a\phi_n$$

$$= \partial_\mu\phi_m - g\varepsilon_{amn}A_\mu^a\phi_n$$

$$= (\partial_\mu\phi + g\mathbf{A}_\mu \times \phi)_m$$

or

$$D_\mu\phi = \partial_\mu\phi + g\mathbf{A}_\mu \times \phi \tag{3.155}$$

in agreement with (3.122) (where, of course, we denoted the potential by $\mathbf{W}$, not $\mathbf{A}$). Finally, the spinor covariant derivative may be written down by putting $M^a = \tau^a/2$, hence

■ $$D_\mu\psi = \partial_\mu\psi - \frac{i}{2}\tau\cdot\mathbf{A}_\mu\psi. \tag{3.156}$$

The 'derivation' of covariant derivatives for an internal gauge group given above is modelled on the derivation of the covariant derivative of a vector in general relativity, where in the case of a curved space–time, it is the space–time axes themselves which vary from point to point. In the case of a (contravariant) vector V^μ, its covariant derivative is

$$D_\nu V^\mu = \partial_\nu V^\mu + \Gamma^\mu_{\lambda\nu}V^\lambda. \tag{3.157}$$

The quantities $\Gamma^{\mu}_{\nu\lambda}$, called the 'connection coefficients', clearly play a similar role to the vector potentials A^a_{μ}. They are called connection coefficients because they *connect* the components of a vector at one point with its components at a nearby point, the vector being transported between the points by 'parallel transport', as explained above. Because of this similarity, some physicists refer to A^a_{μ} as the connection.

Now we know how the generic vector ψ transforms when we do a rotation in isospace; $\psi \to S\psi$. Since $D_{\mu}\psi$ is the covariant derivative of ψ, it transforms in the same way, so we have

$$D_{\mu}\psi \to D'_{\mu}\psi' = SD_{\mu}\psi. \tag{3.158}$$

For simplicity, let us define the matrix

$$A_{\mu} = M^a A^a_{\mu} \tag{3.159}$$

so that (3.153) reads

$$D_{\mu}\psi = (\partial_{\mu} - igA_{\mu})\psi. \tag{3.160}$$

Transforming to a new isoframe gives, in view of (3.158),

$$(\partial_{\mu} - igA'_{\mu})\psi' = S(\partial_{\mu} - igA_{\mu})\psi \tag{3.161}$$

Putting $\psi' = S\psi$ in this equation then gives

■ $$A'_{\mu} = SA_{\mu}S^{-1} - \frac{i}{g}(\partial_{\mu}S)S^{-1}. \tag{3.162}$$

This is the rule for the gauge transformation of the potential. Note the characteristic inhomogeneous term on the right. For the group $U(1)$, $S = e^{-i\Lambda}$, $\partial_{\mu}S = -i(\partial_{\mu}\Lambda)e^{-i\Lambda}$, and (3.162) gives (with $g \to e$ and $M = \cancel{1}$)
$$^{-1}$$

$$A'_{\mu} = A_{\mu} + \frac{1}{e}\partial_{\mu}\Lambda$$

as in (3.74). In the case of $SU(2)$, we have

$$S = \exp\left(\frac{i}{2}\tau\cdot\Lambda\right)$$

so

$$\partial_{\mu}S = \frac{i}{2}\tau\cdot\partial_{\mu}\Lambda\cdot S$$

and, after a little algebra, (3.162) gives (for infinitesimal Λ)

$$\mathbf{A}'_{\mu} = \mathbf{A}_{\mu} - \Lambda \times \mathbf{A}_{\mu} + \frac{1}{g}\partial_{\mu}\Lambda$$

in agreement with (3.124).

It may be noted in passing that the connection coefficients in general relativity also have an inhomogeneous term in their transformation law.

The formula is

$$\Gamma'^{\kappa}_{\lambda\mu} = \frac{\partial x'^{\kappa}}{\partial x^{\alpha}}\frac{\partial x^{\beta}}{\partial x'^{\lambda}}\frac{\partial x^{\gamma}}{\partial x'^{\mu}}\Gamma^{\alpha}_{\beta\gamma} + \frac{\partial^2 x^{\alpha}}{\partial x'^{\lambda}\partial x'^{\mu}}\frac{\partial x'^{\kappa}}{\partial x^{\alpha}}. \tag{3.163}$$

If it were not for the second, inhomogeneous, term, Γ would transform like a tensor.

Now the question presents itself: since A_{μ} transforms inhomogeneously, how do we know whether it may be transformed to *zero* at every point, and therefore have no physical effect? To test this, we perform a series of four infinitesimal displacements round the closed path $ABCDA$, as in Fig. 3.13. We start at A with a vector ψ_A, denoted $\psi_{A,0}$, and transport the vector round the closed path using the rule for 'parallel' transport, using the covariant derivative, then compare the final value of the vector at A, $\psi_{A,1}$ with its initial value $\psi_{A,0}$. If they differ, we take this as a signal that the potential A_{μ} *does* have a physical effect.

Transporting $\psi_{A,0}$ to B will give

$$\psi_B = \psi_{A,0} + D_{\mu}\psi_{A,0}\Delta x^{\mu}$$
$$= (1 + \Delta x^{\mu}D_{\mu})\psi_{A,0};$$

thence to C:

$$\psi_C = \psi_B + \delta x^{\nu}D_{\nu}\psi_B$$
$$= (1 + \delta x^{\nu}D_{\nu})\psi_B$$
$$= (1 + \delta x^{\nu}D_{\nu})(1 + \Delta x^{\mu}D_{\mu})\psi_{A,0};$$

thence to D, and finally back to A:

$$\psi_D = (1 - \Delta x^{\rho}D_{\rho})\psi_C$$
$$\psi_{A,1} = (1 - \delta x^{\sigma}D_{\sigma})\psi_D$$
$$= (1 - \delta x^{\sigma}D_{\sigma})(1 - \Delta x^{\rho}D_{\rho})(1 + \delta x^{\nu}D_{\nu})(1 + \Delta x^{\mu}D_{\mu})\psi_{A,0}$$
$$= \{1 + \delta x^{\mu}\Delta x^{\nu}[D_{\mu}, D_{\nu}]\}\psi_{A,0}. \tag{3.164}$$

Note that what has appeared is the *commutator* of the covariant derivative

Fig. 3.13. A round trip by parallel transport.

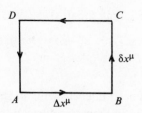

operators. From (3.160) we find

$$[D_\mu, D_\nu] = [\partial_\mu - igA_\mu, \partial_\nu - igA_\nu]$$
$$= -ig\{\partial_\mu A_\nu - \partial_\nu A_\mu - ig[A_\mu, A_\nu]\}. \tag{3.165}$$

If we define the *gauge field* $G_{\mu\nu}$ as

$$G_{\mu\nu} = \partial_\mu A_\nu - \partial_\nu A_\mu - ig[A_\mu, A_\nu], \tag{3.166}$$

we then have

$$[D_\mu, D_\nu] = -igG_{\mu\nu}. \tag{3.167}$$

In (3.164), $\delta x^\sigma \Delta x^\rho$ represents the area of the rectangle, $\Delta S^{\sigma\rho}$, so we rewrite this equation (the suffixes are dummies, so may be relabelled!) as

$$\psi_{A,1} = (1 - ig\Delta S^{\mu\nu}G_{\mu\nu})\psi_{A,0},$$
$$\psi_{A,1} - \psi_{A,0} = \Delta\psi_A = -ig\Delta S^{\mu\nu}G_{\mu\nu}\psi_A, \tag{3.168}$$

and we see that *if the gauge field is non-zero, a journey round a closed path has a physical effect* – the vector ψ is rotated in isospace.

In the case of $U(1)$, an Abelian group, the commutator term is zero, so putting $G \to F$ in (3.166), the gauge field is,

$$F_{\mu\nu} = \partial_\mu A_\nu - \partial_\nu A_\mu$$

which is, of course, just the electromagnetic field (3.81).

In the case of $SU(2)$, the matrices M^a obey the commutation relations (3.147), so

$$G^a_{\mu\nu} = \partial_\mu A^a_\nu - \partial_\nu A^a_\mu + g\varepsilon_{abc}A^b_\mu A^c_\nu \tag{3.169}$$

or, in vector notation,

$$\mathbf{G}_{\mu\nu} = \partial_\mu \mathbf{A}_\nu - \partial_\nu \mathbf{A}_\mu + g\mathbf{A}_\mu \times \mathbf{A}_\nu$$

which agrees with (3.131).

It may easily be seen that, since performing a rotation in isospace gives

$$\psi_{A,0} \to \psi'_{A,0} = S\psi_{A,0}$$
$$\psi_{A,1} \to \psi'_{A,1} = S\psi_{A,1}$$

with the *same* factor S, the field $G_{\mu\nu}$ transforms covariantly:

$$G'_{\mu\nu} = SG_{\mu\nu}S^{-1}. \tag{3.170}$$

It follows that $G_{\mu\nu}$ cannot be transformed away to zero by a gauge transformation; if it is zero in one gauge, it is zero in all.

Returning once more to the general relativistic analogy, the quantity which is analogous to the field tensor $G^a_{\mu\nu}$ is the Riemann–Christoffel curvature tensor $R^\mu_{\rho\sigma\lambda}$ defined by

$$R^\kappa_{\lambda\mu\nu} = \partial_\nu\Gamma^\kappa_{\lambda\mu} - \partial_\mu\Gamma^\kappa_{\lambda\nu} + \Gamma^\rho_{\lambda\mu}\Gamma^\kappa_{\rho\nu} - \Gamma^\rho_{\lambda\nu}\Gamma^\kappa_{\rho\mu}. \tag{3.171}$$

Comparing this with (3.166) or (3.169), we see the structural similarity: the first two terms are derivatives of the connection coefficients (potential) antisymmetric in μ and ν, and the last terms are products of the connection coefficients, also antisymmetric in μ and ν. The way the curvature tensor is introduced is also analogous – on taking a vector V^μ on a round trip by parallel transport, the difference between the initial and final components of the vector is

$$\Delta V^\mu = \tfrac{1}{2} R^\mu_{\rho\sigma\lambda} V^\rho \Delta S^{\sigma\lambda}$$

where $\Delta S^{\sigma\lambda}$ is the area enclosed by the path. This equation parallels (3.168). ΔV^μ is non-zero only if the space is intrinsically curved; for example, on the surface of a sphere (a 2-dimensional space), a vector *will* point in a different direction after a round trip, but on a flat surface it will not. The curvature tensor, being a tensor, has the property that if it is non-zero (has any non-zero components) in one co-ordinate system, it is non-zero in all, and indicates that the space is curved. In general relativity, this means that there is a gravitational field present.

Finally, we derive an interesting identity satisfied by the gauge field (taken from Feynman[24]). Consider the 3-dimensional closed path shown in Fig. 3.14. Start with the vector $\psi_{A,0}$ at A and transport it round the path $ABCDA$, after which its value (from (3.168)) will be

$$\psi_{A,1} = (1 - ig\,\Delta S^{\mu\nu} G_{\mu\nu})\psi_{A,0}$$

where $\Delta S^{\mu\nu} = \delta x^\mu \Delta x^\nu$ is the area of $ABCD$. Now, transporting the vector ψ to P along the line AP of length dx, it will have the value

$$\psi_{P,0} = (1 + dx^\rho D_\rho)\psi_{A,1}.$$

Thence, transporting it round the circuit $PSRQP$, its value will change to

$$\psi_{P,1} = (1 + ig\,\Delta S^{\mu\nu} G_{\mu\nu})\psi_{P,0}$$

where the plus sign arises because the route is in the opposite direction.

Fig. 3.14. A round trip used to derive the Bianchi identity.

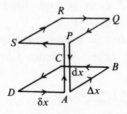

Finally, taking it down to A again, results in the final value

$$
\begin{aligned}
\psi_{A,2} &= (1 - dx^\sigma D_\sigma)\psi_{P,1} \\
&= (1 - dx^\sigma D_\sigma)(1 + ig\,\Delta S^{\mu\nu}G_{\mu\nu})(1 + dx^\rho D_\rho) \\
&\quad \times (1 - ig\Delta S^{\mu\nu}G_{\mu\nu})\psi_{A,0} \\
&= \{1 - ig\,\Delta V^{\rho\mu\nu}[D_\rho, G_{\mu\nu}]\}\psi_{A,0}
\end{aligned}
$$

where $\Delta V^{\rho\mu\nu} = dx^\rho\,\delta x^\mu \Delta x^\nu$ is the volume of the box. Taking into account the fact that the differential operator also acts on $\psi_{A,0}$, we may replace the commutator by the product $D_\rho G_{\mu\nu}$, giving

$$
\psi_{A,2} = (1 - ig\,\Delta V^{\rho\mu\nu}D_\rho G_{\mu\nu})\psi_{A,0}.
$$

The closed route we considered in Fig. 3.14 consists of circuits round the top and bottom faces of the box. There are clearly two similar such circuits enclosing the two other pairs of faces. All six faces are circuited by the path

$$
(ABCDAPSRQPA) + (ADSPABQRCBA) + (APQBADCRSDA),
$$
$$
(3.172)
$$

which would result in a factor

$$
1 - ig\,\Delta V^{\rho\mu\nu}(D_\rho G_{\mu\nu} + D_\mu G_{\nu\rho} + D_\nu G_{\rho\mu})
$$

operating on $\psi_{A,0}$, to give the new vector at A; call it $\psi_{A,3}$. However, as may be easily checked, the path (3.172) traverses each side of the box as many times in one direction as in the opposite one, so the path is equivalent to the reverse of itself, and consequently the vector ψ cannot change, so $\psi_{A,3} = \psi_{A,0}$, and hence

$$
D_\rho G_{\mu\nu} + D_\mu G_{\nu\rho} + D_\nu G_{\rho\mu} = 0. \tag{3.173}
$$

This is the identity we wanted (and it was anticipated in (3.137)). In view of (3.167) it is equivalent to

$$
\sum_{\text{cyclic}} [D_\rho, [D_\mu, D_\nu]] = 0 \tag{3.174}
$$

which is known as the *Jacobi identity*. It is a condition which is identically satisfied by the field tensor. In the Abelian case $U(1)$, it takes the form

$$
\partial_\rho F_{\mu\nu} + \partial_\mu F_{\nu\rho} + \partial_\nu F_{\rho\mu} = 0
$$

which is equation (2.233). This was shown to be equivalent to the homogeneous Maxwell equations.

The identity (3.173) has an analogue in general relativity, known as the *Bianchi identity*, which takes the form

$$
D_\rho R^\kappa_{\lambda\mu\nu} + D_\mu R^\kappa_{\lambda\nu\rho} + D_\nu R^\kappa_{\lambda\rho\mu} = 0. \tag{3.175}
$$

We may similarly refer to (2.233) as the Bianchi identity. The parallels which

Table 3.2. *Parallels between gauge theory and general relativity*

Gauge theory	General relativity
Gauge transformation	Co-ordinate transformation
Gauge group	Group of all co-ordinate transformations
Gauge potential, A_μ	Connection coefficient, $\Gamma^\kappa_{\mu\nu}$
Field strength, $G_{\mu\nu}$	Curvature tensor, $R^\kappa_{\lambda\mu\nu}$
Bianchi identity:	Bianchi identity:
$\sum_{\substack{\rho\mu\nu \\ \text{cyclic}}} D_\rho G_{\mu\nu} = 0$	$\sum_{\substack{\rho\mu\nu \\ \text{cyclic}}} D_\rho R^\kappa_{\lambda\mu\nu} = 0$

we have discussed between gauge theories and general relativity are summarised in Table 3.2.

We close this section by giving a brief account of chromodynamics, which is believed to account for the strong interaction. The notion of the colour quantum number was mentioned in §1.11: quarks of a given type $(u, d, s, \ldots)$ possess an additional label called R, W or B, which takes on three values. The quark Lagrangian is invariant under relabellings described by a group of transformations

$$q \to Mq, \quad q = \begin{pmatrix} q_R \\ q_W \\ q_B \end{pmatrix}. \tag{3.176}$$

M is a 3×3 matrix, and may be either orthogonal or unitary (these matrices form groups – hermitian matrices do not). Which do we choose? We choose $SU(3)$ rather than $O(3)$ for two reasons: (*a*) If the colour group were $O(3)$, the diquark system could be a colour singlet, but diquarks are not found in nature; (*b*) $O(3)$ does not possess asymptotic freedom (see chapter 9) if the number of flavours exceeds two; we know already that the number of flavours is at least five. The corresponding limit for $SU(3)$ is 16 flavours. M is then a unitary matrix, which, because we may subtract out an overall phase, may be chosen to have unit determinant:

$$q \to Uq, \quad U^+U = 1, \quad \det U = 1. \tag{3.177}$$

U may be written in the form

$$U = e^{iH}, \quad H = H^+, \quad \text{Tr}\, H = 0 \tag{3.178}$$

where H is Hermitian, and the zero trace condition follows from $\det U = 0.$ It may be seen that U has eight independent parameters ε_a, and therefore eight generators, which we denote $\lambda_a/2$, so we write

$$U = \exp\left(i\frac{\lambda_a}{2}\varepsilon_a\right)$$

(3.179)

where a summation over a from 1 to 8 is implied, and the eight matrix generators are

$$\lambda_1 = \begin{pmatrix} 0 & 1 & 0 \\ 1 & 0 & 0 \\ 0 & 0 & 0 \end{pmatrix}, \quad \lambda_2 = \begin{pmatrix} 0 & -i & 0 \\ i & 0 & 0 \\ 0 & 0 & 0 \end{pmatrix},$$

$$\lambda_3 = \begin{pmatrix} 1 & 0 & 0 \\ 0 & -1 & 0 \\ 0 & 0 & 0 \end{pmatrix}, \quad \lambda_4 = \begin{pmatrix} 0 & 0 & 1 \\ 0 & 0 & 0 \\ 1 & 0 & 0 \end{pmatrix},$$

$$\lambda_5 = \begin{pmatrix} 0 & 0 & -i \\ 0 & 0 & 0 \\ i & 0 & 0 \end{pmatrix}, \quad \lambda_6 = \begin{pmatrix} 0 & 0 & 0 \\ 0 & 0 & 1 \\ 0 & 1 & 0 \end{pmatrix},$$

$$\lambda_7 = \begin{pmatrix} 0 & 0 & 0 \\ 0 & 0 & -i \\ 0 & i & 0 \end{pmatrix}, \quad \lambda_8 = \frac{1}{\sqrt{3}}\begin{pmatrix} 1 & 0 & 0 \\ 0 & 1 & 0 \\ 0 & 0 & -2 \end{pmatrix}.$$

(3.180)

These are simply the 3-dimensional generalisations of the Pauli matrices, and the $SU(2)$ formula which corresponds to (3.179) is

$$U = \exp\left(i\frac{\tau_i}{2}\varepsilon_i\right)$$

with i summed from 1 to 3. The matrices λ_a obey the commutation relations

$$\left[\frac{\lambda_a}{2},\frac{\lambda_b}{2}\right] = if_{abc}\frac{\lambda_c}{2}$$

(3.181)

(with summation over c from 1 to 8). The quantities if_{abc} are the structure constants of the group (cf. (3.147)) and are totally antisymmetric in their indices. They may be found by explicit calculation using λ_a above, and it turns out that the only non-zero components are

$$\left.\begin{aligned} f_{123} &= 1, \\ f_{147} &= -f_{156} = f_{246} = f_{257} = f_{345} = -f_{367} = \tfrac{1}{2}, \\ f_{458} &= f_{678} = \frac{\sqrt{3}}{2} \end{aligned}\right\}$$

(3.182)

For a fuller treatment of the group $SU(3)$ the reader is referred to the literature. This group originally came to prominence in particle physics in the early 1960s, when it was discovered that the spectrum of particles (which then included strange particles, but not charmed particles) showed an approximate $SU(3)$ symmetry (see §1.8 and §1.9). In quark language, this $SU(3)$ is of the form

$$q \to Uq, \quad q = \begin{pmatrix} u \\ d \\ s \end{pmatrix}. \tag{3.183}$$

In the current jargon, this a flavour group, not a colour group, since it acts on flavour indices only. It is only an approximate symmetry, since for example the s quark is heavier than u and d. By contrast, colour is an exact symmetry. To distinguish the two $SU(3)$ groups of flavour and colour we write them $SU(3)_f$ and $SU(3)_c$.

According to the theory of chromodynamics, colour symmetry is a *gauge symmetry*, so the theory we have developed above may be taken over wholesale. In particular, there is a gauge potential, or rather eight gauge potentials which may be written in matrix form

$$A_\mu = A_\mu^a \frac{\lambda_a}{2} = \tfrac{1}{2} \begin{pmatrix} A_\mu^3 + \dfrac{1}{\sqrt{3}} A_\mu^8 & A_\mu^1 - iA_\mu^2 & A_\mu^4 - iA_\mu^5 \\ A_\mu^1 + iA_\mu^2 & -A_\mu^3 + \dfrac{1}{\sqrt{3}} A_\mu^8 & A_\mu^6 - iA_\mu^7 \\ A_\mu^4 + iA_\mu^5 & A_\mu^6 + iA_\mu^7 & -\dfrac{2}{\sqrt{3}} A_\mu^8 \end{pmatrix}. \tag{3.184}$$

The gauge fields $G_{\mu\nu}^a$ may be written down by referring to (3.169) and substituting for ε_{abc}, which are the structure constants for $SU(2)$ and f_{abc}, the structure constants of $SU(3)$, from (3.182). Hence, for example,

$$\begin{aligned} G_{\mu\nu}^1 &= \partial_\mu A_\nu^1 - \partial_\nu A_\mu^1 + g f_{1bc} A_\mu^b A_\nu^c \\ &= \partial_\mu A_\nu^1 - \partial_\nu A_\mu^1 + g[A_\mu^2 A_\nu^3 - A_\mu^3 A_\nu^2 + \tfrac{1}{2}(A_\mu^4 A_\nu^7 - A_\mu^7 A_\nu^4) \\ &\quad - \tfrac{1}{2}(A_\mu^5 A_\nu^6 - A_\mu^6 A_\nu^5)]. \end{aligned} \tag{3.185}$$

They are gauge fields, and they are massless. They go by the name of *gluons*.

If the explanation of quark confinement is that colour is confined, and therefore that only colour singlet ($SU(3)_c$ singlet) particles may appear in the free state, then gluons will also be confined and therefore unobservable. There should, however, be a combination of gluon fields invariant under $SU(3)_c$, and therefore in principle, observable. This is called a glueball. It has not (yet) been observed.

Summary

[1]A Lagrangian formulation of the motion of a point particle is shown to lead to Newton's second law of motion. [2]A similar treatment is given to the (classical) scalar field. The variational principle gives rise to the

Euler–Lagrange equations, and the action is found for which these give the Klein–Gordon equation. Invariance of the action under space–time translations and rotations leads, via Noether's theorem, to conservation of energy–momentum and angular momentum. [3]A complex scalar field has an $O(2)$ symmetry, and the requirement that this symmetry is local necessitates the introduction of a gauge field A_μ which is shown to obey Maxwell's equations. [4]The Bohm–Aharonov effect is ascribable to a configuration of the gauge field which is only possible in a topologically non-trivial (non-Euclidean) space – in this case, the plane with a hole in. [5]The generalisation of §3 to the case where the scalar field possesses an $O(3)$ symmetry is given, and leads to the Yang–Mills equations. [6]A general treatment of gauge fields along geometrical lines reveals parallels with general relativity, and results in general formulae for covariant derivatives which give the known forms for electromagnetism and the Yang–Mills case, and hold for any symmetry group. A brief account of $SU(3)$ is given, the relevant gauge field being the gluon.

Guide to further reading

Variational principles in classical physics are well described in
(1) H. Goldstein, *Classical Mechanics*, chapters 2 and 11, Reading, Massachusetts: Addison-Wesley Publishing Company, Inc., 1950.

A highly readable, though less mathematical account is
(2) R.P. Feynman, R.B. Leighton & M. Sands, *The Feynman Lectures on Physics*, vol. 2, chapter 19, Reading, Massachusetts: Addison-Wesley Publishing Company, Inc., 1964.

A detailed account of the application of the variational principle to fields is given by
(3) E.L. Hill, *Reviews of Modern Physics*, **23**, 253 (1951).

Shorter accounts may be found in
(4) L.I. Schiff, *Quantum Mechanics*, 2nd edn., section 45, New York: McGraw-Hill Book Company, Inc., 1955.
(5) D.E. Soper, *Classical Field Theory*, New York: John Wiley & Sons, 1976.

Noether's theorem first appeared in
(6) E. Noether, *Nachr. Ges. Wiss. Göttingen*, 171 (1918).

A thorough account of Noether's theorem, in Riemannian as well as flat space is
(7) J.G. Fletcher, *Reviews of Modern Physics*, **32**, 65–87 (1960).

Useful references on the energy–momentum tensor, which also include a consideration of its role in general relativity, are
(8) L.D. Landau & E.M. Lifshitz, *The Classical Theory of Fields*, 3rd English edn, sections 32 and 94, Oxford: Pergamon Press, 1971.
(9) S. Weinberg, *Gravitation and Cosmology: Principles and Applications of the General Theory of Relativity*, pp. 360–3, New York: John Wiley & Sons, Inc., 1972.

The Bohm–Aharonov effect originates in

(10) Y. Aharonov & D. Bohm. *Physical Review*, **115**, 484 (1959).
     The effect was first seen by
(11) R.G. Chambers, *Physical Review Letters*, **5**, 3 (1960).
     A simple account, on which the present account leans heavily, is
(12) R.P Feynman, *et al.*, ref. (2), section 15.5.
     A very lucid discussion of it appears in
(13) Tai Tsun Wu & Chen Ning Yang, *Physical Review D*, **12**, 3845 (1975).
     There have been recent claims that the effect is purely mathematical. For a response to these, and for further references to recent experimental verification of the effect, see
(14) D. Bohm & B.J. Hiley, *Il Nuovo Cimento*, **52A**, 295 (1979).
     For an introduction to homotopy groups, designed for physicists, see
(15) D. Speiser, 'Theory of Compact Lie Groups and Some Applications to Elementary Particle Physics' in F. Gürsey (ed.), *Group Theoretical Concepts and Methods in Elementary Particle Physics*, New York: Gordon & Breach, Science Publishers, 1964.
(16) F. Gürsey, 'Introduction to Group Theory' in C. DeWitt & B. DeWitt (eds.), *Relativity, Groups and Topology*, London: Blackie & Son Ltd; New York: Gordon & Breach, Science Publishers, Inc., 1964.
     The non-Abelian generalisation of electromagnetism was first proposed in
(17) C.N. Yang & R.L. Mills, *Physical Review*, **96**, 191 (1954).
     The application of this gauge principle to particle physics was advocated by
(18) J.J. Sakurai, *Annals of Physics*, **11**, 1 (1960).
(19) A. Salam & J.C. Ward, *Il Nuovo Cimento*, **20**, 419 (1961).
     For a review, see, for example,
(20) J.J. Sakurai, *Currents and Mesons*, Chicago and London: University of Chicago Press, 1969.
     The generalisation of the Yang–Mills idea to arbitrary groups appears in
(21) S.L. Glashow & M. Gell-Mann, *Annals of Physics*, **15**, 437 (1961).
     Parallels between Yang–Mills theories and general relativity were first pointed out by
(22) R. Utiyama, *Physical Review*, **101**, 1597 (1956).
     An early paper on the geometrical aspects of gauge theories is
(23) H.G. Loos, *Journal of Mathematical Physics*, **8**, 2114 (1967);
     they are reviewed in
(24) R.P. Feynman 'Gauge Theories', in R. Balian & C.H. Llewellyn Smith. eds., *Les Houches, Session XXIX, 1976–Intéractions électromagnétiques et faibles à haute energie/Weak and electromagnetic interactions at high energy*, Amsterdam: North-Holland Publishing Company, 1977.
(25) L. O'Raigheartaigh, *Reports on Progress in Physics*, **42**, 159 (1979).
     Chromodynamics is reviewed by Feynman and by Susskind in ref (24).

4

Canonical quantisation and particle interpretation

'I remember that when someone had started to teach me about creation and annihilation operators, that this operator creates an electron, I said "How do you create an electron? It disagrees with conservation of charge"'.

R.P. Feynman
Nobel lecture

4.1   The real Klein–Gordon field

We considered in chapter 2 the simplest relativistic equation, the Klein–Gordon equation, as a single particle equation, and found the following difficulties with it (i) the occurrence of negative energy solutions, (ii) the current j_μ does not give a positive definite probability density ρ, as the Schrödinger equation does. For these reasons we must abandon the interpretation of the Klein–Gordon equation *as a single particle equation*. (Historically, this was the motive which led Dirac to his equation.) Can any sense then be made out of the Klein–Gordon equation? After all, spin 0 particles do exist (π, K, η, etc.) so, surely, there must be some interpretation of the equation which makes sense.

What we shall do first is to consider the Klein–Gordon equation as describing a *field* $\phi(x)$. Since the equation has no classical analogue, $\phi(x)$ is a strictly *quantum* field, but nevertheless we shall begin by treating it as a *classical* field, as we did in the last chapter, and shall find that the negative energy problem does not then exist. We shall then take seriously the fact that $\phi(x)$ is a *quantum* field by recognising that it should be treated as an operator, which is subject to various commutation relations analogous to those in ordinary quantum mechanics. This process is often referred to as 'second quantisation', but I prefer not to use this term. There is, after all, only one quantum theory, not two; what we are doing is quantising a *field*,

129

rather than the motion of a *single particle*, as we do in quantum mechanics. It turns out that the field quantisation has an obvious interpretation as a many-particle theory, which is just what we want. It is a commonplace observation in high energy physics that the number of particles of a given type (say, of spin 0) is not a constant – only consider, for example, the decay $\pi^0 \rightarrow 2\gamma$ – so what right have we to expect that a *single-particle* theory would work? In this new interpretation, $|\phi|^2$ is proportional to the number of particles present. Moreover, after field quantisation, the two difficulties associated with the single-particle Klein–Gordon equation disappear, as we shall see below.

To begin, let us find the energy of the ('classical') Klein–Gordon field. It is obtained from the energy–momentum tensor $\theta^{\mu\nu}$, defined in (3.20), and is

$$H = \int \theta^{00} \, \mathrm{d}^3 x$$

$$= \tfrac{1}{2} \int [(\partial_0 \phi)^2 + \nabla\phi \cdot \nabla\phi + m^2 \phi^2] \mathrm{d}^3 x \tag{4.1}$$

in the case of the real scalar field, with Lagrangian given by (3.10). For the complex scalar field, we have

$$H = \int [(\partial_0 \phi^*)(\partial_0 \phi) + \nabla\phi^* \cdot \nabla\phi + m^2 \phi^* \phi] \mathrm{d}^3 x \tag{4.2}$$

and in each case the energy (Hamiltonian) is positive definite. Thus *the scalar field is not plagued by the negative energy problem* which besets the single-particle theory. But now that we have a satisfactory positive definite energy for the field, we ask, how is this related to the energy of single-particle states? The answer is that *field quantisation* forces us to reinterpret the field as a quantum rather than a classical system. The field $\phi(x)$ is, accordingly, regarded as a *Hermitian operator*, whose Fourier expansion may be written

$$\phi(x) = \int \frac{\mathrm{d}^3 k}{(2\pi)^3 2\omega_k} [a(k)\mathrm{e}^{-ikx} + a^\dagger(k)\mathrm{e}^{ikx}] \tag{4.3}$$

with $\omega_k = (\mathbf{k}^2 + m^2)^{\frac{1}{2}}$. The coefficients $a(k)$ and $a^\dagger(k)$ are also operators. The measure in the integrand has been so chosen because it is relativistically invariant. For the Klein–Gordon field we have the 'mass-shell' condition $k^2 = k_0^2 - \mathbf{k}^2 = m^2 (\hbar = c = 1)$, so an invariant element of phase space is, with $k_0 > 0$ (positive energy condition)

$$\frac{\mathrm{d}^4 k}{(2\pi)^4} 2\pi\delta(k^2 - m^2)\theta(k_0)$$

$$= \frac{d^4k}{(2\pi)^3} \, \delta(k_0^2 - \omega_k^2)\theta(k_0)$$

$$= \frac{d^4k}{(2\pi)^3} \, \delta[(k_0 + \omega_k)(k_0 - \omega_k)]\theta(k_0)$$

$$= \frac{d^4k}{(2\pi)^3} \frac{1}{2k_0} [\delta(k_0 + \omega_k) + \delta(k_0 - \omega_k)]\theta(k_0)$$

$$= \frac{d^3k \, dk_0}{(2\pi)^3 2k_0} \, \delta(k_0 - \omega_k) = \frac{d^3k}{(2\pi)^3 2\omega_k}. \tag{4.4}$$

The quantity $\phi(x, t)$ plays a role in field theory analogous to that played by $\mathbf{x}$, the position vector, in particle mechanics. The quantisation of mechanics follows from the Heisenberg commutation relations

$$\left.\begin{array}{l} [x_i, p_j] = i\delta_{ij} \quad (i, j = 1, 2, 3), \\ [x_i, x_j] = [p_i, p_j] = 0, \end{array}\right\} \tag{4.5}$$

where the momentum p_i is defined canonically as $\partial L / \partial x_i$; $\mathbf{x}$ and $\mathbf{p}$ refer to the position and momentum of the particle, measured at the *same time*. In a scalar field theory, $\phi(\mathbf{x}, t)$ plays a role analogous to $\mathbf{x}(t)$, and describes a system with an infinite number of degrees of freedom, since, at each time, ϕ has an independent value at each point in space. To approach this continuum case, let us divide space up into cells, each of volume δV_r, and let $\phi_r(t)$ be the average value of $\phi(x)$ in cell r at time t. Let the average Lagrangian density in each cell be $\mathcal{L}_r$. Then the momentum variable p_r, conjugate to ϕ_r is

$$p_r(t) = \frac{\partial L}{\partial \dot{\phi}_r(t)} = \delta V_r \frac{\partial \mathcal{L}_r}{\partial \dot{\phi}_r(t)} = \delta V_r \pi_r(t) \tag{4.6}$$

where the field $\pi(x, t)$ is defined by

$$\pi(\mathbf{x}, t) = \frac{\partial \mathcal{L}}{\partial \dot{\phi}(\mathbf{x}, t)} \tag{4.7}$$

and $\pi_r(t)$ is its average value in cell r. Then the Heisenberg commutation relations give

$$\left.\begin{array}{l} [\phi_r(t), p_s(t)] = i\delta_{rs}, \\ [\phi_r(t), \phi_s(t)] = [p_r(t), p_s(t)] = 0. \end{array}\right\} \tag{4.8}$$

Substituting (4.6) into (4.8) gives $[\phi_r(t), \pi_s(t)] = (1/\delta V_s)i\delta_{rs}$. In the continuum limit $\delta V_r \to 0$, and we have

$$\left.\begin{array}{l} [\phi(\mathbf{x}, t), \pi(\mathbf{x}', t)] = i\delta(\mathbf{x} - \mathbf{x}'), \\ [\phi(x, t), \phi(x', t)] = [\pi(\mathbf{x}, t), \pi(\mathbf{x}', t)] = 0. \end{array}\right\} \tag{4.9}$$

These are known as *equal-time commutation relations (ETCR)*, and we now use them to find the commutation relations between $a(k)$ and $a^\dagger(k)$ in equation (4.3). First, from the definition (4.7) and the Lagrangian (3.10), we have

$$\pi(x) = \dot\phi(x). \tag{4.10}$$

Next, it is simple to check that the positive energy (also known as positive frequency) solutions

$$f_k(x) = \frac{1}{[(2\pi)^3 2\omega_k]^{\frac{1}{2}}} e^{-ikx} \tag{4.11}$$

form an orthonormal set

$$\int f_k^*(x) i\overleftrightarrow{\partial}_0 f_{k'}(x) \mathrm{d}^3x = \delta^3(\mathbf{k} - \mathbf{k'}) \tag{4.12}$$

where $\overleftrightarrow{\partial}_0$ is defined by

$$A(t)\overleftrightarrow{\partial}_0 B(t) = A(t)\frac{\partial B(t)}{\partial t} - \frac{\partial A(t)}{\partial t} B(t). \tag{4.13}$$

The field expansion (4.3) may then be written

$$\phi(x) = \int \frac{\mathrm{d}^3k}{[(2\pi)^3 2\omega_k]^{\frac{1}{2}}} [f_k(x)a(k) + f_k^*(x)a^\dagger(k)] \tag{4.14}$$

which may be inverted, using (4.12), to give

$$a(k) = \int \mathrm{d}^3x [(2\pi)^3 2\omega_k]^{\frac{1}{2}} f_k^*(x) i\overleftrightarrow{\partial}_0 \phi(x)$$

$$a^\dagger(k') = \int \mathrm{d}^3x' [(2\pi)^3 2\omega_k]^{\frac{1}{2}} \phi(x') i\overleftrightarrow{\partial}_0 f_{k'}(x') \tag{4.15}$$

From (4.9), (4.10) and (4.15) we then have

$$[a(k), a^\dagger(k')] = -\int \mathrm{d}^3x\, \mathrm{d}^3x' (2\pi)^3 (4\omega_k\omega_{k'})^{\frac{1}{2}}$$

$$\cdot [f_k^*(x) i\overleftrightarrow{\partial}_0 \phi(x), \phi(x') i\overleftrightarrow{\partial}_0 f_{k'}(x')]$$

$$= (2\pi)^3 \int \mathrm{d}^3x\, \mathrm{d}^3x' (4\omega_k\omega_{k'})^{\frac{1}{2}} f_k^*(\mathbf{x}, t)\overleftrightarrow{\partial}_0 f_{k'}(\mathbf{x'}, t)$$

$$\times [\phi(\mathbf{x}, t), \pi(\mathbf{x'}, t)]$$

$$= (2\pi)^3 \int \mathrm{d}^3x (4\omega_k\omega_{k'})^{\frac{1}{2}} f_k^*(x) i\overleftrightarrow{\partial}_0 f_{k'}(x)$$

$$= (2\pi)^3 2\omega_k \delta^3(\mathbf{k} - \mathbf{k'}). \tag{4.16a}$$

Similarly, we have

$$[a(k), a(k')] = 0, \quad [a^\dagger(k), a^\dagger(k')] = 0. \tag{4.16b}$$

The operators $a(k)$ and $a^\dagger(k)$ play a crucial role in the particle interpretation of the quantised field theory. First, construct the operator

$$N(k) = a^\dagger(k)a(k). \tag{4.17}$$

It is simple to show that $N(k)$ and $N(k')$ commute:

$$\begin{aligned}
[N(k), N(k')] &= a^\dagger(k)[a(k), a^\dagger(k')]a(k') \\
&\quad + a^\dagger(k')[a^\dagger(k), a(k')]a(k) \\
&= [a^\dagger(k)a(k) - a^\dagger(k)a(k)]\delta(\mathbf{k} - \mathbf{k}') \\
&= 0,
\end{aligned}$$

so the eigenstates of these operators may be used to form a basis. Let the eigenvalues be denoted by $n(k)$:

$$N(k)|n(k)\rangle = n(k)|n(k)\rangle \tag{4.18}$$

and now use the commutation relations

$$[N(k), a^\dagger(k)] = a^\dagger(k),$$
$$[N(k), a(k)] = -a(k)$$

to find

$$\begin{aligned}
N(k)a^\dagger(k)|n(k)\rangle &= a^\dagger(k)N(k)|n(k)\rangle + a^\dagger(k)|n(k)\rangle \\
&= [n(k) + 1]a^\dagger(k)|n(k)\rangle \tag{4.19}
\end{aligned}$$

and

$$\begin{aligned}
N(k)a(k)|n(k)\rangle &= a(k)N(k)|n(k)\rangle - a(k)|n(k)\rangle \\
&= (n(k) - 1)a(k)|n(k)\rangle. \tag{4.20}
\end{aligned}$$

These equations tell us that if the state $|n(k)\rangle$ has eigenvalue $n(k)$, the states $a^\dagger(k)|n(k)\rangle$ and $a(k)|n(k)\rangle$ are eigenstates of $N(k)$ with respective eigenvalues $n(k) + 1$ and $n(k) - 1$. $N(k)$ is a *particle number operator* (or, to be more precise, a number *density* operator) and to justify the name, note that the field energy, found by substituting (4.3) into (4.1), is

$$\begin{aligned}
H &= \int \frac{d^3k}{(2\pi)^3 2k_0} \frac{k_0}{2}[a^\dagger(k)a(k) + a(k)a^\dagger(k)] \\
&= \int \frac{d^3k}{(2\pi)^3 2k_0} k_0[N(k) + \tfrac{1}{2}] \tag{4.21}
\end{aligned}$$

(with $k_0 = \omega_k$) and, similarly, the field momentum is

$$\mathbf{P} = \int \frac{d^3k}{(2\pi)^3 2k_0} \mathbf{k}[N(k) + \tfrac{1}{2}]. \tag{4.22}$$

These expressions strongly suggest the interpretation that $N(k)$ is the operator for the number of particles with momentum $\mathbf{k}$ and so with energy

k_0 – that is, so long as $N(k)$ never becomes negative! To see that it cannot, merely note that the state $a(k)|n(k)\rangle$ must (as all Hilbert space states must) have non-negative norm:

$$[a(k)|n(k)\rangle]^\dagger[a(k)|n(k)\rangle] = \langle n(k)|a^\dagger(k)a(k)|n(k)\rangle$$
$$= n(k)\langle n(k)|n(k)\rangle \geqslant 0;$$

so that, if $|n(k)\rangle$ has non-negative norm, $n(k)$ must be positive or zero. On the other hand, from equation (4.20), $a(k)$ reduces $n(k)$ by 1, and repeated application will continue to reduce it. The only way to avoid $n(k)$ becoming negative is to have a ground state $|0(k)\rangle$, or $|0\rangle$ for short, with

$$a(k)|0\rangle = 0 \tag{4.23}$$

and so

$$N(k)|0\rangle = a^\dagger(k)a(k)|0\rangle = 0;$$

the ground state (vacuum) contains no particles with momentum **k**. Application of $a^\dagger(k)$ now increases $N(k)$ in steps of 1 at a time, so $N(k)$ *is integral*. This completes the justification for interpreting $N(k)$ as the number operator, and hence for the particle interpretation of the quantised theory.

The student will doubtless recognise the similarity between the above analysis and the quantum mechanical harmonic oscillator. In fact, it is easy to see that our Hamiltonian (4.21) is equivalent to the harmonic oscillator from

$$H = \tfrac{1}{2}P^2(k) + \frac{\omega_k^2}{2}Q^2(k)$$

on putting

$$P(k) = \left(\frac{\omega_k}{2}\right)^{\frac{1}{2}}[a(k) + a^\dagger(k)], \quad Q(k) = \frac{i}{(2\omega_k)^{\frac{1}{2}}}[a(k) - a^\dagger(k)].$$

The Klein–Gordon field, therefore, is equivalent to an infinite sum of oscillators.

The operators $a(k)$ and $a^\dagger(k)$ are called the *annihilation* and *creation* operators for the field quanta. The fact that $N(k)$ is non-negative means, in addition, that the energy of the quantised field, from (4.21), is non-negative, as we found for the classical Klein–Gordon field. It is gratifying that this feature has remained. Actually, of course, the energy contains an infinite contribution from all the oscillator ground states. But, since the zero of energy is arbitrary, this may be subtracted with no physical consequences, and the Hamiltonian redefined as

$$H = \int \frac{d^3k}{(2\pi)^3 2\omega_k}\omega_k N(k), \tag{4.24}$$

with the property

$$\langle 0|H|0\rangle = \int \frac{d^3k}{(2\pi)^3 2k_0} k_0 \langle 0|a^\dagger(k)a(k)|0\rangle$$
$$= 0$$

in view of (4.23). Formally, this is equivalent to writing all annihilation operators to the right of all creation operators; this is called *normal ordering*, and denoted : :. Thus, decomposing $\phi(x)$ into its positive and negative frequency parts, as in equation (4.14),

$$\phi(x) = \phi^{(+)}(x) + \phi^{(-)}(x) \tag{4.25}$$

with

$$\phi^{(+)}(x) = \int \frac{d^3k}{[(2\pi)^3 2\omega_k]^{\frac{1}{2}}} a(k) f_k(x),$$

$$\phi^{(-)}(x) = \int \frac{d^3k}{[(2\pi)^3 2\omega_k]^{\frac{1}{2}}} a^\dagger(k) f_k^*(x),$$

we have

$$:\phi(x)\phi(y): = \phi^{(+)}(x)\phi^{(+)}(y) + \phi^{(-)}(x)\phi^{(+)}(y)$$
$$+ \phi^{(-)}(y)\phi^{(+)}(x) + \phi^{(-)}(x)\phi^{(-)}(y). \tag{4.26}$$

We shall now show that the particles which are quanta of the Klein–Gordon field obey Bose–Einstein statistics. From equation (4.19) we see that the state $a^\dagger(k)|n(k)\rangle$ is proportional to the state $|n(k)+1\rangle$, so we write

$$a^\dagger(k)|n(k)\rangle = c_+(n(k))|n(k)+1\rangle$$

or, to be more precise,

$$a^\dagger(k_i)|n(k_1), n(k_2), \ldots, n(k_i), \ldots\rangle$$
$$= c_+(n(k_i))|n(k_1), n(k_2), \ldots, n(k_i)+1, \ldots\rangle$$

where $c_+(n(k))$ is some coefficient, which is found by the requirement that the states are all normalised:

$$|c_+(n(k))|^2 \langle n(k)+1|n(k)+1\rangle = \langle n(k)|a(k)a^\dagger(k)|n(k)\rangle$$
$$= [n(k)+1]\langle n(k)|n(k)\rangle$$

therefore

$$|c_+(n(k))|^2 = n(k) + 1.$$

So, to within a phase factor, $c_+(n(k)) = [n(k)+1]^{\frac{1}{2}}$. By a similar argument using $a(k)$ and a corresponding coefficient $c_-(n(k))$, we find $c_-(n(k)) = [n(k)]^{\frac{1}{2}}$, so we have

$$a(k_i)|n(k_1), n(k_2), \ldots, n(k_i), \ldots\rangle,$$
$$= [n(k_i)]^{\frac{1}{2}}|n(k_1), n(k_2), \ldots, n(k_i)-1, \ldots\rangle, \tag{4.27}$$

$$a^\dagger(k_i)|n(k_1), n(k_2), \ldots, n(k_i), \ldots\rangle$$
$$= [n(k_i) + 1]^{\frac{1}{2}}|n(k_1), n(k_2), \ldots, n(k_i) + 1, \ldots\rangle. \tag{4.28}$$

The vacuum state contains no particles of any momentum, and we write

$$|0\rangle = |0, 0, \ldots\rangle,$$

and an arbitrary, normalised, state containing $n(k_1)$ particles with momentum k_1, $n(k_2)$ with momentum k_2, etc., may be written

$$|n(k_1), n(k_2), \ldots\rangle = \frac{1}{(n(k_1)!n(k_2)!\ldots)^{\frac{1}{2}}}[a^\dagger(k_1)]^{n(k_1)}[a^\dagger(k_2)]^{n(k_2)}|0\rangle. \tag{4.29}$$

There is, evidently no restriction on $n(k)$; any number of particles may exist in the same momentum state. The particles are therefore *bosons*. It should be appreciated that this conclusion follows directly from the original postulated commutation relations (4.9); these gave rise to equations (4.16) which in essence state that the particles are bosons. It follows that if we want to describe fermions, we must modify these commutation relations. The interesting thing is that, when we come to quantise the Dirac field, we are indeed forced, if we want positive energy, to modify the commutation relations (4.16), so Dirac particles are of necessity fermions. These remarkable conclusions are known as the connection between *spin and statistics*,[10] which was first treated by Pauli.

Finally, the normalisation for 1-particle states follows from (4.16a). With $|k\rangle = a^\dagger(k)|0\rangle$, we have

$$\langle k|k'\rangle = \langle 0|a(k)a^\dagger(k')|0\rangle$$
$$= \langle 0|[a(k), a^\dagger(k')]|0\rangle + \underbrace{\langle 0|a^\dagger(k')a(k)|0\rangle}_{0} \tag{4.30}$$
$$= (2\pi)^3 2k_0 \delta^3(\mathbf{k} - \mathbf{k}').$$

This normalisation is covariant, which is hardly surprising in view of the Lorentz covariant measure in (4.3). An alternative (non-covariant) normalisation, used in many books, is

$$\langle k|k'\rangle = \delta^3(\mathbf{k} - \mathbf{k}'),$$

which results from the commutation relation

$$[a(k), a^\dagger(k)] = \delta^3(\mathbf{k} - \mathbf{k}')$$

in place of (4.16a), which in turn results from an appropriate redefinition of $a(k)$.

Using our normalisation, the 1-particle wave function $\psi(x)$ corresponding to a momentum p is

$$\psi(x) = \langle 0|\phi(x)|p \rangle$$

$$= \int \frac{d^3k}{(2\pi)^3 2\omega_k} [\langle 0|a(k)|p \rangle e^{-ikx} + \langle 0|a^\dagger(k)|p \rangle e^{ikx}]$$

$$= \int \frac{d^3k}{(2\pi)^3 2k_0} [\langle k|p \rangle e^{-ikx}]$$

$$= \int d^3k \, e^{-ikx} \delta^3(k-p)$$

$$= e^{-ipx}. \tag{4.31}$$

We shall now show that this normalisation corresponds to $2p_0$ particles per unit volume – perhaps a surprising result! The point is that we need a Lorentz invariant normalisation. *Non-relativistically*, if S is the probability of finding a particle in a unit volume, then a normalisation of one particle per unit volume $(V = 1)$ is

$$\int_{V=1} S \, d^3r = 1 \tag{4.32}$$

and if $S = \psi^*\psi$, and $\psi \sim e^{ik \cdot r}$, this gives

$$\psi = e^{ik \cdot r}.$$

This argument, however, is not relativistic since (4.32) is not Lorentz invariant. For Lorentz invariance, S must be the time component of a 4-vector S^μ, and the condition then is

$$\int S^0 \, dx^1 \, dx^2 \, dx^3 = 1$$

or, in manifestly covariant form,

$$\frac{1}{4!} \int \varepsilon_{\mu\nu\kappa\lambda} S^\mu \, dx^\nu \, dx^\kappa \, dx^\lambda = 1,$$

which may be written as

$$\int S^\mu n_\mu \, dV = 1$$

where V is a spacelike hypersurface with normal n_μ. This condition is clearly Lorentz invariant. If ψ is a scalar field, however, $\psi^*\psi$ is a scalar, not the time component for a 4-vector. The 4-vector probability current S_μ is

$$S_\mu = i(\psi^*\overleftrightarrow{\partial}_\mu \psi) = 2p_\mu \psi^*\psi;$$

the normalisation condition then is

$$\int_V 2p_0 \psi^* \psi \, d^3 r = 1$$

if there is one particle in the volume V, so (if $V = 1$)

$$\psi = \frac{1}{(2p_0)^{\frac{1}{2}}} e^{i p \cdot r}.$$

We may, however, adopt a normalisation of $2p_0$ particles per unit volume, giving

$$\int 2p_0 \psi^* \psi \, d^3 r = 2p_0$$

$$\psi = e^{i p \cdot r}.$$

This is condition (4.31), which therefore corresponds to $2p_0$ particles per unit volume.

4.2   The complex Klein–Gordon field

We saw in the last chapter that the complex Klein–Gordon field describes a field with electric charge. We consider here the quantisation of this field. Since the classical field is not real, the quantum field will not be Hermitian, and in place of equation (4.14) we have

$$\phi(x) = \int \frac{d^3 k}{(2\pi)^3 2\omega_k} [a(k) e^{-ikx} + b^\dagger(k) e^{ikx}],$$

$$\phi^\dagger(x) = \int \frac{d^3 k}{(2\pi)^3 2\omega_k} [b(k) e^{-ikx} + a^\dagger(k) e^{ikx}]. \tag{4.33}$$

The annihilation and creation operators b and $b^\dagger$ have been introduced as respective coefficients of the positive and negative energy terms. The equal-time commutation relations (4.9), and the Hermitian conjugate equations, are adopted, leading to

$$\left. \begin{aligned} [a(k), a^\dagger(k')] &= (2\pi)^3 2\omega_k \delta^3(\mathbf{k} - \mathbf{k}'), \\ [b(k), b^\dagger(k')] &= (2\pi)^3 2\omega_k \delta^3(\mathbf{k} - \mathbf{k}'), \end{aligned} \right\} \tag{4.34}$$

with all other commutators vanishing. These generalise (4.16). We now have two sets of creation and annihilation operators a and b. What do they correspond to? To answer this, let us substitute the operator expressions (4.33) into the expression (3.62) for the charge of the Klein–Gordon field. This gives (with $c = 1$, replacing ϕ^* by $\phi^\dagger$ and inserting normal ordering)

$$Q = i \int : \phi^\dagger \frac{\partial \phi}{\partial t} - \frac{\partial \phi^\dagger}{\partial t} \phi : d^3 x$$

$$= \int \frac{\mathrm{d}^3k}{(2\pi)^3 2\omega_k} [a^\dagger(k)a(k) - b^\dagger(k)b(k)]. \tag{4.35}$$

On the other hand, the Hamiltonian becomes (cf. (4.24))

$$H = \int \frac{\mathrm{d}^3k}{(2\pi)^3 2\omega_k} [a^\dagger(k)a(k) + b^\dagger(k)b(k)].\,\omega_k \tag{4.36}$$

Hence the $a^\dagger$ and $b^\dagger$ can be interpreted as creation operators for *particles and antiparticles*, which carry opposite charge, but have the same mass. Q, of course, is not positive definite. It is the integral of ρ, which in the single-particle theory we wanted to interpret as the probability density, and were forced to reject because it was not positive definite. We see then, that in the quantised theory, ρ is re-interpreted as the charge density.

It will be useful for future work to know the value of the commutator $[\phi(x), \phi(y)]$. From (4.33) and (4.34) we have, with $k_0 = \omega_k = (\mathbf{k}^2 + m^2)^{\frac{1}{2}}$,

$$[\phi(x), \phi^\dagger(y)] = \frac{1}{(2\pi)^6} \int \int \frac{\mathrm{d}^3k}{2k_0} \frac{\mathrm{d}^3k'}{2k_0'} \{ [a(k), a^\dagger(k')] e^{-ikx+ik'y}$$

$$+ [b^\dagger(k), b(k')] e^{ikx - ik'y} \}$$

$$= \frac{1}{(2\pi)^3} \int \frac{\mathrm{d}^3k}{2k_0} [e^{i\mathbf{k}\cdot(\mathbf{x}-\mathbf{y})} e^{-ik_0(x_0-y_0)} - e^{-i\mathbf{k}\cdot(\mathbf{x}-\mathbf{y})} e^{ik_0(x_0-y_0)}]$$

$$= \frac{-i}{(2\pi)^3} \int \frac{\mathrm{d}^3k}{k_0} e^{i\mathbf{k}\cdot(\mathbf{x}-\mathbf{y})} \sin k_0(x_0 - y_0). \tag{4.37}$$

$$\sin x = \frac{1}{2i}(e^{ix} - e^{-ix})$$

Defining the invariant function $\Delta(x-y)$ by

$$\Delta(x-y) = -i \int \frac{\mathrm{d}^3k}{(2\pi)^3 2\omega_k} (e^{-ik(x-y)} - e^{ik(x-y)}), \tag{4.38}$$

we then have

$$[\phi(x), \phi^\dagger(y)] = i\Delta(x-y). \tag{4.39}$$

At equal times $(x_0 = y_0)$, this gives, from (4.37), the equal-time commutation relation

$$[\phi(x), \phi^\dagger(y)]|_{x^0 = y^0} = 0.$$

It may similarly be verified that, even at unequal times,

$$[\phi(x), \phi(y)] = [\phi^\dagger(x), \phi^\dagger(y)] = 0. \tag{4.40}$$

4.3 The Dirac field

We saw earlier the considerable success of the Dirac equation when treated as a single-particle equation; it predicts, amongst other things, the correct magnetic moment of the electron, and the energy levels of the hydrogen atom (that is, before additional corrections are made). The one serious difficulty was the problem of negative energy states. We saw how Dirac turned this difficulty into a triumph by redefining the vacuum as having all the negative energy states filled (and for this the exclusion principle is crucial), and then predicting antiparticles. The prediction of antiparticles, however, is an abandonment of the Dirac equation as a single-particle equation since it is now required to describe both particles and antiparticles. Drawing on our experience with the Klein–Gordon field, we might guess that the correct way to do this is first to view the Dirac equation as a *field* equation, and then to quantise the field – or, to be more accurate, to recognise that the Dirac field is an intrinsically quantum field, which has no existence in the classical limit $h \to 0$, and therefore to take seriously its status as an *operator*. These same remarks were made in our treatment of the Klein–Gordon field above.

We address ourselves first of all to the question of the energy of the Dirac field. What we shall see is, that in order for the energy to be positive, the Dirac field must obey Fermi–Dirac statistics. To begin, we need a Lagrangian density. It is an easy matter to see that the Dirac equation (2.96),

$$(i\gamma^\mu \partial_\mu - m)\psi = 0,$$

follows from the Euler–Lagrange equation (cf. (3.19)),

$$\frac{\partial \mathscr{L}}{\partial \bar{\psi}} - \partial_\mu \left(\frac{\partial \mathscr{L}}{\partial(\partial_\mu \bar{\psi})} \right) = 0,$$

if we choose

$$
\begin{aligned}
\mathscr{L} &= i\bar{\psi}\gamma^\mu \overset{\leftrightarrow}{\partial}_\mu \psi - m\bar{\psi}\psi \\
&= \frac{i}{2}[\bar{\psi}\gamma^\mu(\partial_\mu \psi) - (\partial_\mu \bar{\psi})\gamma^\mu \psi] - m\bar{\psi}\psi.
\end{aligned}
\tag{4.41}
$$

In this Lagrangian, ψ and $\bar{\psi}$ are treated as dynamically independent fields. From $\mathscr{L}$ we find the canonical momentum field $\pi(x)$

$$\pi(x) = \frac{\partial \mathscr{L}}{\partial \dot{\psi}(x)} = i\psi^\dagger(x). \tag{4.42}$$

(To derive this, the derivative operator in the first line of equation (4.41) is envisaged as operating solely on ψ; the difference between this and the second line of (4.41) is a total divergence, which does not change the action, and is therefore dynamically insignificant.) The Hamiltonian is then

$$
\begin{aligned}
\mathscr{H} &= \pi\dot{\psi} - \mathscr{L} \\
&= \psi^\dagger \gamma^0(-i\gamma^i \partial_i + m)\psi \\
&= \psi^\dagger \gamma^0(i\gamma^0 \partial_0 \psi)
\end{aligned}
$$

$$= \psi^\dagger i \frac{\partial}{\partial t} \psi \tag{4.43}$$

where the Dirac equation itself has been used. This form for the Hamiltonian is, of course, the expected one, but, simply because of this, it is clear that the negative energy and plane wave solutions of the Dirac equation, (see §2.4), give a negative contribution to $\mathcal{H}$, which is therefore *not* positive definite. Hence the negative energy difficulty is not removed by treating the Dirac equation as a field equation, as it was with the Klein–Gordon equation. It is only removed on quantisation.

The general solution to the Dirac equation may be expanded in terms of the plane wave solutions as follows:

$$\left. \begin{aligned} \psi(x) &= \int \frac{\mathrm{d}^3 k}{(2\pi)^3} \frac{m}{k_0} \sum_{\alpha=1,2} [b_\alpha(k) u^{(\alpha)}(k) \mathrm{e}^{-ikx} + d_\alpha^\dagger(k) v^{(\alpha)}(k) \mathrm{e}^{ikx}], \\ \bar{\psi}(x) &= \int \frac{\mathrm{d}^3 k}{(2\pi)^3} \frac{m}{k_0} \sum_{\alpha=1,2} [b_\alpha^\dagger(k) \bar{u}^{(\alpha)}(k) \mathrm{e}^{ikx} + d_\alpha(k) \bar{v}^{(\alpha)}(k) \mathrm{e}^{-ikx}]. \end{aligned} \right\} \tag{4.44}$$

$u^{(1,2)}$ and $v^{(1,2)}$ are the positive and negative energy spinors. As in the comparable expression (4.33) for the Klein–Gordon field, an annihilation operator $b_\alpha(k)$ multiplies the positive energy term, and a creation operator $d_\alpha^\dagger(k)$ the negative energy term. They are noted here as distinct operators, corresponding to a non-Hermitian, i.e. charged, Dirac field. Substituting (4.44) into (4.43) and using the normalisation conditions (2.139) for the spinors u and v gives for the energy

$$\begin{aligned} H &= \int \mathrm{d}^3 x\, \mathcal{H} \\ &= \int \mathrm{d}^3 x\, \psi^\dagger(x) i \frac{\partial}{\partial t} \psi(x) \\ &= \int \mathrm{d}^3 x \sum_{\alpha,\alpha'} \frac{i}{2} \iint \frac{\mathrm{d}^3 k}{(2\pi)^3} \frac{\mathrm{d}^3 k'}{(2\pi)^3} \frac{m^2}{k_0 k_0'} \\ &\quad \times \{ [b_\alpha^\dagger(k) u^{\dagger(\alpha)}(k) \mathrm{e}^{ikx} + d_\alpha(k) v^{\dagger(\alpha)}(k) \mathrm{e}^{-ikx}] \\ &\quad \times [b_{\alpha'}(k') u^{(\alpha')}(k')(-ik_0') \mathrm{e}^{-ik'x} + d_{\alpha'}^\dagger(k') v^{(\alpha')}(k')(ik_0') \mathrm{e}^{ikx}] \\ &\quad - [b_\alpha^\dagger(k) u^{\dagger(\alpha)}(k)(ik_0) \mathrm{e}^{ikx} + d_\alpha(k) v^{\dagger(\alpha)}(k)(-ik_0) \mathrm{e}^{-ikx}] \\ &\quad \times [b_{\alpha'}(k') u^{(\alpha')}(k') \mathrm{e}^{-ik'x} + d_{\alpha'}^\dagger(k') v^{(\alpha')}(k') \mathrm{e}^{ik'x}] \} \\ &= \int \frac{\mathrm{d}^3 k}{(2\pi)^3} \frac{m}{k_0} k_0 \sum_\alpha [b_\alpha^\dagger(k) b_\alpha(k) - d_\alpha(k) d_\alpha^\dagger(k)]. \tag{4.45} \end{aligned}$$

So far no commutation relations have been assumed between bs and ds. It is clear that if they obey commutation relations like (4.34), then the energy in

(4.45) will not be positive definite – the d quanta will contribute a negative amount. The only way to avoid this is to introduce the notion of *anticommutators*, defined by

$$\{A, B\} \equiv AB + BA \tag{4.46}$$

and to postulate the anticommutation relations, first proposed by Jordan and Wigner,

$$\{b_\alpha(k), b_{\alpha'}^\dagger(k')\} = \{d_\alpha(k), d_{\alpha'}^\dagger(k')\} = (2\pi)^3 \frac{k_0}{m} \delta^3(\mathbf{k} - \mathbf{k}')\delta_{\alpha\alpha'},$$

$$\{b_\alpha(k), b_{\alpha'}(k')\} = \{b_\alpha^\dagger(k), b_{\alpha'}^\dagger(k')\} = 0,$$

$$\{d_\alpha(k), d_{\alpha'}(k')\} = \{d_\alpha^\dagger(k), d_{\alpha'}^\dagger(k')\} = 0. \tag{4.47}$$

To subtract out the zero point energy, we normal order the Hamiltonian, *with the additional introduction, in the case of Fermi fields, to change the sign of the term for each interchange of operators.* This gives

$$H = \int d^3x :\psi^\dagger(x) i \frac{\partial}{\partial t} \psi(x):$$

$$= \int \frac{d^3k}{(2\pi)^3} \frac{m}{k_0} k_0 \sum_\alpha [b_\alpha^\dagger(k)b_\alpha(k) + d_\alpha^\dagger(k)d_\alpha(k)]. \tag{4.48}$$

This is now positive definite, but it is easy to see that the anticommutation relations imply Fermi statistics; for example $\{b_\alpha^\dagger(k), b_\alpha^\dagger(k)\} = 0$ implies $b_\alpha^\dagger(k)b_\alpha^\dagger(k) = 0$, hence $b_\alpha^\dagger(k)b_\alpha^\dagger(k)|0\rangle = 0$; it is impossible to have two quanta of the Dirac field in the same state. Hence the use of anticommutators leads directly to the Pauli exclusion principle. Moreover, the total charge is given by

$$Q = \int d^3x :j_0(x):$$

$$= \int d^3x :\psi^\dagger(x)\psi(x):$$

$$= \int \frac{d^3k}{(2\pi)^3} \frac{m}{k_0} \sum_\alpha [b_\alpha^\dagger(k)b_\alpha(k) - d_\alpha^\dagger(k)d_\alpha(k)] \tag{4.49}$$

showing that if $b^\dagger$ creates 'particles', $d^\dagger$ creates 'antiparticles', i.e. of the opposite charge, just as in the case of the charged Klein–Gordon field. These results amount precisely to Dirac's prediction of antiparticles, sketched earlier. Finally, let us calculate the equal-time anticommutator $\{\psi(\mathbf{x}, t), \psi^\dagger(\mathbf{x}', t)\}$; ψ is, of course, a 4-spinor, so, more precisely, we want to calculate $\{\psi_i(\mathbf{x}, t), \psi_j^\dagger(\mathbf{x}', t)\}$, the labels i and j referring to particular spinor components. We shall need to refer to equations (2.145) and (2.146), which are matrix equations, whose (ij) elements read

$$\sum_\alpha u_i^{(\alpha)}(p)\bar{u}_j^{(\alpha)}(p) = \left(\frac{\gamma \cdot p + m}{2m}\right)_{ij},$$

$$\sum_\alpha v_i^{(\alpha)}(p)\bar{v}_j^{(\alpha)}(p) = \left(\frac{\gamma \cdot p - m}{2m}\right)_{ij}.$$

From (4.44), (4.47) and the above equations we then have

$$\{\psi_i(\mathbf{x}, t), \psi_j^\dagger(\mathbf{x}', t)\}$$

$$= \sum_{\alpha,\alpha'} \int\int \frac{\mathrm{d}^3 k\, \mathrm{d}^3 k'}{(2\pi)^6} \frac{m^2}{k_0 k_0'} [u_i^{(\alpha)}(k)\bar{u}_k^{(\alpha')}(k')\gamma_{kj}^0 \{b_\alpha(k), b_{\alpha'}^\dagger(k')\}$$

$$\times\, \mathrm{e}^{-ikx+ik'x'} + v_i^{(\alpha)}(k)\bar{v}_k^{(\alpha')}(k')\gamma_{kj}^0$$

$$\times\, \{d_\alpha^\dagger(k), d_{\alpha'}(k')\}\, \mathrm{e}^{ikx-ik'x'}]$$

$$= \sum_\alpha \int \frac{\mathrm{d}^3 k}{(2\pi)^3} \frac{m}{k_0} [u_i^{(\alpha)}(k)\bar{u}_k^{(\alpha)}(k)\gamma_{kj}^0 \mathrm{e}^{ik\cdot(\mathbf{x}-\mathbf{x}')}$$

$$+ v_i^{(\alpha)}(k)\bar{v}_k^{(\alpha)}(k)\gamma_{kj}^0 \mathrm{e}^{-ik\cdot(\mathbf{x}-\mathbf{x}')}]$$

$$= \int \frac{\mathrm{d}^3 k}{(2\pi)^3 2k_0} \{[(\gamma\cdot k + m)\gamma^0]_{ij}\, \mathrm{e}^{ik\cdot(\mathbf{x}-\mathbf{x}')}$$

$$+ [(\gamma\cdot k - m)\gamma^0]_{ij}\, \mathrm{e}^{-ik\cdot(\mathbf{x}-\mathbf{x}')}\}$$

$$= \int \frac{\mathrm{d}^3 k}{(2\pi)^3 2k_0} 2k_0 \delta_{ij} \mathrm{e}^{ik\cdot(\mathbf{x}-\mathbf{x}')}$$

$$= \delta^3(\mathbf{x} - \mathbf{x}')\delta_{ij}. \tag{4.50}$$

By similar reasoning it follows that

$$\{\psi_i(\mathbf{x}, t), \psi_j(\mathbf{x}', t)\} = 0, \quad \{\psi_i^\dagger(\mathbf{x}, t), \psi_j^\dagger(\mathbf{x}', t)\} = 0. \tag{4.51}$$

4.4 Electromagnetic field

The fundamental fields in nature, we believe, are spinor fields (leptons and quarks) and gauge fields (electromagnetic, weak and gluon fields). Having shown how to quantise the spinor field, we turn now to the gauge fields, and consider, for simplicity the electromagnetic field. There are complications met in quantising gauge fields, and the electromagnetic field exhibits them all; the additional complications of having internal symmetry indices (as the weak and strong gauge fields do) do not, at this stage, pose any particular problems (those problems come later, when we derive the Feynman rules, and for that purpose we shall use the path-integral method).

The origin of the difficulties is that the electormagnetic field, like any massless field, possesses only *two* independent components, but is covariantly described by a 4-vector A_μ. In choosing two of these components as the physical ones, and thence quantising them, we lose manifest covariance. Alternatively, if we wish to keep covariance, we have two redundant components. We shall consider, in turn, examples of these two approaches to quantisation, and shall see the crucial role played by gauge invariance, which tells us that the A_μ are only defined up to a gauge transformation.

We shall need the following information from chapters 2 and 3. The six components of the electromagnetic field may be written as an antisymmetric tensor $F^{\mu\nu}$ (see (2.222)), and the homogeneous Maxwell equations follow identically if it is assumed that $F^{\mu\nu}$ is a 4-dimensional curl (see (2.218–221)),

$$F^{\mu\nu} = \partial^\mu A^\nu - \partial^\nu A^\mu. \tag{4.52}$$

In a vacuum, the other two ('inhomogeneous') Maxwell equations are, from (2.223), $\partial_\mu F^{\mu\nu} = 0$, or

$$\Box A^\nu - \partial^\nu(\partial_\mu A^\mu) = 0. \tag{4.53}$$

These equations follow from a variational principle with the Lagrangian (3.82)

$$\mathcal{L} = -\tfrac{1}{4} F_{\mu\nu} F^{\mu\nu}, \tag{4.54}$$

where A^μ is regarded as the dynamical field. For a given electromagnetic field, A_μ is not unique; the gauge transformation $A_\mu \to A'_\mu = A_\mu + \partial_\mu \Lambda(x)$ leaves $F_{\mu\nu}$ unchanged. By choosing Λ to satisfy

$$\Box \Lambda = -\partial_\mu A^\mu,$$

we obtain $\partial_\mu A'^\mu = 0$. Dropping the prime, the resulting condition (2.229),

$$\partial_\mu A^\mu = 0, \tag{4.55}$$

is called the Lorentz condition. A vector potential satisfying this condition is said to belong to the *Lorentz gauge*. This one condition effectively reduces the number of independent components of A_μ from four to three. However, the Lorentz condition does not make A_μ unique; it is clear that, if A_μ satisfies the Lorentz condition, so will A'_μ as long as $\Box \Lambda(x) = 0$. Then by choosing $\Lambda(x)$ to satisfy

$$\frac{\partial \Lambda}{\partial t} = -\phi,$$

we have $\phi' = 0$, and thence, from (4.55), $\boldsymbol{V} \cdot \boldsymbol{A}' = 0$. Potentials satisfying this additional condition,

$$\phi = 0, \quad \boldsymbol{V} \cdot \boldsymbol{A} = 0, \tag{4.56}$$

are said to belong to the *radiation* (or *Coulomb*) *gauge*. In this gauge there are clearly only two independent components of A^μ. This is the case in the real world, so working in the radiation gauge keeps the physical nature of the electromagnetic field most evident. Let us therefore study quantisation in this gauge.

Radiation gauge quantisation

We proceed, as far as possible, in analogy with the quantisation of the Klein–Gordon field, so first define the conjugate momentum fields

$$\pi^0 = \frac{\partial \mathscr{L}}{\partial \dot{A}_0} = 0, \tag{4.57}$$

$$\pi^i = \frac{\partial \mathscr{L}}{\partial \dot{A}_i} = -\dot{A}^i - \partial^i A^0 = E^i, \tag{4.58}$$

and then impose the usual commutation relations (remembering that π^i is conjugate to $A_i = -A^i$),

$$[A_i(x,t), \pi^j(x',t)] = -[A^i(x,t), E^j(x',t)] = i\delta_{ij}\delta^3(\mathbf{x} - \mathbf{x}'). \tag{4.59}$$

(This equation is, from a strict tensorial point of view, incorrect, since the right-hand side has downstairs indices, and features δ_{ij} instead of g_{ij}, which of course differ in sign for $i,j = 1,2,3$. Numerically, however, the equation is correct.) The condition (4.59) is inconsistent with the radiation gauge condition $\mathbf{V} \cdot \mathbf{A} = 0$, since taking the divergence of both sides gives

$$[\mathbf{V} \cdot \mathbf{A}(\mathbf{x},t), E^i(\mathbf{x}'t)] = i\partial^i\delta^3(\mathbf{x} - \mathbf{x}') \neq 0.$$

The commutation relations (4.59) must therefore be modified.

We replace, for this purpose, δ_{ij} by Δ_{ij}, a rank 2 tensor, symmetric in i and j, and we write $\delta^3(\mathbf{x} - \mathbf{x}')$ in an integral form. Then we have

$$[A^i(\mathbf{x},t), E^j(\mathbf{x}',t)] = -i\Delta^{ij}\frac{1}{(2\pi)^3}\int d^3k e^{i\mathbf{k}\cdot(\mathbf{x} - \mathbf{x}')}$$

and taking the divergence gives

$$[\mathbf{V} \cdot \mathbf{A}(x,t), E^j(x',t)] = \frac{1}{(2\pi)^3}\int d^3k \left(\sum_i k_i \Delta^{ij}\right) e^{i\mathbf{k}\cdot(\mathbf{x} - \mathbf{x}')}.$$

It is clear that the condition for this to vanish is

$$\Delta_{ij} = \delta_{ij} - \frac{k_i k_j}{\mathbf{k}^2},$$

so the correct commutation relations are

$$[A^i(\mathbf{x},t), E^j(\mathbf{x}',t)] = i\int \frac{d^3k}{(2\pi)^3}\left(\delta^{ij} - \frac{k^i k^j}{\mathbf{k}^2}\right)e^{i\mathbf{k}\cdot(\mathbf{x} - \mathbf{x}')}$$

$$= -i\left(\delta^{ij} - \frac{\partial^i\partial^j}{\nabla^2}\right)\delta^3(\mathbf{x} - \mathbf{x}') \tag{4.60}$$

$$[A^i(\mathbf{x}, t), A^j(\mathbf{x}', t)] = [E^i(\mathbf{x}, t), E^j(\mathbf{x}', t)] = 0. \tag{4.61}$$

From these relations we work towards a sensible particle interpretation of the theory, by calculating, as we did in the case of the Klein–Gordon and Dirac fields, the Hamiltonian operator. To begin with, note that, in view of the Lorentz condition (4.55), Maxwell's equations (4.53) become

$$\Box A^\mu = 0 \tag{4.62}$$

and further, since, in the radiation gauge we have $\phi = 0$, this becomes

$$\Box \mathbf{A} = 0. \tag{4.63}$$

This is the Klein–Gordon for a massless field, and we write its solution in terms of the fundamental solutions e^{ikx} and e^{-ikx}. The coefficients must evidently be vectors, called *polarisation vectors*, which we denote $\varepsilon^{(\lambda)}(k)$, where λ is a label, about to be explained. We then write (compare with equation (4.3) for the scalar field)

$$\mathbf{A}(x) = \int \frac{d^3k}{(2\pi)^3 2k_0} \sum_{\lambda=1}^{2} \varepsilon^{(\lambda)}(k)[a^{(\lambda)}(k)e^{-ikx} + a^{(\lambda)\dagger}(k)e^{ikx}] \tag{4.64}$$

with $k^2 = 0, k_0 = |\mathbf{k}|$; λ takes on only two values, since the radiation gauge condition (4.56), $\boldsymbol{\nabla}\cdot\mathbf{A} = 0$, gives

$$\mathbf{k}\cdot\boldsymbol{\varepsilon}^{(\lambda)}(k) = 0. \tag{4.65}$$

So for a given direction of propagation $\mathbf{k}/|\mathbf{k}|$, the $\varepsilon^{(\lambda)}(k)$ are transverse, as shown in Fig. 4.1. It is clear that they may also be chosen to be orthonormal

$$\boldsymbol{\varepsilon}^{(\lambda)}(k)\cdot\boldsymbol{\varepsilon}^{(\lambda')}(k) = \delta_{\lambda\lambda'}. \tag{4.66}$$

It is now straightforward, if a bit tedious, to calculate the commutation relations of the operators $a^{(\lambda)}(k)$ and $a^{(\lambda)\dagger}(k)$. In terms of the functions $f_k(x)$ defined above,

$$f_k(x) = \frac{1}{[(2\pi)^3 2k_0]^{\frac{1}{2}}} e^{-ikx}, \tag{4.11}$$

Fig. 4.1. The polarisation vectors for the electromagnetic field with momentum $\mathbf{k}$.

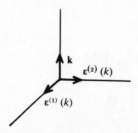

we have

$$\mathbf{A}(x) = \int \frac{\mathrm{d}^3k}{[(2\pi)^3 2k_0]^{\frac{1}{2}}} \sum_\lambda \boldsymbol{\varepsilon}^{(\lambda)}(k)[f_k(x)a^{(\lambda)}(k) + f_k^*(x)a^{(\lambda)\dagger}(k)]. \quad (4.67)$$

Since the $f_k(x)$ and $f_k^*(x)$ form an orthonormal set, it follows that

$$a^{(\lambda)}(k) = \int \mathrm{d}^3x[(2\pi)^3 2k_0]^{\frac{1}{2}} f_k^*(x) \mathrm{i}\overleftrightarrow{\partial}_0 \boldsymbol{\varepsilon}^{(\lambda)}(k) \cdot \mathbf{A}(x),$$

$$a^{(\lambda)\dagger}(k) = -\int \mathrm{d}^3x[(2\pi)^3 2k_0]^{\frac{1}{2}} f_k(x) \mathrm{i}\overleftrightarrow{\partial}_0 \boldsymbol{\varepsilon}^{(\lambda)}(k) \cdot \mathbf{A}(x),$$

and, using (4.60), (4.61), (4.65) and (4.66), we have

$$[a^{(\lambda)}(k), a^{(\lambda')\dagger}(k')] = 2k_0(2\pi)^3 \delta_{\lambda\lambda'}\delta^3(\mathbf{k} - \mathbf{k'}), \quad (4.68)$$
$$[a^{(\lambda)}(k), a^{(\lambda')}(k')] = [a^{(\lambda)\dagger}(k), a^{(\lambda')\dagger}(k')] = 0. \quad (4.69)$$

These commutation relations have the same form as those for the scalar field, and have the same interpretation as annihilation and creation operators for *photons*. The justification for this interpretation follows precisely the lines above, by construction of the number operator $N^{(\lambda)}(k) = a^{(\lambda)\dagger}(k)a^{(\lambda)}(k)$, and the calculation of the field energy. This is

$$H = \frac{1}{2}\int \mathrm{d}^3x(\mathbf{E}^2 + \mathbf{B}^2)$$

$$= \frac{1}{2}\int \mathrm{d}^3x(\dot{\mathbf{A}}^2 + (\nabla \times \mathbf{A})^2),$$

since, in the radiation gauge, $\mathbf{E} = -\dot{\mathbf{A}}$. To calculate $(\nabla \times \mathbf{A})^2$, we write (with summation over repeated indices)

$$\begin{aligned}(\nabla \times A)^2 &= \varepsilon_{ijk}\varepsilon_{imn}\partial_j A_k \partial_m A_n \\ &= (\delta_{jm}\delta_{kn} - \delta_{jn}\delta_{km})\partial_j A_k \partial_m A_n \\ &= (\partial_j A_k)(\partial_j A_k) - (\partial_j A_k)(\partial_k A_j). \end{aligned} \quad (4.70)$$

The second term can be written as a total divergence, since

$$\begin{aligned}\partial_j(A_k\partial_k A_j) &= (\partial_j A_k)(\partial_k A_j) + A_k(\partial_j\partial_k A_j) \\ &= (\partial_j A_k)(\partial_k A_j) + A_k\partial_k(\nabla\cdot\mathbf{A})\end{aligned}$$

and $\nabla\cdot\mathbf{A} = 0$ in the radiation gauge. The integral of a total divergence, however, vanishes, so only the first term in (4.70) is non-zero. Moreover,

$$\partial_j(A_k\partial_j A_k) = (\partial_j A_k)(\partial_j A_k) + A_k\nabla^2 A_k,$$

so, by the same 'total divergence' argument,

$$\int(\partial_j A_k)(\partial_j A_k)\,\mathrm{d}^3x = -\int \mathbf{A}\cdot\nabla^2\mathbf{A}\,\mathrm{d}^3x,$$

and the electromagnetic Hamiltonian is

$$H = \frac{1}{2}\int(\dot{\mathbf{A}}^2 - \mathbf{A}\cdot\nabla^2\mathbf{A})\,\mathrm{d}^3x.$$

Substituting the expansion (4.67) in here, we obtain, after some algebra,

$$H = \sum_\lambda \int \frac{d^3k}{(2\pi)^3 2k_0} \frac{k_0}{2} [a^{(\lambda)}(k)a^{(\lambda)\dagger}(k) + a^{(\lambda)\dagger}(k)a^{(\lambda)}(k)],$$

and on normal-ordering to remove the vacuum energy, we have

$$H = \sum_\lambda \int \frac{d^3k}{(2\pi)^3 k_0} k_0 [a^{(\lambda)\dagger}(k)a^{(\lambda)}(k)]. \tag{4.71}$$

This is the total energy of a collection of photons, with transverse polarisation ($\lambda = 1, 2$). It is, of course, positive definite, and we have a satisfactory quantum theory of the electromagnetic field. It has the merit that only the physical, i.e. transverse, degrees of freedom are quantised, but this is achieved by sacrificing Lorentz invariance, because the radiation gauge condition (4.56) is not Lorentz invariant. This does not mean to say that calculations performed in this gauge do not give relativistically invariant results. On the contrary; we started with a relativistic theory, and the scattering cross sections, for example, which we calculate are also relativistically invariant, but this is not *manifest* in the radiation gauge. For many purposes, it is an advantage to perform the quantisation in a Lorentz invariant manner, for example, in the Lorentz gauge. We now consider this.

Lorentz gauge quantisation

Since our aim is to retain covariance, all four components of A_μ and of π_ν will be brought into play, and will obey the covariant commutation relations

$$[A_\mu(\mathbf{x}, t), \pi_\nu(\mathbf{x}', t)] = ig_{\mu\nu}\delta^3(\mathbf{x} - \mathbf{x}'), \tag{4.72}$$

$$[A_\mu(\mathbf{x}, t), A_\nu(\mathbf{x}', t)] = [\pi_\mu(\mathbf{x}, t), \pi_\nu(\mathbf{x}', t)] = 0, \tag{4.73}$$

where $g_{\mu\nu}$ is the Minkowski metric tensor, and

$$\pi^\mu = \frac{\partial \mathscr{L}}{\partial \dot{A}_\mu}.$$

We immediately meet a problem, however, because, with the Lagrangian (4.54),

$$\pi^0 = \frac{\partial \mathscr{L}}{\partial \dot{A}_0} = 0,$$

so it is impossible to satisfy (4.72) for A_0. In fact, since A_0 commutes with π^0, it is a c-number, not an operator, and we have immediately lost covariance! To find a π^0 that does not vanish we need to change the Lagrangian. The new Lagrangian will not, of course, give Maxwell's equations, so what should it give? The only reasonable answer is Maxwell's equations (4.53) plus a covariant gauge condition, i.e. the Lorentz condition (4.55). Together these yield $\Box A_\mu = 0$, so we look for a Lagrangian which gives this equation of motion, in which the Lorentz gauge is 'already fixed'. It is easy to check that

$$\mathscr{L} = -\tfrac{1}{4}F_{\mu\nu}F^{\mu\nu} - \tfrac{1}{2}(\partial_\mu A^\mu)^2 \tag{4.74}$$

does the job, for we have

$$\frac{\partial\mathscr{L}}{\partial(\partial_\mu A_\nu)} = -\partial^\mu A^\nu + \partial^\nu A^\mu - g^{\mu\nu}\partial_\lambda A^\lambda; \quad \frac{\partial\mathscr{L}}{\partial A_\mu} = 0,$$

and the Euler–Lagrange equations then give

$$\Box A_\mu = 0 \tag{4.75}$$

as desired. The extra term $-\tfrac{1}{2}(\partial \cdot A)^2$ in (4.74) is called a 'gauge-fixing' term. In fact, we may, more generally, add a gauge fixing term $-(\lambda/2)(\partial \cdot A)^2$ to (4.54), giving

$$\mathscr{L} = -\tfrac{1}{4}F_{\mu\nu}F^{\mu\nu} - \frac{\lambda}{2}(\partial_\mu A^\mu)^2, \tag{4.76}$$

which gives the *non-Maxwell* equation of motion

$$\Box A^\mu - (1-\lambda)\partial^\mu(\partial_\lambda A^\lambda) = 0. \tag{4.77}$$

The choice $\lambda = 1$, misleadingly known as the 'Feynman gauge', leads back to (4.75). General gauge-fixing terms prove to be a useful device when we come to quantise non-Abelian gauge fields.

Let us return to the Lagrangian (4.74), and calculate π^0;

$$\pi^0 = \frac{\partial\mathscr{L}}{\partial\dot{A}_0} = -\partial_\mu A^\mu. \tag{4.78}$$

But this is just as bad as before, because it vanishes in the Lorentz gauge! The way out of this dilemma is to postulate that the Lorentz condition does not hold *as an operator identity*. Instead, we impose the weaker requirement that, for physical states $|\psi\rangle$, $\partial_\mu A^\mu$ has vanishing expectation value

$$\langle\psi|\partial_\mu A^\mu|\psi\rangle = 0. \tag{4.79}$$

The consequences of this will be worked out below. First, note that the solution to (4.75) is clearly

$$A_\mu(x) = \int\frac{\mathrm{d}^3k}{(2\pi)^3 2k_0}\sum_{\lambda=0}^3 \varepsilon_\mu^{(\lambda)}(k)\left[a^{(\lambda)}(k)\mathrm{e}^{-\mathrm{i}kx} + a^{(\lambda)\dagger}(k)\mathrm{e}^{\mathrm{i}kx}\right]. \tag{4.80}$$

Here the four polarisation 4-vectors $\varepsilon_\mu^{(0)}, \varepsilon_\mu^{(1)}, \varepsilon_\mu^{(2)}$ and $\varepsilon_\mu^{(3)}$ have a Lorentz-invariant normalisation, $\varepsilon^{(0)}$ being timelike and $\varepsilon^{(1)}$, $\varepsilon^{(2)}$ and $\varepsilon^{(3)}$ spacelike:

$$\begin{aligned}\varepsilon^{(\lambda)}\cdot\varepsilon^{(\lambda')} &= \varepsilon_\mu^{(\lambda)}\varepsilon^{(\lambda')\mu} = \varepsilon_\mu^{(\lambda)}g^{\mu\nu}\varepsilon_\nu^{(\lambda')} \\ &= \varepsilon_0^{(\lambda)}\varepsilon_0^{(\lambda')} - \varepsilon_1^{(\lambda)}\varepsilon_1^{(\lambda')} - \varepsilon_2^{(\lambda)}\varepsilon_2^{(\lambda')} - \varepsilon_3^{(\lambda)}\varepsilon_3^{(\lambda')} \\ &= g^{\lambda\lambda'}.\end{aligned} \tag{4.81}$$

In the frame in which the photon (anticipating the particle interpretation!)

is moving along the third axis, we have $k^\mu = (k, 0, 0, k)$, and

$$\varepsilon^{(0)} = \begin{pmatrix} 1 \\ 0 \\ 0 \\ 0 \end{pmatrix}, \quad \varepsilon^{(1)} = \begin{pmatrix} 0 \\ 1 \\ 0 \\ 0 \end{pmatrix}, \quad \varepsilon^{(2)} = \begin{pmatrix} 0 \\ 0 \\ 1 \\ 0 \end{pmatrix}, \quad \varepsilon^{(3)} = \begin{pmatrix} 0 \\ 0 \\ 0 \\ 1 \end{pmatrix} \tag{4.82}$$

with

$$k \cdot \varepsilon^{(1,2)} = 0. \tag{4.83}$$

Photons with polarisation $\varepsilon^{(0)}$ are called 'scalar' or 'timelike' photons, those with $\varepsilon^{(3)}$ 'longitudinal' and those with $\varepsilon^{(1)}$ and $\varepsilon^{(2)}$ 'transverse'. The scalar and longitudinal photons are clearly the unphysical ones.

We now want fo find $[a^{(\lambda)}(k), a^{(\lambda')\dagger}(k')]$. and we proceed as follows. From the Lagrangian (4.74), the canonical momentum is

$$\pi^\mu = \frac{\partial \mathscr{L}}{\partial \dot{A}_\mu} = F^{\mu 0} - g^{\mu 0}(\partial_\nu A^\nu); \tag{4.84}$$

that is

$$\pi^0 = -\dot{A}^0 + \boldsymbol{V} \cdot \boldsymbol{A}, \quad \pi^i = \partial^i A^0 - \dot{A}^i. \tag{4.85}$$

From (4.73), the spatial derivatives of A_μ commute at equal times, so the non-vanishing commutator (4.72) becomes

$$[\dot{A}_\mu(\boldsymbol{x}, t), A_\nu(\boldsymbol{x}', t)] = i g_{\mu\nu} \delta^3(\boldsymbol{x} - \boldsymbol{x}'). \tag{4.86}$$

Now we substitute into this the expansion (4.80) and obtain

$$[a^{(\lambda)}(k), a^{(\lambda')\dagger}(k')] = -g^{\lambda\lambda'} 2k_0 (2\pi)^3 \delta^3(\boldsymbol{k} - \boldsymbol{k}') \tag{4.87}$$

(which is the 'covariant' version of (4.68)). The other commutators vanish.

The interpretation of the a and $a^\dagger$ as annihilation and creation operators for $\lambda = 1, 2, 3$, that is, for longitudinal and transverse photons, poses no problems and proceeds in the usual way, but the relation for scalar photons

$$[a^{(0)}(k), a^{(0)\dagger}(k')] = -2k_0 (2\pi)^3 \delta^3(\boldsymbol{k} - \boldsymbol{k}') \tag{4.88}$$

will give problems because of the minus sign on the right-hand side. One consequence is that the norm of a one scalar photon state is negative. For if the state is

$$|1\rangle = \int \frac{d^3 k}{(2\pi)^3 2k_0} f(k) a^{(0)\dagger}(k) |0\rangle$$

then we have, using (4.88),

$$\langle 1 | 1 \rangle = \int \frac{d^3 k}{(2\pi)^3 2k_0} \frac{d^3 k'}{(2\pi)^3 2k_0'} f(k) f(k') \langle 0 | a^{(0)}(k') a^{(0)\dagger}(k) | 0 \rangle$$

$$= -\int \frac{d^3k}{(2\pi)^3 2k_0} \frac{d^3k'}{(2\pi)^3 2k'_0} f(k) f^*(k') 2k_0 (2\pi)^3 \delta^3(\mathbf{k} - \mathbf{k}')$$
$$\times \langle 0|0 \rangle$$

$$= -\int \frac{d^3k}{(2\pi)^3 2k_0} |f(k)|^2 \langle 0|0 \rangle. \tag{4.89}$$

Similarly, it follows that a state $|n_t\rangle$ with n_t timelike photons has norm $\langle n_t|n_t \rangle = (-1)^{n_t}$. The Hilbert space of particle states (Fock space) has an *indefinite metric*. This is singularly unpleasant, and raises problems about the quantum mechanical interpretation of these states. Another consequence is that scalar photons give a negative contribution to the energy. A calculation analogous to that leading to equation (4.71) gives

$$H = \int \frac{d^3k}{(2\pi)^3 2k_0} k_0 \left[\sum_{\lambda=1}^{3} a^{(\lambda)\dagger}(k) a^{(\lambda)}(k) - a^{(0)\dagger}(k) a^{(0)}(k) \right]. \tag{4.90}$$

Actually, there is a slight subtlety here, because the number density operator for timelike photons is not $a^{(0)\dagger}(k)a^{(0)}(k)$ but $-a^{(0)\dagger}(k)a^{(0)}(k)$. Consider a state of one timelike photon

$$|1\rangle = \int \frac{d^3k}{(2\pi)^3 2q_0} f(q) a^{(0)\dagger}(q)|0\rangle.$$

We demand $N|1\rangle = |1\rangle$, where N is the integral of the number density operator. We have, inserting the minus sign,

$$N|1\rangle = -\int \frac{d^3k}{(2\pi)^3 2k_0} a^{(0)\dagger}(k) a^{(0)}(k) \int \frac{d^3q}{(2\pi)^3 2q_0} f(q) a^{(0)\dagger}(q)|0\rangle$$

$$= -\int \frac{d^3k}{(2\pi)^3 2k_0} \frac{d^3q}{(2\pi)^3 2q_0} f(q) a^{(0)\dagger}(k) [a^{(0)}(k), a^{(0)\dagger}(q)]|0\rangle$$

$$= \int \frac{d^3k}{(2\pi)^3 2k_0} \frac{d^3q}{(2\pi)^3 2q_0} f(q)(2\pi)^3 2k_0 \delta^3(\mathbf{k} - \mathbf{q}) a^{(0)\dagger}(k)|0\rangle$$

$$= \int \frac{d^3q}{(2\pi)^3 2q_0} f(q) a^{(0)\dagger}(q)|0\rangle$$

$$= |1\rangle.$$

The reader will notice that the only relevant point in this calculation is that the minus sign in the definition of the number density operator cancels that occurring in the commutation relation (4.88). In the light of this, it is clear that the Hamiltonian operator (4.89) cannot have negative eigenvalues. It may, however, have negative expectation values; for example, it is easy to see that

$$\langle 1|H|1\rangle = -\int \frac{d^3k}{(2\pi)^3 2k_0} k_0 |f(k)|^2 \langle 0|0 \rangle.$$

The way to get rid of these unwanted effects is to employ the Lorentz condition (sometimes called the subsidiary condition). It was pointed out above that the condition $\partial_\mu A^\mu = 0$ cannot be taken as an operator identity, because it then conflicts with the commutation relations (4.72). It is even too severe a condition to demand that the physical state $|\psi\rangle$ should satisfy $\partial_\mu A^\mu|\psi\rangle = 0$, for, decomposing into positive and negative frequency parts, we should then have

$$\partial_\mu A^\mu|\psi\rangle = (\partial_\mu A^{(+)\mu} + \partial_\mu A^{(-)\mu})|\psi\rangle = 0,$$

but the negative frequency operator contains creation operators, so not even the vacuum could satisfy this identity. On the other hand, since $A^{(+)\mu}$ contains annihilation operators, we could adopt the less demanding requirement

$$\partial_\mu A^{(+)\mu}|\psi\rangle = 0 \tag{4.91}$$

which the vacuum automatically satisfies. It follows from this that the expectation value of $\partial_\mu A^\mu$ vanishes:

$$\begin{aligned}
\langle\psi|\partial_\mu A^\mu|\psi\rangle &= \langle\psi|\partial_\mu A^{(+)\mu} + \partial_\mu A^{(-)\mu}|\psi\rangle \\
&= \langle\psi|\partial_\mu A^{(-)\mu}|\psi\rangle \\
&= \langle\psi|\partial_\mu A^{(+)\mu}|\psi\rangle^* \\
&= 0.
\end{aligned}$$

It is easy to see that this condition, first formulated by Gupta and Bleuler, resolves the problem of a negative expectation value for the field energy. Substituting equation (4.80) into (4.91) gives

$$\sum_{\lambda=0}^{3} k^\mu \varepsilon_\mu^{(\lambda)} a^{(\lambda)}(k)|\psi\rangle = 0$$

which, in view of (4.83), implies that

$$[k^\mu \varepsilon_\mu^{(0)} a^{(0)}(k) + k^\mu \varepsilon_\mu^{(3)} a^{(3)}(k)]|\psi\rangle = 0;$$

but $k^\mu \varepsilon_\mu^{(0)} = -k^\mu \varepsilon_\mu^{(3)}$, so we have

$$[a^{(0)}(k) - a^{(3)}(k)]|\psi\rangle = 0. \tag{4.92}$$

That is, physical states are *admixtures* of longitudinal and timelike photons, such that (4.92) holds. They do not, as we assumed above, admit the existence of (say) one timelike photon only. It follows immediately from (4.92) that

$$\langle\psi|a^{(0)\dagger}(k)a^{(0)}(k)|\psi\rangle = \langle\psi|a^{(3)\dagger}(k)a^{(3)}(k)|\psi\rangle,$$

and therefore that the contributions of the longitudinal and timelike photons to the Hamiltonian (4.90) *cancel* each other, leaving only the contributions of the (physical) transverse states. It is clear that, *formally*, the effect of the subsidiary condition may be represented by replacing the expectation value of an operator O, $\langle O \rangle = (\Phi, O\Phi)$, by the definition

$$\langle O \rangle = \langle \Phi | \eta O | \Phi \rangle$$

where η is a 'metric operator' equal to $(-1)^{n_t}$. This has the effect of giving the Hamiltonian (4.90) a positive expectation value. By the same token, the definition of the norm of a state vector Φ, $N(\Phi)$, will now be replaced by

$$N(\Phi) = \langle \Phi | \eta | \Phi \rangle$$

which has the effect of giving every state a positive norm. Thus, we see that the problems raised by a fully covariant treatment of the electromagnetic field are not insuperable, provided the Lorentz condition, expressed as an expectation value condition, is taken into account. For a more detailed account of the Gupta–Bleuler formalism, the interested reader is referred elsewhere[5,8,9].

4.5 The massive vector field

The massive spin 1 field is described by the Proca equations (2.238), (2.239):

$$F^{\mu\nu} = \partial^\mu A^\nu - \partial^\nu A^\mu; \quad \partial_\mu F^{\mu\nu} + m^2 A^\nu = 0; \tag{4.93}$$

$$\partial_\mu A^\mu = 0. \tag{4.94}$$

There are three independent components of a massive spin 1 field, and the condition above may serve to eliminate one of the four components of A^μ; we take this to be A^0. The Proca equations follow from the Lagrangian

$$\mathscr{L} = -\tfrac{1}{4} F_{\mu\nu} F^{\mu\nu} + \tfrac{1}{2} m^2 A_\mu A^\mu. \tag{4.95}$$

Note that the positive sign in front of the mass term will result in a term $-\tfrac{1}{2}m^2(\mathbf{A}\cdot\mathbf{A})$ when A^0 is eliminated, so that the mass term has a negative coefficient, as it does for the Klein–Gordon Lagrangian (3.10). The momentum field operators are

$$\pi^\mu = \frac{\partial \mathscr{L}}{\partial \dot{A}_\mu} = \partial^\mu A^0 - \dot{A}^\mu \tag{4.96}$$

from which it follows that $\pi^0 = 0$, but this need not worry us, since we are taking (4.94) to imply that A^0 has been eliminated. Consequently,

$$\pi^i = -\dot{A}^i$$

and the commutation relations

$$[A^i(\mathbf{x},t), \pi_j(\mathbf{x}',t)] = i\delta^i_j \delta^3(\mathbf{x}-\mathbf{x}')$$

become

$$[\dot{A}_i(\mathbf{x},t), A_j(\mathbf{x}',t)] = ig_{ij}\delta^3(\mathbf{x}-\mathbf{x}'). \tag{4.97}$$

The field $A_\mu(x)$ has the expansion

$$A_\mu(k) = \int \frac{d^3k}{(2\pi)^3 2k_0} \sum_{\lambda=1}^{3} \varepsilon_\mu^{(\lambda)}(k)[a^{(\lambda)}(k)e^{-ikx} + a^{(\lambda)\dagger}(k)e^{ikx}] \qquad (4.98)$$

and on inverting this and using (4.97) we find

$$[a^{(\lambda)}(k), a^{(\lambda')\dagger}(k')] = \delta_{\lambda\lambda'} 2k^0 (2\pi)^3 \delta^3(\mathbf{k} - \mathbf{k}'), \qquad (4.99)$$

which is to be compared with (4.87). The analogue, in this case of the scalar photon, is not present, so (4.99) presents no problems, and it may be checked that the Hamiltonian after normal ordering is

$$H = \int \frac{d^3k}{(2\pi)^3 2k_0} k_0 \sum_{\lambda=1}^{3} a^{(\lambda)\dagger}(k)a^{(\lambda)}(k). \qquad (4.100)$$

Thus the particle interpretation of the theory is quite straightforward.

The polarisation vectors $\varepsilon_\mu^{(\lambda)}(k)$ are orthonormal (and spacelike),

$$\varepsilon^{(\lambda)}(k) \cdot \varepsilon^{(\lambda')}(k) = g_{\lambda\lambda'}, \qquad (4.101)$$

and in view of (4.94) are orthogonal to the timelike vector k^μ

$$\varepsilon_\mu^{(\lambda)}(k)k^\mu = 0. \qquad (4.102)$$

In the particle rest-frame, it is easy to see that these conditions are fulfilled by

$$\left.\begin{aligned} k^\mu &= (m,0,0,0), \\ \varepsilon_\mu^{(1)} &= (0,1,0,0), \\ \varepsilon_\mu^{(2)} &= (0,0,1,0), \\ \varepsilon_\mu^{(3)} &= (0,0,0,1). \end{aligned}\right\} \qquad (4.103)$$

It is useful, later on, to know the value of

$$\sum_{\lambda=1}^{3} \varepsilon_\mu^{(\lambda)} \varepsilon_\nu^{(\lambda)} = P_{\mu\nu} \qquad (4.104)$$

which is a rank 2 tensor, whose value in the rest-frame is

$$\text{(rest-frame)} \quad P_{00}=0, \quad P_{11}=P_{22}=P_{33}=1, \quad P_{\mu\nu}=0 \quad (\mu \neq \nu). \qquad (4.105)$$

Let us calculate its value in a moving frame, and choose the particle to be moving along the z axis with momentum $|\mathbf{k}|$, so $k^\mu = (k^0,0,0,|\mathbf{k}|)$. This is obtained from the rest-frame by the boost matrix

$$\Lambda^\mu_{\ \nu} = \frac{1}{m}\begin{pmatrix} k^0 & 0 & 0 & |\mathbf{k}| \\ 0 & 1 & 0 & 0 \\ 0 & 0 & 1 & 0 \\ |\mathbf{k}| & 0 & 0 & k^0 \end{pmatrix}. \qquad (4.106)$$

$P_{\mu\nu}$ transforms as

$$P'_{\mu\nu} = \Lambda^\rho{}_\mu \Lambda^\sigma{}_\nu P_{\rho\sigma} \qquad (4.107)$$

which gives, in the moving frame

$$P_{\mu\nu} = \frac{1}{m^2} \begin{pmatrix} |\mathbf{k}|^2 & 0 & 0 & -k^0|\mathbf{k}| \\ 0 & 1 & 0 & 0 \\ 0 & 0 & 1 & 0 \\ -k^0|\mathbf{k}| & 0 & 0 & (k^0)^2 \end{pmatrix} \qquad (4.108)$$

(The minus signs are due to the interchange of covariant and contravariant indices between (4.106) and (4.107).) It is then straightforward to see that the general form of $P_{\mu\nu}$ is

$$P_{\mu\nu} = \sum_{\lambda=1}^{3} \varepsilon^{(\lambda)}_\mu \varepsilon^{(\lambda)}_\nu = -g_{\mu\nu} + \frac{k_\mu k_\nu}{m^2}. \qquad (4.109)$$

Summary

Canonical quantisation of the real and complex scalar fields, the Dirac field, the electromagnetic field, and the massive spin 1 field are reviewed. [1]For the scalar field, this removes the negative energy problem which beset the single-particle interpretation of the Klein–Gordon equation, and results in a many-particle interpretation of the equation; the particles are necessarily bosons. [2]The complex field, on quantisation, yields particles and antiparticles. [3]Quantisation of the Dirac equation only gives positive definite energy if the resulting particles obey Fermi statistics. This supports Dirac's prediction of antiparticles. [4]Quantisation of the electromagnetic field suffers from difficulties posed by gauge invariance. The quantisation procedure is outlined in both the radiation gauge, in which there appear only the two physical (transverse) polarisation states, and in the Lorentz gauge, in which all four polarisation states appear, the formalism being Lorentz covariant. The resulting difficulties are resolved by the method of Gupta and Bleuler. [5]Quantisation of the massive vector field is straightforward, and leads to the expected particle interpretation.

Guide to further reading

Canonical quantisation of fields of spin 0, $\frac{1}{2}$ and 1 is treated in most books on quantum field theory, for example
(1) C. Itzykson & J.B. Zuber, *Quantum Field Theory*, McGraw-Hill, 1980.
(2) J.D. Bjorken & S.D Drell, *Relativistic Quantum Fields*, McGraw-Hill, 1965.
(3) S.S. Schweber, *An Introduction to Relativistic Quantum Field Theory*, Harper & Row, 1962.
(4) S.S Schweber, H.A. Bethe & F. de Hoffmann, *Mesons and Fields (Vol. 1: Fields)*, Row, Peterson & Co., 1956.

(5) N.N. Bogoliubov & D.V. Shirkov, *Introduction to the Theory of Quantised Fields*, Interscience 1959; 3rd edition,Wiley, 1980.

(6) D. Lurié, *Particles and Fields*, Interscience, 1968.

(7) S. Gasiorowicz, *Elementary Particle Physics*, John Wiley & Sons, Inc., 1966.

(8) F. Mandl, *Introduction to Quantum Field Theory*, Interscience, 1960.

(9) G. Källén, *Quantum Electrodynamics*, Springer-Verlag, 1972.

The connection between spin and statistics is fully reviewed by

(10) R. Jost in M. Fierz & V.F. Weisskopf (eds.), *Theoretical Physics in the Twentieth Century*, Interscience, 1960.

5

Path integrals and quantum mechanics

Thirty-one years ago, Dick Feynman told me about his 'sum over histories' version of quantum mechanics. 'The electron does anything it likes', he said. 'It goes in any direction at any speed, forward or backward in time, however it likes, and then you add up the amplitudes and it gives you the wave-function.' I said to him, 'You're crazy'. But he wasn't.

F.J. Dyson[‡]

A common type of calculation in particle physics is that of a scattering cross section for a particular process, for example electron-electron scattering $e^- e^- \to e^- e^-$. Under the inspiring guidance of Feynman, a short-hand way of expressing – and of thinking about – these quantities has been developed. Thus, in the particular case of $e^- e^-$ scattering, to a first order approximation the process is represented by the 'Feynman diagram' of Fig. 1.14, and the crucial part of this diagram is the 'propagation' of the photon between the two electrons. There are 'Feynman rules' which allow one to associate with each diagram a scattering amplitude, and from the total amplitude (there may be more than one diagram for each process) one calculates the cross section in a straight-forward way. In this chapter and the next one it will be shown how the Feynman rules arise, and, in particular, how to find an expression for the 'propagation' of the virtual particle. In this chapter it will be shown how quantum mechanics can be formulated so that scattering processes may be understood directly in these terms. In the next chapter we shall extend the treatment to scalar and spinor fields, and in chapter 7 to gauge fields. In this chapter we retain Planck's constant $\hbar$ in the relevant formulae.

[‡] In H. Woolf (ed.), *Some strangeness in the proportion*, p. 376. Published by Addison-Wesley, 1980. Quoted with kind permission.

5.1   Path integral formulation of quantum mechanics

In the usual formulation of quantum mechanics, the quantities q and p are replaced by operators which obey Heisenberg commutation relations. The mathematics one invokes is that of operators in Hilbert space. The *path integral formulation* of quantum mechanics, on the other hand, is based directly on the notion of a *propagator* $K(q_f t_f; q_i t_i)$. Given a wave function $\psi(q_i, t_i)$ at time t_i, the propagator gives the corresponding wave function at a later time t_f by an appeal to Huygens' principle:

$$\psi(q_f, t_f) = \int K(q_f t_f; q_i t_i)\psi(q_i t_i)\,dq_i. \tag{5.1}$$

(For simplicity we consider only one spatial dimension.) This equation is quite general and merely expresses causality. According to the usual interpretation of quantum mechanics, $\psi(q_f, t_f)$ is the probability amplitude that the particle is at the point q_f at the time t_f, so $K(q_f t_f, q_i t_i)$ is the probability amplitude for a *transition* from q_i at time t_i to q_f at time t_f. The *probability* that it is observed at q_f at time t_f is

$$P(q_f t_f; q_i t_i) = |K(q_f t_f; q_i t_i)|^2$$

this is a fundamental principle of quantum mechanics, as every student knows.

Let us divide the time interval between t_i and t_f into two, with t as the intermediate time, and q the intermediate point in space, as shown in Fig. 5.1. Repeated application of (5.1) gives

$$\psi(q_f, t_f) = \int\int K(q_f t_f; qt)K(qt; q_i t_i)\psi(q_i, t_i)\,dq_i\,dq,$$

from which it follows that

$$K(q_f t_f; q_i t_i) = \int K(q_f t_f; qt)K(qt; q_i t_i)\,dq, \tag{5.2}$$

so the transition from (q_i, t_i) to (q_f, t_f) may be regarded as the result of

Fig. 5.1. Propagation of a particle from (q_i, t_i) to (q_f, t_f) via an intermediate position (q, t).

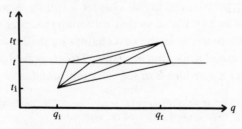

transition from (q_i, t_i) to *all available intermediate points* q followed by transition from (q, t) to (q_f, t_f).

As a simple and familiar illustration of this, consider the 2-slit experiment with electrons, shown in Fig. 5.2. Denote by $K(2A; 1)$ the probability amplitude that the electron passes from the source 1 to the hole $2A$, and by $K(3; 2A)$ the amplitude that it passes from the hole $2A$ to the detectors 3, and so on. Equation (5.2) then gives

$$K(3; 1) = K(3; 2A)K(2A; 1) + K(3; 2B)K(2B; 1)$$

and the intensity pattern on the 'screen' 3 is given by the probability

$$P(3; 1) = |K(3; 1)|^2$$

which will clearly contain interference terms, characteristic of the quantum theory. Note that we cannot say 'the electron travelled either through hole A or hole B' – it travelled, in a sense, over *both paths* (if not detected at the slits).[‡] This notion of *all possible paths* is important in the *path integral* formalism.

We may show that the propagator K is actually the more familiar quantity $\langle q_f t_f | q_i t_i \rangle$. To see this, note that the wave function $\psi(q, t)$ is

$$\psi(q, t) = \langle q | \psi t \rangle_S$$

where the state vector $|\psi t \rangle_S$ in the Schrödinger picture is related to that in the Heisenberg picture $|\psi \rangle_H$ by

$$|\psi t \rangle_S = e^{-iHt/\hbar} |\psi \rangle_H.$$

Let us define the vector

$$|qt \rangle = e^{iHt/\hbar} |q \rangle \tag{5.3}$$

Fig. 5.2. The 2-slit experiment.

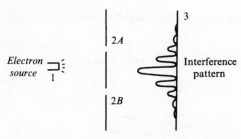

Electron source 1 / 2A / 2B / 3 / Interference pattern

[‡] For an excellent discussion of the 2-slit experiment, see R.P. Feynman, R.B. Leighton & M. Sands, *The Feynman Lectures on Physics*, Vol. 3, Addison-Wesley, 1965.

which we may call, for obvious reasons, a 'moving frame'. We then have

$$\psi(q,t) = \langle qt | \psi \rangle_{\mathrm{H}}. \tag{5.4}$$

Completeness of states enables us to write

$$\langle q_{\mathrm{f}}t_{\mathrm{f}} | \psi \rangle = \int \langle q_{\mathrm{f}}t_{\mathrm{f}} | q_{\mathrm{i}}t_{\mathrm{i}} \rangle \langle q_{\mathrm{i}}t_{\mathrm{i}} | \psi \rangle \, dq_{\mathrm{i}}$$

which, with the help of (5.4), is

$$\psi(q_{\mathrm{f}}, t_{\mathrm{f}}) = \int \langle q_{\mathrm{f}}t_{\mathrm{f}} | q_{\mathrm{i}}t_{\mathrm{i}} \rangle \psi(q_{\mathrm{i}}, t_{\mathrm{i}}) \, dq_{\mathrm{i}}.$$

On comparison with (5.1) we see that

$$\langle q_{\mathrm{f}}t_{\mathrm{f}} | q_{\mathrm{i}}t_{\mathrm{i}} \rangle = K(q_{\mathrm{f}}t_{\mathrm{f}}; q_{\mathrm{i}}t_{\mathrm{i}}), \tag{5.5}$$

as claimed.

The propagator K summarises the quantum mechanics of the system. In the usual formulation of quantum mechanics, given an initial wave function, one can find the final wave function by solving the time-dependent Schrödinger equation. In this formulation, however, the propagator gives the solution directly. The idea now is to express $\langle q_{\mathrm{f}}t_{\mathrm{f}} | q_{\mathrm{i}}t_{\mathrm{i}} \rangle$ as a path integral.

Let us split the time interval between t_{i} and t_{f} into $(n+1)$ equal pieces τ, as in Fig. 5.3. Equation (5.2) now becomes

$$\langle q_{\mathrm{f}}t_{\mathrm{f}} | q_{\mathrm{i}}t_{\mathrm{i}} \rangle = \int \cdots \int dq_1 \, dq_2 \ldots dq_n \langle q_{\mathrm{f}}t_{\mathrm{f}} | q_n t_n \rangle \langle q_n t_n | q_{n-1} t_{n-1} \rangle \cdots$$
$$\cdots \langle q_1 t_1 | q_{\mathrm{i}}t_{\mathrm{i}} \rangle \tag{5.6}$$

and the integral is taken over all possible 'trajectories'; they are not trajectories in the normal sense, since each segment $(q_j t_j; q_{j-1} t_{j-1})$ may be subdivided into smaller segments, so there is no derivative. The paths are really Markov chains.

Let us calculate the propagator over a small segment in the path integral.

Fig. 5.3. Propagation over many paths from $(q_{\mathrm{i}}, t_{\mathrm{i}})$ to $(q_{\mathrm{f}}, t_{\mathrm{f}})$.

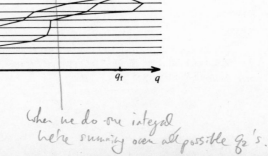

When we do one integral here summing over all possible q_2's.

From (5.3) we have

$$\langle q_{j+1}t_{j+1}|q_jt_j\rangle = \langle q_{j+1}|e^{-iH\tau/\hbar}|q_j\rangle$$

$$= \left\langle q_{j+1}\left|1-\frac{i}{\hbar}H\tau + O(\tau^2)\right|q_j\right\rangle$$

$$= \left[\delta(q_{j+1}-q_j)\right]-\frac{i\tau}{\hbar}\langle q_{j+1}|H|q_j\rangle$$

$$= \frac{1}{2\pi\hbar}\int dp\exp\left[\frac{i}{\hbar}p(q_{j+1}-q_j)\right]$$

$$-\frac{i\tau}{\hbar}\langle q_{j+1}|H|q_j\rangle. \qquad (5.7)$$

The Hamiltonian H is a function of the operators p and q. In the special case where H is of the form

$$H = \frac{p^2}{2m} + V(q) \qquad (5.8)$$

(actually, H can be any function of p + any function of q), the matrix element in (5.7) may easily be calculated. We have

$$\left\langle q_{j+1}\left|\frac{p^2}{2m}\right|q_j\right\rangle = \int dp'\,dp\langle q_{j+1}|p'\rangle\left\langle p'\left|\frac{p^2}{2m}\right|p\right\rangle\langle p|q_j\rangle,$$

[margin note:] $=\frac{p^2}{2m}|p\rangle,$ p *number.*

and now substitute $\langle q_{j+1}|p\rangle = (2\pi\hbar)^{-\frac{1}{2}}\exp(ip'q_{j+1}/\hbar)$, giving

$$\left\langle q_{j+1}\left|\frac{p^2}{2m}\right|q_j\right\rangle = \int\frac{dp'\,dp}{2\pi\hbar}\exp\left[\frac{i}{\hbar}(p'q_{j+1}-pq_j)\right]\frac{p^2}{2m}\delta(p-p')$$

$$= \int\frac{dp}{h}\exp\left[\frac{i}{\hbar}p(q_{j+1}-q_j)\right]\frac{p^2}{2m}. \qquad (5.9)$$

It is to be noted here that p^2 on the left-hand side of (5.9) is an *operator*, whereas on the right-hand side it is a *number*. We could (and perhaps should) have used a notation such as $\hat{p}$ to call attention to the operator nature of p on the left-hand side. But in any case, it is an important result that the formula we have on the right of (5.9) contains no operators. In a similar way,

$$\langle q_{j+1}|V(q)|q_j\rangle = V\left(\frac{q_{j+1}+q_j}{2}\right)\langle q_{j+1}|q_j\rangle$$

$$= V\left(\frac{q_{j+1}+q_j}{2}\right)\delta(q_{j+1}-q_j)$$

$$= \int\frac{dp}{h}\exp\left[\frac{i}{\hbar}p(q_{j+1}-q_j)\right]V(\bar{q}_j), \qquad (5.10)$$

where $\bar{q}_j = \frac{1}{2}(q_j + q_{j+1})$, and $V(q)$ on the left is an operator expression, but the integral on the right contains no operators. Putting (5.9) and (5.10) together, we have

$$\langle q_{j+1}|H|q_j\rangle = \int \frac{\mathrm{d}p}{h}\exp\left[\frac{\mathrm{i}}{\hbar}p(q_{j+1} - q_j)\right]H(p,\bar{q})$$

and, from (5.7),

$$\langle q_{j+1}t_{j+1}|q_jt_j\rangle = \frac{1}{h}\int \mathrm{d}p_j\exp\left\{\frac{\mathrm{i}}{\hbar}[p_j(q_{j+1} - q_j) - \tau H(p_j,\bar{q}_j)]\right\}$$

(5.11)

where p_j is the momentum between t_j and t_{j+1}, or, equivalently, q_j and q_{j+1} – see Fig. 5.4. This gives the propagator over a segment of one possible path. The full propagator is got by substituting this into (5.6), giving, in the continuum limit,

$$\langle q_\mathrm{f}t_\mathrm{f}|q_\mathrm{i}t_\mathrm{i}\rangle = \lim_{n\to\infty}\int \prod_{j=1}^{n}\mathrm{d}q_j\prod_{j=0}^{n}\frac{\mathrm{d}p_i}{h}$$

$$\times \exp\left\{\frac{\mathrm{i}}{\hbar}\sum_{j=0}^{n}[p_j(q_{j+1} - q_j) - \tau H(p_j,\bar{q}_j)]\right\}$$

(5.12)

with $q_0 = q_\mathrm{i}$, $q_{n+1} = q_\mathrm{f}$. This may be written in the symbolic form

$$\blacksquare\qquad \langle q_\mathrm{f}t_\mathrm{f}|q_\mathrm{i}t_\mathrm{i}\rangle = \int \frac{\mathscr{D}q\mathscr{D}p}{h}\exp\frac{\mathrm{i}}{\hbar}\left[\int_{t_\mathrm{i}}^{t_\mathrm{f}}\mathrm{d}t[p\dot{q} - H(p,q)]\right]$$

(5.13)

with $q(t_\mathrm{i}) = q_\mathrm{i}$, $q(t_\mathrm{f}) = q_\mathrm{f}$. In the continuum limit q becomes a function of t, and the integral is a 'functional integral', an integral over all *functions*. It is infinite-dimensional. Expression (5.13) is the path integral expression for the transition amplitude from $(q_\mathrm{i}, t_\mathrm{i})$ to $(q_\mathrm{f}, t_\mathrm{f})$. Each function $q(t)$ and $p(t)$ defines a *path in phase space*. As mentioned above, the more usual approach to quantum mechanics is to solve the Schrödinger equation $\mathrm{i}\hbar(\mathrm{d}|\psi\rangle/\mathrm{d}t) = \hat{H}|\psi\rangle$ where $\hat{H}$ is an operator, subject to some boundary conditions. In the path-integral formulation we have an explicit expression for the transition amplitude, which is clearly very well suited to scattering

Fig. 5.4. Segments of the trajectory in momentum space.

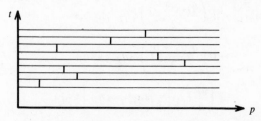

problems. The quantities p and q occurring in the integral are *classical* quantities, *not* operators (c-numbers, not q-numbers). It is not obvious, however, that infinite-dimensional integrals of this type are well-defined mathematically, that is, whether they converge; in other words, exist! We shall assume that they do. The reader interested in the mathematical status of functional integrals is referred to refs. (12) to (17).

There is another form for the transition amplitude, which holds when H is of the form (5.8), since in that case we can perform the p-integration. Equation (5.12) becomes

$$\langle q_f t_f | q_i t_i \rangle = \lim_{n \to \infty} \int \prod_1^n dq_j \prod_0^n \frac{dp_j}{h}$$

$$\times \exp\left\{\frac{i}{\hbar}\left[p_j(q_{j+1} - q_j) - \frac{p_j^2}{2m}\tau - V(\bar{q}_j)\tau \right]\right\}.$$

As far as the p_j integration is concerned, this is of the same form as equation (5A.3) (see the appendix to this chapter), so we get

$$\langle q_f t_f | q_i t_i \rangle = \lim_{n \to \infty} \left(\frac{m}{ih\tau}\right)^{(n+1)/2} \int \prod_1^n dq_j$$

$$\times \exp\left\{\frac{i\tau}{\hbar}\left[\frac{m}{2}\left(\frac{q_{j+1} - q_j}{\tau}\right)^2 - V \right]\right\} \qquad (5.14)$$

and hence, in the continuum limit

■ $$\langle q_f t_f | q_i t_i \rangle = N \int \mathcal{D}q \exp\left[\frac{i}{\hbar}\int_{t_i}^{t_f} L(q, \dot{q})dt\right] \qquad (5.15)$$

where $L = T - V$, the classical Lagrangian. In the limit $n \to \infty$, N becomes infinite, but this does not matter, since we shall always deal with normalised transition amplitudes.

The integrand in (5.15) is the classical action $S = \int L \, dt$. We have proved this equation from the postulates of quantum mechanics, and by assuming that the Hamiltonian is of the form (5.8). Feynman's original approach was to adopt (5.15) as a hypothesis and then prove the Schrödinger equation from it. The disadvantage of this approach is that (5.15) does not hold in general, since (5.8) does not. A counter example has been given by Lee and Yang.[‡] If

$$L = \frac{\dot{q}^2}{2} f(q),$$

which describes a system with a velocity-dependent potential, then the

[‡] T.D. Lee & C.N. Yang, *Physical Review*, **128**, 885 (1962).

momentum is

$$p = \frac{\partial L}{\partial \dot{q}} = \dot{q} f(q)$$

and the Hamiltonian is

$$H = p\dot{q} - L = \tfrac{1}{2}\dot{q}^2 f(q) = \tfrac{1}{2}\frac{p^2}{f(q)}$$

and this is not of the form (5.8). Substituting this into (5.13) and performing the p integrations gives, eventually,

$$\langle q_f t_f | q_i t_i \rangle = N \int \mathscr{D}q \exp\left(\frac{i}{\hbar} S_{eff}\right)$$

where

$$S_{eff} = \int dt \left[L(q, \dot{q}) - \frac{i}{2} \delta(0) \ln f(q) \right].$$

Instead of (5.14), we have an *effective* action, which differs from $S = \int L\,dt$.

In the case of field theories, similarly, the transition from an equation like (5.13) to one like (5.15) may not in general be made, and in particular this is true in the case of non-Abelian gauge-field theories. However, when we come to consider these theories, in chapter 7, we shall for simplicity adopt a 'heuristic' approach to the derivation of the Feynman rules, and work from an equation analogous to (5.15)

5.2   Perturbation theory and the S matrix[‡]

It is our aim to illustrate how the path-integral method is used in the calculation of scattering processes, and we shall consider Rutherford scattering in particular, in §5.3 below. The scattering of one particle on another is described, non-relativistically, by interaction through a potential $V(x)$ (we change the notation, in this section, for the space co-ordinate, from q to x). Since the expression for the transition amplitude is not exactly calculable, we resort, as usual, to perturbation theory. This is valid when the potential $V(x)$ is small, or, more precisely, when the time integral of $V(x, t)$ is small compared with $\hbar$. In that case we may write

$$\exp\left[\frac{-i}{\hbar}\int_{t_i}^{t_f} V(x, t)dt\right] = 1 - \frac{i}{\hbar}\int_{t_i}^{t_f} V(x, t)dt$$
$$- \frac{1}{2!\hbar^2}\left[\int_{t_i}^{t_f} V(x, t)dt\right]^2 + \cdots. \qquad (5.16)$$

[‡] In this section and the following one I have drawn freely from M. Veltman, lectures at Basko-Polje School, 1974 (unpublished).

This is the perturbation expansion. When substituted into the expression (5.15) for the propagator $K(x_f t_f; x_i t_i)$ (see (5.5)) we get a series expansion

$$K = K_0 + K_1 + K_2 + \dots, \tag{5.17}$$

the first term of which is the free propagator K_0:

$$K_0 = N \int \left[\exp\left(\frac{i}{\hbar} S\right) \right] \mathscr{D}x$$

$$= N \int \left[\exp\left(\frac{i}{\hbar} \int \tfrac{1}{2}m\dot{x}^2 \, dt\right) \right] \mathscr{D}x.$$

To evaluate this, we write it in the discrete form (see (5.14))

$$K_0 = \lim_{n\to\infty} \left(\frac{m}{i\hbar\tau}\right)^{(n+1)/2} \int_{-\infty}^{\infty} \prod_{j=1}^{n} dx_j \exp\left[\frac{im}{2\hbar\tau} \sum_{j=0}^{n} (x_{j+1} - x_j)^2\right].$$

The value of this integral is known – see (5A.4),

$$\text{integral} = \frac{1}{(n+1)^{\frac{1}{2}}} \left(\frac{i\hbar\tau}{m}\right)^{n/2} \exp\left[\frac{im}{2\hbar(n+1)\tau}(x_f - x_i)^2\right],$$

so, putting $(n+1)\tau = t_f - t_i$, we have for the free particle propagator

$$K_0(x_f t_f; x_i t_i) = \left(\frac{m}{i\hbar(t_f - t_i)}\right)^{\frac{1}{2}} \exp\left[\frac{im(x_f - x_i)^2}{2\hbar(t_f - t_i)}\right] \quad (t_f > t_i). \tag{5.18}$$

The condition $t_f > t_i$ is clearly crucial for the propagator, since it vanishes, by causality, if $t_f < t_i$, so we may properly put

$$K_0(x_f t_f; x_i t_i) = \theta(t_f - t_i)\left(\frac{m}{i\hbar(t_f - t_i)}\right)^{\frac{1}{2}} \exp\left[\frac{im(x_f - x_i)^2}{2\hbar(t_f - t_i)}\right]. \tag{5.19}$$

Let us now calculate K_1. From (5.14) and (5.16) we have

$$K_1 = \frac{-i}{\hbar} \lim_{n\to\infty} N^{(n+1)/2} \sum_{i=1}^{n} \int \exp\left[\frac{im}{2\hbar\tau} \sum_{j=0}^{n} (x_{j+1} - x_j)^2\right]$$

$$\times V(x_i, t_i) \, dx_1 \dots dx_n,$$

where $N = m/i\hbar\tau$, and we have replaced integration over t by summation over t_i. Noting that V depends on x_i, we now split up the sum in the exponent into two, one going from $j = 0$ to $j = i - 1$, and the other from $j = i$ to $j = n$. We also separate out the integration over x_i, and get

$$K_1 = \lim_{n\to\infty} \frac{-i}{\hbar} \sum_{i=1}^{n} \int dx_i$$

$$\times \left\{ N^{(n-i+1)/2} \int \exp\left[\frac{im}{2\hbar\tau} \sum_{j=i}^{n} (x_{j+1} - x_j)^2\right] dx_{i+1} \dots dx_n \right\}$$

$$\times V(x_i, t_i) \left\{ N^{i/2} \int \exp\left[\frac{im}{2\hbar\tau} \sum_{j=0}^{i-1} (x_{j+1} - x_j)^2\right] dx_1 \dots dx_{i-1} \right\}.$$

The two terms in curly brackets are $K_0(x_f t_f; xt)$ and $K_0(xt; x_i t_i)$, so, replacing $\sum_i \int dx_i$ by $\int dx\, dt$, the above expression becomes

$$K_1(x_f t_f; x_i t_i) = -\frac{i}{\hbar} \int_{t_i}^{t_f} dt \int_{-\infty}^{\infty} K_0(x_f t_f; xt) V(x, t) K_0(xt; x_i t_i) dx.$$

$$(5.20)$$

Now $K_0(x_f t_f; xt)$ vanishes if $t > t_f$ and $K_0(xt; x_i t_i)$ vanishes if $t < t_i$, so the integration in (5.20) may be taken over all values of t, to give

$$K_1(x_f t_f; x_i t_i) = \frac{-i}{\hbar} \int_{-\infty}^{\infty} dt \int K_0(x_f t_f; xt) V(x, t) K_0(xt; x_i t_i) dx. \quad (5.21)$$

This is the first order correction to the free propagator. In a similar way, we may prove that the second order correction is

$$K_2(x_f t_f; x_i t_i) = \left(\frac{-i}{\hbar}\right)^2 \int_{-\infty}^{\infty} dt_1 \int_{-\infty}^{\infty} dt_2$$

$$\times \int_{-\infty}^{\infty} dx_1 \int_{-\infty}^{\infty} dx_2\, K_0(x_f t_f; x_2 x_2) V(x_2, t_2)$$

$$\times K_0(x_2 t_2; x_1 t_1) V(x_1, t_1) K_0(x_1 t_1; x_i t_i). \quad (5.22)$$

Analogous expressions hold for all K_n in the expansion (5.17), so we write

$$K(x_f t_f; x_i t_i) = K_0(x_f t_f; x_i t_i)$$

$$-\frac{i}{\hbar} \int K_0(x_f t_f; x_1 t_1) V(x_1, t_1) K(x_1 t_1; x_i t_i) dx_1\, dt_1$$

$$-\frac{1}{\hbar^2} \int K(x_f t_f; x_1 t_1) V(x_1 t_1) K_0(x_1 t_1; x_2 t_2) V(x_2 t_2)$$

$$\times K_0(x_2 t_2; x_i t_i) dx_1\, dx_2\, dt_1\, dt_2 + \dots . \quad (5.23)$$

Fig. 5.5. The Born series.

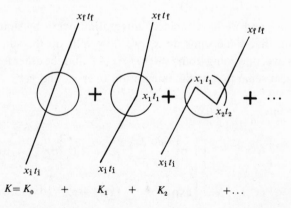

This is the solution to the perturbation series in K, and is called the *Born series*. It may be visualised as in Fig. 5.5. K_0 describes the free propagation of the wave function from $x_i t_i$ to $x_f t_f$; K_1 describes propagation with one interaction with the potential V; and so on.

A noteworthy feature of (5.22) is that it does not include the factor 1/2! present in (5.16). The reason for this is as follows. The two interactions with V occur at different times but are indistinguishable, so we write

$$\frac{1}{2!}\int V(t')V(t'')dt' \, dt'' = \frac{1}{2!}\int [\theta(t'-t'')V(t')V(t'')$$
$$+ \theta(t''-t')V(t')V(t'')] \, dt' \, dt''$$
$$= \int \theta(t_1 - t_2)V(t_1)V(t_2)dt_1 \, dt_2. \qquad (5.24)$$

In a similar way, there is no factor $1/n!$ in the expression for K_n. We shall now show that the free propagator K_0 is simply the Green's function for the Schrödinger equation. To see this, substitute the Born series (5.23) into (5.1), giving

$$\psi(x_f t_f) = \int K(x_f t_f; x_i t_i)\psi(x_i t_i)dx_i$$
$$= \int K_0(x_f t_f; x_i t_i)\psi(x_i t_i)dx_i$$
$$- \frac{i}{\hbar}\int K_0(x_f t_f; xt)V(x, t)K_0(xt; x_i t_i)\psi(x_i t_i)dt \, dx \, dx_i$$
$$+ \dots. \qquad (5.25)$$

Here we have changed from one space dimension to three. Assuming the series above converges, the effect of the unwritten terms is to modify the last K_0 to the full propagator K, so that

$$\psi(x_f t_f) = \int K_0(x_f t_f; x_i t_i)\psi(x_i t_i)dx_i$$
$$- \frac{i}{\hbar}\int K_0(x_f t_f; xt)V(x, t)\psi(xt)dx \, dt. \qquad (5.26)$$

This equation is exact, and is an integral equation for ψ. Now assume that in the distant past, $t_i \rightarrow -\infty$, ψ becomes free – a plane wave ϕ – then the first term on the right-hand side of (5.26) is also a plane wave, since it results from the free propagation of $\psi(x_i t_i)$, and we may write

$$\psi(x_f t_f) = \phi(x_f t_f) - \frac{i}{\hbar}\int K_0(x_f t_f; xt)V(x, t)\psi(xt)dx \, dt. \qquad (5.27)$$

Now $\psi(\mathbf{x}_f t_f)$ obeys the Schrödinger equation

$$\frac{\hbar^2}{2m}\nabla^2_{\mathbf{x}_f}\psi(\mathbf{x}_f t_f) + i\hbar\frac{\partial\psi(\mathbf{x}_f t_f)}{\partial t_f} = V(\mathbf{x}_f t_f)\psi(\mathbf{x}_f t_f). \tag{5.28}$$

Since $\phi(\mathbf{x}_f t_f)$ obeys the free-particle equation (with $V = 0$), then K_0 must obey

$$\frac{\hbar^2}{2m}\nabla^2_{\mathbf{x}_f}K_0(\mathbf{x}_f t_f; \mathbf{x}t) + i\hbar\frac{\partial}{\partial t_f}K_0(\mathbf{x}_f t_f; \mathbf{x}t) = \frac{i}{\hbar}\delta(\mathbf{x}_f - \mathbf{x})\delta(t_f - t) \tag{5.29}$$

which is the equation for the Green's function of (5.28). Note that the presence of the $\delta(t_f - t)$ is to be expected from the $\theta(t_f - t)$ occurring in the definition (5.19) of K_0. The propagator K_0 is then simply the Green's function of the Schrödinger equation, as claimed.

We pass now to the calculation of the scattering amplitude. In the measurement of a scattering process, the experimental conditions are that a particle is free at $t = -\infty$, then scatters, and is free again at $t = +\infty$. This is a source of difficulty, however, because a free particle (that is, one with a definite energy and momentum) is described by a plane wave, which spreads out over all space and time, *including* the centre of the interaction potential $V(\mathbf{x})$, so the particle can never be free! To get round this problem the *adiabatic hypothesis* is invoked; the potential V is switched on and off again slowly, so that $V = 0$ at $t = +\infty$ and the particle is free. V must not be switched on and off too quickly; this would imply, through Fourier transformation, that the time dependence of V results in the scattering centre emitting or absorbing energy, which must not happen.

Returning to the scattering problem, the initial condition is that ψ is a plane wave:

$$\psi_{in}(\mathbf{x}_i t_i) \quad \text{plane wave.}$$

We assume that $V \to 0$ for large negative t, and that t_i is even further in the past. Taking the first Born approximation in (5.25) gives

$$\psi^{(+)}(\mathbf{x}_f t_f) = \int K_0(\mathbf{x}_f t_f; \mathbf{x}_i t_i)\psi_{in}(\mathbf{x}_i t_i)\,d\mathbf{x}_i$$

$$-\frac{i}{\hbar}\int K_0(\mathbf{x}_f t_f; \mathbf{x}t)V(\mathbf{x}, t)K_0(\mathbf{x}t; \mathbf{x}_i t_i)\psi_{in}(\mathbf{x}_i t_i)\,d\mathbf{x}\,d\mathbf{x}_i\,dt. \tag{5.30}$$

The superscript on $\psi^{(+)}(\mathbf{x}_f t_f)$ denotes that it corresponds to a wave which was free at $t = -\infty$, and thus involves the 'retarded' propagator $K_0(\mathbf{x}t; \mathbf{x}'t')$, which vanishes for $t' > t$. It is equally acceptable to write a solution

$\psi^{(-)}(x_1 t_1)$ consisting of a wave which becomes free at $t = \infty$ (ψ_{out}), and an 'advanced' propagator $K_0(\mathbf{x}t; \mathbf{x}'t')$ which vanishes when $t' < t$.

We are interested in the amplitude for detecting a final particle with definite momentum, i.e. a plane wave ψ_{out}. This is called the *scattering amplitude S*, and is the overlap of the wave functions

$$S = \int \psi_{out}^*(\mathbf{x}_f t_f) \psi^{(+)}(\mathbf{x}_f t_f) d\mathbf{x}_f$$

$$= \int \psi_{out}^*(\mathbf{x}_f t_f) K_0(\mathbf{x}_f t_f; \mathbf{x}_i t_i) \psi_{in}(\mathbf{x}_i t_i) d\mathbf{x}_i \, d\mathbf{x}_f$$

$$- \frac{i}{\hbar} \int \psi_{out}^*(\mathbf{x}_f t_f) K_0(\mathbf{x}_f t_f; \mathbf{x}t) V(\mathbf{x}, t) K_0(\mathbf{x}t; \mathbf{x}_i t_i)$$

$$\times \psi_{in}(\mathbf{x}_i t_i) d\mathbf{x}_f \, d\mathbf{x} \, d\mathbf{x}_i \, dt$$

$$= \int \psi_{out}^*(\mathbf{x}_f t_f) \phi(\mathbf{x}_f t_f) d\mathbf{x}_f - \frac{i}{\hbar} \int \psi_{out}^*(\mathbf{x}_f t_f) K_0(\mathbf{x}_f t_f; \mathbf{x}t)$$

$$\times V(\mathbf{x}, t) K_0(\mathbf{x}t; \mathbf{x}_i t_i) \psi_{in}(\mathbf{x}_i t_i) d\mathbf{x}_f \, d\mathbf{x} \, d\mathbf{x}_i \, dt \qquad (5.31)$$

where $\phi(\mathbf{x}_f t_f)$, just like $\psi_{in}(\mathbf{x}_i t_i)$, is a plane wave. If the initial and final momenta are $\mathbf{p}_i = \hbar \mathbf{k}_i, \mathbf{p}_f = \hbar \mathbf{k}_f$, we have, with box normalisation,

$$\psi_{in}(\mathbf{x}t) = \frac{1}{\sqrt{\tau}} \exp\left[\frac{i}{\hbar} (\mathbf{p}_i \cdot \mathbf{x} - E_i t) \right],$$

$$\psi_{out}(\mathbf{x}t) = \frac{1}{\sqrt{\tau}} \exp\left[\frac{i}{\hbar} (\mathbf{p}_f \cdot \mathbf{x} - E_f t) \right] \qquad (5.32)$$

where $E = p^2/2m$ and τ is the volume of the box, which of course is arbitrary. Substituting (5.32) into the first term of (5.31), using

$$\int e^{i\mathbf{q} \cdot \mathbf{x}} d\mathbf{x} = (2\pi)^3 \delta(\mathbf{q},)$$

and putting, for convenience, $\tau = (2\pi)^3$, we get

$$S_{fi} = \delta(\mathbf{k}_i - \mathbf{k}_f) - \frac{i}{\hbar} \int \psi_{out}^*(\mathbf{x}_f t_f) K_0(\mathbf{x}_f t_f; \mathbf{x}t) V(\mathbf{x}, t)$$

$$\times K_0(\mathbf{x}t; \mathbf{x}_i t_i) \psi_{in}(\mathbf{x}_i t_i) d\mathbf{x}_f \, d\mathbf{x} \, d\mathbf{x}_i \, dt. \qquad (5.33)$$

The scattering amplitude is then seen to be one element of a matrix S, whose (fi) element appears above; this is the *scattering matrix* or *S matrix*. The first term corresponds to no interaction giving momentum conservation and a unit S matrix. Genuine interactions are represented by the second term in (5.33), and the amplitude that a particular 'out' state results from a

particular 'in' state is

$$A = -\frac{i}{\hbar} \int \psi_{out}^*(\mathbf{x}_f t_f) K_0(\mathbf{x}_f t_f; \mathbf{x}t) V(\mathbf{x}, t) K_0(\mathbf{x}t; \mathbf{x}_i t_i)$$

$$\times \psi_{in}(\mathbf{x}_i t_i) \, d\mathbf{x}_f \, d\mathbf{x} \, d\mathbf{x}_i \, dt. \qquad (5.34)$$

We now have an expression for the scattering amplitude in terms of the free propagator K_0 and the interaction potential V. Equation (5.34) may be translated into a set of simple rules for the scattering amplitude; these rules are known as the *Feynman rules*.

We may represent amplitude (5.34) (which is a first order approximation), by the diagram

$$V \qquad \mathbf{x}_f t_f$$

$$(5.35)$$

$$\mathbf{x}_i t_i$$

It is clear that the rules for translating this diagram into the expression for the scattering amplitude may be summarised by making the correspondence

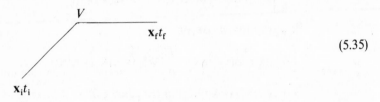

$$\overline{} \quad K_0(\mathbf{x}_2 t_2; \mathbf{x}_1 t_1) \qquad (5.36)$$
$$\mathbf{x}_1 t_1 \qquad\qquad \mathbf{x}_2 t_2$$

$$V$$
$$\mathbf{x}t \qquad -\frac{i}{\hbar} V(\mathbf{x}t); \quad \text{integration over } x \text{ and } t$$

In addition, we multiply by ψ_{in} and ψ_{out}^* at the ends of the diagram and integrate over the two relevant spatial variables. Thus, the amplitude for the second order process

$$\mathbf{x}_i t_i$$
$$V$$
$$V \qquad \mathbf{x}'t'$$
$$\mathbf{x}t \qquad\qquad \mathbf{x}_f t_f$$

$$(5.37)$$

is

$$A^{(2)} = \left(\frac{-i}{\hbar}\right)^2 \int \psi_{out}^*(\mathbf{x}_f t_f) K_0(\mathbf{x}_f t_f; \mathbf{x}'t') V(\mathbf{x}', t') K_0(\mathbf{x}'t'; \mathbf{x}t) V(\mathbf{x},t)$$

$$\times K_0(\mathbf{x}t; \mathbf{x}_i t_i) \, d\mathbf{x}_i \, d\mathbf{x} \, dt \, d\mathbf{x}' \, dt' \, d\mathbf{x}_f.$$

The rules (5.36) are called the Feynman rules. In the non-relativistic quantum mechanics we are dealing with at present, these rules are hardly necessary to do calculations, but in quantum field theory, to be considered in the next chapter, they are a great aid to calculation.

The rules (5.36) are written in co-ordinate space. In many calculations, however, it is more convenient to work in momentum space, and in the remainder of this section we will derive the corresponding Feynman rules in momentum space. Let $\mathscr{K}(\mathbf{p}, t; \mathbf{p}_0 t_0)$ be the amplitude that a particle with momentum $\mathbf{p}_0$ at time t_0 be later observed to have momentum $\mathbf{p}_1$ at time t_1. It is given by

$$\mathscr{K}(\mathbf{p}_1 t_1; \mathbf{p}_0 t_0) = \int \exp\left(-\frac{i}{\hbar}\mathbf{p}_1\cdot\mathbf{x}_1\right) K(\mathbf{x}_1 t_1; \mathbf{x}_0 t_0)$$

$$\times \exp\left(\frac{i}{\hbar}\mathbf{p}_0\cdot\mathbf{x}_0\right) d\mathbf{x}_0\, d\mathbf{x}_1. \tag{5.38}$$

The free propagator $K_0(\mathbf{x}_1 t_1; \mathbf{x}_0 t_0)$ is given by the 3-dimensional generalisation of (5.19), i.e.

$$K_0(\mathbf{x}_1 t_1; \mathbf{x}_0 t_0) = \theta(t_1 - t_0)\left[\frac{m}{i\hbar(t_1 - t_0)}\right]^{3/2} \exp\left[\frac{im}{2\hbar}\frac{(\mathbf{x}_0 - \mathbf{x}_1)^2}{t_1 - t_0}\right]. \tag{5.39}$$

$\mathscr{K}_0$ is then

$$\mathscr{K}_0(\mathbf{p}_1 t_1; \mathbf{p}_0 t_0) = \theta(t_1 - t_0)\left(\frac{m}{i\hbar(t_1 - t_0)}\right)^{3/2}$$

$$\times \int \exp\left[\frac{i}{\hbar}(\mathbf{p}_0\cdot\mathbf{x}_0 - \mathbf{p}_1\cdot\mathbf{x}_1)\right]$$

$$\times \exp\left[\frac{im}{2\hbar}\frac{(\mathbf{x}_0 - \mathbf{x}_1)^2}{t_1 - t_0}\right] d\mathbf{x}_0\, d\mathbf{x}_1.$$

To evaluate this integral we introduce the variables

$$\mathbf{x} = \mathbf{x}_0 - \mathbf{x}_1, \quad \mathbf{X} = \mathbf{x}_0 + \mathbf{x}_1, \quad \mathbf{p} = \mathbf{p}_0 - \mathbf{p}_1, \quad \mathbf{P} = \mathbf{p}_0 + \mathbf{p}_1$$

so that

$$2(\mathbf{p}_0\cdot\mathbf{x}_0 - \mathbf{p}_1\cdot\mathbf{x}_1) = \mathbf{P}\cdot\mathbf{x} + \mathbf{p}\cdot\mathbf{X}.$$

The Jacobian of the transformation is $(\frac{1}{2})^3 = \frac{1}{8}$, so we have

$$\mathscr{K}_0(\mathbf{p}_1, t_1; \mathbf{p}_0, t_0) = \theta(t_1 - t_0)\left(\frac{\alpha}{i\pi}\right)^{3/2}\frac{1}{8}\int \exp\left(\frac{i}{2\hbar}\mathbf{p}\cdot\mathbf{X}\right) d\mathbf{X}$$

$$\times \int \exp\left(\frac{i}{2\hbar}\mathbf{P}\cdot\mathbf{x}\right) e^{i\alpha x^2}\, d\mathbf{x}$$

where $\alpha = m/2\hbar(t_1 - t_0)$. The first integral is $8(2\pi\hbar)^3 \delta \mathbf{p} = 8(2\pi\hbar)^3 \delta(\mathbf{p}_0 - \mathbf{p}_1)$, so

$$\mathcal{K}_0(\mathbf{p}_1 t_1; \mathbf{p}_0 t_0) = (2\pi\hbar)^3 \theta(t' - t^0)\delta(\mathbf{p}_0 - \mathbf{p}_1)\left(\frac{\alpha}{i\pi}\right)^{3/2}$$

$$\times \int \exp\left(\frac{i}{2\hbar}\mathbf{p}\cdot\mathbf{x} + i\alpha\mathbf{x}^2\right) d\mathbf{x}.$$

The integral may be evaluated by appealing to equation (5A.3), giving

$$\mathcal{K}_0(\mathbf{p}_1 t_1; \mathbf{p}_0 t_0) = (2\pi\hbar)^3 \theta(t_1 - t_0)\delta(\mathbf{p}_0 - \mathbf{p}_1)\exp\left[\frac{-i\mathbf{P}^2(t_1 - t_0)}{8m\hbar}\right].$$

Note that the delta function implies that $\mathbf{p}_1 = \mathbf{p}_0$, and there is only propagation when momentum is conserved. Moreover, we then have $\mathbf{P}^2 = 4\mathbf{p}_0^2$, so finally

$$\mathcal{K}_0(\mathbf{p}_1 t_1; \mathbf{p}_0 t_0) = (2\pi\hbar)^3 \theta(t_1 - t_0)\delta(\mathbf{p}_0 - \mathbf{p}_1)$$

$$\times \exp\left[\frac{-i\mathbf{p}_0^2(t_1 - t_0)}{2m\hbar}\right]. \tag{5.40}$$

This propagator, as already noted, gives the amplitude for observing a particle with momentum $\mathbf{p}_1$ at time t_1, if one has been observed with momentum $\mathbf{p}_0$ at time t_0. The Fourier transform of this quantity is, of course, $K_0(\mathbf{x}_1 t_1; \mathbf{x}_0 t_0)$, given by the inverse of (5.38), which yields, on substituting (5.40),

$$K_0(\mathbf{x}_1 t_1; \mathbf{x}_0 t_0) = \frac{1}{(2\pi\hbar)^6}\int \exp\left(\frac{i}{\hbar}\mathbf{p}_1\cdot\mathbf{x}_1\right)\mathcal{K}_0(\mathbf{p}_1 t_1; \mathbf{p}_0 t_0)$$

$$\times \exp\left(-\frac{i}{\hbar}\mathbf{p}_0\cdot\mathbf{x}_0\right)d\mathbf{p}_1\, d\mathbf{p}_0$$

$$= \theta(t_1 - t_0)\frac{1}{(2\pi\hbar)^3}$$

$$\times \int \exp\left\{\frac{i}{\hbar}\left[\mathbf{q}\cdot(\mathbf{x}_1 - \mathbf{x}_0) - \frac{\mathbf{q}^2}{2m}(t_1 - t_0)\right]\right\}d\mathbf{q}. \tag{5.41}$$

We shall use this expression in the calculation of Coulomb scattering in the next section.

Finally, let us take the Fourier transform of the t dependence, so as to treat time and space in a symmetric manner. This is necessary for relativistic examples. The propagator we require is then

$$k_0(\mathbf{p}_1 E_1; \mathbf{p}_0 E_0) = \int \exp\left(\frac{i}{\hbar}E_1 t_1\right)\mathcal{K}_0(\mathbf{p}_1 t_1; \mathbf{p}_0 t_0) \times$$

$$\times \exp\left(-\frac{i}{\hbar}E_0 t_0\right) dt_0\, dt_1$$

$$= (2\pi\hbar)^3 \delta(\mathbf{p}_0 - \mathbf{p}_1) \int \theta(\tau) \exp\left(\frac{-i p_1^2}{2m\hbar}\tau\right)$$

$$\times \exp\left[\frac{i}{\hbar}(E_1 t_1 - E_0 t_0)\right] dt_0\, dt_1 \qquad (5.42)$$

where $\tau = t_1 - t_0$. Regarding τ and t_0 as the independent variables, this gives

$$k_0(\mathbf{p}_1 E_1; \mathbf{p}_0 E_0) = (2\pi\hbar)^3 \delta(\mathbf{p}_0 - \mathbf{p}_1) \int_{-\infty}^{\infty} \exp\left[\frac{i}{\hbar}(E_1 - E_0)t_0\right] dt_0$$

$$\times \int_{-\infty}^{\infty} \theta(\tau) \exp\left[\frac{i}{\hbar}\left(E - \frac{p_1^2}{2m}\right)\tau\right] d\tau.$$

The first of these integrals is $(2\pi\hbar)\delta(E_1 - E_0)$. The presence of the $\theta(\tau)$ function in the second integral means that it is of the form

$$\int_0^{\infty} e^{i\omega\tau}\, d\tau$$

which, if ω is real, does not converge. To make it converge, ω must be replaced by $\omega + i\varepsilon$, where ε is small and positive. The value of the integral is then $i/(\omega + i\varepsilon)$.[‡] Substituting for ω, we have finally

$$k_0(\mathbf{p}_1 E_1; \mathbf{p}_0 E_0) = (2\pi\hbar)^4 \delta(\mathbf{p}_0 - \mathbf{p}_1)\delta(E_0 - E_1)\frac{i\hbar}{E - \dfrac{p_1^2}{2m} + i\varepsilon}.$$

$$(5.43)$$

As may have been anticipated, this propagator yields energy conservation as well as momentum conservation. The limit $\varepsilon \to 0$ should be understood in equation (5.43). An important observation to make is that the energy E is not necessarily $p^2/2m$ for a particle described by wave mechanics. E and p are independent variables (used to define Fourier transforms from t and x space). It is only for a classical point particle, described in quantum theory by a wave packet of vanishing size, that $E = p^2/2m$. In this limiting case, the propagator above has a pole. Propagation takes place in general, however, for any value of E and p.

It is now straightforward, if tedious, to show that if we introduce the

---

[‡] Equivalently, the Fourier transform of $\theta(t)$ is given by

$$\theta(t) = \lim_{\varepsilon \to 0^+} \frac{1}{2\pi} \int e^{i\omega t} \frac{1}{\omega - i\varepsilon}\, d\omega$$

Fourier transform of the potential $V(\mathbf{x}, t)$ by

$$V(\mathbf{x}, t) = \int \exp\left[\frac{i}{\hbar}(\mathbf{q}\cdot\mathbf{x} - Wt)\right] v(\mathbf{q}, W)\, d\mathbf{q}\, dW \qquad (5.44)$$

then the amplitude (5.34) may be expressed in terms of k_0 and v, and summarised by the momentum-space diagram

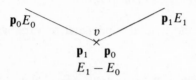

$$\begin{array}{cc} \mathbf{p}_1 & \mathbf{p}_0 \\ E_1 & - E_0 \end{array}$$

whose meaning is given by the *Feynman rules*

$$\left.\begin{array}{l} \dfrac{\mathbf{p}, E}{\rule{3cm}{0.4pt}} \qquad\qquad \dfrac{1}{(2\pi\hbar)^4}\dfrac{i\hbar}{E - \dfrac{p^2}{2m} + i\varepsilon}, \\[2em] \begin{array}{c} \diagdown\ \diagup \\ \times \\ \mathbf{q}, W \end{array}\qquad\qquad \dfrac{-i}{\hbar}(2\pi\hbar)^4 v(\mathbf{q}, W). \\[1em] \qquad\qquad\qquad\text{with energy and momentum} \\ \qquad\qquad\qquad\text{conservation.} \end{array}\right\} \qquad (5.45)$$

These are the Feynman rules in momentum space. The expression for the scattering amplitude A contains $\psi_{\text{out}}^*, \psi_{\text{in}}$ and integration over relevant variables.

5.3   Coulomb scattering

Let us now apply the theory developed above to the well-known problem of the scattering of charged spinless particles in a Coulomb field (Rutherford scattering). The scattering amplitude in the first Born approximation is given by (5.34)

$$A = \frac{-i}{\hbar}\int \psi_{\text{out}}^*(\mathbf{x}_1 t_1) K_0(\mathbf{x}_1 t_1; \mathbf{x}t) V(\mathbf{x}, t) K_0(\mathbf{x}t; \mathbf{x}_0 t_0)$$

$$\times \psi_{\text{in}}(\mathbf{x}_0 t_0)\, d\mathbf{x}_1\, d\mathbf{x}\, d\mathbf{x}_0\, dt$$

where $V(\mathbf{x}, t)$ represents the Coulomb potential. Now substitute for K_0 from (5.41), and for ψ_{out} and ψ_{in} from (5.32), giving

$$A = \frac{-i}{\hbar\tau}\frac{1}{(2\pi\hbar)^6}\int \exp\left[\frac{-i}{\hbar}\left(\mathbf{p}_f\cdot\mathbf{x}_1 - \frac{p_f^2}{2m}t_1\right)\right]$$

$$\times \exp\left\{\frac{i}{\hbar}\left[\mathbf{q}\cdot(\mathbf{x}_1 - \mathbf{x}) - \frac{q^2}{2m}(t_1 - t)\right]\right\} \times$$

$$\times \, V(\mathbf{x}, t) \exp\left\{\frac{i}{\hbar}\left[\mathbf{q}' \cdot (\mathbf{x} - \mathbf{x}_0) - \frac{q'^2}{2m}(t - t_0)\right]\right\}$$

$$\times \exp\left[\frac{i}{\hbar}\left(\mathbf{p}_i \cdot \mathbf{x}_0 - \frac{p_i^2}{2m}t_0\right)\right] d\mathbf{x}_1 \, d\mathbf{x} \, d\mathbf{x}_0 \, dt \, d\mathbf{q} \, d\mathbf{q}'.$$

Integration over $\mathbf{x}_1$ and $\mathbf{x}_0$ gives the delta functions

$$(2\pi\hbar)^3 \delta(\mathbf{p}_f - \mathbf{q}) \quad \text{and} \quad (2\pi\hbar)^3 \delta(\mathbf{q}' - \mathbf{p}_1).$$

Integration over $\mathbf{q}$ and $\mathbf{q}'$ then eliminates the terms in t_1 and t_0, giving

$$A = \frac{-i}{\hbar\tau} \int \exp\left\{\frac{i}{\hbar}\left[(\mathbf{p}_i - \mathbf{p}_f) \cdot \mathbf{x} - (E_i - E_f)t\right]\right\} V(\mathbf{x}, t) \, d\mathbf{x} \, dt$$

with $E_{i,f} = p_{i,f}^2/2m$. For the Coulomb potential $V = Ze^2/4\pi\varepsilon_0 r$ and integration over t then gives

$$A = \frac{-i}{\hbar\tau} 2\pi\delta\left(\frac{E_i - E_f}{\hbar}\right) \frac{Ze^2}{4\pi\varepsilon_0} \int \exp\left[\frac{i}{\hbar}(\mathbf{p}_i - \mathbf{p}_f) \cdot \mathbf{x}\right] \frac{1}{r} \, d\mathbf{x}.$$

The last integral does not converge at infinity, so a factor e^{-ar} is introduced, and, on letting $a \to 0$, the value of the integral is $4\pi\hbar^2/q^2$, where $\mathbf{q} = \mathbf{p}_i - \mathbf{p}_f$, so

$$A = \frac{-i}{\hbar\tau} \frac{2\pi Ze^2\hbar^2}{\varepsilon_0 q^2} \delta\left(\frac{E_i - E_f}{\hbar}\right). \tag{5.46}$$

This is the scattering amplitude, from which we want to calculate the scattering cross section σ. $|A|^2$ gives the probability that a particle emerges with momentum p_f. Assuming a box normalisation volume τ, then

$$|A|^2 \frac{\tau \, d\mathbf{p}_f}{(2\pi\hbar)^3}$$

gives the probability that a particle emerges with momentum between $\mathbf{p}_f$ and $\mathbf{p}_f + d\mathbf{p}_f$. If the interaction lasts an effective time T, then

$$\frac{|A|^2}{T} \tau \frac{d\mathbf{p}_f}{(2\pi\hbar)^3}$$

is the number of particles per second emerging in this momentum range. To get the cross section, we divide by the incident flux and integrate over $\mathbf{p}_f$. The incident particles travel with speed p_i/m and there are $1/\tau$ of them per unit volume, so the flux is $p_i/\tau m$ particle per second per unit area. The cross section is then

$$\sigma = \int \frac{|A|^2 \tau m}{T} \frac{\tau \, d\mathbf{p}_f}{(2\pi\hbar)^3}. \tag{5.47}$$

$|A|^2$ involves $|\delta(E_i - E_f/\hbar)|^2$. What is this? We appeal to the definition of $\delta(x)$:

$$\left| \delta\left(\frac{E_i - E_f}{\hbar} \right) \right|^2 = \lim_{T \to \infty} \left| \frac{1}{2\pi} \int_{-T/2}^{T/2} \exp[i(E_i - E_f)t/\hbar] \, dt \right|^2$$

$$= \lim_{T \to \infty} \left| \frac{\sin[(E_i - E_f)T/2\hbar]}{\pi(E_i - E_f)/\hbar} \right|^2$$

$$= \frac{T}{2\pi} \delta\left(\frac{E_i - E_f}{\hbar} \right)$$

$$= \frac{T\hbar}{2\pi} \delta(E_i - E_f) \tag{5.48}$$

where we have used the formula $\lim_{\alpha \to \infty} (\sin^2 \alpha x / \alpha x^2) = \pi \delta(x)$. Collecting together equations (5.46–8) gives

$$\sigma = \frac{mZ^2 e^4}{4\pi^2 \varepsilon_0^2} \int \frac{1}{p_i} \frac{1}{q^4} \delta(E_f - E_i) \, d^3 p_f$$

$$= \frac{mZ^2 e^4}{4\pi^2 \varepsilon_0^2} \int \frac{1}{p_i} \frac{1}{q^4} p_f^2 \, dp_f \delta(E_f - E_i) \, d\Omega.$$

Now use $E = p^2/2m$ to put $p_i = (2mE_i)^{\frac{1}{2}}$ and $p_f^2 \, dp_f = (2m^3 E_f)^{\frac{1}{2}} \, dE_f$, and integrate over E_f to give

$$\sigma = \frac{m^2 Z^2 e^4}{4\pi^2 \varepsilon_0^2} \int \frac{1}{q^4} \, d\Omega$$

where, because of the delta function, $p_i = p_f = p$. Hence $q^2 = 4p^2 \sin^2(\theta/2)$ where θ is the angle between $\mathbf{p}_i$ and $\mathbf{p}_f$. Finally, putting $p = mv$ we have the differential cross section

$$\frac{d\sigma}{d\Omega} = \left(\frac{Ze^2}{8\pi\varepsilon_0 mv^2} \right)^2 \frac{1}{\sin^4(\theta/2)} \tag{5.49}$$

which is the Rutherford formula.

5.4 Functional calculus: differentiation

Quantities like the propagator

$$\langle x_f t_f | x_i t_i \rangle = \int \mathscr{D}x \exp\left[\frac{i}{\hbar} \int_{t_i}^{t_f} L(x, \dot{x}) dt \right]$$

are *functional integrals*: the integration is taken over all functions $x(t)$. The left-hand side is a number, so the integral associates with each function $x(t)$, a *number*. The integral is called a *functional*, and clearly depends on *the value of the function* $x(t)$ *at all points.* We may write this in short hand:

functional: function → number. (5.50)

A *function,* for example $f(t) = t + 2t$, has a value (a *number*) for each value of the independent parameter, which is also a number. Given a value for t, we calculate the value of f. In short hand:

function: number → number. (5.51)

In mathematical notation, numbers belong to the space of *reals* $\mathbb{R}$, so a function defines a mapping

function: $\mathbb{R} \to \mathbb{R}$. (5.52)

Sometimes, of course, a function may be a vector quantity, like an electric field **E**, and therefore belong to $\mathbb{R}^3$; and it associates this electric field with every point of 3-dimensional space, and so is a mapping $\mathbb{R}^3 \to \mathbb{R}^3$. On the other hand, a scalar function $\phi(\mathbf{x})$ clearly defines a mapping $\mathbb{R}^3 \to \mathbb{R}$. In general, then, we have the definition

function $= \mathbb{R}^n \to \mathbb{R}^m$. (5.53)

Functions are continuous – to be precise, they are n-times differentiable. In physics, we generally concern ourselves with functions which are infinitely differentiable. The underlying co-ordinate space is a *manifold M* (for example $\mathbb{R}$, or $\mathbb{R}^3$ for 3-dimensional Euclidean space), and a function is denoted $C^n(M)$; and in the case of infinitely differentiable functions, $C^\infty(M)$. A functional, then, from (5.50), defines a mapping

functional: $C^\infty(M) \to \mathbb{R}$. (5.54)

It should by now be obvious, but nonetheless important to note, that a functional is *not* a function of a function, which is, of course, a function. It is common to denote a functional F of a function f by using square brackets, $F[f]$.

We now define functional differentiation. By analogy with ordinary differentiation, the derivative of the functional $F[f]$ with respect to the function $f(y)$ is defined by

$$\frac{\delta F[f(x)]}{\delta f(y)} = \lim_{\varepsilon \to \infty} \frac{F[f(x) + \varepsilon\delta(x-y)] - F[f(x)]}{\varepsilon}. \quad (5.55)$$

Let us take a specific example. Consider the functional

$$F[f] = \int f(x)\,dx. \quad (5.56)$$

Then

$$\frac{\delta F[f]}{\delta f(y)} = \lim_{\varepsilon \to \infty} \frac{1}{\varepsilon} \left\{ \int [f(x) + \varepsilon \delta(x - y)]\, dx - \int f(x)\, dx \right\}$$

$$= \int \delta(x - y)\, dx$$

$$= 1. \qquad (5.57)$$

As a second example, consider

$$F_x[f] = \int G(x, y) f(y)\, dy. \qquad (5.58)$$

Here, x on the left-hand side is to be regarded as a parameter. Then

$$\frac{\delta F_x[f]}{\delta f(z)} = \lim_{\varepsilon \to \infty} \frac{1}{\varepsilon} \left(\int \{G(x, y)[f(y) + \varepsilon \delta(y - z)]\}\, dy \right.$$

$$\left. - \int G(x, y) f(y)\, dy \right)$$

$$= \int G(x, y) \delta(y - z)\, dy$$

$$= G(x - z). \qquad (5.59)$$

5.5   Further properties of path integrals

We have shown that the transition amplitide from $q_i t_i$ to $q_f t_f$ is given by

$$\langle q_f t_f | q_i t_i \rangle = N \int \mathscr{D}q \exp\left[\frac{i}{\hbar} \int_{t_i}^{t_f} dt L(q, \dot{q}) \right]$$

in the case where $H = (p^2/2m) + V(q)$, which is sufficiently general for the present purposes, and the boundary conditions of the problem are

$$q(t_f) = q_f, \quad q(t_i) = q_i.$$

This type of boundary condition may be appropriate in the motion of classical particles, but it is not what we meet in field theory. Its analogue there would be, for example, $\psi(t_i) = \psi_i$, $\psi(t_f) = \psi_f$. But what really happens is that particles are *created* (for example, by collision), they interact, and are *destroyed* by observation (i.e. by detection). For example, in measuring the differential cross section $d\sigma/d\Omega$ for πN scattering, the pion is created by an NN collision, and it is destroyed when it is detected.

The act of creation may be represented as a source, and that of destruction by a sink, which is, in a manner of speaking, a source. The boundary conditions of the problem may then be represented as in Fig. 5.6; the vacuum at $t = -\infty$ evolves into the vacuum at $t \to \infty$, via the creation, interaction and destruction of a particle, through the agency of a source. We want to know the *vacuum-to-vacuum transition amplitude in the presence of a source*. This formulation, using the language of sources, is due to Schwinger[18]. The source $J(t)$ is represented by modifying the Lagrangian

$$L \to L + \hbar J(t)q(t). \tag{5.60}$$

If $|0, t\rangle^J$ is the ground state (vacuum) vector (in the moving frame) in the presence of the source, i.e. for a system described by (5.60), then the transition amplitude is

$$Z[J] \propto \langle 0, \infty | 0, -\infty \rangle^J \tag{5.61}$$

where a proportionality factor has been omitted. $Z[J]$ is a functional of J, and we now derive an expression for it, i.e. for the transition amplitude up to a constant factor. The salient feature is the presence of the ground state; how do we arrive at it?

The situation is represented by looking at the time axis in Fig. 5.7. What follows leans heavily on ref. (11). The source $J(t)$ is non-zero only between times t and t' $(t < t')$. T is an earlier time than t, and T' a later time than t', so the transition amplitude is

$$\langle Q'T'|QT\rangle^J = N \int \mathscr{D}q \exp\left[\frac{i}{\hbar} \int_T^{T'} dt(L + \hbar Jq)\right]. \tag{5.62}$$

Fig. 5.6. Representation of the vacuum–vacuum transition amplitude in the presence of a source.

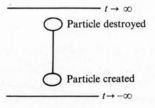

Fig. 5.7. Rotation of the time axis in calculating the vacuum–vacuum transition amplitude.

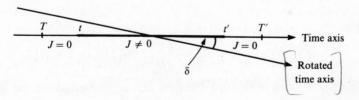

We may write

$$\langle Q'T'|QT\rangle^J = \int dq'\, dq\, \langle Q'T'|q't'\rangle\langle q't'|qt\rangle^J\langle qt|QT\rangle. \quad (5.63)$$

Referring to (5.3) we then have

$$\langle Q'T'|q't'\rangle = \left\langle Q'\left|\exp\left(-\frac{i}{\hbar}HT'\right)\exp\left(\frac{i}{\hbar}Ht'\right)\right|q'\right\rangle$$

$$= \sum_m \phi_m(Q')\phi_m^*(q')\exp\left[\frac{i}{\hbar}E_m(t'-T')\right], \quad (5.64)$$

where $\phi_m(q)$ are a complete set of energy eigenstates. Similarly,

$$\langle qt|QT\rangle = \sum_n \phi_n(q)\phi_n^*(Q)\exp\left[-\frac{i}{\hbar}E_n(t-T)\right]. \quad (5.65)$$

Now substitute these equations into (5.63). By taking the limit $T' \to \infty e^{-i\delta}$, $T \to -\infty e^{-i\delta}$, with δ an arbitrary angle $\leqslant \pi/2$ (see Fig. 5.7), we see that *only the ground state (vacuum) contribution survives*. This is the feature we want. We then have

$$\lim_{\substack{T'\to\infty e^{-i\delta}\\T\to-\infty e^{-i\delta}}} \langle Q'T'|QT\rangle^J = \phi_0^*(Q)\phi_0(Q')\exp\left[-\frac{i}{\hbar}E_0(T'-T)\right]$$

$$\times \int dq'\, dq\, \phi_0^*(q',t')\langle q't'|qt\rangle^J\phi_0(q,t)$$

or

$$\int dq'\, dq\, \phi_0^*(q',t')\langle q't'|qt\rangle^J\phi_0(q,t)$$

$$= \lim_{\substack{T'\to\infty e^{-i\delta}\\T\to-\infty e^{-i\delta}}} \frac{\langle Q'T'|QT\rangle^J}{\phi_0^*(Q)\phi_0(Q')\exp\left[-\frac{i}{\hbar}E_0(T'-T)\right]}. \quad (5.66)$$

The left-hand side is the ground state expectation value of the transition amplitude. The times t' and $-t$ may be taken as large as one likes, so the left-hand side becomes $\langle 0,\infty|0,-\infty\rangle^J$. The denominator on the right-hand side is simply a numerical factor, so we have

$$\langle 0,\infty|0,-\infty\rangle^J \propto \lim_{\substack{T'\to\infty e^{-i\delta}\\T\to-\infty e^{-i\delta}}} \langle Q'T'|QT\rangle^J \quad (5.67)$$

with

$$\langle Q'T'|QT\rangle^J = N\int \mathscr{D}Q\exp\left\{\frac{i}{\hbar}\int_T^{T'} dt[L(Q,\dot{Q})+\hbar JQ]\right\}.$$

Finally, instead of rotating the time axis as we have done, the ground state contribution may be isolated by adding a small negative imaginary part to the Hamiltonian in (5.64) and (5.65); adding $-\frac{1}{2}i\varepsilon q^2$ will achieve the result. This is equivalent to adding $+\frac{1}{2}i\varepsilon q^2$ to L, so we finally define $Z[J]$ in (5.61) by

$$\blacksquare \qquad Z[J] = \int \mathscr{D}q \exp\left[\frac{i}{\hbar}\int_{-\infty}^{\infty} dt(L + \hbar Jq + \tfrac{1}{2}i\varepsilon q^2)\right]$$

$$\propto \langle 0, \infty | 0, -\infty \rangle^J. \tag{5.68}$$

This expression for the transition amplitude will be taken over when we consider field theory, in the next chapter. Meanwhile, we shall prove one more relation, involving the functional derivatives of Z with respect to $J(t)$.

To begin, instead of $\langle q_f t_f | q_i t_i \rangle$, consider $\langle q_f t_f | q(t_{n_1}) | q_i t_i \rangle$, where $t_f > t_{n_1} > t_i$, and it should be remembered that $q(t_{n_1})$ is an operator. Consider equation (5.6), and choose t_{n_1} to be one of the times $t_1, \ldots, t_n$. Then

$$\langle q_f t_f | q(t_{n_1}) | q_i t_i \rangle = \int dq_1 \ldots dq_n \langle q_f t_f | q_n t_n \rangle \langle q_n t_n | q_{n-1} t_{n-1} \rangle$$

$$\ldots \langle q_{n_1} t_{n_1} | q(t_{n_1}) | q_{n_1-1} t_{n_1-1} \rangle \ldots \langle q_1 t_1 | q_i t_i \rangle.$$

The expression $\langle q_{n_1} t_{n_1} | q(t_{n_1}) | q_{n_1-1} t_{n_1-1} \rangle$ may clearly be replaced by $q(t_{n_1}) \langle q_{n_1} t_{n_1} | q_{n_1-1} t_{n_1-1} \rangle$, where this time $q(t_{n_1})$ is a scalar. The rest of the argument is analogous to that leading from (5.6) to (5.13), so we have, finally,

$$\langle q_f t_f | q(t_1) | q_i t_i \rangle = \int \frac{\mathscr{D}q \mathscr{D}p}{h} q(t_1)$$

$$\times \exp\left\{\frac{i}{\hbar}\int_{t_i}^{t_f} [p\dot{q} - H(p,q)]dt\right\}. \tag{5.69}$$

Next, suppose we want to find

$$\langle q_f t_f | q(t_{n_1}) q(t_{n_2}) | q_i t_i \rangle.$$

If $t_{n_1} > t_{n_2}$ we have

$$\langle q_f t_f | q(t_{n_1}) q(t_{n_2}) | q_i t_i \rangle$$

$$= \int dq_1 \ldots dq_n \langle q_f t_f | q_n t_n \rangle \ldots \langle q_{n_1} t_{n_1} | q(t_{n_1}) | q_{n_1-1} t_{n_1-1} \rangle$$

$$\ldots \langle q_{n_2} t_{n_2} | q(t_{n_2}) | q_{n_2-1} t_{n_2-1} \rangle \ldots \langle q_1 t_1 | q_i t_i \rangle,$$

giving, finally,

$$\langle q_f t_f | q(t_1) q(t_2) | q_i t_i \rangle = \int \frac{\mathscr{D}q \mathscr{D}p}{h} q(t_1) q(t_2)$$

$$\times \exp\left[\frac{i}{\hbar}\int_{t_i}^{t_f} (p\dot{q} - H)dt\right] \tag{5.70}$$

if $t_1 > t_2$. If, on the other hand, $t_2 > t_1$, this is not true; in that case, the right-hand side of (5.70) is equal to

$$\langle q_f t_f | q(t_2) q(t_1) | q_i t_i \rangle.$$

In general, then, the right-hand side of (5.70) is equal to

$$\langle q_f t_f | T[q(t_1) q(t_2)] | q_i t_i \rangle$$

where the *time ordering operator* T has the definition

$$T[A(t_1)B(t_2)] = \begin{cases} A(t_1)B(t_2) & \text{if } t_1 > t_2, \\ B(t_2)A(t_1) & \text{if } t_2 > t_1. \end{cases} \tag{5.71}$$

T has the effect of putting earlier times to the right. The result we have found generalises to

$$\langle q_f t_f | T[q(t_1) q(t_2) \ldots q(t_n)] | q_i t_i \rangle = \int \frac{\mathscr{D}q \mathscr{D}p}{h} q(t_1) q(t_2) \ldots q(t_n)$$

$$\times \exp \left\{ \frac{i}{\hbar} \int_{t_i}^{t_f} [p\dot{q} - H(p,q)] dt \right\}. \tag{5.72}$$

In the case where H is of the form (5.8), this becomes

$$\langle q_f t_f | T[q(t_1) q(t_2) \ldots q(t_n)] | q_i t_i \rangle = N \int \mathscr{D}q \, q(t_1) q(t_2) \ldots q(t_n)$$

$$\times \exp \left(\frac{i}{\hbar} \int_{t_i}^{t_f} L \, dt \right). \tag{5.73}$$

However, from the definition of $Z[J]$ in (5.68), its functional derivative with respect to J is

$$\frac{\delta Z[J]}{\delta J(t_1)} = i \int \mathscr{D}q \, q(t_1) \exp \left[\frac{i}{\hbar} \int_{-\infty}^{\infty} dt(L + \hbar Jq + \tfrac{1}{2} i \varepsilon q^2) \right]$$

and hence

$$\frac{\delta^n Z[J]}{\delta J(t_1) \ldots \delta J(t_n)} = i^n \int \mathscr{D}q \, q(t_1) \ldots q(t_n)$$

$$\times \exp \left[\frac{i}{\hbar} \int_{-\infty}^{\infty} dt(L + \hbar Jq + \tfrac{1}{2} i \varepsilon q^2) \right] \tag{5.74}$$

which gives, on putting $J = 0$,

$$\left. \frac{\delta^n Z[J]}{\delta J(t_1) \ldots \delta J(t_n)} \right|_{J=0} = i^n \int \mathscr{D}q \, q(t_1) \ldots q(t_n)$$

$$\times \exp \left[\frac{i}{\hbar} \int_{-\infty}^{\infty} dt(L + \tfrac{1}{2} i \varepsilon q^2) \right]. \tag{5.75}$$

Comparing the right-hand side of this equation with that of (5.73) above, we note that the difference lies in the $\frac{1}{2}i\varepsilon q^2$ term; but know, from what was said above, that it is this term that has the effect of isolating the ground state contribution, so we finish up with the vacuum expectation value of the time-ordered product:

$$\left. \frac{\delta^n Z[J]}{\delta J(t_1)\dots\delta J(t_n)}\right|_{J=0} \propto i^n \langle 0, \infty | T[q(t_1)\dots q(t_n)]|0, -\infty\rangle. \qquad (5.76)$$

This is the second result we wanted, and to which we shall have recourse in the next chapter.

Appendix: some useful integrals
We begin by stating a well-known formula

$$\int_{-\infty}^{\infty} e^{-\alpha x^2}\,dx = \left(\frac{\pi}{\alpha}\right)^{\frac{1}{2}}. \qquad (5A.1)$$

To prove it, is clearly the same as proving that

$$\int_{-\infty}^{\infty}\int_{-\infty}^{\infty} e^{-\alpha(x^2+y^2)}\,dx\,dy = \frac{\pi}{\alpha}, \qquad (5A.2)$$

and this is shown by going over to polar co-ordinates (r,θ):

$$\int_0^{2\pi}\int_0^{\infty} e^{-\alpha r^2} r\,dr\,d\theta = 2\pi \int_0^{\infty} e^{-\alpha r^2} r\,dr$$
$$= \pi \int_0^{\infty} e^{-\alpha r^2}\,d(r^2)$$
$$= \frac{\pi}{\alpha}.$$

Thus, (5A.1) is proved.

Now we pass from the integration of Gaussians to that of quadratic forms

$$\int_{-\infty}^{\infty} e^{-ax^2+bx+c}\,dx \equiv \int_{-\infty}^{\infty} e^{q(x)}\,dx.$$

Let $\bar{x}$ be the value of x giving a minimum of q:

$$\bar{x} = \frac{b}{2a}, \quad q(\bar{x}) = \frac{b^2}{4a} + c.$$

This allows us to 'uncomplete' the square:

$$q(x) = q(\bar{x}) - a(x - \bar{x})^2.$$

Hence

$$\int_{-\infty}^{\infty} e^{q(x)} dx = e^{q(\bar{x})} \int_{-\infty}^{\infty} e^{-a(x-\bar{x})^2} dx$$

$$= e^{q(\bar{x})} \left(\frac{\pi}{a}\right)^{\frac{1}{2}}$$

from (5A.1). Finally, we have

$$\int_{-\infty}^{\infty} \exp(-ax^2 + bx + c)dx = \exp\left(\frac{b^2}{4a} + c\right)\left(\frac{\pi}{a}\right)^{\frac{1}{2}}. \qquad (5A.3)$$

Lastly, we show that

$$\int_{-\infty}^{\infty} \exp\{i\lambda[(x_1 - a)^2 + (x_2 - x_1)^2 + \ldots + (b - x_n)^2]\}dx_1 \ldots dx_n$$

$$= \left[\frac{i^n \pi^n}{(n+1)\lambda^n}\right]^{\frac{1}{2}} \exp\left[\frac{i\lambda}{n+1}(b-a)^2\right]. \qquad (5A.4)$$

It is proved by induction; we assume it is true for *n*, and show it is true for *n* + 1. We have

$$\int_{-\infty}^{\infty} \exp\{i\lambda[(x_1 - a)^2 + \ldots + (b - x_{n+1})^2]\}dx_1 \ldots dx_{n+1}$$

$$= \left[\frac{i^n \pi^n}{(n+1)\lambda^n}\right]^{\frac{1}{2}} \int_{-\infty}^{\infty} \exp\left[\frac{i\lambda}{n+1}(x_{n+1} - a)^2\right]$$

$$\times \exp[i\lambda(b - x_{n+1})^2]dx_{n+1}$$

$$= \left[\frac{i^n \pi^n}{(n+1)\lambda^n}\right]^{2} \int_{-\infty}^{\infty} \exp i\lambda$$

$$\times \left[\frac{1}{n+1}(x_{n+1} - a)^2 + (b - x_{n+1})^2\right]dx_{n+1}.$$

Putting $x_{n+1} - a = y$, the term in square brackets becomes

$$\frac{1}{n+1}y^2 + (b - a - y)^2 = \frac{n+2}{n+1}y^2 - 2y(b-a) + (b-a)^2$$

$$= \frac{n+2}{n+1}\left[y - \frac{n+1}{n+2}(b-a)\right]^2$$

$$+ \frac{1}{n+2}(b-a)^2.$$

Now we put $\lambda - (n + 1/n + 2)(b - a) = z$ and find that the integral is

$$\left(\frac{i^n \pi^n}{(n+1)\lambda^n}\right)^{\frac{1}{2}} \int_{-\infty}^{\infty} \exp\left[i\lambda\frac{n+2}{n+1}z^2 + \frac{i\lambda}{n+2}(b-a)^2\right]dz$$

$$= \left[\frac{i^{n+1} \pi^{n+1}}{(n+2)\lambda^{n+1}}\right]^{\frac{1}{2}} \exp\left[\frac{i\lambda}{n+2}(b-a)^2\right]$$

which is (5A.4) with $n + 1$ instead of n. It only remains to show that the formula holds when $n = 1$. In this case the integral is

$$I = \int_{-\infty}^{\infty} \exp\{i\lambda[(x - a)^2 + (b - x)^2]\} \, dx$$

$$= \exp\left[i\frac{\lambda(a - b)^2}{2}\right]\left(\frac{i\pi}{2\lambda}\right)^{\frac{1}{2}}$$

where in the last step (5A.3) has been used, but with a in that equation imaginary. This value for I is the same as the value obtained from (5A.4) by putting $n = 1$. We have therefore proved (5A.4) for all n.

Summary

[1]Feynman's path integral formulation of quantum mechanics is explained, and [2]a perturbation series (the Born series) developed. The S matrix (for scattering of 'particles') is defined and it is shown how the resulting transition amplitudes may be obtained by reference to the 'Feynman rules'. These are written down in co-ordinate space and momentum space. [3]It is shown how the case of Coulomb scattering results in Rutherford's formula. [4]A brief account of functional differentiation is followed by [5]a demonstration that the scattering amplitude, written as a vacuum-to-vacuum transition amplitude in the presence of a source J, is a functional integral of J, and relates the vacuum expectation values of time-ordered products of operators to corresponding functional derivatives of this functional integral. The appendix proves the integrals needed in the course of the chapter.

Guide to further reading

The first papers on path-integral quantisation were
(1) P.A.M. Dirac, *Physikalisches Zeitschrift der Sowjet Union*, **3**, 64 (1933).
(2) R.P. Feyman, *Reviews of Modern Physics*, **20**, 367 (1948).

These are both reprinted in
(3) J. Schwinger (ed.), *Quantum Electrodynamics*, Dover Publications, 1958.

An expanded account is to be found in
(4) R.P. Feynman & A.R. Hibbs, *Quantum Mechanics and Path Integrals*, McGraw-Hill, 1965.

There are by now a number of good reviews of path-integral quantisation as used in physics, among which are the following.
(5) M.S. Marinov, *Physics Reports*, **60**, 1 (1980).
(6) C. Dewitt-Morette, A. Maheshwari & B. Nelson, *Physics Reports*, **50**, 255 (1979).
(7) L.S. Schulman, *Techniques and Applications of Path Integration*, John Wiley & Sons, 1981.
(8) T.D. Lee, *Particle Physics and Introduction to Field Theory*, chapter 19, Harwood Academic Publishers, 1981.

(9) J.R. Klauder in G.J. Papadopoulos & J.T. Devreese (eds.), *Path Integrals and their Applications in Quantum, Statistical and Solid State Physics*, Plenum Press, 1978.
(10) D.J. Blokhinstev & B.M. Barbashov, *Soviet Physics Uspekhi*, **15**, 193 (1972).
(11) E.S. Abers & B.W. Lee, *Physics Reports*, **9C**, 1 (1973).

Good introductions to the mathematical aspects of functional integration are
(12) I.M. Gel'fand & A.M. Yaglom, *Journal of Mathematical Physics*, **1**, 4 (1960).
(13) M. Kac, *Probability and Related Topics in Physical Sciences*, Interscience, 1959.
(14) J.B. Keller & D.W. McLaughlin, *American Mathematical Monthly*, **82**, 451 (1975).

For more rigorous treatment, see
(15) S.P. Gudder, *Stochastic Methods in Quantum Mechanics*, North Holland Publishing Co., 1979.
(16) B. Simon, *Functional Integration and Quantum Physics*, Academic Press, 1979.
(17) M.C. Reed in G. Velo & A.S. Wightman (eds.), *Constructive Quantum Field Theory* (Lecture Notes in Physics, 25), Springer-Verlag, 1973.

Schwinger's philosophy of sources is explained in, for example,
(18) J. Schwinger, *Particles and Sources*, Gordon & Breach, 1969.

6

Path-integral quantisation and Feynman rules: scalar and spinor fields

In this chapter we shall quantise scalar and spinor fields by path-integral quantisation, in analogy with the treatment of quantum mechanics in the last chapter. This will enable us to find the propagators for the scalar and spinor fields. We shall then introduce interactions, treat them perturbatively, and find the Feynman rules. After considering spinor fields in more detail, we conclude by calculating the pion–nucleon scattering cross section.

6.1 Generating functional for scalar fields[‡]

Suppose the scalar field $\phi(x)$ has a source, in the sense of §5.5, $J(x)$, then, analogously to expression (5.68), we may define the vacuum-to-vacuum transition amplitude in the presence of the source J as

$$Z[J] = \int \mathcal{D}\phi \exp\left\{ i \int d^4x \left[\mathcal{L}(\phi) + J(x)\phi(x) + \frac{i}{2}\varepsilon\phi^2 \right] \right\}$$

$$\propto \langle 0, \infty | 0, -\infty \rangle^J. \tag{6.1}$$

Here we have made the substitution $\mathcal{D}q(t) \to \mathcal{D}\phi(x^\mu)$, and have put $\hbar = 1$. $\mathcal{L}$ is the Klein–Gordon Lagrangian (3.10). So, instead of dividing time up into segments, we divide space and time up, and Minkowski space is broken down into 4-dimensional cubes of volume δ^4 in each of which ϕ is taken to be constant

$$\phi \sim \phi(x_i, y_j, z_k, t_l).$$

Derivatives are approximated by, for example,

$$\left.\frac{\partial\phi}{\partial x}\right|_{i,j,k,l} \approx \frac{1}{\delta}[\phi(x_i + \delta, y_j, z_k, t_l) - \phi(x_i, y_j, z_k, t_l)]$$

[‡] In this section and for much of this chapter I have drawn on the lectures of J. Wess, Karlsruhe University, 1974 (unpublished), and V.N. Popov, CERN preprint TH 2424, December, 1977 (unpublished).

Now let us replace the four indices (i, j, k, l) formally by one index n, and write

$$\mathscr{L}\left(\phi(x_i, y_j, z_k, t_l), \partial_\mu \phi(x_i, y_j, z_k, t_l)\right) = \mathscr{L}(\phi_n, \partial_\mu \phi_n)$$
$$= \mathscr{L}_n.$$

If i, j, k and l each take on N values, then n takes on N^4 values, so the action $S = \int \mathscr{L} \, \mathrm{d}^4 x$ becomes

$$S \approx \sum_{n=1}^{N^4} \delta^4 \mathscr{L}_n.$$

The vacuum-to-vacuum amplitude $Z[J]$ is then

$$Z[J] = \lim_{N \to \infty} \int \prod_{n=1}^{N^4} \mathrm{d}\phi_n(x) \exp \mathrm{i} \sum_{n=1}^{N^4} \delta^4 \left(\mathscr{L}_n + \phi_n J_n + \frac{\mathrm{i}}{2} \varepsilon \phi_n^2 \right). \quad (6.2)$$

Let us now calculate this for a free particle (field), for which

$$\mathscr{L}_0 = \tfrac{1}{2}(\partial_\mu \phi \, \partial^\mu \phi - m^2 \phi^2).$$

The corresponding vacuum-to-vacuum amplitude is (taking the limit $N \to \infty$)

$$Z_0[J] = \int \mathscr{D}\phi \exp\left(\mathrm{i} \int \{\tfrac{1}{2}[\partial_\mu \phi \, \partial^\mu \phi - (m^2 - \mathrm{i}\varepsilon)\phi^2] + \phi J\} \, \mathrm{d}^4 x \right). \quad (6.3)$$

We use the identity

$$\int \partial_\mu \phi \, \partial^\mu \phi \, \mathrm{d}^4 x = \int \partial_\mu (\phi \partial^\mu \phi) \mathrm{d}^4 x - \int \phi \, \square \, \phi \, \mathrm{d}^4 x,$$

and convert the first term on the right to a surface integral, using the 4-dimensional version of Gauss's theorem. This surface term vanishes if $\phi \to 0$ at infinity, so we have

$$\int \partial_\mu \phi \, \partial^\mu \phi \, \mathrm{d}^4 x = - \int \phi \, \square \, \phi \, \mathrm{d}^4 x, \quad (6.4)$$

giving

■ $$Z_0[J] = \int \mathscr{D}\phi \exp\left\{ -\mathrm{i} \int [\tfrac{1}{2}\phi(\square + m^2 - \mathrm{i}\varepsilon)\phi - \phi J] \mathrm{d}^4 x \right\}. \quad (6.5)$$

(Note that the field ϕ in this generating functional does *not* obey the Klein–Gordon equation (3.8).) To evaluate $Z_0[J]$, let us change ϕ to

$$\phi(x) \to \phi(x) + \phi_0(x). \quad (6.6)$$

Using the fact that

$$\int \phi_0(\square + m^2 - \mathrm{i}\varepsilon)\phi \, \mathrm{d}^4 x = \int \phi(\square + m^2 - \mathrm{i}\varepsilon)\phi_0 \, \mathrm{d}^4 x,$$

which follows from an argument analogous to that leading to (6.4), we have, under (6.6),

$$\int [\tfrac{1}{2}\phi(\Box + m^2 - i\varepsilon)\phi - \phi J]\,d^4x$$

$$\to \int [\tfrac{1}{2}\phi(\Box + m^2 - i\varepsilon)\phi + \phi(\Box + m^2 - i\varepsilon)\phi_0$$
$$+ \tfrac{1}{2}\phi_0(\Box + m^2 - i\varepsilon)\phi_0 - \phi J - \phi_0 J]\,d^4x.$$

If ϕ_0 is now chosen to satisfy

$$(\Box + m^2 - i\varepsilon)\phi_0(x) = J(x) \tag{6.7}$$

then this becomes

$$\int [\tfrac{1}{2}\phi(\Box + m^2 - i\varepsilon)\phi - \tfrac{1}{2}\phi_0 J]\,d^4x. \tag{6.8}$$

Now the solution to (6.7) is

$$\phi_0(x) = - \int \Delta_F(x - y)J(y)\,d^4y \tag{6.9}$$

where $\Delta_F(x - y)$ is the so-called Feynman propagator, obeying

$$(\Box + m^2 - i\varepsilon)\Delta_F(x) = -\delta^4(x). \tag{6.10}$$

Substituting (6.9) into (6.8), we see that the exponent in (6.5) is $-i$ times

$$\tfrac{1}{2}\int \phi(\Box + m^2 - i\varepsilon)\phi\,d^4x + \tfrac{1}{2}\int J(x)\Delta_F(x - y)J(y)\,d^4x\,d^4y. \tag{6.11}$$

So $Z_0[J]$ now takes the form (where dx stands for d^4x, and similarly for y)

$$Z_0[J] = \exp\left[-\frac{i}{2}\int J(x)\Delta_F(x - y)J(y)\,dx\,dy \right]$$
$$\times \int \mathscr{D}\phi \exp\left[-\frac{i}{2}\int \phi(\Box + m^2 - i\varepsilon)\phi\,dx \right]. \tag{6.12}$$

The superiority of this expression to (6.5) lies in the fact that here $Z_0[J]$ has separated into two factors, one depending on ϕ only, and the other on J only. In fact, the integral involving ϕ is actually a *number*, let us call it N, since the integral has been taken over all functions ϕ. Finally, then, we have

$$\blacksquare \qquad Z_0[J] = N \exp\left[-\frac{i}{2}\int J(x)\Delta_F(x - y)J(y)\,dx\,dy \right]. \tag{6.13}$$

Since we are only interested in normalised transition amplitudes, the value of N, in the applications we consider, is irrelevant.

The aim of this section, to derive equation (6.13) for the vacuum-to-vacuum transition amplitude, has now been achieved. In the next section we shall show how the same equation is derived using rather higher brow mathematical techniques of functional integration. Before finishing this section, however, we shall consider briefly the Feynman propagator $\Delta_F(x)$, defined by (6.10). It is easy to see that $\Delta_F(x)$ has a Fourier representation

$$\Delta_F(x) = \frac{1}{(2\pi)^4} \int d^4k \frac{e^{-ikx}}{k^2 - m^2 + i\varepsilon}. \tag{6.14}$$

Note that the presence of the $i\varepsilon$ term, which was put in originally (see (6.1)) to ensure vacuum-to-vacuum boundary conditions, dictates the path of integration round the poles at $k_0 = \pm(\mathbf{k}^2 + m^2)^{\frac{1}{2}}$. In fact, the poles are at $k_0^2 = \mathbf{k}^2 + m^2 - i\varepsilon$, and therefore at

$$k_0 = \pm(\mathbf{k}^2 + m^2)^{\frac{1}{2}} \mp i\delta = \pm E \mp i\delta. \tag{6.15}$$

This is shown in Fig. 6.1, where the integration path of k_0 is along its real axis, as shown. In the limit $\delta \to 0$, i.e. $\varepsilon \to 0$, which is implied in the expression (6.14), the poles reach the real axis, and in this case the integration path is as shown in Fig. 6.2.

There is another way of incorporating the vacuum-to-vacuum boundary conditions, which is to rotate the time axis, instead of through the small

Fig. 6.1. Integration path along the real k_0 axis in the definition of $\Delta_F(x)$.

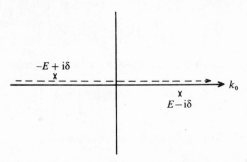

Fig. 6.2. New integration path on taking the limit $\varepsilon \to 0 (\delta \to 0)$.

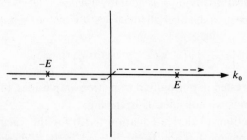

angle δ, shown in Fig. 5.7, through an angle of $\pi/2$, so that $t \to -i\infty$. Defining

$$x_4 = it = ix_0 \tag{6.16}$$

this limit is $x_4 \to \infty$. This space–time, with an imaginary time axis, is *Euclidean*, for the invariant interval is

$$ds^2 = (dx^0)^2 - (dx^1)^2 - (dx^2)^2 - (dx^3)^2$$

$$= - \sum_{\mu=1}^{4} (dx^\mu)^2.$$

Defining, in addition,

$$k_4 = -ik_0, \tag{6.17}$$

giving, in the Euclidean space,

$$k^2 = -(k_1^2 + k_2^2 + k_3^2 + k_4^2) = -k_E^2 \tag{6.18}$$

and

$$d^4k_E = d^3k \, dk_4 = -i \, d^4k,$$

the Feynman propagator is

$$\Delta_F(x) = \frac{-i}{(2\pi)^4} \int d^4k_E \frac{e^{-ikx}}{k_E^2 + m^2}. \tag{6.19}$$

Here there is no problem of choosing a contour, for the poles are not on the real axis, but at $k_4 = \pm i(\mathbf{k}^2 + m^2)^{\frac{1}{2}}$. Referring to equation (6.3) for $Z_0[J]$, noting that $d^4x = -i \, d^4x_E$ and that $(\partial^\mu \phi)^2 = -(\partial_E^\mu \phi)^2$, the Euclidean transition amplitude is

$$Z_{0_E}[J] = \int \mathscr{D}\phi \exp\left(-\int \{\tfrac{1}{2}[(\partial_E^\mu \phi)^2 + m^2\phi^2] - \phi J\} d^4x_E \right). \tag{6.20}$$

The exponent in the integrand is negative definite, so the integrand converges; the role of the ε term in (6.3) was, in effect, to make the integrand converge.

6.2 Functional integration

We shall now generalise the usual formulae for Gaussian integration over a finite number of variables, to formulae for functional integrals, and then show how (6.12) (or (6.13)) follows from (6.5). To begin, we have, from (5A.1),

$$\int e^{-\frac{1}{2}ax^2} dx = \left(\frac{2\pi}{a} \right)^{\frac{1}{2}} \tag{6.21}$$

(the limits $-\infty$ and ∞ are to be understood in this, and all subsequent, integrals). Now we take the product of n such integrals

$$\int \exp\left(-\tfrac{1}{2}\sum_n a_n x_n^2\right) dx_1 \ldots dx_n = \frac{(2\pi)^{n/2}}{\prod_{i=1}^{n} a_i^{\frac{1}{2}}}. \tag{6.22}$$

Let A be a diagonal matrix with elements $a_1, \ldots, a_n$, and let x be an n-vector $(x_1, \ldots, x_n)$. Then the exponent above is the inner product

$$\sum_n a_n x_n^2 = (x, Ax)$$

and the determinant of A is

$$\det A = a_1 a \ldots a_n = \prod_{i=1}^{n} a_i.$$

Equation (6.22) then becomes

$$\int e^{-\frac{1}{2}(x, Ax)} d^n x = (2\pi)^{n/2} (\det A)^{-\frac{1}{2}}. \tag{6.23}$$

Since this holds for any diagonal matrix, it also holds for any real symmetric matrix. Defining the measure

$$(dx) = d^n x (2\pi)^{-n/2},$$

equation (6.23) becomes

$$\int e^{-\frac{1}{2}(x, Ax)} (dx) = (\det A)^{-\frac{1}{2}}. \tag{6.24}$$

This equation may be extended to quadratic forms

$$Q(x) = \tfrac{1}{2}(x, Ax) + (b, x) + c. \tag{6.25}$$

The minimum of Q lies at $\bar{x} = -A^{-1}b$ and

$$Q(x) = Q(\bar{x}) + \tfrac{1}{2}[x - \bar{x}, A(x - \bar{x})]$$

so we have

$$\int \exp\{-[\tfrac{1}{2}(x, Ax) + (b, x) + c]\}(dx)$$
$$= \exp[\tfrac{1}{2}(b, A^{-1}b) - c](\det A)^{-\frac{1}{2}}. \tag{6.26}$$

This equation is analogous to (5A.3).

Let us make a small digression to the case of Hermitian matrices. Squaring (6.21) gives

$$\int e^{-\frac{1}{2}a(x^2 + y^2)} dx \, dy = \frac{2\pi}{a}.$$

Putting $z = x + iy$, $z^* = x - iy$, $dx\,dy = -\frac{1}{2}i\,dz^*\,dz$, we have

$$\int e^{-az^*z} \frac{dz^*}{(2\pi i)^{\frac{1}{2}}} \frac{dz}{(2\pi i)^{\frac{1}{2}}} = \frac{1}{a}. \tag{6.27}$$

The generalisation of this formula is that if A is a positive definite Hermitian matrix and we define the measure $(dz) = d^n z (2\pi i)^{-n/2}$, then

$$\int e^{-(z^*, Az)}(dz^*)(dz) = (\det A)^{-1}. \tag{6.28}$$

The formulae written down so far are rigorously true; they are simply the result of generalising the integration space from one dimension to a finite-dimensional vector space. We now assume that we can generalise the above formulae to an infinite-dimensional function space. This actually needs a careful mathematical justification, but, assuming that this can be made, then, if the generalisation is to the case of a single real variable $f(t)$, the inner product (f, f) is

$$(f, f) = \int [f(t)]^2 \, dt.$$

In our case, we are concerned with real functions of space–time, $\phi(x^\mu)$, so we have

$$(\phi, \phi) = \int [\phi(x)]^2 \, d^4x. \tag{6.29}$$

The generalisation of equation (6.24) is

$$\int \mathscr{D}\phi \exp\left[-\frac{1}{2} \int \phi(x) A \phi(x) \, dx \right] = (\det A)^{-\frac{1}{2}}; \tag{6.30}$$

A may be, in general, a differential operator. If ϕ is a complex field, we instead generalise (6.28) to give

$$\int \mathscr{D}\phi^* \mathscr{D}\phi \exp\left[-\int \phi^*(x) A \phi(x) \, dx \right] = (\det A)^{-1}. \tag{6.31}$$

We are now in a position to prove (6.12) from (6.5). The exponent of the integrand of (6.5) is a quadratic form, so we employ equation (6.26) (or, rather, its functional generalisation), with $A = i(\Box + m^2 - i\varepsilon)$, $b = -iJ$, $c = 0$, giving

$$Z_0[J] = \exp\left[\frac{i}{2} \int J(x)(\Box + m^2 - i\varepsilon)^{-1} J(y) \, dx \, dy \right]$$
$$\times [\det i(\Box + m^2 - i\varepsilon)]^{-\frac{1}{2}}.$$

The determinant term is defined by (6.30), and we have, from (6.10),

$$(\Box + m^2 - i\varepsilon)^{-1} = -\Delta_F(x - y), \qquad (6.32)$$

so

$$Z_0[J] = \exp\left[-\frac{i}{2} \int J(x)\Delta_F(x - y)J(y)\mathrm{d}x\,\mathrm{d}y \right]$$

$$\times \int \mathscr{D}\phi \exp\left[-\frac{i}{2} \int \phi(\Box + m^2 - i\varepsilon)\phi\,\mathrm{d}x \right]$$

which is equation (6.12). We recall that the last factor is just a number N, so $Z_0[J]$ can be written in the form (6.13).

6.3   Free particle Green's functions

We shall now show that the amplitude $Z_0[J]$ is the 'generating functional' for the free particle Green's functions, which will be defined as we proceed. We begin by expanding equation (6.13), to give

$$Z_0[J] = N\left\{ 1 - \frac{i}{2} \int J(x)\Delta_F(x - y)J(y)\,\mathrm{d}x\,\mathrm{d}y \right.$$

$$+ \frac{1}{2!}\left(\frac{i}{2}\right)^2 \left[\int J(x)\Delta_F(x - y)J(y)\mathrm{d}x\,\mathrm{d}y \right]^2$$

$$\left. + \frac{1}{3!}\left(\frac{-i}{2}\right)^3 \left[\int J(x)\Delta_F(x - y)J(y)\mathrm{d}x\,\mathrm{d}y \right]^3 + \cdots \right\}. \qquad (6.33)$$

Introducing the Fourier transform of $J(x)$ by

$$J(x) = \int J(p)\mathrm{e}^{-ipx}\mathrm{d}^4p, \qquad (6.34)$$

we have, using (6.14),

$$-\frac{i}{2}\int J(x)\Delta_F(x - y)J(y)\mathrm{d}^4x\,\mathrm{d}^4y$$

$$= \frac{-i}{2(2\pi)^4} \int \frac{J(p_1)\mathrm{e}^{-i(p_1+k)x}\mathrm{e}^{-i(p_2-k)y}J(p_2)}{k^2 - m^2 + i\varepsilon}$$

$$\times \mathrm{d}^4p_1\,\mathrm{d}^4p_2\,\mathrm{d}^4k\,\mathrm{d}^4x\,\mathrm{d}^4y$$

$$= -\frac{i}{2}(2\pi)^4 \int \frac{J(-k)J(k)}{k^2 - m^2 + i\varepsilon}\mathrm{d}^4k \qquad (6.35)$$

in which we have integrated over x and y to give two delta functions, and then over p_1 and p_2. We may represent this diagrammatically by using the following rules of correspondence (Feynman rules in momentum space).

$$\left.\begin{array}{l} \dfrac{p}{} \quad \dfrac{1}{(2\pi)^4} \dfrac{i}{p^2 - m^2 + i\varepsilon}, \\[3mm] \times\!\!\dfrac{p}{J} \quad i(2\pi)^4 J(p). \end{array}\right\} \tag{6.36}$$

These Feynman rules may be compared with those in (5.45). Since the non-relativistic limit of $p^2 - m^2$ is $2m(T - \mathbf{p}^2/2m)$ ($T =$ kinetic energy), the propagator above is clearly the (relativistic) propagator for *one particle*. The expression (6.35) then corresponds to the diagram

$$\tfrac{1}{2} \underset{J \qquad\qquad J}{\times\!\!-\!\!\!-\!\!\!-\!\!\!-\!\!\!-\!\!\times} \tag{6.37}$$

The vacuum-to-vacuum amplitude (6.33) may then be written in terms of Feynman diagrams (ignoring N)

$$Z_0 = 1 + \frac{1}{2}\, \times\!\!-\!\!\!-\!\!\!-\!\!\times + \frac{1}{2!}\left(\frac{1}{2}\right)^2 \begin{array}{c} \times\!\!-\!\!\!-\!\!\times \\ \times\!\!-\!\!\!-\!\!\times \end{array}$$

$$+ \frac{1}{3!}\left(\frac{1}{2}\right)^3 \begin{array}{c} \times\!\!-\!\!\!-\!\!\times \\ \times\!\!-\!\!\!-\!\!\times \\ \times\!\!-\!\!\!-\!\!\times \end{array} + \cdots$$

$$= 1 + \frac{1}{2}\ \underset{-\infty}{\overset{\infty}{\Bigg|}} \quad + \frac{1}{2!}\left(\frac{1}{2}\right)^2\ \underset{-\infty}{\overset{\infty}{\Bigg|\Bigg|}}$$

$$+ \cdots \tag{6.38}$$

where, in the last line, we have resorted to a pictorial representation reminiscent of Fig. 5.6. We shall now show that the implied suggestion is correct; in other words, it is correct to interpret this series as the propagation of one particle between sources, the propagation of two particles between sources, and so on. We therefore have a *many-particle theory*, consistent with our original philosophy of using a field, which yielded particles on quantisation. Each term in the above series is a Green's function so $Z_0[J]$ is a *generating functional* for the Green's functions of the theory.

To understand how to interpret the power series expansion of a functional, let us first recall the formula for the power series expansion of a function, say $F(y_1, \ldots, y_k)$ of k variables $y_1, \ldots, y_k$. It is

$$F\{y\} = F(y_1, \ldots, y_k)$$

$$= \sum_{n=0}^{\infty} \sum_{i_1=0}^{k} \cdots \sum_{i_n=0}^{k} \frac{1}{n!}\, T_n(i_1, \ldots, i_n) y_{i_1}, \ldots, y_{i_n}$$

where

$$T_n = \frac{\partial^n F\{y\}}{\partial y_1 \dots \partial y_n}\bigg|_{y=0}. \tag{6.39}$$

Going over to the case of continuously many variables, $i \to x_i$, y_i $(i = 1, \dots, k) \to y(x)$, $-\infty < x < \infty$, and $\sum_i \to \int dx$, and we obtain the power series expansion of a functional

$$F[y] = \sum_{n=0}^{\infty} \int dx_1 \dots dx_n \frac{1}{n!} T_n(x_1, \dots, x_n) y(x_1) \dots y(x_n)$$

where

$$T_n(x_1, \dots, x_n) = \frac{\delta}{\delta y(x)} \dots \frac{\delta}{\delta y(x_n)} F[y]\bigg|_{y=0}. \tag{6.40}$$

$F[y]$ is called the *generating functional* of the functions $T_n(x_1, \dots, x_n)$.

Now we return to $Z_0[J]$. It is convenient here to settle the question of its normalisation. $Z[J]$ is the vacuum-to-vacuum transition amplitude in the presence of the source, J, so it is sensible to normalise it to $Z[J = 0] = 1$. In that case, referring to (5.61), we may *define* $Z[J]$ by

$$Z[J] = \langle 0, \infty | 0, -\infty \rangle^J \tag{6.41}$$

which automatically obeys

$$Z[0] = 1. \tag{6.42}$$

Equations (6.5) and (6.13) must then be rewritten as

$$Z_0[J] = \frac{\int \mathcal{D}\phi \exp\left\{ -i \int [\tfrac{1}{2}\phi(\Box + m^2 - i\varepsilon)\phi - \phi J] \, dx \right\}}{\int \mathcal{D}\phi \exp\left[-i \int \tfrac{1}{2}\phi(\Box + m^2 - i\varepsilon)\phi \, dx \right]} \tag{6.43}$$

and

$$Z_0[J] = \exp\left[-\frac{i}{2} \int J(x)\Delta_F(x - y)J(y) \, dx \, dy \right]. \tag{6.44}$$

These new definitions clearly obey (6.42). Then $Z_0[J]$, as defined by (6.44), is clearly the generating functional for

$$\tau(x_1, \dots, x_n) = \frac{1}{i^n} \frac{\delta^n Z_0[J]}{\delta J(x_1) \dots \delta J(x_n)}\bigg|_{J=0}. \tag{6.45}$$

At this point we refer back to equation (5.76), which was derived for its use here. The analogous equation in the case of fields, and using our new normalisation, is

$$\frac{\delta^n Z_0[J]}{\delta J(x_1) \dots J(x_n)}\bigg|_{J=0} = i^n \langle 0 | T(\phi(x_1) \dots \phi(x_n)) | 0 \rangle \tag{6.46}$$

or

$$\langle 0|T(\phi(x_1)\dots\phi(x_n))|0\rangle = \frac{1}{i^n}\frac{\delta^n Z_0[J]}{\delta J(x_1)\dots\delta J(x_n)}\bigg|_{J=0}. \qquad (6.47)$$

Comparing (6.45) and (6.47) we have

$$\tau(x_1,\dots,x_n) = \langle 0|T(\phi(x_1)\dots(x_n))|0\rangle. \qquad (6.48)$$

These quantities, the vacuum expectation values of time-ordered products of field operators, are called the *Green's functions* or *n*-point *functions* of the theory. They are closely related to the S matrix elements – we shall see this connection later, when we introduce interactions. We then have

$$Z_0[J] = \sum_{n=0}^{\infty}\frac{(-i/2)^n}{n!}\int dx_1\dots dx_n J(x_1)\dots J(x_n)\tau(x_1,\dots,x_n) \qquad (6.49)$$

which expresses Z as the generating functional of the Green's functions τ. This equation corresponds to the pictorial equation (6.38).

Let us now calculate some *n*-point functions. We start with the 2-point function

$$\tau(x,y) = -\frac{\delta^2 Z_0[J]}{\delta J(x)\delta J(y)}\bigg|_{J=0} \qquad (6.50)$$

with $Z_0[J]$ given by (6.44). (Recall again that we are still concerned with a *free* field theory, since we started with the free Lagrangian, and thus with (6.3). The expressions we shall find below refer, therefore, to the free particle Green's functions. The corresponding Green's functions for interacting fields will differ from these, and will be found later.)

From (6.44) we have

$$\frac{1}{i}\frac{\delta Z_0[J]}{\delta J(x)} = \frac{1}{i}\frac{\delta}{\delta J(x)}\exp\left[-\frac{i}{2}\int dx_1\,dx_2 J(x_1)\Delta_F(x_1-x_2)J(x_2)\right]$$

$$= -\int\Delta_F(x-x_1)J(x_1)\,dx_1$$

$$\times\exp\left[-\frac{i}{2}\int dx_1\,dx_2 J(x_1)\Delta_F(x_1-x_2)J(x_2)\right],$$

$$\frac{1}{i}\frac{\delta}{\delta J(x)}\frac{1}{i}\frac{\delta}{\delta J(y)}Z_0[J] = i\Delta_F(x-y)\exp\left(-\frac{i}{2}\int J\Delta_F J\right)$$

$$+\int\Delta_F(x-x_1)J(x_1)\,dx_1\int\Delta_F(y-x_1)J(x_1)\,dx_1$$

$$\times\exp\left(-\frac{i}{2}\int J\Delta_F J\right). \qquad (6.51)$$

Here we have abbreviated the notation in the exponent, for simplicity. Finally, putting $J = 0$,

$$\frac{1}{i}\frac{\delta}{\delta J(x)}\frac{1}{i}\frac{\delta}{\delta J(y)}Z_0[J]\bigg|_{J=0} = i\Delta_F(x - y),$$

or

$$\tau(x, y) = i\Delta_F(x - y). \tag{6.52}$$

What is the physical significance of this? From (6.48) the 2-point function is

$$\begin{aligned}\tau(x, y) &= \langle 0|T(\phi(x)\phi(y))|0\rangle \\ &= \langle 0|\theta(x_0 - y_0)\phi(x)\phi(y) \\ &\quad + \theta(y_0 - x_0)\phi(y)\phi(x)|0\rangle.\end{aligned} \tag{6.53}$$

From (4.14), we may decompose ϕ into its positive and negative frequency parts

$$\phi(x) = \phi^{(+)}(x) + \phi^{(-)}(x) \tag{6.54}$$

with

$$\phi^{(+)}(x) = \int \frac{d^3k}{[(2\pi)^3 2\omega_k]^{\frac{1}{2}}} f_k(x)a(k)$$

$$\phi^{(-)}(x) = \int \frac{d^3k}{[(2\pi)^3 2\omega_k]^{\frac{1}{2}}} f_k^*(x)a^\dagger(k)$$

and where $f_k(x)$ is given by (4.11). Because $a(k)$ and $a^\dagger(k)$ are annihilation and creation operators, the only terms to survive the vacuum expectation value in (6.53) are $\phi^{(+)}(x)\phi^{(-)}(y)$, so that

$$\begin{aligned}\tau(x, y) &= \theta(x_0 - y_0)\langle 0|\phi^{(+)}(x)\phi^{(-)}(y)|0\rangle \\ &\quad + \theta(y_0 - x_0)\langle 0|\phi^{(+)}(y)\phi^{(-)}(x)|0\rangle.\end{aligned} \tag{6.55}$$

The first term is the amplitude for creating a particle at y, at time y_0, and destroying it at x, at time x_0 $(> y_0)$. The second term is the amplitude for creating a particle at x, at time x_0, and destroying it at y, at time y_0 $(> x_0)$. These are represented schematically in Fig. 6.3. We shall now verify that the sum of these terms is the Feynman propagator $i\Delta_F(x - y)$.

Fig. 6.3. Interpretation of equation (6.54). (See text.)

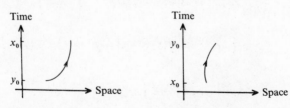

To prove this, let us first re-express $\Delta_F(x - y)$. From (6.14)

$$\Delta_F(x) = \int \frac{d^4k}{(2\pi)^4} \frac{e^{-ikx}}{k^2 - m^2 + i\varepsilon}$$

$$= \int \frac{d^3k \, dk_0}{(2\pi)^4} \frac{e^{-ikx}}{k_0^2 - (\mathbf{k}^2 + m^2) + i\varepsilon}$$

$$= \int \frac{d^3k \, dk_0}{(2\pi)^4} \frac{e^{-ikx}}{2\omega_k} \left(\frac{1}{k_0 - \omega_k + i\delta} - \frac{1}{k_0 + \omega_k - i\delta} \right)$$

where $\omega_k^2 = \mathbf{k}^2 + m^2$. The integration path over k_0 is shown in Fig. 6.1. The exponential contains $e^{-ik_0 x_0}$, so, when $x_0 > 0$, we may complete the integral in the lower half of the k_0 plane, and the contribution on the semicircle at infinity will vanish; the path encloses the pole at $k_0 = \omega_k - i\delta$. On the other hand, when $x_0 < 0$, we complete in the upper half plane and enclose the pole at $k_0 = -\omega_k + i\delta$. Applying Cauchy's theorem then gives

$$\Delta_F(x) = \int \frac{d^3k}{(2\pi)^3} \frac{e^{ik \cdot x}}{2\omega_k}$$
$$\times [\theta(x_0)(-i)e^{-i\omega_k x_0} - \theta(-x_0)i e^{i\omega_k x_0}].$$

In the second integral, we may change $\mathbf{k}$ to $-\mathbf{k}$ without affecting the value of the integral, so, finally,

$$\Delta_F(x - y) = -i \int \frac{d^3k}{(2\pi)^3} \frac{1}{2\omega_k}$$
$$\times [\theta(x_0 - y_0)e^{-ik(x-y)} + \theta(y_0 - x_0)e^{ik(x-y)}]. \tag{6.56}$$

This is the form we want for $\Delta_F(x - y)$. Now we substitute (6.54) into (6.55), giving

$$\tau(x, y) = \int \frac{d^3k \, d^3k'}{(2\pi)^3 (2\omega_k 2\omega_{k'})^{\frac{1}{2}}} [\theta(x_0 - y_0)\langle 0|a(k)f_k(x)a^\dagger(k')f_{k'}^*(y)|0\rangle$$
$$+ \theta(y_0 - x_0)\langle 0|a(k)f_k(y)a^\dagger(k')f_{k'}^*(x)|0\rangle]$$

$$= \int \frac{d^3k \, d^3k'}{(2\pi)^6 2\omega_k 2\omega_{k'}} [\theta(x_0 - y_0)e^{-i(kx - k'y)}\langle 0|a(k)a^\dagger(k')|0\rangle$$
$$+ \theta(y_0 - x_0)e^{-i(ky - k'x)}\langle 0|a(k)a^\dagger(k')|0\rangle]$$

where we have used (4.11) for $f_k(x)$. Now we use the commutation relation (4.16a), to find

$$\tau(x, y) = \int \frac{d^3k}{(2\pi)^3 2\omega_k} [\theta(x_0 - y_0)e^{-ik(x-y)} + \theta(y_0 - x_0)e^{ik(x-y)}]$$

$$= i\Delta_F(x - y) \tag{6.57}$$

from (6.56). We have now proved (6.52) again, and have in the process found an interpretation for the 2-point function in terms of the creation, propagation and destruction of our particle between two points.

We have found an expression for the 2-point function. What is the 1-point function? It is clear that

$$\tau(x) = \langle 0|T\phi(x)|0\rangle = \langle 0|\phi(x)|0\rangle = \frac{1}{i}\frac{\delta Z_0[J]}{\delta J(x)}\bigg|_{J=0} = 0. \qquad (6.58)$$

Let us now find the 3-point function.

$$\tau(x_1, x_2, x_3) = \frac{1}{i}\frac{\delta}{\delta J(x_1)}\frac{1}{i}\frac{\delta}{\delta J(x_2)}\frac{1}{i}\frac{\delta}{\delta J(x_3)}Z_0[J]\bigg|_{J=0}.$$

We saw above (equation (6.51)) that

$$\frac{1}{i}\frac{\delta}{\delta J(x_2)}\frac{1}{i}\frac{\delta}{\delta J(x_3)}Z_0[J]$$

$$= i\Delta_F(x_2 - x_3)\exp\left(-\frac{i}{2}\int J\Delta_F J\right)$$

$$+ \int \Delta_F(x_2 - x)J(x)\,dx \int \Delta_F(x_3 - x)J(x)\,dx \exp\left(-\frac{i}{2}\int J\Delta_F J\right).$$

Further differentiation then gives

$$\frac{1}{i}\frac{\delta}{\delta J(x_1)}\frac{1}{i}\frac{\delta}{\delta J(x_2)}\frac{1}{i}\frac{\delta}{\delta J(x_3)}Z_0[J]$$

$$= -i\Delta_F(x_2 - x_3)\int \Delta_F(x_1 - x)J(x)\,dx \exp\left(-\frac{i}{2}\int J\Delta_F J\right)$$

$$- i\Delta_F(x_2 - x_1)\int \Delta_F(x_3 - x)J(x)\,dx \exp\left(-\frac{i}{2}\int J\Delta_F J\right)$$

$$- i\Delta_F(x_3 - x_1)\int \Delta_F(x_2 - x)J(x)\,dx \exp\left(-\frac{i}{2}\int J\Delta_F J\right)$$

$$- \int \Delta_F(x_2 - x)J(x)\,dx \int \Delta_F(x_3 - x)J(x)$$

$$\times \int \Delta_F(x_1 - x)J(x)\,dx \exp\left(-\frac{i}{2}\int J\Delta_F J\right). \qquad (6.59)$$

Putting $J = 0$ reduces this expression to zero, so

$$\tau(x_1, x_2, x_3) = \langle |T(\phi(x_1)\phi(x_2)\phi(x_3))|0\rangle = 0. \qquad (6.60)$$

To find the 4-point function we carry on in the same way, by differentiating (6.59) once more and putting $J = 0$. We have

$$\frac{1}{i}\frac{\delta}{\delta J(x_1)}\cdots\frac{1}{i}\frac{\delta}{\delta J(x_4)}Z_0[J]$$

$$= -\Delta_F(x_2 - x_3)\Delta_F(x_1 - x_4)\exp\left(-\frac{i}{2}\int J\Delta_F J\right)$$

$$-\Delta_F(x_2 - x_1)\Delta_F(x_3 - x_4)\exp\left(-\frac{i}{2}\int J\Delta_F J\right)$$

$$-\Delta_F(x_3 - x_1)\Delta_F(x_2 - x_4)\exp\left(-\frac{i}{2}\int J\Delta_F J\right)$$

$+$ (terms which vanish when $J = 0$),

so the 4-point function is

$$\begin{aligned}
\tau(x_1, x_2, x_3, x_4) &= \langle 0|T(\phi(x_1)\phi(x_2)\phi(x_3)\phi(x_4))|0\rangle \\
&= -[\Delta_F(x_1 - x_2)\Delta_F(x_3 - x_4) \\
&\quad + \Delta_F(x_1 - x_3)\Delta_F(x_2 - x_4) \\
&\quad + \Delta_F(x_1 - x_4)\Delta_F(x_2 - x_3)].
\end{aligned} \tag{6.61}$$

This is simply the sum of products of 2-point functions, and may be written

$$(6.62)$$

Going to higher orders, it is clear that if n is odd, the n-point function always vanishes,

$$\tau(x_1, x_2, \ldots, x_{2n+1}) = 0, \tag{6.63}$$

and it turns out that, if n is even, the n-point function is a sum of products of 2-point functions,

$$\tau(x_1, x_2, \ldots, x_{2n}) = \sum_{\text{perms}} \tau(x_{p_1}, x_{p_2})\ldots(x_{p_{2n-1}}, x_{p_{2n}}), \tag{6.64}$$

where

$$\tau(x, y) = i\Delta_F(x - y).$$

This important result, when derived using the 'canonical' method, employing the operator commutator relations, is known as *Wick's theorem*.

In this section we have derived the Green's functions in a scalar free field theory. The interesting case, however, is the one for which interactions are present; how do we calculate the Green's functions then? When we know this, we are one step nearer calculating a scattering amplitude for a real physical process!

6.4   Generating functionals for interacting fields

The Lagrangian

$$\mathscr{L} = \tfrac{1}{2}\partial_\mu\phi\partial^\mu\phi - \tfrac{1}{2}m^2\phi^2 - \frac{g}{4!}\phi^4 = \mathscr{L}_0 + \mathscr{L}_{\text{int}} \qquad (6.65)$$

describes a scalar field which interacts with itself, through the term in ϕ^4. We shall first show how to find the Green's functions for a general interaction $\mathscr{L}_{\text{int}}$, and then in the next section apply the formulae to the case of ϕ^4 theory. The normalised generating functional is

$$Z[J] = \frac{\displaystyle\int \mathscr{D}\phi \exp\left(iS + i\int J\phi\,dx\right)}{\displaystyle\int \mathscr{D}\phi e^{iS}} \qquad (6.66)$$

with $S = \int \mathscr{L}\,dx$. It is clear that, when $\mathscr{L}_{\text{int}} = 0$, this becomes the expression (6.43), which we were able to show is the same as (6.44). Equation (6.44) is in a form suitable for functional differentiation with respect to J, and therefore for finding the Green's functions. We want to find the expression which corresponds to (6.44) in the case of interacting fields. We proceed by finding the differential equation satisfied by $Z[J]$, and then solving it in terms of $Z_0[J]$.

First note that, from (6.44),

$$\frac{1}{i}\frac{\delta}{\delta J(x)}Z_0[J] = -\int \Delta_F(x-y)J(y)\,dy\exp\left(-\frac{i}{2}\int J\Delta_F J\,dx\,dy\right),$$

so, since Δ_F is minus the inverse of $\Box + m^2$,

$$(\Box + m^2)\frac{1}{i}\frac{\delta}{\delta J(x)}Z_0[J] = J(x)Z_0[J]. \qquad (6.67)$$

This is the differential equation satisfied by $Z_0[J]$.

Now we have, from (6.66),

$$\frac{1}{i}\frac{\delta Z[J]}{\delta J(x)} = \frac{\displaystyle\int \exp\left(iS + i\int J\phi\,dx\right)\phi(x)\mathscr{D}\phi}{\displaystyle\int e^{iS}\mathscr{D}\phi}. \qquad (6.68)$$

We define the functional

$$\hat{Z}[\phi] = \frac{e^{iS}}{\displaystyle\int e^{iS}\mathscr{D}\phi}. \qquad (6.69)$$

Then

$$Z[J] = \int \hat{Z}[\phi]\exp\left[i\int J(x)\phi(x)\,dx\right]\mathscr{D}\phi. \qquad (6.70)$$

This is the functional analogue of the Fourier transform. Now we take the functional derivative of $\hat{Z}[\phi]$, noting that

$$S = \int (\tfrac{1}{2}\partial_\mu\phi\partial^\mu\phi - \tfrac{1}{2}m^2\phi^2 + \mathscr{L}_{\text{int}})\mathrm{d}^4x$$

$$= -\int [\tfrac{1}{2}\phi(\square + m^2)\phi - \mathscr{L}_{\text{int}}]\mathrm{d}^4x. \tag{6.71}$$

We obtain

$$\mathrm{i}\frac{\delta\hat{Z}[\phi]}{\delta\phi(x)} = \mathrm{i}\frac{\delta}{\delta\phi}\left\{\exp-\mathrm{i}\int[\tfrac{1}{2}\phi(\square + m^2)\phi - \mathscr{L}_{\text{int}}]\mathrm{d}^4x\right\}$$

$$\times\left[\int e^{\mathrm{i}S}\mathscr{D}\phi\right]^{-1}$$

$$= (\square + m^2)\phi(x)\hat{Z}[\phi] - \frac{\partial\mathscr{L}_{\text{int}}}{\partial\phi}\hat{Z}[\phi]$$

$$= (\square + m^2)\phi(x)\hat{Z}[\phi] - \mathscr{L}'_{\text{int}}(\phi)\hat{Z}[\phi] \tag{6.72}$$

where the prime on $\mathscr{L}_{\text{int}}$ means differentiation with respect to the argument. Now we multiply both sides of (6.72) by $\exp \mathrm{i}\int J(x)\phi(x)\,\mathrm{d}x$ and integrate over ϕ. The right-hand side gives

$$\frac{\int(\square + m^2)\phi(x)\exp\left(\mathrm{i}S + \mathrm{i}\int J\phi\,\mathrm{d}x\right)\mathscr{D}\phi}{\int e^{\mathrm{i}S}\mathscr{D}\phi}$$

$$-\frac{\int\mathscr{L}'_{\text{int}}(\phi)\exp\left(\mathrm{i}S + \mathrm{i}\int J\phi\,\mathrm{d}x\right)\mathscr{D}\phi}{\int e^{\mathrm{i}S}\mathscr{D}\phi}$$

$$= (\square + m^2)\frac{1}{\mathrm{i}}\frac{\delta Z[J]}{\delta J(x)} - \mathscr{L}'_{\text{int}}\left[\frac{1}{\mathrm{i}}\frac{\delta}{\delta J}\right]Z[J] \tag{6.73}$$

where (6.68) has been used, and the argument of $\mathscr{L}'_{\text{int}}$ has been changed from ϕ to $(1/\mathrm{i})(\delta/\delta J)$, since it operates on $Z[J]$. The left-hand side of (6.72) gives

$$\mathrm{i}\int\frac{\delta\hat{Z}[\phi]}{\delta\phi}\exp\left(\mathrm{i}\int J\phi\,\mathrm{d}x\right)\mathscr{D}\phi$$

$$= \mathrm{i}\exp\left(\mathrm{i}\int J\phi\,\mathrm{d}x\right)\hat{Z}[\phi]\Big|_{\phi\to\infty}$$

$$+ \int J(x)\hat{Z}[\phi]\exp\left(\mathrm{i}\int J\phi\,\mathrm{d}x\right)\mathscr{D}\phi$$

$$= J(x)Z[J] \tag{6.74}$$

from (6.70). Equating (6.73) and (6.74) gives

$$(\Box + m^2)\frac{1}{i}\frac{\delta Z[J]}{\delta J(x)} - \mathscr{L}'_{\text{int}}\left(\frac{1}{i}\frac{\delta}{\delta J(x)}\right)Z[J] = J(x)Z[J]. \qquad (6.75)$$

We must solve this equation for $Z[J]$. In the free field case, $\mathscr{L}_{\text{int}} = 0$ and the equation reduces to (6.67), for $Z_0[J]$. We shall now show that the solution to (6.75) is

$$Z[J] = N\exp\left[i\int\mathscr{L}_{\text{int}}\left(\frac{1}{i}\frac{\delta}{\delta J}\right)dx\right]Z_0[J] \qquad (6.76)$$

where N is a normalising factor. The proof is in two stages.

Proof
(a) We first prove the identity

$$\exp\left[-i\int\mathscr{L}_{\text{int}}\left(\frac{1}{i}\frac{\delta}{\delta J(y)}\right)dy\right]J(x)\exp\left[+i\int\mathscr{L}_{\text{int}}\left(\frac{1}{i}\frac{\delta}{\delta J(y)}\right)dy\right]$$

$$= J(x) - \mathscr{L}'_{\text{int}}\left(\frac{1}{i}\frac{\delta}{\delta J(x)}\right). \qquad (6.77)$$

This follows by observing that the functional analogue of

$$\left[x_i, \frac{1}{i}\frac{\partial}{\partial x_j}\right] = i\delta_{ij}$$

is

$$\left[J(x), \frac{1}{i}\frac{\delta}{\delta J(y)}\right] = i\delta(x - y).$$

Repeated application of this equation gives

$$\left[J(x), \left(\frac{1}{i}\frac{\delta}{\delta J(y)}\right)^n\right] = i\delta(x - y)\left(\frac{1}{i}\frac{\delta}{\delta J(y)}\right)^{n-1}$$

$$-\frac{1}{i}\frac{\delta}{\delta J(y)}\left[J(x), \left(\frac{1}{i}\frac{\delta}{\delta J(y)}\right)^{n-1}\right]$$

$$\vdots$$

$$= i\delta(x - y)n\left(\frac{1}{i}\frac{\delta}{\delta J(y)}\right)^{n-1}. \qquad (6.78)$$

By expanding the function

$$F(\phi) = F(0) + \phi F'(0) + \frac{\phi^2}{2!}F''(0) + \ldots = \sum_{n=0}^{\infty}\frac{\phi^n}{n!}F^{(n)}(0)$$

and making the replacement $\phi \to (1/i)(\delta/\delta J)$, it then follows from (6.78) that

$$\left[J(x),\int F\left(\frac{1}{i}\frac{\delta}{\delta J(y)}\right)dy\right]=iF'\left(\frac{1}{i}\frac{\delta}{\delta J(x)}\right). \tag{6.79}$$

Now we use the Hausdorff formula

$$e^A Be^{-A}=B+[A,B]+\frac{1}{2!}[A,[A,B]]+\cdots \tag{6.80}$$

where A and B are operators, and put $A=-i\int\mathscr{L}_{int}((1/i)(\delta/\delta J(y)))dy$ and $B=J(x)$. Since, in this case, A commutes with $[A,B]$ (from (6.79)), only the first two terms on the right-hand side of (6.80) appear, and (6.77) is proved.

(b) We must now show that (6.76) is the solution of (6.75). From (6.76) and (6.77)

$$J(x)Z[J]=NJ(x)\exp\left[i\int\mathscr{L}_{int}\left(\frac{1}{i}\frac{\delta}{\delta J(y)}\right)dy\right]Z_0[J]$$

$$=N\exp\left[i\int\mathscr{L}_{int}\left(\frac{1}{i}\frac{\delta}{\delta J(y)}\right)dy\right]$$

$$\times\left[J(x)-\mathscr{L}'_{int}\left(\frac{1}{i}\frac{\delta}{\delta J(x)}\right)\right]Z_0[J].$$

The first of these terms is transformed using (6.67), and in the second, the order of $e^{i\int\mathscr{L}_{int}}$ and $\mathscr{L}'_{int}$ may be interchanged, giving

$$J(x)Z[J]=N\exp\left[i\int\mathscr{L}_{int}\left(\frac{1}{i}\frac{\delta}{\delta J(y)}\right)dy\right](\Box+m^2)\frac{1}{i}\frac{\delta Z_0}{\delta J(x)}$$

$$-N\mathscr{L}'_{int}\left(\frac{1}{i}\frac{\delta}{\delta J(x)}\right)\exp\left[i\int\mathscr{L}_{int}\left(\frac{1}{i}\frac{\delta}{\delta J(y)}\right)dy\right]Z_0[J]$$

$$=(\Box+m^2)\frac{1}{i}\frac{\delta Z[J]}{\delta J(x)}-\mathscr{L}'_{int}\left[\frac{1}{i}\frac{\delta}{\delta J(x)}\right]Z[J],$$

where (6.76) was used. This is equation (6.75). QED.

We are now in a position to calculate the Green's functions in the interacting field case, which we proceed to do, as usual in quantum theory, by perturbation theory.

6.5 ϕ^4 theory

Generating functional

As we saw in (6.65), the interaction Lagrangian in ϕ^4 theory is

$$\mathscr{L}_{int}=-\frac{g}{4!}\phi^4. \tag{6.81}$$

The normalised generating functional $Z[J]$ is

$$Z[J] = \frac{\exp\left[i\int \mathscr{L}_{\text{int}}\left(\frac{1}{i}\frac{\delta}{\delta J(z)}\right)dz\right]\exp\left[-\frac{i}{2}\int J(x)\Delta_F(x-y)J(y)dx\,dy\right]}{\left\{\exp\left[i\int \mathscr{L}_{\text{int}}\left(\frac{1}{i}\frac{\delta}{\delta J(z)}\right)dz\right]\exp\left[-\frac{i}{2}\int J(x)\Delta_F(x-y)J(y)dx\,dy\right]\right\}\Big|_{J=0}}.$$

(6.82)

The only way of treating $\exp(i\int \mathscr{L}_{\text{int}})$ is as a power series in the coupling constant g, i.e. by perturbation theory. Substituting (6.81) into (6.82) and expanding in powers of g, the numerator of $Z[J]$ is

$$\left[1 - \frac{ig}{4!}\int\left(\frac{1}{i}\frac{\delta}{\delta J(z)}\right)^4 dz + O(g^2)\right]$$

$$\times \exp\left[-\frac{i}{2}\int J(x)\Delta_F(x-y)J(y)dx\,dy\right].$$

To order g^0, we just have the free particle generating functional $Z_0[J]$. To order g, we proceed as follows:

$$\frac{1}{i}\frac{\delta}{\delta J(z)}\exp\left[-\frac{i}{2}\int J(x)\Delta_F(x-y)J(y)dx\,dy\right]$$

$$= -\int \Delta_F(z-x)J(x)dx\exp\left[-\frac{i}{2}\int J(x)\Delta_F(x-y)J(y)dx\,dy\right]$$

$$\left(\frac{1}{i}\frac{\delta}{\delta J(z)}\right)^2\exp\left[-\frac{i}{2}\int J(x)\Delta_F(x-y)J(y)dx\,dy\right]$$

$$= \left\{i\Delta_F(0) + \left[\int \Delta_F(z-x)J(x)dx\right]^2\right\}$$

$$\times \exp\left[-\frac{i}{2}\int J(x)\Delta_F(x-y)J(y)dx\,dy\right]$$

$$\left(\frac{1}{i}\frac{\delta}{\delta J(z)}\right)^3\exp\left[-\frac{i}{2}\int J(x)\Delta_F(x-y)J(y)dx\,dy\right]$$

$$= \left\{3[-i\Delta_F(0)]\int \Delta_F(z-x)J(x)dx - \left[\int \Delta_F(z-x)J(x)dx\right]^3\right\}$$

$$\times \exp\left[-\frac{i}{2}\int J(x)\Delta_F(x-y)J(y)dx\,dy\right]$$

$$\left(\frac{1}{i}\frac{\delta}{\delta J(z)}\right)^4\exp\left[-\frac{i}{2}\int J(x)\Delta_F(x-y)J(y)dx\,dy\right]$$

$$= \left\{-3[\Delta_F(0)]^2 + 6i\Delta_F(0)\left[\int \Delta_F(z-x)J(x)dx\right]^2\right.$$

$$+\left[\int\Delta_F(z-x)J(x)\mathrm{d}x\right]^4\bigg\}\exp\left[-\frac{i}{2}\int J(x)\Delta_F(x-y)\mathrm{d}x\,\mathrm{d}y\right]$$
$$(6.83)$$

We may write this expression diagramatically. Let

$$x\,\text{———————}\,y\to\Delta_F(x-y)\qquad\qquad(6.84)$$

represent the free particle propagator. $\Delta_F(0)=\Delta_F(x-x)$ is then represented by a closed loop

$$\bigcirc\to\Delta_F(0).\qquad\qquad(6.85)$$

Equation (6.83) may then be written

$$\left(\frac{1}{i}\frac{\delta}{\delta J(z)}\right)^4\exp\left(-\frac{i}{2}\int J\Delta_F J\right)=\{-3\;\infty\;+6i\;\underline{\bigcirc}\;+\times\}$$

$$\times\exp\left(-\frac{i}{2}\int J\,\Delta_F J\right).\qquad(6.86)$$

The meeting of four lines at a point in the three diagrams in (6.86) is clearly a consequence of the fact that $\mathcal{L}_{\mathrm{int}}$ contains ϕ^4. Moreover, the coefficients 3, 6 and 1 follow from rather simple considerations of symmetry. The first term on the right-hand side of (6.86), for example, results from joining up the two pairs of lines in the third term, in all possible ways; and there are three ways to do this. The second term is got by joining any two lines of the third term, and there are six ways to do this. These numerical coefficients are known as *symmetry factors*. The first term, with two closed loops, is known as a *vacuum graph*, because it has no external lines. The term with one closed loop has two external lines (i.e. two Js) and the last term four external lines (four Js). It is now an easy matter to write down the denominator of (6.82). Putting $J=0$ eliminates the second and third terms in (6.86), so we have

$$\left[\exp\left(i\int\mathcal{L}_{\mathrm{int}}\right)\exp\left(-\frac{i}{2}\int J\,\Delta_F J\right)\right]\bigg|_{J=0}$$

$$=1-\frac{ig}{4!}\int(-3\;\infty\;)\mathrm{d}z.\qquad\qquad(6.87)$$

The complete generating functional, given by equation (6.82), is, to order g,

$$Z[J]=\frac{\left[1-\frac{ig}{4!}\int\left(-3\;\infty\;+6i\;\underline{\bigcirc}\;+\times\right)\mathrm{d}z\right]\exp\left(-\frac{i}{2}\int J\Delta_F J\right)}{1-\frac{ig}{4!}\int(-3\;\infty\;)\mathrm{d}z}$$

$$=\left[1-\frac{ig}{4!}\int\left(6i\;\underline{\bigcirc}\;+\times\right)\mathrm{d}z\right]\exp\left(-\frac{i}{2}\int J\,\Delta_F J\right),\qquad(6.88)$$

where the denominator has been expanded by the binomial theorem. The interesting thing that has happened is that the *vacuum diagram has disappeared* in $Z[J]$. It turns out that this is true to all orders in perturbation theory, and is a general property of *normalised* generating functionals.

2-point function
The 2-point function is defined by

$$\tau(x_1, x_2) = -\frac{\delta^2 Z[J]}{\delta J(x_2)\delta J(x_1)}\bigg|_{J=0}. \tag{6.89}$$

By looking at (6.88), it is seen that the first term in Z will give $i\Delta_F(x_1 - x_2)$ in τ; this is the free particle propagator. The term in $\times$ in (6.88) contains four Js and so gives no contribution to the 2-point function. The term in $\underline{\quad\bigcirc\quad}$ is

$$\frac{g}{4}\Delta_F(0)\int dx\,dy\,dz\,\Delta_F(z-x)J(x)\Delta_F(z-y)J(y)\exp\left(-\frac{i}{2}\int J\,\Delta_F J\right).$$

On differentiation we get

$$\frac{1}{i}\frac{\delta}{\delta J(x_1)}\bigg(\qquad\bigg) = -\frac{ig}{4}\Delta_F(0)2\int dy\,dz\,\Delta_F(z-x)\Delta_F(z-y)J(y)$$

$$\times \exp\left(-\frac{i}{2}\int J\,\Delta_F J\right) + \cdots,$$

$$\frac{1}{i}\frac{\delta}{\delta J(x_2)}\frac{1}{i}\frac{\delta}{\delta J(x_1)}\bigg(\qquad\bigg) = -\frac{g}{2}\Delta_F(0)\int dz\,\Delta_F(z-x_1)\Delta_F(z-x_2)$$

$$\times \exp\left(-\frac{i}{2}\int J\,\Delta_F J\right) + \dots,$$

where the omitted terms vanish when $J = 0$. We then have

$$\tau(x_1, x_2) = i\Delta_F(x_1 - x_2) - \frac{g}{2}\Delta_F(0)\int dz\,\Delta_F(z-x_1)\Delta_F(z-x_2) + O(g^2) \tag{6.90}$$

$$= i\frac{\quad}{\quad} - \frac{g}{2}\frac{\quad\bigcirc\quad}{\quad} + O(g^2). \tag{6.91}$$

To order g, this represents the effect of interaction on the free particle propagation. The free particle propagator is, from (6.14),

$$\Delta_F(x-y) = \frac{1}{(2\pi)^4}\int \frac{e^{-ik\cdot(x-y)}}{k^2 - m^2 + i\varepsilon}d^4k,$$

and its Fourier transform contains a pole at $k^2 = m^2$. This identifies the mass of the particle as m. We shall now show that the effect of the interaction is to change the value of the physical mass away from m. Indeed, the second term in (6.90) is

$$-\tfrac{1}{2}g\,\Delta_F(0)\int \Delta_F(x_1-z)\Delta_F(x_2-z)\mathrm{d}z$$

$$= -\frac{g\,\Delta_F(0)}{2\,(2\pi)^8}\int \frac{\mathrm{e}^{-\mathrm{i}p\cdot(x_1-z)}}{p^2-m^2+\mathrm{i}\varepsilon}\cdot\frac{\mathrm{e}^{-\mathrm{i}q\cdot(x_2-z)}}{q^2-m^2+\mathrm{i}\varepsilon}\mathrm{d}^4p\,\mathrm{d}^4q\,\mathrm{d}^4z$$

$$= -\frac{g\,\Delta_F(0)}{2\,(2\pi)^4}\int \frac{\mathrm{e}^{-\mathrm{i}p\cdot(x_1-x_2)}}{(p^2-m^2+\mathrm{i}\varepsilon)^2}\delta^4(p-q)\mathrm{d}^4p\,\mathrm{d}^4q$$

$$= -\frac{g\,\Delta_F(0)}{2\,(2\pi)^4}\int \frac{\mathrm{e}^{-\mathrm{i}p\cdot(x_1-x_2)}}{(p^2-m^2+\mathrm{i}\varepsilon)^2}\mathrm{d}^4p,$$

giving, for the 2-point function (6.90),

$$\tau(x_1,x_2) = \frac{\mathrm{i}}{(2\pi)^4}\int \frac{\mathrm{e}^{-\mathrm{i}p\cdot(x_1-x_2)}}{p^2-m^2+\mathrm{i}\varepsilon}\left[1+\frac{\tfrac{1}{2}\mathrm{i}g\,\Delta_F(0)}{p^2-m^2+\mathrm{i}\varepsilon}\right]\mathrm{d}^4p. \qquad (6.92)$$

Formally, the term in square brackets above can be written as

$$\left[1-\frac{\tfrac{1}{2}\mathrm{i}g\,\Delta_F(0)}{p^2-m^2+\mathrm{i}\varepsilon}\right]^{-1},$$

so we have

$$\tau(x_1,x_2) = \frac{\mathrm{i}}{(2\pi)^4}\int \frac{\mathrm{e}^{-\mathrm{i}p\cdot(x_1-x_2)}}{p^2-m^2+\tfrac{1}{2}\mathrm{i}g\,\Delta_F(0)+\mathrm{i}\varepsilon}\mathrm{d}^4p. \qquad (6.93)$$

The Fourier transform of $\tau(x_1,x_2)$ will now possess a pole at p^2 equal to

$$m^2-\tfrac{1}{2}\mathrm{i}g\,\Delta_F(0) \equiv m^2+\delta m^2 = m_r^2 \qquad (6.94)$$

where

$$\delta m^2 = -\tfrac{1}{2}\mathrm{i}g\,\Delta_F(0); \qquad (6.95)$$

m_r is now identified as the *physical mass*, or *renormalised mass*. The change in (mass)^2, δm^2, is a quadratically divergent quantity, since $\Delta_F(0)$ contains four powers of p in the numerator (d^4p) and two in the denominator. It happens, then, that the renormalisation of the mass is by an infinite quantity, but that is a distinct circumstance; the essence of renormalisation is that the physical quantity (mass, in this case) is not the same as the parameter in the Lagrangian, if an interaction is present. More will be said about renormalisation in chapter 9.

4-point function

Let us proceed, finally, to the 4-point function, given by (equation (6.45))

$$\tau(x_1,x_2,x_3,x_4) = \frac{\delta^4 Z[J]}{\delta J(x_1)\delta J(x_2)\delta J(x_3)\delta J(x_4)}\bigg|_{J=0}, \qquad (6.96)$$

where $Z[J]$ is given by (6.88). The first (order g^0) term in τ is the same as that found in (6.60)

$$- [\Delta_F(x_1 - x_2)\Delta_F(x_3 - x_4) + \Delta_F(x_1 - x_3)\Delta_F(x_2 - x_4)$$
$$+ \Delta_F(x_1 - x_4)\Delta_F(x_2 - x_3)]$$
$$= -(= + || + \times \,) = -3(=). \qquad (6.97)$$

This is the free particle 4-point function and will not contribute to the scattering. The next term in $Z[J]$, of order g, gives, as may be checked,

$$\frac{g}{4}\frac{\delta^4}{\delta J(x_1)\delta J(x_2)\delta J(x_3)\delta J(x_4)}$$

$$\times \left\{ -\!\!\bigcirc\!\!- \exp\left[-\frac{i}{2}\int J(x)\Delta_F(x - y)J(y)\mathrm{d}x\,\mathrm{d}y \right] \right\}\bigg|_{J=0}$$

$$= \frac{g}{4}\frac{\delta^4}{\delta J(x_1)\delta J(x_2)\delta J(x_3)\delta J(x_4)}$$

$$\times \left\{ \Delta_F(0)\int \mathrm{d}x\,\mathrm{d}y\,\Delta_F(x - z)\Delta_F(y - z)J(y)J(x) \right.$$

$$\left. \times \exp\left(-\frac{i}{2}\int J\Delta_F J \right) \right\}\bigg|_{J=0}$$

$$= \frac{-ig}{2}\Delta_F(0)\int \mathrm{d}z\,[\Delta_F(z - x_1)\Delta_F(z - x_2)\Delta_F(x_3 - x_4)$$

$$+ \Delta_F(z - x_1)\Delta_F(z - x_3)\Delta_F(x_2 - x_4)$$

$$+ \Delta_F(z - x_1)\Delta_F(z - x_4)\Delta_F(x_2 - x_3)$$

$$+ \Delta_F(z - x_2)\Delta_F(z - x_3)\Delta_F(x_1 - x_4)$$

$$+ \Delta_F(z - x_2)\Delta_F(z - x_4)\Delta_F(x_1 - x_3)$$

$$+ \Delta_F(z - x_3)\Delta_F(z - x_4)\Delta_F(x_1 - x_2)]$$

$$= -3ig\left[\frac{\bigcirc}{} \right]. \qquad (6.98)$$

The diagram above does service for the six equivalent terms in the expression above it. Each of these terms contributes twice, so the 'symmetry factor' of the diagram is 12.

The final term in $Z[J]$, of order g, gives

$$\frac{-ig}{4!}\frac{\delta^4}{\delta J(x_1)\delta J(x_2)\delta J(x_3)\delta J(x_4)}$$

$$\times \left\{ \times \exp\left[-\frac{i}{2}\int J(x)\Delta_F(x - y)J(y)\mathrm{d}x\,\mathrm{d}y \right] \right\}\bigg|_{J=0}$$

$$= \frac{-ig}{4!} \frac{\delta^4}{\delta J(x_1)\delta J(x_2)\delta J(x_3)\delta J(x_4)}$$

$$\times \left\{ \left[\int \Delta_F(z-x)J(x)dx \right]^4 dz \exp\left(-\frac{i}{2}\int J\,\Delta_F J \right) \right\}\Big|_{J=0}$$

$$= -ig \int \Delta_F(x_1-z)\Delta_F(x_2-z)\Delta_F(x_3-z)\Delta_F(x_4-z)dz$$

$$= -ig \bowtie . \tag{6.99}$$

The complete 4-point function, to order g, is then

$$\tau(x_1,x_2,x_3,x_4) = -3\left[\underline{\quad\quad} \right] - 3ig\left(\underline{\overset{\bigcirc}{\quad\quad}} \right) - ig\left(\bowtie \right)$$

$$= -3\left[\underline{\quad\quad} \right] \frac{-ig}{4!}\left[12 \times 6 \left(\underline{\overset{\bigcirc}{\quad\quad}} \right) + 24\left(\bowtie \right) \right]. \tag{6.100}$$

The first term, of order g^0, does not contribute to the scattering. The numerical coefficients above are easily derived by simple combinatorics, and this suggests a rather direct way to write down all the diagrams of a given order. Let us consider, for example, all diagrams contributing to the 4-point function of order g, in $g\phi^4/4!$ theory. We deduce them as follows. First of all, we are considering a $g\phi^4$ theory, so to order g^n we have the n vertices

$$\times \times \times \cdots \times \tag{6.101}$$

Corresponding to the 4-point function so we draw four external points

$$\begin{array}{ll} x_1 \bullet\!\!\!-\!\!\!- & -\!\!\!-\!\!\!\bullet x_3 \\ x_2 \bullet\!\!\!-\!\!\!- & -\!\!\!-\!\!\!\rightarrow x_4 \end{array}. \tag{6.102}$$

The 4-point function in $g\phi^4$ theory to order g is then constructed from the following *prediagram*

$$\begin{array}{ll} x_1 \underline{\quad\quad} \times \underline{\quad\quad} & x_3 \\ x_2 \underline{\quad\quad} \underline{\quad\quad} & x_4 \end{array}. \tag{6.103}$$

Fig. 6.4. First order terms in the 4-point function.

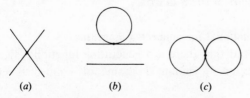

(a)   (b)   (c)

(This is called a prediagram to distinguish it from a real Feynman diagram.) We now join up all the lines. There are three topologically distinct types of Feynman diagram which result, drawn in Fig. 6.4. The multiplicities are calculated as follows. To get diagram (a) (in Fig. 6.4) join x_1 up to one of the legs of the vertex in (6.103). There are 4 ways to do this. Now join x_2 up to one of the remaining three legs – there are 3 ways. There are $4! = 24$ ways to complete the diagram (a), which is the coefficient in equation (6.100). Next, to make diagram (b) join x_1 directly to one of the other external points x_2, x_3, x_4. There are 3 ways to do this. Choose one leg of the vertex and join it up to one of the two remaining external points. There are 4×2 ways to do this. Join one of the three remaining legs of the vertex to the one remaining point. There are 3 ways to do this. Join the remaining two legs together. The total multiplicity is $3 \times 4 \times 2 \times 3 = 12 \times 6$, as in (6.100). The reader will easily convince himself that the multiplicity of diagram (c) is $3 \times 3 = 9$. The reason this diagram does not appear in (6.100) is that it is a multiple of the vacuum diagram $\bigcirc\!\!\bigcirc$ (see above) and $Z[J]$, being properly normalised, produces no vacuum diagrams.

In summary, the Feynman rules for ϕ^4 scalar field theory, in co-ordinate space, are

$$
\left\{
\begin{array}{lcc}
\text{line} & x \underline{\hspace{1cm}\times\hspace{0.3cm}} y & \Delta_F(x - y) \\
\text{vertex} & \times z & -ig \\
& \text{integration over } z & \\
\text{symmetry factor} & & S/4!
\end{array}
\right\} .
\tag{6.104}
$$

In calculations of realistic processes, involving, for example, electrons and photons, the particles are not identical and there is then no symmetry factor to contend with. We postpone, until we consider these real scattering processes, the derivation of the S matrix from the Green's function.

It is however, convenient to mention here that, of the two order g diagrams in (6.100), the first one, $\underline{\hspace{0.6cm}Q\hspace{0.6cm}}$, only contributes to the trivial (diagonal) part of the S matrix, so is not interesting. It describes the two particles moving independently, and the effect of the interaction is to modify the propagator of one of them. This graph is called *disconnected*. The other order g graph, $\times$, is *connected*, since every line in it is connected to every other line. Only connected Feynman diagrams contribute to $S - 1$, i.e. to the non-trivial part of the S matrix.

6.6   Generating functional for connected diagrams

Now it turns out that there is a generating functional W, which generates only connected Feynman diagrams or connected Green's functions. It is related to Z by

$$Z[J] = e^{iW[J]} \qquad (6.105)$$

or

$$W[J] = -i \ln Z[J]. \qquad (6.106)$$

We shall now show, by considering the 2-point and 4-point functions in ϕ^4 theory, that $W[J]$ generates no disconnected graphs. We have, firstly,

$$\frac{\delta^2 W}{\delta J(x_1)\delta J(x_2)} = \frac{i}{Z^2} \frac{\delta Z}{\delta J(x_1)} \frac{\delta Z}{\delta J(x_2)} - \frac{i}{Z} \frac{\delta^2 Z}{\delta J(x_1)\delta J(x_2)}. \qquad (6.107)$$

When $J = 0$, we have

$$\left. \frac{\delta Z[J]}{\delta J(x)} \right|_{J=0} = 0, \quad Z[0] = 1 \qquad (6.108)$$

so

$$\left. \frac{\delta^2 W}{\delta J(x_1)\delta J(x_2)} \right|_{J=0} = -i \left. \frac{\delta^2 Z}{\delta J(x_1)\delta J(x_2)} \right|_{J=0} = i\tau(x_1, x_2) \qquad (6.109)$$

showing that W generates the propagator, to any order in g. This is as we expected, since the propagator has no disconnected part. To find the 4-point function, we differentiate (6.107) twice more and set $J = 0$, to give

$$\left. \frac{\delta^4 W}{\delta J(x_1)\delta J(x_2)\delta J(x_3)\delta J(x_4)} \right|_{J=0}$$

$$= i\Bigg[\frac{1}{Z^2} \frac{\delta^2 Z}{\delta J(x_1)\delta J(x_2)} \frac{\delta^2 Z}{\delta J(x_3)\delta J(x_4)}$$

$$+ \frac{1}{Z^2} \frac{\delta^2 Z}{\delta J(x_1)\delta J(x_3)} \frac{\delta^2 Z}{\delta J(x_2)\delta J(x_4)}$$

$$+ \frac{1}{Z^2} \frac{\delta^2 Z}{\delta J(x_1)\delta J(x_4)} \frac{\delta^2 Z}{\delta J(x_2)\delta J(x_3)}$$

$$- \frac{1}{Z} \frac{\delta^4 Z}{\delta J(x_1)\delta J(x_2)\delta J(x_3)\delta J(x_4)} \Bigg]\Bigg|_{J=0}$$

$$= i[\tau(x_1, x_2)\tau(x_3, x_4) + \tau(x_1, x_3)\tau(x_2, x_4)$$

$$+ \tau(x_1, x_4)\tau(x_2, x_3) - \tau(x_1, x_2, x_3, x_4)]. \qquad (6.110)$$

We have to show that this contains no disconnected diagrams. The most convenient way to do this is to substitute the diagrams themselves into (6.110). Working, as always to order g, and substituting equations (6.91) and (6.100) into (6.110) gives (in a notation which is meant to be self-explanatory – 1 stands for x_1, etc.)

$$\frac{\delta^4 W}{\delta J(x_1)\delta J(x_2)\delta J(x_3)\delta J(x_4)}\bigg|_{J=0}.$$

$$= \mathrm{i}\Bigg[\left(\mathrm{i}\,\overline{\underset{1\quad\;\;2}{}} - \frac{g}{2}\,1\!\bigcirc\!2\right)\left(\mathrm{i}\,\overline{\underset{3\quad\;\;4}{}} - \frac{g}{2}\,3\!\bigcirc\!4\right)$$

$$+\left(\mathrm{i}\,\overline{\underset{1\quad\;\;3}{}} - \frac{g}{2}\,1\!\bigcirc\!3\right)\left(\mathrm{i}\,\overline{\underset{2\quad\;\;4}{}} - \frac{g}{2}\,2\!\bigcirc\!4\right)$$

$$+\left(\mathrm{i}\,\overline{\underset{1\quad\;\;4}{}} - \frac{g}{2}\,1\!\bigcirc\!4\right)\left(\mathrm{i}\,\overline{\underset{2\quad\;\;3}{}} - \frac{g}{2}\,2\!\bigcirc\!3\right)$$

$$+\left(\frac{1\;\;2}{3\;\;4} + \frac{1\;\;3}{2\;\;4} + \frac{1\;\;4}{2\;\;3}\right)$$

$$+\frac{\mathrm{i}g}{2}\left(\frac{1\!\bigcirc\!2}{3\;\;4} + \frac{1\!\bigcirc\!3}{2\;\;4} + \frac{1\!\bigcirc\!4}{2\;\;3} + \frac{3\!\bigcirc\!4}{1\;\;2}\right.$$

$$\left.+\frac{2\!\bigcirc\!4}{1\;\;3} + \frac{2\!\bigcirc\!3}{1\;\;4}\right)$$

$$+\frac{\mathrm{i}g}{4!}\left({}_3^1\!\!\times\!\!{}_4^2 + {}_2^1\!\!\times\!\!{}_4^3 + \ldots(24\text{ terms})\right)\Bigg]$$

$$=\frac{-g}{4!}\left({}_3^1\!\!\times\!\!{}_4^2 + {}_2^1\!\!\times\!\!{}_4^3 + \ldots(24\text{ terms})\right)$$

$$= -g\times. \tag{6.111}$$

We see, indeed, that only the connected diagram survives.

Finally, let us briefly consider the *n*-point function. It is

$$\tau(x_1,\ldots,x_n) = \frac{1}{\mathrm{i}^n}\frac{\delta^n Z[J]}{\delta J(x_1)\ldots\delta J(x_n)}\bigg|_{J=0}. \tag{6.112}$$

We then define the *irreducible n-point function* $\phi(x_1,\ldots,x_n)$ by

$$\phi(x_1,\ldots,x_n) = \frac{1}{\mathrm{i}^n}\frac{\delta^n W[J]}{\delta J(x_1)\ldots\delta J(x_n)}\bigg|_{J=0}. \tag{6.113}$$

To justify the name, note that we have, from (6.100) and (6.111),

$$\left.\begin{array}{l}\tau(x_1,\ldots,x_4) = -\mathrm{i}g\times - 3\mathrm{i}g\,\overline{\!\bigcirc\!} - 3\,\overline{}\\[4pt] \mathrm{i}\phi(x_1,\ldots,x_4) = -\mathrm{i}g\times\end{array}\right\} \tag{6.114}$$

and from (6.110)

$$\mathrm{i}\phi(x_1,\ldots,x_4) = \tau(x_1,\ldots,x_4) - \tau(x_1,x_2)\tau(x_3,x_4)$$
$$- \tau(x_1,x_3)\tau(x_2,x_4) - \tau(x_1,x_4)\tau(x_2,x_3).$$

We have, however, $\tau(x_1, x_2) = i\phi(x_1, x_2)$, so

$$\tau(x_1, \ldots, x_4) = i\phi(x_1, \ldots, x_4) - \sum_p \phi(x_{i_1}, x_{i_2})\phi(x_{i_3}, x_{i_4}) \qquad (6.115)$$

where $\sum_p$ stands for the sum over all possible partitions of the indices $(1 \ldots 4)$ into classes $(i_1, i_2)(i_3, i_4)$. Equation (6.115) is now the same as (6.114). The 4-point function is decomposed into an 'irreducible' (or 'connected') part $\times$ and reducible parts $\underline{\quad\quad}$ and $\underline{\quad O\quad}$, in which, in general, a subset of the incident particle scatters into a subset of the final particles, quite independently of the rest. For the 4-point function we have,

$$\text{(diagram)} = \text{(diagram)} \ \text{irred.} + \underline{\quad\quad} \qquad (6.116)$$

which gives, to first order in g,

$$\text{(diagram)} = \text{(diagram)} + \underline{\quad\quad} + \underline{\quad O\quad} \cdot \qquad (6.117)$$

This generalises to n-point functions, for example

$$\text{(diagram)} = \text{(diagram irred.)} + \text{(diagram)} + \text{(diagram)} \cdot \qquad (6.118)$$

6.7 Fermions and functional methods

We saw in §4.3 that there is a connection between spin and statistics, so that fermion fields obey the anticommutation relations

$$\{\psi(x), \psi(y)\}|_{x^0 = y^0} = 0.$$

(Actually, the restriction $x^0 = y^0$ is unnecessary; the fields anticommute at all times.) In the canonical approach to field theory, $\psi(x)$ are regarded as operators, so we deal with a set of anticommuting operators. In the functional approach, however, the generating functional for the Green's functions is written as a functional integral over the fields, which are regarded as *classical* functions – c-numbers. To extend functional methods to Fermi fields, therefore, demands that, in the functional integral, they are regarded as *anticommuting c-numbers*. To most physicists this notion is strange, if not contradictory, but in the mathematical literature it goes back to 1855, where it appears in a paper by Hermann Grassmann on linear algebra. The generators C_i of an n-dimensional *Grassmann algebra* obey

$$\{C_i, C_j\} \equiv C_i C_j + C_j C_i = 0 \tag{6.119}$$

where $i = 1, 2, \ldots, n$. In particular,

$$C_i^2 = 0. \tag{6.120}$$

The expansion of a function $f(C_i)$ only contains a finite number of terms. For example, in the 1-dimensional algebra

$$f(C) = a + bC,$$

since the quadratic term vanishes, by (6.120).

Now consider the notion of differentiation. Because of the anticommutation relations there are two types, left and right. Their definitions follow from these examples; the *left derivative* of the product $C_1 C_2$ is

$$\frac{\partial^L}{\partial C_i}(C_1 C_2) = \delta_{i_1} C_2 - \delta_{i_2} C_1 \tag{6.121a}$$

and the *right derivative* is

$$\frac{\partial^R}{\partial C_i}(C_1 C_2) = \delta_{i_2} C_1 - \delta_{i_1} C_2. \tag{6.121b}$$

It is clear that the derivative operators must obey

$$\left\{\frac{\partial}{\partial C_i}, C_j\right\} = \delta_{ij}. \tag{6.122}$$

For example, in the 1-dimensional algebra this gives

$$\left\{\frac{d}{dC}, C\right\} = 1$$

and this relation may be checked by applying it to $f(C)$ in equation (6.121). It may also be verified, by performing a similar exercise, that

$$\left\{\frac{\partial}{\partial C_i}, \frac{\partial}{\partial C_j}\right\} = 0. \tag{6.123}$$

Then we have $(\partial/\partial C_i)^2 = 0$, which implies that there is no inverse to differentiation! This will make the definition of integration a bit tricky. A simple way of 'deducing' the rules for integration is to observe that if it has the same effect on a function as differentiation has, then we can get round the dilemma. So, since in the 1-dimensional case we have $df/dC = b$, we must also have $\int dC f(C) = b$. This requires

$$\int dC = 0, \quad \int dC \cdot C = 1,$$

which generalises to the n-dimensional case

$$\int dC_i = 0, \quad \int dC_i C_i = 1. \tag{6.124}$$

Now let η and $\bar{\eta}$ be independent Grassmann quantities, so that

$$\int d\eta = \int d\bar{\eta} = 0, \quad \int d\eta\,\eta = \int d\bar{\eta}\,\bar{\eta} = 1.$$

Because $\eta^2 = \bar{\eta}^2 = 0$, we have

$$e^{-\bar{\eta}\eta} = 1 - \bar{\eta}\eta$$

and hence

$$\int d\bar{\eta}\,d\eta\,e^{-\bar{\eta}\eta} = \int d\bar{\eta}\,d\eta - \int d\bar{\eta}\,d\eta\,\bar{\eta}\eta$$

$$= 0 + \int d\bar{\eta}\,d\eta\,\eta\bar{\eta}$$

$$= 1.$$

We now look for the generalisation of this formula to higher dimensions: let us consider the 2-dimensional case,

$$\eta = \begin{pmatrix} \eta_1 \\ \eta_2 \end{pmatrix}, \quad \bar{\eta} = \begin{pmatrix} \bar{\eta}_1 \\ \bar{\eta}_2 \end{pmatrix}.$$

The exponent $\bar{\eta}\eta$ (which should properly be written $\bar{\eta}^T\eta$, where T stands for transpose) is

$$\bar{\eta}\eta = \bar{\eta}_1\eta_1 + \bar{\eta}_2\eta_2$$

so

$$(\bar{\eta}\eta)^2 = (\bar{\eta}_1\eta_1 + \bar{\eta}_2\eta_2)(\bar{\eta}_1\eta_1 + \bar{\eta}_2\eta_2)$$

$$= \bar{\eta}_1\eta_1\bar{\eta}_2\eta_2 + \bar{\eta}_2\eta_2\bar{\eta}_1\eta_1$$

$$= 2\bar{\eta}_1\eta_1\bar{\eta}_2\eta_2$$

and higher powers of $\bar{\eta}\eta$ are zero, so we have

$$e^{-\bar{\eta}\eta} = 1 - (\bar{\eta}_1\eta_1 + \bar{\eta}_2\eta_2) + \bar{\eta}_1\eta_1\bar{\eta}_2\eta_2.$$

Applying the integration rules above, we then see that

$$\int d\bar{\eta}\,d\eta\,e^{-\bar{\eta}\eta} = \int d\bar{\eta}_1\,d\bar{\eta}_2\,d\eta_1\,d\eta_2\,\bar{\eta}_1\eta_1\bar{\eta}_2\eta_2$$

$$= 1, \tag{6.125}$$

as in the 1-dimensional case.

Now let us change variables, putting

$$\eta = M\alpha, \quad \bar{\eta} = N\bar{\alpha} \tag{6.126}$$

where M and N are 2×2 matrices, and α and $\bar{\alpha}$ are the new independent Grassmann quantities. We have

$$\eta_1\eta_2 = (M_{11}\alpha_1 + M_{12}\alpha_2)(M_{21}\alpha_1 + M_{22}\alpha_2)$$

$$= (M_{11}M_{22} - M_{12}M_{21})\alpha_1\alpha_2$$

$$= (\det M)\alpha_1\alpha_2.$$

However, in order to preserve the integration rules

$$\int d\eta_1 \, d\eta_2 \eta_1 \eta_2 = \int d\alpha_1 \, d\alpha_2 \alpha_1 \alpha_2$$

we must require

$$d\eta_1 \, d\eta_2 = (\det M)^{-1} \, d\alpha_1 \, d\alpha_2 \tag{6.127}$$

in contrast to the normal rule for a change of variable. Substituting (6.126) into (6.125), and remembering (6.127), we now have

$$(\det MN)^{-1} \int d\bar{\alpha} \, d\alpha \, e^{-\bar{\alpha} M^T N \alpha} = 1.$$

But since $\det MN = \det M^T N$, this gives, putting $MN = A$,

$$\int d\bar{\alpha} \, d\alpha \, e^{-\bar{\alpha} A \alpha} = \det A. \tag{6.128}$$

This formula, or rather its generalisation to the infinite-dimensional case, will be used in the next chapter, in finding the Feynman rules for gauge fields.

To describe Fermi fields, we now make the transition to an *infinite-dimensional Grassmann algebra*, whose generators may be denoted $C(x)$. They obey the relations

$$\left.\begin{aligned}
&\{C(x), C(y)\} = 0, \\[4pt]
&\frac{\partial^{\mathrm{L,R}} C(x)}{\partial C(y)} = \delta(x - y), \\[4pt]
&\int dC(x) = 0; \quad \int C(x) \, dC(x) = 1.
\end{aligned}\right\} \tag{6.129}$$

Integrals like (6.128) then become functional integrals over complex Grassmann variables. As in the case of scalar fields, we shall treat these formulae with the confidence that, one day, a rigorous mathematical justification for them will be found. In this spirit, we write down an expression for the generating functional for free Dirac fields, by analogy with our treatment of scalar fields – see equation (6.1). Since the Lagrangian for the Dirac field is

$$\mathcal{L} = i\bar{\psi}\gamma^\mu \partial_\mu \psi - m\bar{\psi}\psi,$$

the normalised generating functional for free Dirac fields is

$$Z_0[\eta, \bar{\eta}] = \frac{1}{N} \int \mathcal{D}\bar{\psi} \mathcal{D}\psi \exp\left\{ i \int [\bar{\psi}(x)(i\gamma \cdot \partial - m)\psi(x) \right.$$
$$\left. + \bar{\eta}(x)\psi(x) + \bar{\psi}(x)\eta(x)] \, dx \right\} \tag{6.130}$$

where the integral over x is 4-dimensional, and

$$N = \int \mathscr{D}\bar{\psi}\mathscr{D}\psi \exp\left[i \int \bar{\psi}(x)(i\gamma\cdot\partial - m)\psi(x)\,dx \right]. \tag{6.131}$$

Here $\bar{\eta}(x)$ represents the source term for $\psi(x)$, and $\eta(x)$ the source for $\bar{\psi}(x)$. It is now our aim to express this in a form analogous to equation (6.13), so that we can perform functional differentiation, and calculate Green's functions and S-matrix elements. To simplify the appearance of the formulae, we introduce the notation

$$S^{-1} = i\gamma^\mu\partial_\mu - m. \tag{6.132}$$

Then

$$Z_0[\eta,\bar{\eta}] = \frac{1}{N} \int \mathscr{D}\bar{\psi}\mathscr{D}\psi \exp\left[i \int (\bar{\psi}S^{-1}\psi + \bar{\eta}\psi + \bar{\psi}\eta)\,dx \right]. \tag{6.133}$$

Putting

$$Q(\psi,\bar{\psi}) = \bar{\psi}S^{-1}\psi + \bar{\eta}\psi + \bar{\psi}\eta,$$

we now find the value of ψ which minimises Q. It is

$$\psi_m = -S\eta, \quad \bar{\psi}_m = -\bar{\eta}S$$

(where we have assumed that S^{-1} possesses an inverse; it will be shown below that it does), and the minimum value of Q is

$$Q_m = Q(\psi_m, \bar{\psi}_m) = -\bar{\eta}S\eta.$$

We then have

$$Q = Q_m + (\bar{\psi} - \bar{\psi}_m)S^{-1}(\psi - \psi_m)$$

and

$$Z_0 = \frac{1}{N} \int \mathscr{D}\bar{\psi}\mathscr{D}\psi \exp\left\{ i \int [Q_m + (\bar{\psi} - \bar{\psi}_m)S^{-1}(\psi - \psi_m)]\,dx \right\}$$

$$= \frac{1}{N}\exp\left[-i \int \bar{\eta}(x)S\eta(y)\,dx\,dy \right]\det(iS^{-1}).$$

In the last step, e^{iQ_m} has been placed outside the integral, since Q_m does not depend on ψ or $\bar{\psi}$, and equation (6.128) has been used, duly extended to the functional case. Moreover, it is clear that $N = \det(iS^{-1})$, so, finally,

$$Z_0[\eta,\bar{\eta}] = \exp\left[-i \int \bar{\eta}(x)S(x-y)\eta(y)\,dx\,dy \right]. \tag{6.134}$$

It is easy to show that S exists. It is given by

$$S(x) = (i\gamma\cdot\partial + m)\Delta_F(x) \tag{6.135}$$

where $\Delta_F(x)$ is the Feynman propagator. With equation (6.132) we have

$$S^{-1}S = (i\gamma \cdot \partial - m)(i\gamma \cdot \partial + m)\Delta_F(x)$$
$$= (-\Box - m^2)\Delta_F(x)$$
$$= \delta^4(x).$$

We may now find the free propagator for the Dirac field. By analogy with equation (6.50), it is defined by

$$\tau(x, y) = -\left.\frac{\delta^2 Z_0[\eta, \bar{\eta}]}{\delta\eta(x)\delta\bar{\eta}(y)}\right|_{\eta=\eta=0}$$
$$= -\frac{\delta}{\delta\eta(x)}\frac{\delta}{\delta\bar{\eta}(y)}\left\{-i\int\bar{\eta}(x)S(x-y)\eta(y)\,dx\,dy\right\}\bigg|_{\eta=\bar{\eta}=0}$$
$$= iS(x-y). \tag{6.136}$$

Let us summarise our formulae for the free propagators of scalar and spinor fields. For scalar fields, with Lagrangian (up to a total divergence)

$$\mathscr{L}_0 = \tfrac{1}{2}\partial_\mu\phi\partial^\mu\phi - \tfrac{1}{2}m^2\phi^2 = -\tfrac{1}{2}\phi(\Box + m^2)\phi$$

we found the 2-point function

$$\tau(x, y) = i\Delta_F(x - y) \tag{6.52}$$

where the Feynman propagator Δ_F obeys

$$(\Box + m^2)\Delta_F(x - y) = -\delta^4(x - y). \tag{6.10}$$

For spinor fields, the Lagrangian is (see (6.132))

$$\mathscr{L}_0 = i\bar{\psi}\gamma^\mu\partial_\mu\psi - m\bar{\psi}\psi = \bar{\psi}S^{-1}\psi$$

and the 2-point function is i times the propagator

$$\tau(x, y) = iS(x - y).$$

In each case, it is seen that *the propagator is the inverse of the operator appearing in the quadratic term in the Lagrangian*. (The factor $\tfrac{1}{2}$ in the scalar Lagrangian is immaterial, and appears because ϕ is real; for complex ϕ, it is absent.) It is possible to take this as a *definition* of a propagator, and this is what we shall do when we consider gauge fields.

Finally, it will be convenient here to point out a further consequence of fields obeying Fermi statistics. It follows from the relation obeyed by the

Fig. 6.5. Modification to the scalar field propagator by a closed fermion loop.

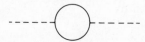

differential operators for the Grassmann fields. By a generalisation of (6.123), we have

$$\frac{\delta^2}{\delta\eta(x)\delta\eta(y)} = -\frac{\delta^2}{\delta\eta(y)\delta\eta(x)} \tag{6.137}$$

where η is a fermion source, and the operation of differentiation refers to either left or right differentiation. For left differentiation we have

$$\frac{\delta}{\delta\eta(x_1)}[\eta(x)\eta(y)] = \delta^4(x_1 - x)\eta(y) - \delta^4(x_1 - y)\eta(x).$$

What we want to show is that these rules result in a factor of -1 for each closed fermion loop in a Feynman diagram. For example, a spinor field coupled to a scalar field will give a correction to the free scalar propagator shown in Fig. 6.5, which contains a closed fermion loop and two interaction vertices. The appropriate 2-point function will be derived from the generating functional with interactions included. By a generalisation of (6.76), this will take the form

$$Z[\eta,\bar{\eta}] = \exp\left[i\int \mathscr{L}_{\text{int}}\left(\frac{1}{i}\frac{\delta}{\delta\eta}, \frac{1}{i}\frac{\delta}{\delta\bar{\eta}}\right)dx\right]Z_0[\eta,\bar{\eta}] \tag{6.138}$$

with Z_0 given by (6.134). The third term in its expansion is

$$-\frac{1}{2}\int dx\,dy\,dx'\,dy'\bar{\eta}(x)S(x-y)\eta(y)\bar{\eta}(x')S(x'-y')\eta(y').$$

The loop in Fig. 6.5 will contribute a term of the form

$$\frac{\delta^2}{\delta\bar{\eta}_i(z)\delta\eta_j(z)}\frac{\delta^2}{\delta\bar{\eta}_k(z')\delta\eta_l(z')}Z[\eta,\bar{\eta}]$$

(i,j,k and l are spinor indices). On substituting (6.138) for $Z[\eta,\bar{\eta}]$ and applying (6.137), this term is seen to be

$$+ S_{il}(z-z')S_{kj}(z'-z).$$

The overall sign would be $-$ if the fields obeyed Bose statistics, hence the -1 factor for the fermion loop.

6.8 The S matrix and reduction formula

We have seen how to calculate the Green's functions for an interacting theory, but we now want to calculate quantities which we measure directly in experiments. The commonest types of process which concern the elementary particle physicist are firstly scattering processes, in which a *cross section* for a particular reaction is measured, and secondly a decay of one particle into two or more, in which a partial lifetime is measured. The calculation of both these quantities, the cross section and the lifetime, is carried out by first calculating the quantum mechanical *amplitude* that the process takes place. Once we have the amplitude, the rest of the calculation is fairly straightforward. In this section we show how to calculate the amplitude, which we call the scattering amplitude, and show how a particular scattering amplitude is related by a simple formula, to a corresponding Green's function. We apply this to the case of the pion–nucleon interaction, and in the next section show how to obtain the scattering cross section. Let us consider, quite generally, a process in which an initial configuration of particles α, ends up as a final configuration β. We denote the scattering amplitude for this $S_{\beta\alpha}$, and call it the ($\beta\alpha$) 'matrix element' of the *scattering matrix* or S *matrix*. S is the collection of all $S_{\beta\alpha}$. The states α and β are defined asymptotically, at times $t \to -\infty$ and $t \to \infty$ respectively, so we define

$$S_{\beta\alpha} = \langle \beta, t \to \infty \,|\, \alpha, t \to -\infty \rangle. \tag{6.139}$$

In the absence of interactions with long-range forces, these asymptotic states consist of *free particles*, which is a great simplification. Long-range interactions, like electromagnetism, bring complications which we prefer to leave aside. An alternative notation is to define the 'in' and 'out' states

$$|\alpha\rangle_{\text{in}} = |\alpha, t \to -\infty\rangle, \quad |\beta\rangle_{\text{out}} = |\beta, t \to +\infty\rangle.$$

Suppose the 'in' state consists of two scalar particles, with momenta $\mathbf{p}_1$ and $\mathbf{p}_2$; then

$$|\alpha\rangle_{\text{in}} = a_{\text{in}}^\dagger(\mathbf{p}_1) a_{\text{in}}^\dagger(\mathbf{p}_2)|0\rangle,$$

and the creation operators $a_{\text{in}}^\dagger$ obey the usual free particle commutation relation (4.16a) with a_{in}.

The definition

$$S_{\beta\alpha} = {}_{\text{out}}\langle \beta |\alpha\rangle_{\text{in}}$$

is clearly equivalent to

$$a_{\text{out}}(p) = S^\dagger a_{\text{in}}(p)S,$$
$$a_{\text{out}}^\dagger(p) = S^\dagger a_{\text{in}}^\dagger(p)S,$$

which in turn is equivalent to the relation between the free field operators

$$\phi_{\text{out}}(x) = S^\dagger \phi_{\text{in}}(x)S. \tag{6.140}$$

We want to find an expression for the operator S. To begin, we take the seemingly irrelevant step of considering the field operator $\phi(x)$ at a time intermediate between $-\infty$ and $+\infty$; in other words, when it is subject to interactions. From the Lagrangian

$$\mathcal{L} = \tfrac{1}{2}\partial_\mu \phi \partial^\mu \phi - \frac{m^2}{2}\phi^2 + \mathcal{L}_{\text{int}}$$

ϕ obeys the equation of motion

$$(\Box_x + m^2)\phi(x) = \frac{\partial \mathscr{L}_{\text{int}}}{\partial \phi(x)}, \tag{6.141}$$

or, denoting the Klein–Gordon operator by

$$K_x = (\Box_x + m^2), \tag{6.142}$$

we have

$$K_x \phi(x) = \frac{\partial \mathscr{L}_{\text{int}}}{\partial \phi(x)}. \tag{6.143}$$

Let us now solve this equation.[‡] Denoting its Green's function by $G(x - y)$, we have, by definition,

$$(\Box_y + m^2)G(y - x) = \delta^4(y - x), \tag{6.144}$$

so on multiplying (6.141) by $G(y - x)$, and the above equation by $\phi(y)$, subtracting, and integrating over y, we have

$$\int d^4 y [G(y - x)(\Box_y + m^2)\phi(y) - \phi(y)(\Box_y + m^2)G(y - x)]$$

$$= \int G(y - x)\frac{\partial \mathscr{L}_{\text{int}}}{\partial \phi(y)} d^4 y - \phi(x).$$

The left-hand side of this equation is

$$\int d^3 y \, dy_0 \left\{ [G(y - x)\nabla^2 \phi(y) - \phi(y)\nabla^2 G(y - x)] \right.$$

$$\left. - \left[G(y - x)\frac{\partial^2}{\partial y_0^2}\phi(y) - \phi(y)\frac{\partial^2}{\partial y_0^2}G(y - x) \right] \right\}.$$

The 'space' part of this expression vanishes by Green's theorem:

$$\int d^3 y \, dy_0 (G\nabla^2 \phi - \phi \nabla^2 G) = \int dS^y \, dy_0 \cdot (G\nabla\phi - \phi\nabla G)$$

$$= 0,$$

so long as ϕ and $\nabla\phi$ vanish on the boundary at spatial infinity. As for the 'time' part, we have

$$G\frac{\partial^2}{\partial y_0^2}\phi - \phi\frac{\partial^2}{\partial y_0^2}G = \frac{\partial}{\partial y_0}\left(G\frac{\partial}{\partial y_0}\phi - \phi\frac{\partial}{\partial y_0}G \right)$$

$$= \frac{\partial}{\partial y_0}(G\overleftrightarrow{\partial_0}\phi),$$

[‡] The solution is equation (6.149) or (6.150), below. Readers who are familiar with, or too impatient to follow, the proof, are invited to skip to the solution.

so, collecting these results together gives

$$\phi(x) = \left(\int_{y_0^+} - \int_{y_0^-} \right) d^3y \, G(y - x) \overleftrightarrow{\partial}_0 \phi(y)$$

$$+ \int_{y_0^-}^{y_0^+} dy \, G(y - x) \frac{\partial \mathcal{L}_{\text{int}}}{\partial \phi(y)} \tag{6.145}$$

where we have carried out the time integration between times y_0^- and y_0^+, or, more generally, over the spacelike hypersurfaces σ^- and σ^+. Equation (6.145) is the solution to (6.141), but it is not yet in the form we want. From the theory of differential equations, we know that the general solution to (6.141) will consist of the sum of the solution of the homogeneous (*free field*) equation, plus the integral of the Green's function times the inhomogeneous term. It is obvious that this last is the second term in (6.145); we shall show that the first term represents a free field, but exactly *what* free field depends on the *boundary conditions*. In fact, the Green's function, satisfying (6.144), is itself not unique; for uniqueness, boundary conditions must be imposed. We therefore define the *advanced* and *retarded Green's functions* by

$$\left. \begin{array}{l} \Delta_{\text{ret}}(x) = 0 \quad \text{for } x^2 > 0, x_0 < 0, \\ \Delta_{\text{adv}}(x) = 0 \quad \text{for } x^2 < 0, x_0 > 0, \end{array} \right\} \tag{6.146}$$

$$(\Box + m^2) \Delta_{\substack{\text{ret} \\ \text{adv}}}(x) = \delta^4(x). \tag{6.147}$$

Now for any two Green's functions, G_1 and G_2, we have

$$G_1(x - y)(\Box_x + m^2)G_2(x - z) = \delta(x - z)G_1(x - y)$$

$$G_2(x - z)(\Box_x + m^2)G_1(x - y) = \delta(x - y)G_2(x - z),$$

hence, subtracting and integrating over x,

$$G_1(z - y) - G_2(y - z) = \left(\int_{x_0^+} - \int_{x_0^-} \right) d^3x \, G_2(x - z) \overleftrightarrow{\partial}_0 G_1(x - y).$$

Applying this to the retarded and advanced Green's functions gives

$$\Delta_{\text{ret}}(x) = \Delta_{\text{adv}}(-x). \tag{6.148}$$

Now substitute $G = \Delta_{\text{adv}}$ into equation (6.145), remembering that y_0^+ is a later time, and y_0^- an earlier time than x_0; we find

$$\phi(x) = - \int_{y_0^-} d^3y \, \Delta_{\text{ret}}(x - y) \overleftrightarrow{\partial}_0 \phi(y) + \int dy \, \Delta_{\text{ret}}(x - y) \frac{\partial \mathcal{L}}{\partial \phi(y)}.$$

This is the conventional result using retarded Green's functions, where $\phi(x)$ is expressed in terms of its boundary values at an *earlier* time. Now let $y_0^- \to -\infty$, and denote the first term on the right-hand side by

$$\phi_{-\infty}(x) = - \lim_{y_0 \to -\infty} \int_{y_0} d^3y \, \Delta_{\text{ret}}(x - y) \overleftrightarrow{\partial}_{y_0} \phi(y).$$

It is easy to see that this satisfies the free Klein–Gordon equation, for

$$(\Box + m^2)\phi_{-\infty}(x) = - \lim_{y_0 \to -\infty} \int d^3y \delta^4(x - y) \overleftrightarrow{\partial}_{y_0} \phi(y),$$

and the right-hand side clearly vanishes as $y_0 \to -\infty$, for any given x_0. So we may identify $\phi_{-\infty}(x)$ with the 'in' field $\phi_{in}(x)$, and write the solution to equation (6.141) as

$$\phi(x) = \phi_{in}(x) + \int dy \Delta_{ret}(x - y) \frac{\partial \mathcal{L}}{\partial \phi(y)}. \tag{6.149}$$

It is clear, moreover, that we may equally well take an alternative solution, using the advanced Green's function, and the boundary conditions at $y_0 = +\infty$:

$$\phi(x) = \phi_{out}(x) + \int dy \Delta_{adv}(x - y) \frac{\partial \mathcal{L}}{\partial \phi(y)}. \tag{6.150}$$

Using equation (6.143) we express these solutions in the form

$$\phi(x) = \phi_{in}(x) + \int dy \Delta_{ret}(x - y) K_y \phi(y),$$
$$\phi(x) = \phi_{out}(x) + \int dy \Delta_{adv}(x - y) K_y \phi(y). \tag{6.151}$$

It is obvious that, in some sense, we require

$$\phi(x) \xrightarrow[t \to \pm\infty]{} \phi_{\substack{out \\ in}}(x),$$

as an *asymptotic condition*. It has been shown, however, that if this condition is taken as it stands, as an operator condition, there is no scattering. This is called the *strong* asymptotic condition, and is obviously not what we want. The correct form was found by Lehmann, Symanzik and Zimmermann (LSZ)[‡] and is known as the *weak asymptotic condition*:

$$\lim_{t \to \pm\infty} \langle a | \phi(x) | b \rangle = \langle a | \phi_{\substack{out \\ in}}(x) | b \rangle \tag{6.152}$$

where $|a\rangle$ and $|b\rangle$ are arbitrary states in Hilbert space. The point is, of course, that although the strong condition implies the weak one, the converse is not true. It is our job now to find an expression for the operator S, defined through equation (6.140)

$$\phi_{out}(x) = S^\dagger \phi_{in}(x) S, \tag{6.140}$$

[‡] H. Lehmann, K. Symanzik & W. Zimmermann, *Nuovo Cimento*, **1**, 425 (1955).

using equations (6.151) and (6.152). First, we define the functional

$$I[J] = T \exp\left[i \int J(x)\phi(x)\,dx \right] \tag{6.153}$$

where T is the time-ordering operator. This implies that

$$\frac{1}{i}\frac{\delta I[J]}{\delta J(x)} = T(\phi(x)I[J]).$$

Comparing this equation (and corresponding higher derivatives) with (6.46), it follows that the vacuum expectation value of $I[J]$ is $Z[J]$:

$$Z[J] = \langle 0|I[J]|0 \rangle. \tag{6.154}$$

Now we multiply equation (6.151) by $I[J]$ and T, bearing in mind the asymptotic condition, to give

$$\frac{1}{i}\frac{\delta I[J]}{\delta J(x)} = I[J]\phi_{in}(x) + \int dy\,\Delta_{ret}(x-y)K_y\frac{1}{i}\frac{\delta I[J]}{\delta J(y)},$$

$$\frac{1}{i}\frac{\delta I[J]}{\delta J(x)} = \phi_{out}(x)I[J] + \int dy\,\Delta_{adv}(x-y)K_y\frac{1}{i}\frac{\delta I[J]}{\delta J(y)},$$

and hence, on subtraction,

$$\phi_{out}I - I\phi_{in} = i \int dy\,\Delta(x-y)K_y\frac{\delta I[J]}{\delta J(y)} \tag{6.155}$$

where

$$\Delta(x) = \Delta_{adv}(x) - \Delta_{ret}(x).$$

Substituting for ϕ_{out} from (6.140) and noting that S is a unitary operator, $SS^\dagger = S^\dagger S = 1$, we then have

$$[\phi_{in}(x), SI[J]] = i \int dy\,\Delta(x-y)K_y\frac{\delta(SI[J])}{\delta J(y)}. \tag{6.156}$$

This equation, unappetising as it looks, is not difficult to solve. Indeed, a simple re-arrangement of the Baker–Campbell–Hausdorff formula (6.80)

$$e^B A e^{-B} = A + [B, A],$$

where A and B are operators whose commutator is a c-number, gives

$$[A, e^B] = [A, B]e^B.$$

On noting that

$$[\phi_{in}(x), \phi_{in}(y)] = i\Delta(x-y),$$

the following solution to (6.156) suggests itself:

$$SI[J] = \exp\left[\int \phi_{in}(z)K\frac{\delta}{\delta J(z)}\,dz \right]F[J] \tag{6.157}$$

where $F[J]$ is some functional of J. Indeed, we have

$$[\phi_{in}(x), SI] = i \int \Delta(x-y) K \frac{\delta}{\delta J(y)} dy \exp\left[\int \phi_{in}(z) K \frac{\delta}{\delta J(z)} dz\right] F[J]$$

$$= \exp\left[\int \phi_{in}(z) K \frac{\delta}{\delta J(y)} dz\right] i \int \Delta(x-y) K \frac{\delta F[J]}{\delta J(y)} dy$$

and

$$\frac{\delta(SI[J])}{\delta J(y)} = \exp\left[\int \phi_{in}(z) K \frac{\delta}{\delta J(z)} dz\right] \frac{\delta F[J]}{\delta J(y)},$$

so (6.157) solves (6.156). To find $F[J]$, we take the vacuum expectation value of (6.157). This implies first that we shall have to normal-order the exponential. But having done that, note that

$$\langle 0|:e^A:|0\rangle = 1$$

for any operator A, so we have

$$\langle 0|SI[J]|0\rangle = F[J].$$

On the other hand,

$$\langle 0|SI[J]|0\rangle = \langle 0|I[J]|0\rangle = Z[J]$$

so $F[J]$ is simply $Z[J]$, and we have

$$SI[J] = :\exp\left[\int \phi_{in}(z) K \frac{\delta}{\delta J(z)} dz\right]: Z[J].$$

Finally, noting that $I[J] \to 1$ as $J \to 0$, we obtain

$$S = :\exp\left[\int \phi_{in}(z) K \frac{\delta}{\delta J(z)} dz\right]: Z[J]\Big|_{J=0} \tag{6.158}$$

This is the *reduction formula* in functional form. How it works out, as we shall see below, is that, on expanding the exponential, a typical term contains $(\delta/\delta J)^n Z[J]|_{J=0}$, which is, of course, the n-point Green's function. For each external particle, we have an operator $K = (\Box + m^2)$, which reduces the propagator to a delta function; we then multiply by the free particle wave function ϕ_{in}. In the above formula, of course, ϕ_{in} is not a wave function, it is a field operator – note that S is an operator. What we shall show is that when we find a particular *S-matrix element*, it is indeed given by this simple prescription. To be more explicit, we first find the variational derivatives

$$\frac{1}{i} \frac{\delta}{\delta J(x_1)} \frac{1}{i} \frac{\delta}{\delta J(x_2)} \cdots \frac{1}{i} \frac{\delta}{\delta J(x_n)} Z[J]\Big|_{J=0}$$

$$= G(x_1, \ldots, x_n),$$

which are the n-point Green's functions, then multiply by

$$\prod_i (\Box_{x_i} + m^2),$$

which removes the external legs from the Green's function, and finally multiply by

$$\prod_i \phi(x_i),$$

the external free particle wave functions. Collecting this together, the n-particle S-matrix element is

$$S_n(x_1, \ldots, x_n) = \prod_i \phi(x_i)(\Box_{x_i} + m)G(x_1, \ldots, x_n). \tag{6.159}$$

Of course, it must be remembered that this formula refers to scalar fields, and interesting physical processes involve spinor fields. In this case the Klein–Gordon operator $K_x = (\Box_x + m^2)$ is replaced by the Dirac operator $D_x = (\gamma \cdot \partial + m)$. We now have all the formulae we need to calculate the pion–nucleon scattering amplitude.

$\rightarrow (i\gamma - m)$?

6.9   Pion–nucleon scattering amplitude

In the presence of scalar and spinor fields, of respective masses m and M, the formula (6.158) becomes generalised to

$$S =\, :\exp\left[\int \left(\phi_{in}\vec{K}\frac{\delta}{\delta J} + \bar{\psi}_{in}\vec{D}\frac{\delta}{\delta \bar{\eta}} - \frac{\delta}{\delta \eta}\overleftarrow{\bar{D}}\psi_{in} \right)dx \right]: Z[J, \eta, \bar{\eta}]\Big|_0 \tag{6.160}$$

where

$$D = i\gamma^\mu \partial_\mu - M, \quad \bar{D} = -i\gamma^\mu \partial_\mu - M \tag{6.161}$$

and the arrows indicate that K, D and $\bar{D}$ operate on the ϕ, ψ and $\bar{\psi}$ brought in from the functional differentiation of Z, and *not* on ϕ_{in}, $\bar{\psi}_{in}$ and ψ_{in}. The pion is actually a pseudoscalar particle, so its interaction with the nucleon system is given by

$$\mathscr{L}_{int} = ig\bar{\psi}\gamma_5\vec{\tau}\psi \cdot \vec{\phi}. \tag{6.162}$$

The bilinear $\bar{\psi}\gamma_5\psi$ is a pseudoscalar – see equation (2.126) – making $\mathscr{L}_{int}$ a scalar. The vectors $\bar{\psi}\gamma_5\vec{\tau}\psi$ and $\vec{\phi}$ are in isospace, since pions are isovectors. Defining

$$\phi^\mp = \frac{1}{\sqrt{2}}(\phi_1 \pm i\phi_2), \quad \phi^0 = \phi_3,$$

this gives

$$\mathscr{L}_{int} = i\sqrt{2}g(\bar{p}\gamma_5 n\phi^+ + \bar{n}\gamma_5 p\phi^-) + ig(\bar{p}p - \bar{n}n)\phi^0.$$

Suppose we are considering $\pi^+ p$ scattering. Then the only interaction term of interest is

$$\mathscr{L}_{\text{int}} = i\sqrt{2}g(\bar{p}\gamma_5 n\phi^+ + \bar{n}\gamma_5 p\phi^-),$$

and the complete Lagrangian is

$$\mathscr{L} = \mathscr{L}_0 + \mathscr{L}_{\text{int}} = i\bar{\psi}\gamma^\mu\partial_\mu\psi \mp M\bar{\psi}\psi + \tfrac{1}{2}\partial^\mu\vec{\phi}\partial_\mu\vec{\phi} - \tfrac{1}{2}m^2\vec{\phi}^2$$
$$+ i\sqrt{2}g(\bar{p}\gamma_5 n\phi^+ + \bar{n}\gamma_5 p\phi^-). \tag{6.163}$$

Here the terms in the free part, $\mathscr{L}_0$, are isoscalars, and, in the interaction term, n and p describe the destruction of n and p particles, or creation of their antiparticles; ϕ^+ describes destruction of π^+ or creation of π^-, and ϕ^- destruction of π^- or creation of π^+.

The generating functional $Z[J,\eta,\bar{\eta}]$ is given by the equation analogous to (6.76):

$$Z[J,\eta,\bar{\eta}] = \frac{\exp\left\{i\int\left(i\sqrt{2}g\dfrac{\delta}{\delta\eta(z)}\gamma_5\dfrac{\delta}{\delta\bar{\eta}(z)}\dfrac{\delta}{\delta J(z)}\right)dz\right\}Z_0[J,\eta,\bar{\eta}]}{\left[\quad''\quad\right]\bigg|_{J=\eta=\bar{\eta}=0}}$$

$$\tag{6.164}$$

with

$$Z_0[J,\eta,\bar{\eta}]$$

$$= \frac{\int \mathscr{D}\phi\,\overset{\psi}{\mathscr{D}\bar{\eta}}\,\overset{\bar{\psi}}{\mathscr{D}\eta}\exp\left[i\int(\bar{\psi}S^{-1}\psi - \tfrac{1}{2}\phi K\phi + \bar{\eta}\psi + \bar{\psi}\eta + J\phi)\,dx\right]}{\left[\quad''\quad\right]\bigg|_{J=\eta=\bar{\eta}=0}}$$

$$= \exp\left\{-i\int[\tfrac{1}{2}J(x)\Delta_F(x-y)J(y) + \bar{\eta}(x)S(x-y)\eta(y)]\,dx\,dy.\right\}$$

$$\tag{6.165}$$

Here J and η refer to the sources of whichever pion or nucleon field we are concerned with. This will become clear below, when we use these expressions. First, however, we shall calculate the S-matrix element for $\pi^+ p$ scattering, and show that it is obtained from the corresponding Green's function in the way described at the end of the last section.

The kinematics are shown in Fig. 6.6. The initial proton has momentum p and spin s, the initial pion momentum k, and the final values of these quantities are denoted by a prime. We want to calculate the scattering amplitude

$$S_{\text{fi}} = \langle p', s'; k'|S|p, s, k\rangle$$

with

$$S =: \exp \int \phi_{in}\overleftarrow{K}\frac{\delta}{\delta J} + \overline{\psi}_{in}\overrightarrow{D}\frac{\delta}{\delta\overline{\eta}} - \frac{\delta}{\delta\eta}\overleftarrow{D}\psi_{in}: Z\bigg|_0 .$$

The operators $K(\delta/\delta J)$, $D(\delta/\delta\overline{\eta})$ and $(\delta/\delta\eta)\overline{D}$ may be brought outside the bracket, and act directly on Z. Writing the initial and final states as

$$|p, s; k\rangle = b_s^\dagger(p)a^\dagger(k)|0\rangle$$

$$\langle p', s'; k| = \langle 0|b_{s'}(p')a(k'),$$

and expanding the exponential, we have, up to first order,

$$S_{fi} = \langle 0|b_{s'}(p')a(k')$$

$$\times \left[1 + : \int \phi_{in}\overleftarrow{K}\frac{\delta}{\delta J} + \overline{\psi}_{in}\overrightarrow{D}\frac{\delta}{\delta\overline{\eta}} - \frac{\delta}{\delta\eta}\overleftarrow{D}\psi_{in}: + \cdots\right]$$

$$\times b_s^\dagger(p)a^\dagger(k)|0\rangle.$$

The zeroth order term only contributes if $p = p'$, $s = s'$ and $k = k'$, and that corresponds to no interaction, which is not interesting. We shall now show that the first order term vanishes. The general idea is to change the order of the creation and annihilation operators, remembering that $b_s(p)|0\rangle = a(k)|0\rangle = 0$. Since the boson and fermion operators commute with each other, we may consider these terms separately. In that case, substituting for ϕ_{in}, which we simply write as ϕ, the free field operator, and using equation (4.14), the boson term is

$$\langle 0|a(k') \int dx \frac{d^3q}{[(2\pi)^3 2\omega_q]^{\frac{1}{2}}} [f_q(x)a(q) + f_q^*(x)a^\dagger(q)]a^\dagger(k')|0\rangle \overleftarrow{K}\frac{\delta}{\delta J}$$

$$= \langle 0|a(k') \int dx \frac{d^3q}{[(2\pi)^3 2\omega_q]^{\frac{1}{2}}} f_q(x)[a(q), a^\dagger(k')]$$

$$+ \int dx \frac{d^3q}{[(2\pi)^3 2\omega_q]^{\frac{1}{2}}} f_q^*(x)[a(k'), a^\dagger(q)]a^\dagger(k')|0\rangle \overleftarrow{K}\frac{\delta}{\delta J}$$

$$= 0 .$$

where we have used the fact that the commutators displayed are c-numbers, hence in the first term we have $a(k')$ acting on $|0\rangle$, and in the second, $a^\dagger(k')$ acting on $\langle 0|$. So the first order term is zero, as claimed.

Fig. 6.6. Kinematics of $p\pi^+$ scattering. The solid line denotes the proton, the dashed line the pion.

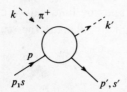

Now consider the second order term. The coefficient of $\frac{1}{2}(\vec{K}[\delta/\delta J])^2$ is, in the bosonic sector,

$$
\langle 0|a(k') \int dx_1\, dx_2 \frac{d^3q}{[(2\pi)^3 2\omega_q]^{\frac{1}{2}}} \frac{d^3q'}{[(2\pi)^3 2\omega_{q'}]^{\frac{1}{2}}} [f_q(x_1) f_{q'}(x_2) a(q) a(q')
$$
$$
+ f_q(x_1) f_{q'}^*(x_2) a^\dagger(q') a(q) + f_q^*(x_1) f_{q'}(x_2) a^\dagger(q) a(q')
$$
$$
+ f_q^*(x_1) f_{q'}^*(x_2) a^\dagger(q) a^\dagger(q')] a^\dagger(k)|0\rangle. \tag{6.166}
$$

We have, here, carried out the normal-ordering prescription of putting annihilation operators to the right of creation operators. As above, we then write suitably chosen products of a and $a^\dagger$ as (*c*-number) commutators, so it is immediately clear that the first and last terms vanish. As for the second and third terms, observe first that

$$
\langle 0|a(k')a^\dagger(q')a(q)a^\dagger(k)|0\rangle = \langle 0|[a(k'), a^\dagger(q')][a(q), a^\dagger(k)]|0\rangle
$$
$$
= (2\pi)^6 4\omega_{k'}\omega_k \delta^3(k'-q')\delta^3(q-k),
$$
$$
\langle 0|a(k')a^\dagger(q)a(q')a^\dagger(k)|0\rangle = (2\pi)^6 4\omega_{k'}\omega_k \delta^3(k'-q)\delta^3(k-q'),
$$

so expression (6.166) becomes, using (4.11),

$$
\int dx_1\, dx_2 \frac{d^3q\, d^3q'}{[4\omega_q\omega_{q'}]^{\frac{1}{2}}} 4\omega_k\omega_{k'} \frac{1}{[4\omega_q\omega_{q'}]^{\frac{1}{2}}} [e^{-i(qx_1 - q'x_2)}
$$
$$
\times \delta^3(k'-q')\delta^3(q-k) + e^{i(qx_1 - q'x_2)}\delta^3(k'-q)\delta^3(k-q')]
$$
$$
= \int dx_1\, dx_2 (e^{-ikx_1}e^{ik'x_2} + e^{ikx_1}e^{-ikx_2}).
$$

This whole term is the coefficient of

$$
\frac{1}{2} K_{x_1} \frac{\delta}{\delta J(x_1)} K_{x_2} \frac{\delta}{\delta J(x_2)} Z[J,\eta,\bar{\eta}]\Big|_{J=0}.
$$

Because of the symmetry under $x_1 \leftrightarrow x_2$, the bosonic contribution to the *S*-matrix element is

$$
\int dx_1\, dx_2\, e^{ik'x_2} K_{x_2} \frac{\delta}{\delta J(x_1)} \frac{\delta}{\delta J(x_2)} Z[J,\eta,\bar{\eta}]\Big|_{J=0} \overleftarrow{K}_{x_1} e^{-ikx_1} \tag{6.167}
$$

where the arrow on K_{x_1} indicates that it acts on the functional derivation of Z.

We now turn to the fermionic sector, and see a very similar pattern. We have to evaluate

$$
\int dx_3\, dx_4 \langle 0|b_{s'}(p'){:}\left[\psi(x_4) D_{x_4} \frac{\delta}{\delta\bar{\eta}(x_4)} - \bar{\psi}(x_4)\frac{\delta}{\delta\eta(x_4)}\bar{D}_{x_4} \right]
$$
$$
\times \left[\bar{\psi}(x_3) D_{x_3} \frac{\delta}{\delta\bar{\eta}(x_3)} - \psi(x_3)\frac{\delta}{\delta\eta(x_3)}\bar{D}_{x_3} \right]{:}\, b_s^\dagger(p)|0\rangle.
$$

On substituting for $\psi(x)$ and $\bar{\psi}(x)$ from (4.44), and turning products of operators into anticommutators, it is seen that the $:\bar{\psi}(x_4)\bar{\psi}(x_3):$ and $:\psi(x_4)\psi(x_3):$ terms vanish, while the $:\bar{\psi}(x_4)\psi(x_3):$ term gives

$$- \langle 0|b_{s'}(p') \int \frac{d^3q\,d^3q'}{(2\pi)^6} \frac{M^2}{q_0'q_0} \sum_{\alpha\alpha'}$$

$$\times :[b_\alpha^\dagger(q)\bar{u}^\alpha(q)e^{iqx_4} + d_\alpha(q)\bar{v}^\alpha(q)e^{-iqx_4}]$$

$$\times [b_{\alpha'}(q')u^{\alpha'}(q')e^{-iq'x_3} + d_{\alpha'}^\dagger(q')v^{\alpha'}(q')e^{iq'x_3}]:b_s(p)|0\rangle$$

$$= - \int \frac{d^3q\,d^3q'}{(2\pi)^6} \frac{M^2}{q_0'q_0} \sum_{\alpha\alpha'} \langle 0|b_{s'}(p')b_\alpha^\dagger(q)b_{\alpha'}(q')b_s^\dagger(p)|0\rangle$$

$$\times \bar{u}^\alpha(q)u^{\alpha'}(q')e^{iqx_4}e^{-iq'x_3}$$

$$= - \bar{u}^{s'}(p')u^s(p)e^{ip'x_4}e^{-ipx_3}.$$

The other non-vanishing term comes from $:\psi(x_4)\bar{\psi}(x_3):$, which gives the same as above, with $x_3 \leftrightarrow x_4$, and an additional minus sign, coming from the definition of normal-ordering (see the remark above equation (4.48)). These terms are, respectively, the coefficients of $\frac{1}{2}D_{x_4}(\delta/\delta\bar{\eta}(x_4))(\delta/\delta\bar{\eta}(x_3))\bar{D}_{x_3}$ and $\frac{1}{2}(\delta/\delta\eta(x_4))\bar{D}_{x_4}D_{x_3}(\delta/\delta\bar{\eta}(x_3))$. On making the interchange $x_3 \leftrightarrow x_4$ in the second term, and observing that the Ds commute, but $\delta/\delta\eta$ and $\delta/\delta\bar{\eta}$ anticommute (cf. (6.137)), thus introducing yet another minus sign, we can then write the fermionic contribution to the S-matrix element as

$$\int dx_3\,dx_4 e^{ip'x}\bar{u}^{s'}(p')D_{x_4} \frac{\delta}{\delta\eta(x_3)} \frac{\delta}{\delta\bar{\eta}(x_4)} Z[J,\eta,\bar{\eta}]\bigg|_{\eta=\bar{\eta}=0}$$

$$\times \overleftarrow{\bar{D}}_{x_3}u^s(p)e^{-ipx_3}. \tag{6.168}$$

Combining, finally, the pion and nucleon contributions, we may write the pion–nucleon scattering amplitude as

$$S_{fi} = \int dx_1\ldots dx_4 e^{ik'x_2}e^{ip'x_4}\bar{u}^{s'}(p')\overleftarrow{K}_{x_2}\overleftarrow{D}_{x_4}\tau(x_1,\ldots,x_4)\overleftarrow{\bar{D}}_{x_3}\overleftarrow{K}_{x_1}$$

$$\times u^s(p)e^{-ipx_3}e^{-ikx_1} \tag{6.169}$$

where $\tau(x_1,x_2,x_3,x_4)$ is the *4-point function* or *Green's function*:

$$\tau(x_1,x_2,x_3,x_4) = \frac{\delta}{\delta J(x_1)} \frac{\delta}{\delta J(x_2)} \frac{\delta}{\delta\eta(x_3)} \frac{\delta}{\delta\eta(x_4)} Z[J,\eta,\bar{\eta}]\bigg|_{J=\eta=\bar{\eta}=0} \tag{6.170}$$

This is the content of the reduction formula for this particular scattering process; the scattering amplitude is obtained from the Green's function by multiplying by the Klein–Gordon, or Dirac, operators, as appropriate, for the external particles, then multiplying by the wave function of the particles at positions $x_1, \ldots, x_4$, and integrating over $x_1, \ldots, x_4$. Although exhibited here for the case of pion–nucleon scattering, this recipe is general.

To get an explicit expression we need to know the 4-point function, obtained by functional differentiation of the generating functional given by (6.164). Fortunately, however, we do not need to perform this task explicitly, for we can invoke the rules we discovered in §6.5, making the obvious adaptation from ϕ^4 to $i\bar{\psi}\gamma_5\psi\phi$ for the interaction term. We first draw a *prediagram* for the propagation of (πN) to (πN), with two interaction terms, as in Fig. 6.7. We are only interested in constructing connected Feynman diagrams, since only these give a contribution to the scattering. The pion lines from the interaction vertices must be connected to the external pion lines, and a pair of nucleon lines from the vertices must be joined together. There are clearly two ways of doing this, indicated by the dotted lines in Fig. 6.7. The two resulting Feynman diagrams are redrawn in Fig. 6.8. Let us recall that these diagrams, in our convention, are to be read from left to right (with the incoming particles on the left). Many writers draw them to be read from bottom to top, and in fact the most logical way is to draw them from right to left, the same way as the expression for the matrix element $\langle \text{out}|S|\text{in}\rangle$. In the general case, both diagrams of Fig. 6.8 will contribute to the scattering and the cross section will show an interference term so characteristic of quantum phenomena, but in the case of $\pi^+ p$ scattering, Fig. 6.8(a) will not contribute, since the intermediate nucleon state, N^{++} with charge 2, does not exist. The virtual N in Fig. 6.8(b) is, in this case, a neutron. Recalling that the coupling constant

Fig. 6.7. Prediagram for πN scattering to second order, showing the two possible ways of forming a Feynman diagram.

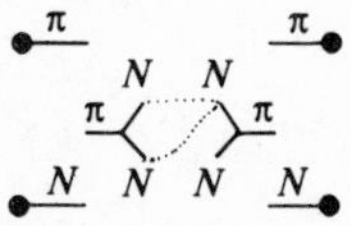

Fig. 6.8. Connected Feynman diagrams for pion–nucleon scattering, to second order.

(a)   (b)

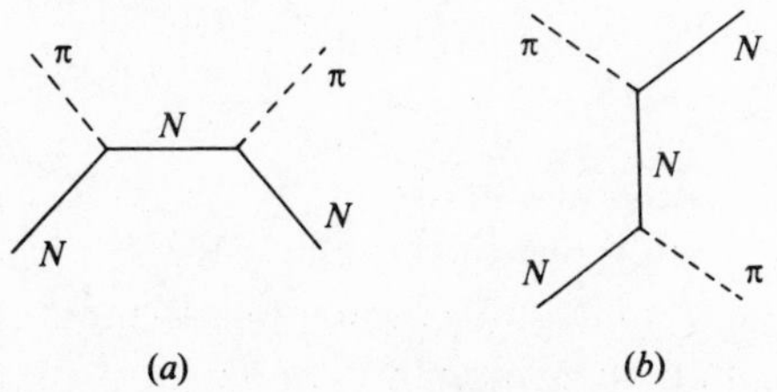

for the $(\pi^+ pn)$ vertex (equation 6.163)) is $\sqrt{2}g$, and that the vertex contains an $(i\gamma_5)$ factor, we may immediately write the 4-point function for Fig. 6.8(b), redrawn in Fig. 6.9, as

$$\tau(x_1, x_2, x_3, x_4) = 2g^2 \int dy_1\, dy_2 i\Delta_F(x_2 - y_2)iS(x_4 - y_1)i\gamma_5$$

$$\times iS(y_1 - y_2)i\gamma_5 iS(y_2 - x_3)i\Delta_F(x_1 - y_1). \qquad (6.171)$$

This is to be inserted into (6.169). Recalling that

$$K_x \Delta_F(x - y) = (\Box_x + m^2)\Delta_F(x - y) = -\delta^4(x - y),$$
$$D_x S(x - y) = (i\gamma \cdot \partial_x - M)S(x - y) = \delta^4(x - y),$$
$$S(x - y)\overleftarrow{D}_x = S(x - y)(\overleftarrow{-i\gamma \cdot \partial_x} - M) = \delta^4(x - y),$$

we obtain

$$S_{\text{fi}} = -2ig^2 \int dx_1 \dots dx_4\, dy_1\, dy_2 e^{ik'x_2} e^{ip'x_4} \bar{u}^{s'}(p')\delta^4(x_2 - y_2)$$

$$\times \delta^4(x_4 - y_1)\gamma_5 S(y_1 - y_2)\gamma_5 \delta^4(y_2 - x_3)$$

$$\times \delta^4(y_1 - x_1)u^s(p)e^{-ipx_3}e^{-ikx_1}$$

$$= -2ig^2 \int dy_1\, dy_2 e^{i(p'-k)y_1} e^{i(k'-p)y_2} \bar{u}^{s'}(p')\gamma_5 S(y_2 - y_1)\gamma_5 u^s(p).$$

Combining equations (6.135) for $S(x)$, and (6.14) for $\Delta_F(x)$, we have

$$S(y_1 - y_2) = \frac{1}{(2\pi)^4} \int dq \frac{\gamma^\mu q_\mu + M}{q^2 - M^2} e^{-iq(y_1 - y_2)}. \qquad (6.172)$$

Inserting this into the above equation, we may first carry out the integration over y_1 and y_2:

$$\int dy_1\, dy_2 e^{i(p'-k)y_1} e^{i(k'-p)y_2} e^{-iq(y_1-y_2)}$$

$$= (2\pi)^8 \delta^4(p' - k - q)\delta^4(k' - p + q)$$
$$= (2\pi)^8 \delta^4(q - p + k')\delta^4(p' + k' - p - k)$$
$$= (2\pi)^8 \delta^4(q - p + k')\delta^4(P_f - P_i),$$

Fig. 6.9. $\pi^+ p$ scattering.

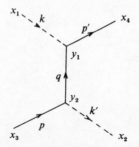

showing that the intermediate neutron has 4-momentum q, and 4-momentum is conserved at each vertex, as well as overall. The integration over q may now be carried out, as long as q is replaced by $p - k' = p' - k$. Therefore

$$S_{fi} = -i\delta^4(P_f - P_i)2g^2(2\pi)^4\bar{u}^{s'}(p')\gamma_5 \frac{\gamma\cdot(p-k')+M}{(p-k')^2 - M^2}\gamma_5 u^s(p).$$

(6.173)

Now using the Dirac equation (2.140) in momentum space and the properties of the γ matrices, this expression may be simplified to give

$$S_{fi} = i\delta^4(P_f - P_i)2g^2(2\pi)^4\bar{u}^{s'}(p')\gamma\cdot k' u^s(p)\frac{1}{2pk' - m^2}.$$

(6.174)

From the scattering amplitude, the next task is to find the scattering cross section, and this is dealt with in the next section. Before finishing this section, however, we shall summarise briefly the *Feynman rules* for calculating the scattering amplitude for an interaction involving scalar (or pseudoscalar) and spinor particles. It will immediately be seen that they yield the expression (6.173) for the particular case of $\pi^+ p$ scattering to lowest order perturbation theory.

1. To nth order, perturbation theory corresponds to a diagram with n vertices. The amplitude for a particular process (i.e. with particular ingoing and outgoing external lines) to a particular order is obtained by adding the amplitudes of all topologically inequivalent connected diagrams. Fig. 6.8 shows the two diagrams for (pseudo-) scalar spinor scattering to second order. Fig. 6.10 shows

Fig. 6.10. Some fourth order diagrams for scalar/pseudoscalar-spinor (e.g. pion–nucleon) scattering. The last diagram is disconnected and is not counted.

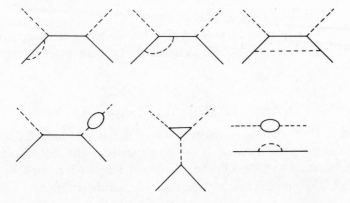

Table 6.1. *Feynman rules for scalar (or pseudoscalar) and spinor theories*

Factor in S matrix	Diagrammatic representation
$u(p)$	—————————incoming fermion (external line)
$\bar{u}(p)$	—————————outgoing fermion (external line)
$ig(\gamma_5)(2\pi)^4\delta^4(p+k-p')$	vertex
$\dfrac{i}{(2\pi)^4}\dfrac{1}{\gamma\cdot p - M}\left(\text{with }\int d^4p\right)$	————>———fermion propagator (internal line)
$\dfrac{i}{(2\pi)^4}\dfrac{1}{p^2 - m^2}\left(\text{with }\int d^4p\right)$	----->----scalar propagator (internal line)

some fourth order diagrams. Spinor lines are continuous, scalar lines dotted.

2. For each incoming spinor particle write $u(p)$ $(u(-p)$ for its antiparticle), and for each outgoing spinor particle $\bar{u}(p)$.

3. For each vertex write ig (for scalar interaction) or $ig\gamma_5$ (for pseusoscalar), where g is the relevant coupling constant read off from the interaction Lagrangian, and multiply by $(2\pi)^4\delta^4$ (incoming momenta).

4. For each spinor propagator (internal line) of momentum p write

$$\frac{1}{(2\pi)^4}\frac{i}{(\gamma\cdot p - M)}d^4p.$$

5. For each (pseudo-) scalar propagator write

$$\frac{1}{(2\pi)^4}\frac{i}{p^2 - m^2}d^4p.$$

6. Integrate over internal momenta.

These rules are summarised in Table 6.1.

6.10   Scattering cross section

We now show how to calculate the scattering cross section from the scattering amplitude. First, we define the Lorentz invariant amplitude M by

$$\langle p'_1, p'_2, \ldots, |S - 1|p_1, p_2, \ldots \rangle$$
$$= (2\pi)^4\delta^4(P_f - P_i)iM(p'_1, p'_2, \ldots, p_1, p_2, \ldots), \qquad (6.175)$$

recalling that what we calculated above did not include the identity contribution to the S matrix. It also only referred to the connected part of $S - 1$, but this we shall take to be understood. Further, for simplicity, let us consider the simplest case of 2-particle $\rightarrow$ 2-particle scattering.

Now, in general, the initial particles will not be sharp momentum states p_1, p_2, but will be wave packets:

$$\text{initial state } |i\rangle = \int \frac{d^3k_1}{(2\pi)^3 2k_1^0} \frac{d^3k_2}{(2\pi)^3 2k_2^0} f(k_1)g(k_2)|k_1, k_2\rangle$$

which, however, have momenta peaked respectively about p_1 and p_2:

$$k_1 \approx p_1, \quad k_2 \approx p_2.$$

The final state is

$$|f\rangle = |p_1', p_2'\rangle$$

so the transition amplitude is (with $d\tilde{k} = d^3k/(2\pi)^3 2k^0$)

$$\int d\tilde{k}_1 \, d\tilde{k}_2 f(k_1)g(k_2)\langle p_1' p_2'|S - 1|k_1 k_2\rangle$$

$$= (2\pi)^4 i \int d\tilde{k}_1 \, d\tilde{k}_2 f(k_1)g(k_2)\delta(p_1' + p_2' - k_1 - k_2)$$

$$\times M(p_1', p_2', k_1, k_2)$$

and the transition probability is

$$W = (2\pi)^8 \int d\tilde{k}_1 \, d\tilde{k}_2 \, d\tilde{q}_1 \, d\tilde{q}_2 f(k_1)g(k_2)f^*(q_1)g^*(q_2)$$

$$\times \delta(p_1' + p_2' - k_1 - k_2)\delta(p_1' + p_2' - q_1 - q_2)$$

$$\times M(p_1', p_2', k_1, k_2)M^*(p_1', p_2', q_1, q_2).$$

Because of the first delta function, the second one may be written as $\delta(k_1 + k_2 - q_1 - q_2)$. Also, since f and g are peaked around p_1 and p_2, we may approximate M above by $M(p_1', p_2', p_1, p_2)$. Finally, to get W into a comprehensible form, we introduce the Fourier transforms of f and g

$$\tilde{f}(x) = \int d\tilde{q} e^{iqx} f(q)$$

where, as above,

$$d\tilde{q} = \frac{d^3q}{(2\pi)^3 2q_0}.$$

So

$$|\tilde{f}(x)|^2 = \int d\tilde{k}_1 \, d\tilde{q}_1 e^{i(k_1 - q_1)x} f(k_1)f^*(q_1)$$

and similarly

$$|\tilde{g}(x)|^2 = \int d\tilde{k}_2 \, d\tilde{q}_2 e^{i(k_2 - q_2)x} g(k_2)g^*(q_2);$$

$\tilde{f}(x)$ and $\tilde{g}(x)$ are the wave functions of the incoming particles. Noting that

$$(2\pi)^4 \delta(k_1 + k_2 - q_1 - q_2) = \int e^{i(k_1 + k_2 - q_1 - q_2)x}\,\mathrm{d}^4 x,$$

we have, gathering these facts together,

$$W = \int \mathrm{d}^4 x |\tilde{f}(x)|^2 |\tilde{g}(x)|^2 (2\pi)^4 \delta(p_1' + p_2' - p_1 - p_2)$$
$$\times |M(p_1', p_2', p_1, p_2)|^2. \tag{6.176}$$

The first factor is the overlap of the wave functions, a precondition for scattering. This factor is unity when the initial state is a momentum eigenstate. The second factor is the familiar expression of Fermi's Golden Rule. We can then write the transition probability per unit volume per unit time as

$$\frac{\mathrm{d}W}{\mathrm{d}V\,\mathrm{d}t} = |\tilde{f}(x)|^2 |\tilde{g}(x)|^2 (2\pi)^4 \delta(p_1' + p_2' - p_1 - p_2)$$
$$\times |M(p_1', p_2', p_1, p_2)|^2.$$

Using our covariant normalisation, which corresponds to $2p_0$ particles per unit volume (see §4.1), we have, for our wave packets, that the number of particles '1' per unit volume is $|\tilde{f}(x)|^2 2p_1^0$, and the corresponding number of particles '2' is $|\tilde{g}(x)|^2 2p_2^0$. Now assume that particle 2 is initially at rest, $p_2^0 = m_2$. The incident flux is the relative velocity $v = |\mathbf{p}_1|/p_{10}$ times the particle density $2p_{10}|\tilde{f}(x)|^2$, which is $2|\mathbf{p}_1||\tilde{f}(x)|^2$. The target density, similarly, is $2m_2|\tilde{g}(x)|^2$.

Now the scattering cross section $\mathrm{d}\sigma$ is defined in terms of $\mathrm{d}W/\mathrm{d}V\,\mathrm{d}t$, the number of transitions per unit time and unit volume, by

$$\frac{\mathrm{d}W}{\mathrm{d}V\,\mathrm{d}t} = (\text{incident flux}) \times (\text{target density}) \times \mathrm{d}\sigma$$

which gives

$$\mathrm{d}\sigma = (2\pi)^4 \delta^4(p_1' + p_2' - p_1 - p_2)\frac{1}{4m_2|\mathbf{p}_1|}|M|^2.$$

There is a Lorentz invariant generalisation of the quantity $m_2|\mathbf{p}_1|$, viz.

$$B = [(p_1 \cdot p_2)^2 - m_1^2 m_2^2]^{\frac{1}{2}}$$
$$= m_2|\mathbf{p}_1| \quad \text{in laboratory frame.}$$

As it stands, this expression for the cross section refers to an undefined final state, subject only to the condition $p_1' + p_2' = p_1 + p_2$. What is measured in the laboratory, however, is the differential cross section, for scattering into a particular solid angle $\mathrm{d}\Omega$; and therefore with final momentum in a range

$d\mathbf{p}_1$. We therefore write the cross section for the final momenta in the momentum-space interval $d^3 p'_1 \, d^3 p'_2$ as

$$d\sigma = \frac{(2\pi)^4}{4B} \frac{d^3 p'_1}{(2\pi)^3 2(p'_1)_0} \frac{d^3 p'_2}{(2\pi)^3 2(p'_2)_0} \delta^4(p_f - p_i)|M|^2. \tag{6.177}$$

In the case in which the initial particles have spins s_1 and s_2, there will be a summation over spin states, so for an initially unpolarised state we replace $|M|^2$ by

$$|M|^2 \to \frac{1}{(2s_1 + 1)(2s_2 + 1)} \sum_{s_i, s_f} |M_{fi}|^2.$$

The formulae above hold for bosons. When (massive) fermions are involved, the normalisation of our Dirac spinors is equivalent to p_0/m particles per unit volume, so if the target is a fermion, the target density is $|\tilde{g}(x)|^2$. For the final state fermion, the Lorentz invariant phase space $d^3 p/(2\pi)^3 2p_0$ is replaced by $(m/p_0)[d^3 p/(2\pi)^3]$. Putting these things together, the cross section for pion–nucleon scattering is given by

$$d\sigma = \frac{1}{(2\pi)^2} \frac{d^3 p'_1}{2E'_1} \frac{d^3 p'_2}{E'_2/M} \frac{M}{2B} \delta^4(p_f - p_i)\tfrac{1}{2}\sum_{\text{spin}} |M_{fi}|^2$$

$$= \frac{1}{32\pi^2} \frac{d^3 p'_1}{E'_1} \frac{d^3 p'_2}{E'_2} \frac{M^2}{B} \delta(E'_1 + E'_2 - E_i)\delta^3(p'_1 + p'_2 - p_i)$$

$$\times \sum_{\text{spin}} |M_{fi}|^2.$$

Let us now work in the centre of mass system

$$\mathbf{p}'_1 = -\mathbf{p}'_2 \equiv \mathbf{p}_f.$$

Then the phase space integral becomes

$$I = \int \frac{d^3 p'_1}{E'_1} \frac{d^3 p'_2}{E'_2} \delta(E'_1 + E'_2 - E_i)\delta^3(p'_1 + p'_2 - p_i)$$

$$= \int \frac{d^3 p_f}{E'_1 E'_2} \delta(E'_1 + E'_2 - E_i)$$

$$= \int \frac{p_f^2 \, dp_f \, d\Omega_f}{E'_1 E'_2} \delta(E'_1 + E'_2 - E_i).$$

Invoking the formula

$$\delta(f(x)) = [f'(x_0)]^{-1}\delta(x - x_0), \quad \text{where } f(x_0) = 0,$$

we then see that

$$I = \frac{p_f}{E_i} \int d\Omega_f,$$

so the differential cross section is

$$\frac{d\sigma}{d\Omega} = \frac{1}{32\pi^2} \frac{M^2 p_f}{BE_i} \sum_{\text{spin}} |M_{fi}|^2.$$

In the centre of mass system, $B = p_f(E_1' + E_2') = p_f W$ where $W = E_i$ is the total energy of the system, so

$$\frac{d\sigma}{d\Omega} = \frac{1}{32\pi^2} \left(\frac{M}{W}\right)^2 \sum_{\text{spin}} |M_{fi}|^2. \tag{6.178}$$

The invariant amplitude M_{fi} is, by comparing equations (6.174) and (6.175),

$$M_{fi} = 2g^2 \bar{u}^{s'}(p')\gamma \cdot k' u^s(p) \frac{1}{2p \cdot k' - m^2}. \tag{6.179}$$

We now have to evaluate a quantity of the form

$$\sum_{\text{spins}} |\bar{u}' A u|^2$$

where $u' = u^{s'}(p')$ and A is a Dirac matrix operator. We have

$$(\bar{u}' A u)^* = u'^T \gamma^{0*} A^* u^*$$
$$= u^\dagger A^\dagger \gamma^{0\dagger} u'$$
$$= \bar{u} \bar{A} u'$$

with $\bar{A} = \gamma^0 A^\dagger \gamma^0$. Hence

$$|\bar{u}' A u|^2 = (\bar{u}' A u)^* (\bar{u}' A u)$$
$$= \bar{u} \bar{A} u' \bar{u}' A u$$
$$= \bar{u}'_k A_{kl} u_l \bar{u}_m \bar{A}_{mi} u'_i$$
$$= u'_i \bar{u}'_k A_{kl} u_l \bar{u}_m \bar{A}_{mi}.$$

Summing over spin, we now appeal to equation (2.145)

$$\sum_\alpha u_i^\alpha(p) \bar{u}_j^\alpha(p) = \left(\frac{\gamma \cdot p + M}{2M}\right)_{ij},$$

to give

$$\sum_{\text{spins}} |\bar{u}' A u|^2 = \text{Tr}\left(\frac{\gamma \cdot p' + M}{2M}\right) A \left(\frac{\gamma \cdot p + M}{2M}\right) \bar{A} \tag{6.180}$$

where the trace is over the Dirac matrices. We also know that (see (2.147) and (2.151))

$$\text{Tr}(\gamma \cdot a)(\gamma \cdot b) = 4a \cdot b$$
$$\text{Tr}(\gamma \cdot a)(\gamma \cdot b)(\gamma \cdot c)(\gamma \cdot d) = -\text{Tr}(\gamma \cdot b)(\gamma \cdot a)(\gamma \cdot c)(\gamma \cdot d)$$
$$+ 2a \cdot b \, \text{Tr}(\gamma \cdot c)(\gamma \cdot d)$$

and that the trace of an odd number of γ matrices vanishes ((2.150)). This gives

$$\sum_{\text{spin}} |M_{\text{fi}}|^2 = 4g^4 \left(\frac{1}{2p \cdot k' - m^2} \right)^2 \frac{1}{4M^2} \text{Tr}(\gamma \cdot p' + M)$$
$$\times \gamma \cdot k'(\gamma \cdot p + M)\gamma \cdot k'$$
$$= \frac{g^2}{M^2} \left(\frac{1}{2p \cdot k' - m^2} \right)^2 4\{2(p \cdot k')(p' \cdot k')$$
$$+ m^2[M^2 - (p \cdot p')]\}.$$

In the centre-of-mass system

$$|\mathbf{p}| = |\mathbf{p}'| = |\mathbf{k}| = |\mathbf{k}'| = q$$
$$p = ((q^2 + M^2)^{\frac{1}{2}}, \mathbf{q}), \quad k = ((q^2 + m^2)^{\frac{1}{2}}, -\mathbf{q})$$
$$p' = ((q^2 + M^2)^{\frac{1}{2}}, \mathbf{q}'), \quad k' = ((q^2 + m^2)^{\frac{1}{2}}, -\mathbf{q}').$$

At low energies, $m, M \gg q$ and $p \cdot k' \approx Mm$, $p' \cdot k' \approx Mm$, $p \cdot p' \approx M^2$, giving

$$\sum_{\text{spin}} |M_{\text{fi}}|^2 \approx \frac{8g^4}{(2M - m)^2}$$

and hence, since $W \approx M + m$ in equation (6.178),

$$\frac{d\sigma}{d\Omega} = \frac{g^4}{4\pi^2} \left(\frac{M}{M + m} \right)^2 \frac{1}{(2M - m)^2} \approx \frac{g^4}{16\pi^2} \cdot \frac{1}{M^2} \qquad (6.181)$$

if m is neglected in comparison with M.

If we know g we can arrive at a prediction, from first order perturbation theory, of the pion–nucleon scattering cross section; g may be found by two distinct methods, and they give very different values. First, it may be found from the deuteron binding energy. The Yukawa potential between nucleons a distance r apart is

$$V = \frac{ge^{-\eta r}}{r}$$

where $\eta = \lambda^{-1}$, where λ is the pion Compton wavelength, 1.4×10^{-15} m. At a separation of $r = 2.8 \times 10^{-15}$ m, the exponential term may be neglected, and the r dependence of the potential is the same as the Coulomb law. We know from the theory of the deuteron that the potential well between two nucleons at a distance of 2.8×10^{-15} m is about 20 MeV deep. On the other hand, the electrostatic interaction is only about 0.5 MeV at the same distance. Hence

$$\frac{g^2}{e^2} \approx \frac{20}{0.5} = 40$$

where e is the electric charge of the proton in e.s.u. Hence

$$\frac{g^2}{\hbar c} \approx 40 \times \frac{e^2}{\hbar c} \approx 0.3. \tag{6.182}$$

Putting this value in (6.181), we predict

$$\begin{aligned}
\sigma &= 4\pi \left(\frac{d\sigma}{d\Omega} \right) \\
&= 4\pi \left(\frac{g^2}{4\pi} \right)^2 \frac{1}{M^2} \\
&= \frac{1}{4\pi} \left(\frac{g^2}{\hbar c} \right)^2 \left(\frac{\hbar}{Mc} \right)^2 \\
&\simeq 120\,\mu b^{\ddagger}
\end{aligned} \tag{6.183}$$

which is an order of magnitude smaller than the low energy pion–nucleon scattering cross section $\approx$ a few mb.

A second determination of g can be made from a knowledge of the phase shifts in π–N scattering. By using a *derivative* coupling (m = pion mass)

$$\mathscr{L} = \frac{f}{m} \bar{\psi} \gamma_5 \gamma^\mu \psi \partial_\mu \phi \tag{6.184}$$

it may be shown using a static theory with cut-off, that a resonant behaviour is to be expected for the π–N scattering amplitude. This is, of course, observed (see, for example, Fig. 1.2). Comparison of the theory with the measured phase shifts gives a value for $f^{\ddagger\ddagger}$:

$$\frac{f^2}{4\pi} = 0.08. \tag{6.185}$$

On the other hand, it may be shown[‡‡‡] that to first order, the pseudovector coupling (6.184) and the pseudoscalar coupling (6.162) are equivalent if $g = (2M/m)f$ (M = nucleon mass). This gives

$$\frac{g^2}{4\pi} \approx 15. \tag{6.186}$$

The corresponding cross section is

$$\sigma = 4\pi \frac{d\sigma}{d\Omega} \approx 48\,b, \tag{6.187}$$

which is far in excess of the experimental value.

---

‡ $1\,b = 10^{-28}\,m^2$.

‡‡ See, for example, J.J. Sakurai, *Invariance Principles and Elementary Particles*, p. 215, Princeton University Press, 1964.

‡‡‡ See, for example, S.S. Schweber, H.A. Bethe & F. de Hoffmann, *Mesons and Fields*, section 26, vol. 1, Row, Peterson & Co. 1956.

In conclusion, neither value for g produces agreement with experiment. The value (6.186), which tends to be taken more seriously by particle physicists, is so high that a perturbation series in g^2 will diverge. It may reasonably be objected that this makes the theory meaningless. From a contemporary perspective, however, g is not a fundamental parameter. The basic strong interaction is the one between quarks and gluons, which is a gauge interaction. We shall start to consider these interactions in the next chapter, and in chapter 9 will see that the coupling constant concerned actually changes, depending on the energy at which it is measured. The present exercise, then, is seen to be a rather academic one, but has served the useful purpose of illustrating the application of the Feynman rules.

Summary

[1]The generating functional $Z[J]$ for scalar fields is written down, and converted into a form involving the Feynman propagator Δ_F, which is also written down in Euclidean space. [2]Functional integration is introduced and it is shown how the results of §6.1 may be derived by use of it. [3]$Z_0[J]$ is shown to be the generating functional for the free particle Green's functions and the relationship is shown between the n-point functions and the vacuum expectation value of the time-ordered product of n fields. Wick's theorem is proved. [4]Interactions are introduced and the relation between $Z[J]$ in the interacting case, and $Z_0[J]$ in the free particle case, is shown. [5]These results are applied to ϕ^4 theory, whose 2- and 4-point functions are calculated to first order. Feynman diagrams are introduced and the distinction between connected and disconnected diagrams made. [6]The generating functional for connected diagrams is found. [7]Grassmann numbers are introduced and it is shown how the generating functional for a spinor field may be expressed using elements of a Grassmann algebra. [8]The S matrix is defined, and the reduction formula, giving S in terms of functional derivatives of Z, obtained. Using this formula, [9]the pion–nucleon scattering amplitude is obtained to second order, and the Feynman rules summarised for the case of scalar and spinor fields. [10]The cross section for $\pi^+ p$ is obtained.

Guide to further reading

Early papers on functional methods in quantum field theory are
(1) P.T. Matthews & A. Salam, *Nuovo Cimento*, **2**, 120 (1955).
(2) P.W. Higgs, *Nuovo Cimento*, **4**, 1262 (1956).
(3) H. Umezawa & A. Visconti, *Nuovo Cimento*, **1**, 1079 (1955).

Good accounts appear in
(4) C. Nash, *Relativistic Quantum Fields*, Academic Press, 1978.

(5) D.J. Amit, *Field Theory, the Renormalisation Group, and Critical Phenomena*, McGraw-Hill, 1978.

(6) H.M. Fried, *Functional Methods and Models in Field Theory*, MIT Press, 1972.

For an account of Grassmann algebras, see

(7) D. Fearnley-Sander, *American Mathematical Monthly*, **86**, 809 (1979).

The application of Grassmann algebras to Fermi fields is discussed in

(8) F.A. Berezin, *The Method of Second Quantisation*, Academic Press, 1966.

(9) P. Ramond, *Field Theory: A Modern Primer*, Benjamin/Cummings, 1981.

(10) T.D. Lee, *Particle Physics and Introduction to Field Theory*, Harwood Academic Publishers, 1981.

(11) C. Itzykson & J.B. Zuber, *Quantum Field Theory*, McGraw-Hill, 1980.

The reduction formula and the S matrix are treated in

(12) K. Symanzik in *Lectures in Theoretical Physics*, Vol. III (Boulder 1960), Interscience, 1961.

(13) J.D. Bjorken & S.D. Drell, *Relativistic Quantum Fields*, McGraw-Hill, 1965.

(14) L.D. Faddeev & A.A. Slavnov, *Gauge Fields: Introduction to Quantum Theory*, Benjamin/Cummings, 1980.

and in refs. (5), (6) and (11) above.

7

Path integral quantisation: gauge fields

7.1   Propagators and gauge conditions in QED

We saw in the last chapter how the Feynman rules could be 'read off' from the generating functional – more particularly, from the Lagrangian of the theory. As remarked in section 6.7 (p. 220), the propagator is the inverse of the operator appearing in the quadratic part of the Lagrangian. Our aim now is to extend this work to the gauge fields, and in this section we restrict our attention to quantum electrodynamics. It turns out that, because of the freedom to make gauge transformations, we encounter difficulties which were not present with scalar and spinor fields. We exhibit these difficulties in both the canonical and path-integral formalism, the former being more transparent, the latter more succinct and formal. It is, however, the path-integral method which has the greater power, and in the next section we shall show how to apply it to the case of non-Abelian gauge fields.

We begin this section by outlining the difficulty in finding the propagator of the electromagnetic field, using the canonical formalism.

Photon propagator – canonical formalism

We shall proceed, in our study of gauge fields, from a generating functional which in the case of quantum electrodynamics takes the form

$$Z[J] = \int \mathscr{D}A_\mu \exp i \int (\mathscr{L} + J^\mu A_\mu)\,dx \tag{7.1}$$

where J^μ is the external current source, and, in the absence of matter,

$$\mathscr{L} = -\tfrac{1}{4}F_{\mu\nu}F^{\mu\nu}. \tag{7.2}$$

This was our method of procedure for scalar fields (see (6.1)) and for spinor fields (see (6.130)), and it stemmed from equation (5.15) in quantum mechanics. We should recall, however, that the more fundamental expression for the transition amplitude (generating functional) is equation (5.13), which reduces to (5.15) when H is quadratic in p with a constant coefficient (as in (5.8)). It is true also in field theory that the only *safe* path integral is the one based on the canonical formalism, that is, involving

$$\int\int \mathscr{D}\phi\mathscr{D}\pi \exp i \int (\phi\pi - \mathscr{H}) \, dx.$$

After performing the integration over π, the canonical momentum, the simple 'Lagrangian' path integral is not recovered unless there are no constraints and $\mathscr{H}$ is quadratic in the π with constant coefficients. There are many field theories in which these conditions are not satisfied, and they include the non-linear sigma model, non-Abelian gauge theories and gravity. In non-Abelian gauge theories, however, the Faddeev–Popov technique gets one out of trouble, so at the expense of some rigour it is justifiable to proceed from expressions like (7.1), introducing the Faddeev–Popov ghost fields when the need arises. In the case of gravity, the Faddeev–Popov technique *fails* to get one out of trouble – but we are not intending to quantise gravity. We shall therefore use a simple 'Lagrangian' path integral for the generating functional in what follows. Students interested in seeing the method based on the canonical path integral will find it in refs. (2), (4) and (5).[‡]

After these cautionary remarks, we return to equations (7.1) and (7.2). As we saw in §3.3, variation of the Lagrangian (7.2) gives Maxwell's equations

$$\partial_\mu F^{\mu\nu} = \partial_\mu(\partial^\mu A^\nu - \partial^\nu A^\mu) = 0$$

or

$$(g_{\mu\nu}\Box - \partial_\mu\partial_\nu)A^\mu = 0. \tag{7.3}$$

After partial integration, and discarding of surface terms, $\mathscr{L}$ may be written

$$\mathscr{L} = \tfrac{1}{2}A^\mu[g_{\mu\nu}\Box - \partial_\mu\partial_\nu]A^\nu. \tag{7.4}$$

According to our prescription, the photon propagator $D_{\mu\nu}$ is the inverse of this operator

$$(g_{\mu\nu}\Box - \partial_\mu\partial_\nu)D^{\nu\lambda}(x - y) = \delta_\mu^\lambda\delta^4(x - y). \tag{7.5}$$

(Alternatively, and equivalently, we see that $D^{\nu\lambda}$ is the Green's function of Maxwell's equation (7.3).) Now we multiply (7.5) by the operator ∂^μ, giving

$$(0 \cdot \partial_\nu)D^{\nu\lambda} = \partial^\lambda\delta^4(x - y),$$

from which we see that $D^{\nu\lambda}$ has *no inverse*, and is therefore formally *infinite*. This is because the operator $(g_{\mu\nu}\Box - \partial_\mu\partial_\nu)$ has no inverse; if applied to $\partial^\mu\Lambda$,

---

[‡] I am grateful to Professor P.W. Higgs for emphasising these points to me.

for example, it gives zero:

$$(g_{\mu\nu}\Box - \partial_\mu\partial_\nu)\partial^\mu\Lambda = (\partial_\nu\Box - \Box\partial_\nu)\Lambda = 0,$$

so the operator has a zero eigenvalue, and therefore no inverse.

Our straightforward attempt to find the photon propagator has failed. Some light is thrown on the reason for this failure by appealing to the path-integral method.

Photon propagator – path-integral method

Here we simply consider the generating functional

$$Z = \int \mathscr{D}A_\mu \, e^{i\int \mathscr{L}\,\mathrm{d}x}.$$

$\mathscr{L}$ is invariant under gauge transformations $A_\mu \to A_\mu + \partial_\mu\Lambda$, but the integration is taken over *all* A_μ, including those that are related only by a gauge transformation. This clearly gives an infinite contribution to Z, and therefore to the Green's functions, obtained by functional differentiation of Z.

Gauge-fixing terms

What must we do to get a finite value for Z? The obvious answer is to fix a particular gauge, so that the integral over A_μ does *not* extend over values simply related by a gauge transformation. To be definite, we impose the Lorentz gauge condition $\partial_\mu A^\mu = 0$. This gives the Lagrangian

$$\mathscr{L} = \tfrac{1}{2}A^\mu g_{\mu\nu}\Box A^\nu. \tag{7.6}$$

The operator $g_{\mu\nu}\Box$ has an inverse, known as the *Feynman propagator*

$$D_{\mathrm{F}}(x,y)_{\mu\nu} = -g_{\mu\nu}\Delta_{\mathrm{F}}(x,y;m=0) \tag{7.7}$$

The Lagrangian is then written

$$\mathscr{L} = -\tfrac{1}{4}F_{\mu\nu}F^{\mu\nu} - \tfrac{1}{2}(\partial_\mu A^\mu)^2$$
$$= \mathscr{L}_0 + \mathscr{L}_{\mathrm{GF}} \tag{7.8}$$

where

$$\mathscr{L}_{\mathrm{GF}} = -\tfrac{1}{2}(\partial_\mu A^\mu)^2 \tag{7.9}$$

is known as the 'gauge-fixing' term.

It may be helpful to reproduce the above line of reasoning in momentum space. The operator $g_{\mu\nu}\Box - \partial_\mu\partial_\nu$ becomes $-g_{\mu\nu}k^2 + k_\mu k_\nu$, and it is easy to see that this has no inverse. If it did, it would have to be of the form $Ag^{\nu\lambda} + Bk^\nu k^\lambda$, with $(-k^2 g_{\mu\nu} + k_\mu k_\nu)(Ag^{\nu\lambda} + Bk^\nu k^\lambda) = \delta_\mu^\lambda$. This implies

$$-Ak^2\delta_\mu^\lambda + Ak_\mu k^\lambda = \delta_\mu^\lambda,$$

which has no solution. On the other hand, in the Lorentz gauge, the operator $-g_{\mu\nu}k^2$ clearly has an inverse $-g^{\nu\lambda}(1/k^2)$, and the Feynman propagator is

$$D_{\mathrm{F}}(k)_{\mu\nu} = -\frac{g_{\mu\nu}}{k^2}. \tag{7.10}$$

More generally, instead of (7.8), we may add to the Lagrangian an arbitrary amount of $\mathscr{L}_{\mathrm{GF}}$:

$$\begin{aligned}
\mathscr{L} &= -\tfrac{1}{4}F_{\mu\nu}F^{\mu\nu} - \frac{1}{2\alpha}(\partial_\mu A^\mu)^2 \\
&= \tfrac{1}{2}A^\mu\left[g_{\mu\nu}\Box + \left(\frac{1}{\alpha}-1\right)\partial_\mu\partial_\nu\right]A^\nu
\end{aligned} \tag{7.11}$$

with α finite. The 'quadratic' operator in momentum space is

$$-k^2 g_{\mu\nu} + \left(1-\frac{1}{\alpha}\right)k_\mu k_\nu$$

whose inverse gives the propagator

$$D(k)_{\mu\nu} = -\frac{1}{k^2}\left[g_{\mu\nu} + (\alpha-1)\frac{k_\mu k_\nu}{k^2}\right]. \tag{7.12}$$

Particular nomenclatures associated with α are

$$\left.\begin{aligned}
\alpha \to 1:&\quad \text{Feynman propagator (Feynman gauge),} \\
\alpha \to 0:&\quad \text{Landau gauge.}
\end{aligned}\right\} \tag{7.13}$$

The physics, of course, is unaffected by the value of α.

Propagator for transverse photons

The propagator (7.10) applies to photons in the Lorentz gauge. This is a covariant gauge in which photons have a polarisation vector ε^μ, with $k_\mu\varepsilon^\mu = 0$, so that only three of the four states are independent. *Physical* photons, however, are described by only *two* polarisation states – those with transverse polarisation. In this subsection, we calculate the propagator for transverse (physical) photons. From §4.4, we see that we are working in the radiation (or Coulomb) gauge, in which $\mathbf{V}\cdot\mathbf{A} = 0$, $\phi = 0$.

Recall that the relation between the propagator and the 2-point function is

$$\langle 0| T(A_\mu(x)A_\nu(y))|0\rangle = iD_{\mu\nu}(x-y). \tag{7.14}$$

We calculate the left-hand side by substituting for $A_\mu(x)$ from (4.64) (or from (4.80), but with the sum over $\lambda = 1, 2$ only). Hence

$$\langle 0| T(A_\mu(x)A_\nu(y))|0\rangle$$

$$= \left\langle 0 \left| \int \frac{\mathrm{d}^3 k}{(2\pi)^3 2k_0} \frac{\mathrm{d}^3 k'}{(2\pi)^3 2k_0'} \sum_{\lambda, \lambda'=1}^{2} \varepsilon_\mu^{(\lambda)}(k) \varepsilon_\nu^{(\lambda')}(k') \right. \right.$$

$$\times \{ [a^{(\lambda)}(k) \mathrm{e}^{-\mathrm{i}kx} + a^{(\lambda)\dagger}(k) \mathrm{e}^{\mathrm{i}kx}]$$

$$\times [a^{(\lambda')}(k') \mathrm{e}^{-\mathrm{i}k'y} + a^{(\lambda')\dagger}(k') \mathrm{e}^{\mathrm{i}k'y}] \theta(x_0 - y_0)$$

$$+ [a^{(\lambda')}(k') \mathrm{e}^{-\mathrm{i}k'y} + a^{(\lambda')\dagger}(k') \mathrm{e}^{\mathrm{i}k'y}]$$

$$\left. \left. \times [a^{(\lambda)}(k) \mathrm{e}^{-\mathrm{i}kx} + a^{(\lambda)\dagger}(k) \mathrm{e}^{\mathrm{i}kx}] \theta(y_0 - x_0) \} \right| 0 \right\rangle$$

$$\doteq \left\langle 0 \left| \int \frac{\mathrm{d}^3 k \, \mathrm{d}^3 k'}{(2\pi)^6 2k_0 2k_0'} \sum_{\lambda, \lambda'=1}^{2} \varepsilon_\mu^{(\lambda)}(k) \varepsilon_\nu^{(\lambda')}(k') \right. \right.$$

$$\times [a^{(\lambda)}(k) a^{(\lambda')\dagger}(k') \mathrm{e}^{\mathrm{i}(k'y - kx)} \theta(x_0 - y_0)$$

$$\left. \left. + a^{(\lambda')}(k') a^{(\lambda)\dagger}(k) \mathrm{e}^{\mathrm{i}(kx - k'y)} \theta(y_0 - x_0)] \right| 0 \right\rangle.$$

The two terms in $aa^\dagger$ may be replaced by their commutators, given by (4.68). The delta functions then enable one to integrate over k' and sum over λ', giving

$$\langle 0 | T(A_\mu(x) A_\nu(y)) | 0 \rangle = \int \frac{\mathrm{d}^3 k}{(2\pi)^3 2k_0} \sum_{\lambda=1}^{2} \varepsilon_\mu^{(\lambda)}(k) \varepsilon_\nu^{(\lambda)}(k)$$

$$\times [\mathrm{e}^{\mathrm{i}k(y-x)} \theta(x_0 - y_0) + \mathrm{e}^{\mathrm{i}k(x-y)} \theta(y_0 - x_0)]. \tag{7.15}$$

Now, from equations (6.14) and (6.56), the Feynman propagator for massless particles is

$$\Delta_\mathrm{F}(x, m=0) = -\mathrm{i} \int \frac{\mathrm{d}^3 k}{(2\pi)^3 2k_0} [\theta(x_0) \mathrm{e}^{-\mathrm{i}kx} + \theta(-x_0) \mathrm{e}^{\mathrm{i}kx}]$$

$$= \int \frac{\mathrm{d}^4 k}{(2\pi)^4} \frac{\mathrm{e}^{-\mathrm{i}kx}}{k^2 + \mathrm{i}\varepsilon}.$$

Hence we may write

$$\langle 0 | T(A_\mu(x) A_\nu(y)) | 0 \rangle = \mathrm{i} \int \frac{\mathrm{d}^4 k}{(2\pi)^4} \frac{\mathrm{e}^{-\mathrm{i}kx}}{k^2 + \mathrm{i}\varepsilon} \sum_{\lambda=1}^{2} \varepsilon_\mu^{(\lambda)}(k) \varepsilon_\nu^{(\lambda)}(k)$$

and the propagator for transverse photons is then

$$D_{\mu\nu}^{\mathrm{tr}}(x - y) = \int \frac{\mathrm{d}^4 k}{(2\pi)^4} \frac{\mathrm{e}^{-\mathrm{i}k(x-y)}}{k^2 + \mathrm{i}\varepsilon} \sum_{\lambda=1}^{2} \varepsilon_\mu^{(\lambda)}(k) \varepsilon_\nu^{(\lambda)}(k). \tag{7.16}$$

What is $\sum_{\lambda=1}^{2} \varepsilon_\mu^{(\lambda)}(k) \varepsilon_\nu^{(\lambda)}(k)$? Now ε_μ is orthogonal to k_μ, which is lightlike; ε_μ is therefore spacelike. We introduce a timelike vector $\eta_\mu = (1, 0, 0, 0)$, which is orthogonal to ε_μ in the radiation gauge. We then form a tetrad from $\varepsilon_\mu^{(1,2)}$,

η_μ and one other spacelike vector, denoted $\bar{k}$:

$$\bar{k}^\mu = \frac{k^\mu - (k \cdot \eta)\eta^\mu}{[(k \cdot \eta)^2 - k^2]^{\frac{1}{2}}}.$$

It is straightforward to verify that $\bar{k}$ is spacelike:

$$\bar{k}^2 = \frac{k^2 - 2(k \cdot \eta)^2 + (k \cdot \eta)^2 \eta^2}{(k \cdot \eta)^2 - k^2} = -1,$$

where we have used $\eta^2 = 1$; and that $\bar{k}$ is orthogonal to ε:

$$\bar{k} \cdot \varepsilon = \frac{k \cdot \varepsilon - (k \cdot \eta)(\eta \cdot \varepsilon)}{[(k \cdot \eta)^2 - k^2]^{\frac{1}{2}}} = 0,$$

since $k \cdot \varepsilon = 0$ and $\eta \cdot \varepsilon = 0$. Having constructed a tetrad, we now have

$$g_{\mu\nu} = \eta_\mu \eta_\nu - \sum_{\lambda=1}^{2} \varepsilon_\mu^{(\lambda)}(k)\varepsilon_\nu^{(\lambda)}(k) - \bar{k}_\mu \bar{k}_\nu,$$

hence

$$\sum_{\lambda=1}^{2} \varepsilon_\mu^{(\lambda)} \varepsilon_\nu^{\lambda}(k) = -g_{\mu\nu} + \eta_\mu \eta_\nu - \bar{k}_\mu \bar{k}_\nu$$

$$= -g_{\mu\nu} - \frac{k_\mu k_\nu}{(k \cdot \eta)^2 - k^2} + \frac{(k \cdot \eta)(k_\mu \eta_\nu + \eta_\mu k_\nu)}{(k \cdot \eta)^2 - k^2}$$

$$- \frac{k^2 \eta_\mu \eta_\nu}{(k \cdot \eta)^2 - k^2}. \tag{7.17}$$

This is the desired expression for $\sum_{\lambda=1}^{2} \varepsilon_\mu^{(\lambda)}(k)\varepsilon_\nu^{(\lambda)}(k)$, and when substituted in (7.15) it gives an explicit expression for the propagator for transverse photons.

7.2 Non-Abelian gauge fields and the Faddeev–Popov method

We now want to extend what we have done to non-Abelian gauge fields (Yang–Mills fields). Our aim is to discover the general rules for finding the gauge-field propagator. We proceed by developing the formal path-integral method referred to in the last section, based on making Z finite. This method was first devised by Faddeev and Popov. Because it is not particularly transparent, we shall first give a heuristic derivation of the Faddeev–Popov formula, applied to electrodynamics. We shall then give a more rigorous derivation, which will be applied to Yang–Mills fields.

Heuristic derivation

We saw above that Z is infinite because the functional integration extends over all A_μ, even over those related by a gauge transformation, under which the integrand is invariant. Let us write each A_μ as

$$A_\mu \sim \bar{A}_\mu, \Lambda(x). \tag{7.18}$$

In words, A_μ is the class of all potentials that can be reached from a *fixed* $\bar{A}_\mu$ by a gauge transformation, given by some function $\Lambda(x)$. Then the integral for Z may be broken down:

$$Z = \int \mathscr{D} A_\mu e^{iS} \sim \int \mathscr{D} \bar{A}_\mu e^{iS} \int \mathscr{D} \Lambda. \tag{7.19}$$

It is the last factor, $\int \mathscr{D} \Lambda$, which is an 'overcounting' and causes the divergence. Let us modify it by introducing a convergence factor

$$\int \mathscr{D} \Lambda \to \int \mathscr{D} \Lambda \, e^{-(i/2\alpha)F^2}$$

where $F = F(A_\mu)$ is some function of A_μ, for example $F = \partial^\mu A_\mu$. The integral now converges, but Z now depends on F – which we do want! So we modify the integral further

$$\int \mathscr{D} \Lambda \to \int \mathscr{D} F \, e^{-(i/2\alpha)F^2} = \int \mathscr{D} \Lambda \det\left(\frac{\partial F}{\partial \Lambda}\right) e^{-(i/2\alpha)F^2}$$

so that Z now becomes (without the source term)

$$Z = \int \mathscr{D} A_\mu \int \mathscr{D} \Lambda \exp\left[i \int \left(\mathscr{L} - \frac{1}{2\alpha} F^2 \right) dx \right] \det \frac{\partial F}{\partial \Lambda}. \tag{7.20}$$

What is $\partial F / \partial \Lambda$? Under a gauge transformation, $F \to F + M\Lambda$, where M will in general be a differential operator, so

$$\frac{\partial F}{\partial \Lambda} = M.$$

There will then be a factor $\det M$ in the expression for Z. We now appeal to equation (6.128) – or rather, its infinite-dimensional generalisation – to express $\det M$ as an integral over *Grassmann scalar fields* $\bar{\eta}, \eta$. Changing A into iA we have

$$\det M = \int \mathscr{D} \eta \, \mathscr{D} \bar{\eta} \exp\left(-i \int \bar{\eta} M \eta \, dx \right) \tag{7.21}$$

so that (removing the bar from $\bar{A}_\mu$)

$$Z = \frac{1}{N} \int \mathscr{D} A_\mu \, \mathscr{D} \eta \, \mathscr{D} \bar{\eta} \exp\left[i \int \left(\mathscr{L} - \frac{1}{2\alpha} F^2 - \bar{\eta} M \eta \right) dx \right]$$

$$\equiv \frac{1}{N} \int \mathscr{D} A_\mu \, \mathscr{D} \eta \, \mathscr{D} \bar{\eta} \exp\left(i \int \mathscr{L}_{\text{eff}} \, dx \right) \tag{7.22}$$

where, as usual, N is a normalising factor, and the effective Lagrangian $\mathscr{L}_{\text{eff}}$ is

$$\mathscr{L}_{\text{eff}} = \mathscr{L} - \frac{1}{2\alpha} F^2 - \bar{\eta} M \eta$$

$$\equiv \mathscr{L} + \mathscr{L}_{\text{GF}} + \mathscr{L}_{\text{FPG}}. \tag{7.23}$$

$\mathscr{L}_{GF}$ is the gauge-fixing term discussed above (equation (7.9)) and $\mathscr{L}_{FPG}$ is the 'Faddeev–Popov ghost' term, η being the ghost field, a scalar field with Fermi statistics. In the case of QED, the ghost contribution in (7.22) integrates out, and so has no physical effect (at least in the Lorentz gauge). We shall see below that, in the Yang–Mills case, there is a coupling between the ghost and gauge fields, so the ghost cannot be integrated out. It therefore occurs in Feynman diagrams, but only in internal loops, not as an external particle. Let us now turn to the non-Abelian case and repeat the line of argument above, but in a more rigorous manner.

More rigorous derivation

We need to introduce a convergence factor into Z, or, alternatively, to take out a multiplicative infinity. The key lies in referring to a particular gauge, and seeing if the integration 'separates out' as in (7.19) above. Let us write the gauge condition as

$$F^a[A_\mu^b] = 0 \tag{7.24}$$

where a and b are internal indices. We shall consider the case in which F is a linear function (or functional) of A. More generally, we could have $F^a[A_\mu^b, \phi] = 0$, where ϕ is a scalar field.

The integration over Λ in (7.19) is clearly an integration over the group space. Suppose the symmetry group of the problem is G, and two elements $g, g' \in G$, then we may define the *Hurwitz measure* over G as an integration measure invariant under group transformations

$$dg = d(g'g). \tag{7.25}$$

To understand this, consider the case of finite groups. Then a function $\phi(g)$ is a set of numbers $\phi(g_1), \phi(g_2), \ldots, \phi(g_n)$. We then define the 'integral' of ϕ over the group as the sum of these numbers

$$\int \phi(g)\,dg = \sum_{g \in G} \phi(g).$$

This is then invariant (more precisely, 'left invariant') under the group transformations

$$\int \phi(g'g)\,dg = \int \phi(g)\,dg$$

since, as g runs over the group, so does $g'g$, and each number ϕ is counted once and only once. Writing $g'g = g''$, then $g = g'^{-1}g''$, and interpreting the following formulae for the case of continuous groups, we have

$$\int \phi(g'')d(g'^{-1}g'') = \int \phi(g)\,dg = \int \phi(g'')dg''$$

or, relabelling,

$$\int \phi(g) \mathrm{d}(g'g) = \int \phi(g) \mathrm{d}g$$

and hence

$$\mathrm{d}(g'g) = \mathrm{d}g$$

as in (7.25). An invariant measure may always be found for compact groups.[‡] For example, for the rotation group it is $\sin \beta \, \mathrm{d}\alpha \, \mathrm{d}\beta \, \mathrm{d}\gamma$, where α, β and γ are the Euler angles. A_μ, defined by (3.159), and redefined by $A \to A/g$, transforms according to (3.162) (with $S \to U$):

$$A'_\mu = U A_\mu U^{-1} - \mathrm{i}(\partial_\mu U) U^{-1} \tag{7.26}$$

where $U = \exp [\mathrm{i}\omega^a(x) T^a]$ and T^a are the generators of G in the regular representation. Infinitesimally, we have

$$U(\omega) = 1 + \mathrm{i}\omega^a T^a + O(\omega^2)$$

so the group measure $\mathrm{d}g$, for $g \approx 1$, may be expressed as

$$\mathrm{d}g = \prod_a \mathrm{d}\omega^a = \mathrm{d}\boldsymbol{\omega}. \tag{7.27}$$

In addition, from (3.124) we have, with $\Lambda \to \omega$,

$$\begin{aligned} A'^a_\mu &\equiv (A_\omega)^a_\mu \\ &= A^a_\mu + f^{abc} A^b_\mu \omega^c + \partial_\mu \omega^a \end{aligned} \tag{7.28}$$

where f^{abc} are the structure constants of the group G ($= \varepsilon^{abc}$ in the case of $SU(2)$). Now consider the quantity

$$\Delta^{-1}[A] = \int \mathscr{D}\boldsymbol{\omega} \, \delta(F[A_\omega]) = \int \mathscr{D}g \, \delta(F[A_g]) \tag{7.29}$$

where $\mathscr{D}\boldsymbol{\omega} = \prod_x \mathrm{d}\boldsymbol{\omega}(x) \, (\mathscr{D}g = \prod_x \mathrm{d}g(x))$ and $\delta(F[A_\omega])$ is a product of Dirac delta functions, one at each point of space–time – it is really a 'δ-functional'. A is short hand for A^a_μ. Now $\Delta^{-1}[A]$ is gauge invariant:

$$\begin{aligned} \Delta^{-1}[A_g] &= \int \mathscr{D}g' \, \delta(F[A_{g'g}]) \\ &= \int \mathscr{D}(g'g) \delta(F[A_{g'g}]) \\ &= \int \mathscr{D}(g'') \delta(F[A_{g''}]) \\ &= \Delta^{-1}[A] \end{aligned} \tag{7.30}$$

[‡] See, for example, M. Hamermesh, *Group Theory and its Application to Physical Problems*, p. 313, Addison-Wesley, 1962, or J.D. Talman, *Special Functions: A Group Theoretic Approach*, section 9–2, Benjamin, 1968.

where we have used (7.25). We may therefore write (putting ω for g)

$$1 = \Delta[A] \int \mathscr{D}\omega \, \delta(F[A_\omega]). \tag{7.31}$$

We now insert this into the expression for the path integral, giving

$$\int \mathscr{D}A_\mu e^{iS} = \int \mathscr{D}A_\mu \Delta[A_\mu] \int \mathscr{D}\omega \, \delta(F[(A_\mu)_\omega]) e^{iS}. \tag{7.32}$$

Now we perform a gauge transformation in the integrand, from $(A_\mu)_\omega$ to A_μ, using (7.30). This gives

$$\int \mathscr{D}A_\mu e^{iS} = \int \mathscr{D}A_\mu \Delta[A_\mu] \int \mathscr{D}\omega \, \delta(F[A]) e^{iS} \tag{7.33}$$

since S is invariant. The factors following $\mathscr{D}\omega$ are now independent of ω, so $\int \mathscr{D}\omega$ gives a multiplicative divergence, which may be removed by *redefining* Z as

$$Z = \int \mathscr{D}A_\mu \Delta[A_\mu] \delta(F[A_\mu]) e^{iS[A]}. \tag{7.34}$$

We must now evaluate $\Delta[A_\mu]$. For infinitesimal transformations,

$$
\begin{aligned}
F^a[A_\omega] &= F^a[A] + \frac{\partial F^a}{\partial A_\mu^b} \delta A_\mu^b \\[2mm]
&= F^a[A] + \frac{\partial F^a}{\partial A_\mu^b} (\delta^{bd}\partial_\mu + f^{bcd}A_\mu^c)\omega^d \\[2mm]
&= F^a[A] + \frac{\partial F^a}{\partial A_\mu^b} (D_\mu\omega)^b
\end{aligned}
\tag{7.35}
$$

where we have used (7.28) and (3.122). So, if $F^a[A] = 0$, then from (7.29) we have

$$\Delta^{-1}[A] = \int \mathscr{D}\omega^a \delta\left[\frac{\partial F^a}{\partial A_\mu^b} D_\mu^{bc}\omega^c(y) \right]. \tag{7.36}$$

Now we denote the argument of the delta function by M:

$$
\begin{aligned}
\frac{\partial F^a}{\partial A_\mu^b} D_\mu^{bc} \delta^4(x-y) &= M_{ac}(x,y)\delta^4(x-y) \\[2mm]
&= \langle a, x | M | c, y \rangle \\[2mm]
&\sim \frac{\delta F^a[A(x)]}{\delta \omega^c(y)}.
\end{aligned}
\tag{7.37}
$$

Then

$$\Delta^{-1}[A] \sim (\det M)^{-1}. \tag{7.38}$$

Proof[‡]

If M has eigenfunctions f^i and eigenvalues λ^i, then (no summation over i)

$$\sum_{b,y} \langle a, x|M|b, y \rangle f_b^i(y) = \lambda^i f_a^i(x)$$

$$\sum_{b,y} M_{ab}(x, y)\delta^4(x - y)f_b^i(y) = \lambda^i f_a^i(x)$$

$$\sum_b M_{ab}(x, x)f_b^i(x) = \lambda^i f_a^i(x). \tag{7.39}$$

Now we expand $\omega^c(y)$ in terms of the eigenfunctions of M:

$$\omega^c(y) = \sum_i \omega^i f_c^i(y).$$

Hence, from (7.39),

$$\sum_c M_{ac}(y, y)\omega^c(y) = \sum_i \omega^i \lambda^i f_a^i(y)$$

and (7.36) then reads

$$\Delta^{-1}[A] = \int \mathscr{D}\omega^a \prod_a \delta\left[\sum_i \omega^i \lambda^i f_a^i(y)\right]. \tag{7.40}$$

Putting

$$u_a = \sum_i \omega^i \lambda^i f_a^i,$$

equation (7.40) is

$$\begin{aligned}
\Delta^{-1}[A] &= \int \mathscr{D}u_a \frac{\partial(\omega^1 \omega^2 \omega^3 \ldots)}{\partial(u_1 u_2 u_3 \ldots)} \prod_a \delta(u_a) \\
&= \frac{\partial(\omega^1 \omega^2 \omega^3 \ldots)}{\partial(u_1 u_2 u_3 \ldots)}\bigg|_{u=0} \\
&= (\lambda_1 \lambda_2 \ldots)^{-1}|f_a^i|^{-1} \\
&\propto (\lambda_1 \lambda_2 \ldots)^{-1}.
\end{aligned}$$

Hence $\Delta[A] \propto$ product of eigenvalues of $M = \det M$;

$$\Delta[A] \sim \det\left(\frac{\delta F^a(x)}{\delta \omega^b(y)}\right)^{A(x)}\bigg|_{\omega=0} = \det M \tag{7.41}$$

which is (7.38). QED. It will be noted that we need pay no attention to normalisation factors.

It is common to consider the Lorentz gauge

$$F^a = \partial^\mu F_\mu^a \tag{7.42}$$

$$\underset{A}{\uparrow}$$

[‡] I am grateful to Professor G. Rickayzen for this proof.

but more generally we may consider

$$F^a = \partial^\mu A^a_\mu + C^a(x) \tag{7.43}$$

where C is an arbitrary function. The generating functional is then

$$Z = \int \mathscr{D}A_\mu \Delta[A_\mu] \delta(F[A] - C) e^{iS[A]}. \tag{7.44}$$

Since C is independent of A, $\Delta[A]$ is the same as before (equation (7.41)), and the left-hand side is independent of $C(x)$. We may then include in Z any weighting factor we wish; it only changes the normalisation of Z. So we include the factor

$$\exp\left[-\frac{i}{2\alpha} \int C^2_a(x)\, d^4x \right]$$

giving

$$Z = N \int \mathscr{D}A_\mu \Delta[A] \exp\left[i \int \left(\mathscr{L} - \frac{1}{2\alpha} F[A]^2 \right) dx \right]. \tag{7.45}$$

Note that $F[A]$ could be any functional of A, despite the fact that we referred to the Lorentz condition above. Now putting $\Delta[A] = \det iM$ and using equation (6.128) in the form

$$\det iM = \int \mathscr{D}\eta \mathscr{D}\bar{\eta} \exp\left(-i \int \bar{\eta}^a M_{ab} \eta^b \, dx \right), \tag{7.46}$$

we get

$$Z = N \int \mathscr{D}A_\mu \mathscr{D}\eta \mathscr{D}\bar{\eta} \exp\left[i \int \left(\mathscr{L} - \frac{1}{2\alpha} F^2 - \bar{\eta}^a M_{ab} \eta^b \right) dx \right]. \tag{7.47}$$

We may write this in the form

$$Z = N \int \mathscr{D}A_\mu \mathscr{D}\eta \mathscr{D}\bar{\eta} \exp\left(i \int \mathscr{L}_{\text{eff}} \, dx \right) \tag{7.48}$$

where the effective Lagrangian is

$$\mathscr{L}_{\text{eff}} = \mathscr{L} - \frac{1}{2\alpha} F^2 - \bar{\eta}^a M_{ab} \eta^b$$

$$= \mathscr{L} + \mathscr{L}_{\text{GF}} + \mathscr{L}_{\text{FPG}}. \tag{7.49}$$

$\mathscr{L}_{\text{GF}}$ is the gauge-fixing term, encountered above, and $\mathscr{L}_{\text{FPG}}$ is the Faddeev–Popov ghost term. The Grassmann fields $\eta, \bar{\eta}$ are called ghost fields, since their unphysical spin statistics allows them only to appear in closed loops in Feynman diagrams, never as external fields.

After this heavy formal reasoning, we may now find the Feynman rules for QED and non-Abelian gauge fields. They depend on the gauge. We shall consider first the Lorentz gauge and then the so-called axial gauge.

Feynman rules in the Lorentz gauge

First, let us consider electrodynamics, in which there is no internal symmetry index. We have

$$F = \partial^\mu A_\mu.$$

From (7.28) we have

$$\delta A_\mu = \partial_\mu \omega$$

therefore

$$\delta F = \Box \omega.$$

Hence from (7.37)

$$M = \frac{\delta F}{\delta \omega} = \Box$$

and

$$Z = N \int \mathscr{D} A_\mu \exp \left\{ i \int \left[\mathscr{L}_0 - \frac{1}{2\alpha}(\partial \cdot A)^2 \right] dx \right\}$$
$$\times \int \mathscr{D}\eta \mathscr{D}\bar\eta \exp \left(i \int \bar\eta \Box \eta \, dx \right) \tag{7.50}$$

where $\mathscr{L}_0$ is the gauge-field Lagrangian, $-\frac{1}{4} F^a_{\mu\nu} F^{\mu\nu a}$. The last integral in the ghost fields contributes only an overall constant factor to Z. The photon propagator is obtained by finding the inverse of the quadratic term in $\mathscr{L}_0 - (1/2\alpha)(\partial \cdot A)^2$, which has been found in (7.12) above. The 2-point function is then

$$\begin{array}{c} k \\ \mu \ \text{\wavy} \ \nu \end{array} \qquad -\frac{i}{k^2} \left[g_{\mu\nu} + (\alpha - 1)\frac{k_\mu k_\nu}{k^2} \right]. \tag{7.51}$$

Next, we consider Yang–Mills fields. The gauge condition

$$F^a = \partial^\mu A^a_\mu$$

together with (7.28)

$$\delta A^a_\mu = f^{abc} A^b_\mu \omega^c + \partial_\mu \omega^a$$

gives

$$\delta F^a = f^{abc} \partial^\mu (A^b_\mu \omega^c) + \Box \omega^a$$

and hence

$$M_{ab} = \frac{\delta F^a}{\delta \omega^b} = - f^{abc} \partial^\mu A^c_\mu - f^{abc} A^c_\mu \partial^\mu + \delta^{ab} \Box.$$

Re-introducing the coupling constant g, which was dropped in (7.26), the ghost part of the action is

$$\int \mathscr{D}\eta\,\mathscr{D}\bar\eta\exp\left(-i\int\bar\eta^a\frac{\delta F^a}{\delta\omega^b}\eta^b\,dx\right)$$

$$=\int \mathscr{D}\eta\,\mathscr{D}\bar\eta\exp\left\{-i\int\bar\eta^a\square\eta^a\,dx+igf^{abc}\int[(\bar\eta^a\partial^\mu\eta^b)A^c_\mu\right.$$

$$\left.+\partial^\mu A^c_\mu(\bar\eta^a\eta^b)]dx\right\}.\tag{7.52}$$

This gives the ghost propagator and the ghost gauge-field coupling

$$(i\square)^{-1}=\frac{i}{k^2}\delta^{ab},$$

$$-gf^{abc}p_\mu.\tag{7.53}$$

(Here we have not included the $(2\pi)^4$ and momentum conservation δ-functions. They are to be understood – cf. Table 6.1). The ghost propagates like a scalar particle but has Fermi statistics.

The gauge field term F^a is like that above, so the gauge propagator is like the photon propagator

$$-\frac{i}{k^2}\left[g_{\mu\nu}+(\alpha-1)\frac{k_\mu k_\nu}{k^2}\right]\delta^{ab}.\tag{7.54}$$

Besides the fact that the ghost couples to the gauge field, the other complicating feature of the non-Abelian case is that the gauge field *couples to itself*. The physical significance of this was discussed in §3.5. This means that we have at least cubic terms in A appearing in the Lagrangian. To obtain the pure gauge-field Lagrangian, we define the field by (3.166)

$$G_{\mu\nu}=\partial_\mu A_\nu-\partial_\nu A_\mu-ig[A_\mu,A_\nu]$$

and the Lagrangian is then

$$\mathscr{L}=-\tfrac{1}{2}\mathrm{Tr}\,G_{\mu\nu}G^{\mu\nu}.\tag{7.55}$$

It may, for example, be checked that in the case of $SU(2)$, this gives $-\tfrac{1}{4}\mathbf{W}_{\mu\nu}\cdot\mathbf{W}^{\mu\nu}$, as in (3.132). On expansion this is

$$\mathscr{L}=-\tfrac{1}{4}G^a_{\mu\nu}G^{a\mu\nu}$$

$$=-\tfrac{1}{4}[\partial_\mu A^a_\nu-\partial_\nu A^a_\mu+gf^{abc}A^b_\mu A^c_\nu]$$

$$\times [\partial^\mu A^{\nu a} - \partial^\nu A^{\mu a} + g f^{amn} A^{\mu m} A^{\nu n}]$$
$$= -\tfrac{1}{4}[\text{quadratic term} + 2g f^{abc} A_\mu^b A_\nu^c (\partial^\mu A^{\nu a} - \partial^\nu A^{\mu a})$$
$$+ g^2 f^{abc} f^{amn} A_\mu^b A_\nu^c A^{\mu m} A^{\nu n}]. \tag{7.56}$$

The quadratic term gives the propagator, which we have already found. The cubic and quartic terms give the couplings

$$-g f^{abc}[(r_\mu - q_\mu)g_{\nu\rho} + (p_\nu - r_\nu)g_{\mu\rho}$$
$$+ (q_\rho - p_\rho)g_{\mu\nu}], \tag{7.57}$$

$$-g^2 [f^{abe} f^{cde}(g_{\mu\rho}g_{\nu\sigma} - g_{\mu\sigma}g_{\nu\rho})$$
$$+ f^{ace} f^{bed}(g_{\mu\sigma}g_{\rho\nu} - g_{\mu\nu}g_{\rho\sigma})$$
$$+ f^{ade} f^{bce}(g_{\mu\nu}g_{\sigma\rho} - g_{\mu\rho}g_{\sigma\nu})]. \tag{7.58}$$

The coupling of spin $\tfrac{1}{2}$ matter fields with gauge fields follows immediately by replacing ∂_μ in the free matter Lagrangian by the covariant derivative

$$D_\mu = \partial_\mu - ig A_\mu^a T^a \tag{7.59}$$

where T^a are $n \times n$ Hermitian matrix generators of the group, the matter field ψ forming the basis for an n-dimensional representation. This is easily seen to give the Feynman rules

matter propagator $\dfrac{i\delta_{ab}}{\gamma \cdot p - m}$,

gauge field–matter coupling $-ig\gamma_\mu(T^c)_{ab}$.

$$\tag{7.60}$$

This completes our search for the Feynman rules for non-Abelian gauge fields and their coupling with matter, in the Lorentz gauge. A crucial feature of these rules is the Faddeev–Popov ghost. As we have mentioned before, ghosts only occur in the internal parts of Feynman diagrams, and have the wrong spin-statistics relation. Their existence, however, is crucial, for without them unitarity would be violated, as we shall see below. Nevertheless, the reader would be forgiven for regarding them as something of a mathematical artifice, and asking, in consequence, if there is a gauge in

Yang–Mills theory, in which their contribution may be integrated out, as it is in electrodynamics. In fact there is: it is called the axial gauge, and we conclude this section with a brief consideration of it.

Gauge-field propagator in the axial gauge
The axial gauge is defined by the condition

$$t^\mu A_\mu^a = 0, \quad t^\mu t_\mu = -1 \tag{7.61}$$

where t is a spacelike vector. The gauge-fixing term is then

$$F^a = t^\mu A_\mu^a \tag{7.62}$$

and under the gauge transformation (7.28) we have

$$\delta F^a = f^{abc}\omega^b t^\mu A_\mu^c + t^\mu \partial_\mu \omega^a$$
$$= t^\mu \partial_\mu \omega^a.$$

Hence

$$\frac{\delta F^a}{\delta \omega^b} = \delta^{ab} t^\mu \partial_\mu. \tag{7.63}$$

This does not contain A_μ^a, so, inserting it into the ghost part of the action, we see that the ghost decouples from the gauge field, and its contribution may be integrated out. This covenience is bought at a price, however, for the gauge-field propagator turns out to be quite complicated in this gauge. The effective Lagrangian, not including the ghost term, is

$$\mathscr{L} + \mathscr{L}_{GF} = -\tfrac{1}{4} G_{\mu\nu}^a G^{a\mu\nu} - \frac{1}{2\alpha}(t^\mu A_\mu^a)^2.$$

The corresponding quadratic part of the action is, after partial integration,

$$\frac{1}{2} \int A^{\mu a}\left(\Box g_{\mu\nu} - \partial_\mu \partial_\nu - \frac{1}{\alpha} t_\mu t_\nu \right) A^{\nu a}\, dx.$$

The operator in brackets is, in momentum space,

$$-k^2 g_{\mu\nu} + k_\mu k_\nu - \frac{1}{\alpha} t_\mu t_\nu.$$

It is straightforward to check that this has the inverse

$$-\frac{1}{k^2}\left[g^{\mu\nu} + \frac{(t^2 + \alpha k^2)k^\mu k^\nu}{(k\cdot t)^2} - \frac{k^\mu t^\nu + t^\mu k^\nu}{(k\cdot t)} \right]. \tag{7.64}$$

In the limit $\alpha \to 0$, we have the correspondence

$$\underset{\mu}{\overset{a \quad k \quad b}{\text{〰〰〰〰}}}\underset{\nu}{} \quad -\frac{i}{k^2}\left[g^{\mu\nu} + \frac{t^2}{(k\cdot t)^2}k^\mu k^\nu - \frac{k^\mu t^\nu + t^\mu k^\nu}{k\cdot t} \right]\delta^{ab}. \tag{7.65}$$

7.3   Self-energy operator and vertex function

Having got the Feynman rules, we can now calculate the amplitude for any process to any order in perturbation theory. A question which then naturally arises is, are there any implications of gauge invariance for physical processes which hold true to *all* orders in perturbation theory – in other words, which are *exactly* true? It turns out that there are, and they are called the Ward identities in QED; they have generalisations to the non-Abelian case. The generalised Ward identities are crucial in proving the renormalisability of gauge theories, and renormalisability, in turn, is crucial in order that these theories make sense and are believable. The Ward identities and their generalisations are stated in terms of complete propagators and vertex functions, and our job in this section is to define these things.

We saw in the last chapter how perturbation theory was applied to calculating the 2- and 4-point functions, and therefore to scattering processes, and saw how the mass of a particle, defined as a pole in the 2-point function, was no longer the bare mass m, but $m + \delta m$, with $\delta m^2 = -\frac{1}{2} i g \Delta_F(0)$ in ϕ^4 theory. We want now to consider the problem of summing to all orders, and therefore (assuming that perturbation theory makes physical sense!) of obtaining exact Green's functions. We continue to refer, for the present, to ϕ^4 theory, though, of course, we shall broaden our outlook when we consider gauge invariance – ϕ^4 is not a gauge theory!

Let us begin by recalling some results from chapter 6 and introducing some new notation. Z is the generating functional for the n-point functions $\tau(x_1, \ldots, x_n)$ or Green's functions $G^{(n)}(x_1, \ldots, x_n)$:

$$\tau(x_1, \ldots, x_n) = G^{(n)}(x_1, \ldots, x_n) = \frac{1}{i^n} \frac{\delta^n Z[J]}{\delta J(x_1) \ldots \delta J(x_n)} \bigg|_{J=0}. \tag{7.66}$$

These contain connected (or irreducible) and disconnected (reducible) parts, for example (see (6.100)),

$$G^{(4)} = -3 \overline{} - 3ig \overset{O}{\overline{}} - ig \times + O(g^2); \tag{7.67}$$

the first two of these terms are reducible, the last term is irreducible. Only the connected Green's functions contribute to the scattering, and these are generated by W (where $Z = e^{iW}$), so that the connected Green's functions are

$$i\phi(x_1, \ldots, x_n) = G_c^{(n)}(x_1, \ldots, x_n) = \frac{1}{i^{n-1}} \frac{\delta^n W[J]}{\delta J(x_1) \delta J(x_n)} \bigg|_{J=0}. \tag{7.68}$$

In ϕ^4 theory, referring to (7.67),

$$G_c^{(4)} = -ig \times + O(g^2). \tag{7.69}$$

We saw that $G^{(n)}$ may be expressed in terms of G_c entirely; it is $G_c^{(n)}$ plus products of $G_c^{(m)}$, connected m-point functions of lower order, $m < n$. This was a useful simplification, based on a classification of diagrams. We now introduce a further classification. Ignoring numerical factors such as i, the (connected) 2-point function is, to *all orders*,

$$+ O(g^4). \tag{7.70}$$

These are all connected (irreducible) diagrams, and we want to devise a method of summing them. The sum is called the *complete* or *dressed* propagator and is denoted

$$x\text{———}\bigcirc\text{———}y = G_c^{(2)}(x, y). \tag{7.71}$$

The effect of the terms of order g and higher is to change the physical mass away from the 'bare mass' m, and hence to give rise to 'self-energy'. The graphs above all contribute to self-energy. Consider the first graph of order g^2, . It may be written as a product

The first and last factors are merely external propagators, which are common to all the graphs, so we define *truncated* graphs by multiplying the external legs by inverse propagators. We denote them then by dashed lines, for example,

The second graph in g^2 becomes , and the third graph ----$\ominus$----. We may similarly deal with the g^3 graphs, and those of higher orders. Of the three graphs of order g^2, then, the first is a product of graphs of lower order, but the other two are not; this is because the first graph contains a propagator. It is called a 1-*particle reducible* graph. It may be cut into two by cutting one internal line. This is clearly not true of the second two graphs, which are therefore called 1-*particle irreducible* (1PI) graphs. Based on this classification, we define the *proper self-energy* part as the sum of 1PI graphs, denoted as follows:

$$- - - - -\bigcirc\!\!\!\!\!\! \diagdown - - - - - = \frac{1}{i}\Sigma(p)$$

$$= \underset{p}{- - -}\bigcirc\underset{p}{- - -} + \underset{p}{-}\overset{\text{\Large 8}}{\underset{p}{-}} + \underset{p}{- -}\ominus\underset{p}{- -}$$

$$+ - - - -\overset{\text{\large 8}}{-} - - - - + \cdots. \tag{7.72}$$

The complete propagator (7.70) (in momentum space) may therefore be written in terms of the bare propagator $G_0(p) = i/(p^2 - m^2)$ and the proper self-energy function $\Sigma(p)$ as follows

$$G_c^{(2)}(p) = G_0(p) + G_0(p)\frac{\Sigma(p)}{i}G_0(p)$$

$$+ G_0(p)\frac{\Sigma(p)}{i}G_0(p)\frac{\Sigma(p)}{i}G_0(p) + \cdots$$

$$= G_0\left(1 + \frac{\Sigma}{i}G_0 + \frac{\Sigma}{i}G_0\frac{\Sigma}{i}G_0 + \cdots\right)$$

$$= G_0\left(1 - \frac{\Sigma}{i}G_0\right)^{-1}$$

$$= \left[G_0^{-1}(p) - \frac{1}{i}\Sigma(p)\right]^{-1}$$

$$= \frac{i}{p^2 - m^2 - \Sigma(p)}; \tag{7.73}$$

or, in diagrams

$$-\!\!\!-\!\!\!\bigcirc\!\!\!-\!\!\!- \;=\; -\!\!\!-\!\!\!-\!\!\!- \;+\; -\!\!\!-\!\!\!\bigcirc\!\!\!\!\!\!\diagdown\!\!\!-\!\!\!- \;+$$

$$+ \;-\!\!\!-\!\!\!\bigcirc\!\!\!\!\!\!\diagdown\!\!\!\!-\!\!\!\bigcirc\!\!\!\!\!\!\diagdown\!\!\!-\!\!\!- \;+\; \cdots. \tag{7.74}$$

Defining the physical mass m_{phys} by the pole in the complete propagator

$$G_c^{(2)}(p) = \frac{i}{p^2 - m_{\text{phys}}^2} \tag{7.75}$$

gives, on comparison with (7.73),

$$m_{\text{phys}}^2 = m^2 + \Sigma(p) \tag{7.76}$$

which justifies the appellation 'self-energy' term to Σ. It represents the change in mass from the 'bare' to the 'physical' value, calculated to all orders in perturbation theory.

From (7.73), we have

$$G_c^{(2)}(p)^{-1} = G_0(p)^{-1} - \frac{1}{i}\Sigma(p), \tag{7.77}$$

so the inverse of the 2-point function contains, apart from the inverse bare propagator, *only 1PI graphs*. This is an example of a *vertex function*, and may be generalised.

The 2-*point vertex function* $\Gamma^{(2)}(p)$ is defined by

$$G_c^{(2)}(p)\Gamma^{(2)}(p) = i. \tag{7.78}$$

Together with (7.77) this gives

$$\Gamma^{(2)}(p) = p^2 - m^2 - \Sigma(p) \tag{7.79}$$

We shall now show that there is a generating functional for $\Gamma^{(n)}(p)$. It is denoted $\Gamma[\phi]$ and is defined by the following *Legendre transformation* on $W[J]$:

$$W[J] = \Gamma[\phi] + \int dx J(x)\phi(x). \tag{7.80}$$

This gives

$$\frac{\delta W[J]}{\delta J(x)} = \phi(x), \quad \frac{\delta \Gamma[\phi]}{\delta \phi(x)} = -J(x). \tag{7.81}$$

Hence we have for the propagator

$$G(x, y) = -\frac{\delta^2 W[J]}{\delta J(x)\delta J(y)} = -\frac{\delta\phi(x)}{\delta J(y)}. \tag{7.82}$$

Let us now define the kernel

$$\Gamma(x, y) = \frac{\delta^2\Gamma[\phi]}{\delta\phi(x)\delta\phi(y)} = -\frac{\delta J(x)}{\delta\phi(y)}. \tag{7.83}$$

These are clearly the inverses of each other, for

$$\int dz G(x, z)\Gamma(z, y) = -\int dz\frac{\delta^2 W[J]}{\delta J(x)\delta J(z)}\frac{\delta^2\Gamma[\phi]}{\delta\phi(z)\delta\phi(y)}$$

$$= \int dz\frac{\delta\phi(x)}{\delta J(z)}\frac{\delta J(z)}{\delta\phi(y)}$$

$$= \frac{\delta\phi(x)}{\delta\phi(y)}$$

$$= \delta(x - y). \tag{7.84}$$

Actually, the Fourier transform of this equation would be expected to be

$$G(p, -p)\Gamma(p, -p) = 1$$

rather than equation (7.78), which has i on the right-hand side (and where we have put $G(p, -p) = G(p))$. This slight irregularity may be properly accounted for by defining Fourier transforms in the following way:

$$G(x, y) = \frac{1}{(2\pi)^8} \int dp \, dq \, e^{i(px + qy)} \tilde{G}(p, q), \tag{7.85}$$

$$\tilde{G}(p, q) = \int dx \, dy \, e^{-i(px + qy)} G(x, y), \tag{7.86}$$

and, from translation invariance,

$$\tilde{G}(p, q) = (2\pi)^4 \delta(p + q) G(p, q). \tag{7.87}$$

(The integrals and delta functions and scalar products above are all, of course, 4-dimensional.) On the other hand, for the 2-point vertex functions we have

$$\Gamma(x, y) = \frac{i}{(2\pi)^8} \int dp \, dq \, e^{i(px + qy)} \tilde{\Gamma}(p, q), \tag{7.88}$$

$$\tilde{\Gamma}(p, q) = -i \int dx \, dy \, e^{-i(px + qy)} \Gamma(x, y), \tag{7.89}$$

$$\tilde{\Gamma}(p, q) = (2\pi)^4 \delta(p + q) \Gamma(p, q). \tag{7.90}$$

Equations (7.84) and (7.78) are now consistent.

Now we differentiate equation (7.84),

$$\int dz \frac{\delta^2 W}{\delta J(x) \delta J(z)} \frac{\delta^2 \Gamma}{\delta\phi(z)\delta\phi(z')} = -\delta(x - z'),$$

with respect to $J(x'')$, using

$$\frac{\delta}{\delta J(x'')} = \int dz'' \frac{\delta\phi(z'')}{\delta J(x'')} \frac{\delta}{\delta\phi(z'')}$$

$$= -\int dz'' G(x'', z'') \frac{\delta}{\delta\phi(z'')}.$$

This gives

$$\int dz \frac{\delta^3 W}{\delta J(x) \delta J(x'') \delta J(z)} \frac{\delta^2 \Gamma}{\delta\phi(z)\delta\phi(z')}$$

$$- \int dz \, dz'' \frac{\delta^2 W}{\delta J(x) \delta J(y)} G(x'', z'') \frac{\delta^3 \Gamma}{\delta\phi(z)\delta\phi(z')\delta\phi(z'')} = 0,$$

hence

$$\int dz \frac{\delta^3 W}{\delta J(x) \delta J(x'') \delta J(z)} \Gamma(z, z')$$

$$+ \int dz \, dz'' G(x, z) G(x'', z'') \frac{\delta^3 \Gamma}{\delta\phi(z)\delta\phi(z')\delta\phi(z'')} = 0.$$

Now we multiply by $G(x', z')$ and integrate over z', using (7.84), to give

$$\frac{\delta^3 W}{\delta J(x)\delta J(x')\delta J(x'')} = -\int dz\, dz'\, dz''\, G(x, z)G(x', z')G(x'', z'')$$

$$\times \frac{\delta^3 \Gamma}{\delta\phi(z)\delta\phi(z')\delta\phi(z'')}. \tag{7.91}$$

This says that the connected 3-point function is the same as the irreducible (1PI) 3-point vertex, with exact propagators in the external lines. It may be represented diagramatically by Fig. 7.1. Equation (7.91) may also be inverted, using (7.84), to give

$$\frac{\delta^3 \Gamma}{\delta\phi(y)\delta\phi(y')\delta\phi(y'')}$$

$$= -\int dx\, dx'\, dx''\, \Gamma(x, y)\Gamma(x', y')\Gamma(x'', y'')\frac{\delta^3 W}{\delta J(x)\delta J(x')\delta J(x'')}. \tag{7.92}$$

Since $\Gamma(x, y)$, etc., are the inverse propagators, the right-hand side of this equation is the truncated connected 3-point function, which is therefore the same as the 1PI 3-point function.

Differentiating (7.91) once more gives an equation for the connected 4-point function, which is represented in Fig. 7.2. The connected 4-point function is seen to contain a 1PI part (the first term on the right), with exact propagators in the external lines, and three 1-particle reducible parts, related to each other by 'crossing'.

The 1-particle irreducible function Γ are useful for stating the generalised Ward identities, and also in connection with spontaneous symmetry breaking, as we shall see in the next chaper.

The relations we have derived, or alternatively Figs. 7.1 and 7.2, show that the theory can be expressed either in terms of $W[J]$ or $\Gamma[\phi]$, where

Fig. 7.1. Two representations of equation (7.91).

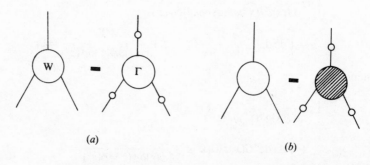

(a)   (b)

$\delta W/\delta J = \phi, \delta\Gamma/\delta\phi = -J$, and

$$W = \Gamma + \int J\phi. \tag{7.93}$$

This Legendre transformation has a simple geometrical interpretation, and is also commonly used in thermodynamics. In fact, it would appear that there are fairly close analogies between quantum field theory and statistical mechanics (thermodynamics) which are revealed by this formulation of field theory. We close this section by considering these two topics in turn.

Geometrical interpretation of the Legendre transformation

For simplicity, consider the case of functions of only one independent variable, x. Suppose we have such a function

$$f = f(x); \tag{7.94}$$

for example, f and x may be physical quantities and (7.94) is the relation between them; x is the independent variable. Suppose now that we want to change the independent variable to $\mathrm{d}f/\mathrm{d}x = u$. The question is then: what dependent variable should we choose, so as not to lose any of the information contained in (7.94)? Consider a graphical representation, and let $f(x)$ be represented as in Fig. 7.3(a). This corresponds to (7.94). Now, when the independent variable is $u = \mathrm{d}f/\mathrm{d}x$, if the dependent variable is f, we have the graph in Fig. 7.3(b). Does this contain all the information in Fig. 7.3(a)? To see if it does, try to reconstruct $f(x)$ from Fig. 7.3(b). We arrive at Fig. 7.3(c), which shows that the reconstruction cannot be done uniquely – some information has been lost. So we ask again: if $u = \mathrm{d}f/\mathrm{d}x$ is to be taken as the independent variable, what should we choose as the dependent variable?

Fig. 7.2. See text.

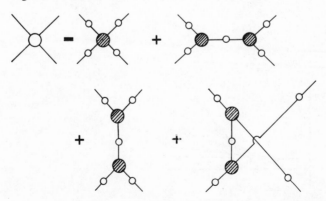

A *tangent* to the curve in Fig. 7.3(a) has the equation

$$f = ux + g$$

where g is the intercept on the f axis, and u is the gradient. So, if u is the independent variable, choosing g as the dependent variable,

$$g = f - ux \qquad (7.95)$$

means that, from a knowledge of $g = g(u)$, Fig. 7.3(a) may be reconstructed as the *envelope of tangents*, as in Fig. 7.3(d). (7.95) is an example of the Legendre transformation. The generalisation to more than one dimension is trivial: if $f(x, y)$ is a given function, and we wish to change the independent variables from (x, y) to (u, y) where $u = \mathrm{d}f/\mathrm{d}x$, then the correct

Fig. 7.3. Geometrical interpretation of the Legendre transformation.

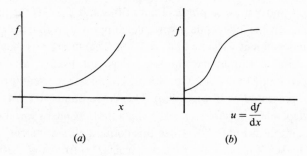

(a)   (b)

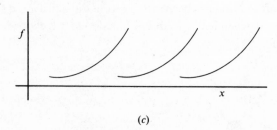

(c)

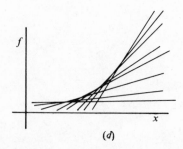

(d)

function of u and y to choose is

$$g(u, y) = f(x, y) - ux. \tag{7.96}$$

Equation (7.93) is of this form (though of course it is a *functional* equation), for we may write it as

$$\Gamma[\phi] = W[J] - \int J\phi \tag{7.97}$$

where $\phi = \delta W/\delta J$.

Thermodynamic analogy

In a thermodynamic system, the internal energy U is a function of entropy S and volume V, $U = U(S, V)$. It may, however, sometimes not be convenient to express the energy of a system in terms of a function of S, since the entropy is not easily measured! The temperature T is, on the other hand, very easily measured (and is an intensive quantity, while S is extensive). What is the corresponding function of T which provides a description of the system? It is the free energy F, given by

$$F = U - TS$$

or

$$F(T, V) = U(S, V) - TS \tag{7.98}$$

where

$$T = \left(\frac{\partial U}{\partial S}\right)_V$$

as follows from

$$dU = TdS - PdV.$$

It is clear that equations (7.97) and (7.98) have the same mathematical form, but actually the analogies between field theory and thermodynamics – approached through statistical mechanics – go much deeper than this. They start with the partition function Z, which is analogous to the generating functional for Green's functions, also denoted Z. The partition function is related to the free energy F by

$$Z = e^{-F/NkT}.$$

These analogies are summarised in Table 7.1.

Another example is magnetisation.[12] Let Z be the partition function of a system in the presence of an external field H. Then it is also the generating functional for the Euclidean Green's function

$$Z[H] = \sum \frac{1}{N!} \int dx_1 \dots dx_N H(x_1) \dots H(x_N) G^{(N)}(x_1, \dots, x_N).$$

Table 7.1. *Analogies between field theory and statistical mechanics*

Field theory	Statistical mechanics
Z	Z partition function
$Z = e^{iW}$	$Z = e^{-F/NkT}$
$W[J] = \Gamma[\phi] + \int J\phi$	$F(T) = U(S) - TS$

Similarly,

$$W[H] = \ln Z[H]$$

is the generating functional for connected Green's functions. Also $\delta W/\delta H$ is the 'magnetisation' M

$$M(x) = \frac{\delta W[H]}{\delta H(x)}$$

and 1PI vertices are generated by $\Gamma[M]$ where

$$\Gamma[M] + W[H] = \int dx\, H(x)M(x).$$

As a final example of a Legendre transformation we may cite the relation between the Lagrangian and Hamiltonian formulation of classical mechanics. For point particles, L is a function of $x, \dot{x}$ and t, and H a function of x, p and t, and we have

$$H(x, p) = -L(x, \dot{x}) + \dot{x}p, \quad p = \frac{\partial L}{\partial \dot{x}}.$$

7.4   Ward–Takahashi identities in QED

The Ward identity and its generalisation by Takahashi are exact relations between 1PI vertex functions and propagators, true to all orders of perturbation theory. They follow from the gauge invariance of QED, and play a key role in the proof of the renormalisability of this theory. We shall prove these identities starting from the generating functional Z for a system of photons and electrons, given by

$$Z = N \int \mathscr{D}A_\mu \mathscr{D}\bar{\psi}\mathscr{D}\psi \exp\left(i\int \mathscr{L}_{\text{eff}}\, dx\right), \tag{7.99}$$

$$\mathscr{L}_{\text{eff}} = -\tfrac{1}{4}F_{\mu\nu}F^{\mu\nu} + i\bar{\psi}\gamma^\mu(\partial_\mu + ieA_\mu)\psi - m\bar{\psi}\psi$$
$$- \frac{1}{2\alpha}(\partial^\mu A_\mu)^2 + J^\mu A_\mu + \bar{\eta}\psi + \bar{\psi}\eta. \tag{7.100}$$

This effective Lagrangian contains the free field photon part, the free field electron part, with the ordinary derivative replaced by the covariant derivative to account for the interaction with the electromagnetic field, a gauge-fixing term for the Lorentz gauge, and source terms for A_μ, ψ and $\bar\psi$. What is missing is the Faddeev–Popov ghost term. Since, as we have seen, the ghost does not couple to physical fields (in this gauge), its contribution to Z is only an overall constant, and may be taken to have been absorbed into N.

We recall that without the gauge-fixing term (and source terms), the Lagrangian is gauge invariant. This made Z infinite and the search for a photon propagator doomed. To find a finite propagator, we were forced to introduce the gauge-fixing term (and ghost term, which in the Abelian case we may ignore). This however, means that $\mathscr{L}_{eff}$ is not gauge invariant. The physical consequences of the theory, expressed in terms of Green's functions, however, cannot depend on the gauge, so Z *must* be gauge invariant. This is a non-trivial requirement, and leads to a differential equation for Z, which we now find.

On performing an infinitesimal gauge transformation

$$A_\mu \to A_\mu + \partial_\mu \Lambda, \quad \psi \to \psi - ie\Lambda\psi$$
$$\bar\psi \to \bar\psi + ie\Lambda\bar\psi \tag{7.101}$$

(cf. (3.67) and (3.74), with $\Lambda \to e\Lambda$; the electron has charge $e(<0)$), the first three terms in (7.100) are invariant, but the rest are not, so the integrand of Z picks up a factor:

$$\exp\left\{ i\int dx\left[-\frac{1}{\alpha}(\partial^\mu A_\mu)\Box\Lambda + J^\mu\partial_\mu\Lambda - ie\Lambda(\bar\eta\psi - \bar\psi\eta) \right]\right\},$$

which, since Λ is infinitesimal, may be written

$$1 + i\int dx\left[-\frac{1}{\alpha}(\partial^\mu A_\mu)\Box - \partial^\mu J_\mu - ie(\bar\eta\psi - \bar\psi\eta) \right]\Lambda(x), \tag{7.102}$$

where we have integrated by parts to remove the derivative operator from Λ. Invariance of Z implies that this operator (7.102), when acting on Z, is merely the identity. Since Λ is an arbitrary function, this implies that

$$\left[-\frac{1}{\alpha}\Box(\partial^\mu A_\mu) - \partial^\mu J_\mu - ie(\bar\eta\psi - \bar\psi\eta) \right]Z = 0.$$

Making the substitutions

$$\psi \to \frac{1}{i}\frac{\delta}{\delta\bar\eta}, \quad \bar\psi \to \frac{1}{i}\frac{\delta}{\delta\eta}, \quad A_\mu \to \frac{1}{i}\frac{\delta}{\delta J^\mu},$$

we find the functional differential equation

$$\left[\frac{i}{\alpha}\Box\,\partial^\mu\frac{\delta}{\delta J^\mu} - \partial^\mu J_\mu - e\left(\bar\eta\frac{\delta}{\delta\bar\eta} - \eta\frac{\delta}{\delta\eta}\right)\right]Z[\eta,\bar\eta,J] = 0. \tag{7.103}$$

Putting $Z = e^{iW}$, this may be written as an equation for W:

$$-\frac{\Box}{\alpha}\partial^\mu\frac{\delta W}{\delta J^\mu} - \partial^\mu J_\mu - ie\left(\bar\eta\frac{\delta W}{\delta\bar\eta} - \eta\frac{\delta W}{\delta\eta}\right) = 0, \tag{7.104}$$

where $W = W[\eta,\bar\eta,J]$. Finally, we convert this into an equation for the vertex function Γ, given by

$$\Gamma[\psi,\bar\psi,A_\mu] = W[\eta,\bar\eta,J_\mu] - \int dx(\bar\eta\psi + \bar\psi\eta + J^\mu A_\mu) \tag{7.105}$$

which implies that

$$\left.\begin{aligned}
\frac{\delta\Gamma}{\delta A_\mu(x)} &= -J^\mu(x), & \frac{\delta W}{\delta J_\mu(x)} &= A^\mu(x), \\[2mm]
\frac{\delta\Gamma}{\delta\psi(x)} &= -\bar\eta(x), & \frac{\delta W}{\delta\bar\eta(x)} &= \psi(x), \\[2mm]
\frac{\delta\Gamma}{\delta\bar\psi(x)} &= -\eta(x), & \frac{\delta W}{\delta\eta(x)} &= \bar\psi(x).
\end{aligned}\right\} \tag{7.106}$$

Equation (7.104) then becomes

$$-\frac{\Box}{\alpha}\partial^\mu A_\mu(x) + \partial_\mu\frac{\delta\Gamma}{\delta A_\mu(x)} - ie\psi\frac{\delta\Gamma}{\delta\psi(x)} + ie\bar\psi\frac{\delta\Gamma}{\delta\bar\psi(x)} = 0. \tag{7.107}$$

Now functionally differentiate this equation twice, with respect to $\bar\psi(x_1)$ and $\psi(y_1)$, and put $\bar\psi = \psi = A_\mu = 0$. The first term vanishes, and we obtain

$$-\partial^\mu_x\frac{\delta^3\Gamma[0]}{\delta\bar\psi(x_1)\delta\psi(y_1)\delta A^\mu(x)} = ie\delta(x - x_1)\frac{\delta^2\Gamma[0]}{\delta\bar\psi(x_1)\delta\psi(y_1)}$$

$$-ie\delta(x - y_1)\frac{\delta^2\Gamma[0]}{\delta\bar\psi(x_1)\delta\psi(y_1)}. \tag{7.108}$$

The left-hand side of this equation is the derivative of the (1PI) electron–photon vertex, and the two terms on the right are the inverses of exact propagators. The content of (7.108) becomes clear if we express it in momentum space. We therefore define the proper vertex function $\Gamma_\mu(p,q,p')$ by

$$\int dx\,dx_1\,dy_1\,e^{i(p'x_1 - py_1 - qx)}\frac{\delta^3\Gamma[0]}{\delta\bar\psi(x_1)\delta\psi(y_1)\delta A^\mu(x)}$$

$$= ie(2\pi)^4\delta(p' - p - q)\Gamma_\mu(p,q,p'). \tag{7.109}$$

On the other hand, $\delta^2\Gamma/\delta\bar{\psi}\,\delta\psi$ is, as we have seen, the inverse propagator, which we denote S'_F (to distinguish it from S_F, the bare propagator), so

$$\int dx_1\,dy_1\,e^{i(p'x_1-py_1)}\frac{\delta^2\Gamma[0]}{\delta\bar{\psi}(x_1)\delta\psi(y_1)}=(2\pi)^4\delta(p'-p)iS'^{-1}_F(p). \quad (7.110)$$

Multiplying (7.108) by $\exp i(p'x_1-py_1-qx)$ and integrating over x, x_1 and y_1 then gives

$$q^\mu\Gamma_\mu(p,q,p+q)=S'^{-1}_F(p+q)-S'^{-1}_F(p). \quad (7.111)$$

This is known as the *Ward–Takahashi identity* and it may be represented pictorially, as in Fig. 7.4. Taking the limit $q_\mu\to 0$ yields the *Ward identity*

$$\blacksquare\quad \frac{\partial S'^{-1}_F}{\partial p^\mu}=\Gamma_\mu(p,0,p). \quad (7.112)$$

As stated above, this relation holds to all orders in perturbation theory. It is instructive, however, to examine it to the lowest two orders; they are shown in Figs. 7.5 and 7.6. To lowest order, S'_F is simply the bare propagator S_F, so

$$S^{-1}_F(p)=\gamma_\mu p^\mu-m$$

and

$$\frac{\partial S^{-1}_F(p)}{\partial p^\mu}=\gamma_\mu. \quad (7.113)$$

Fig. 7.4. Ward–Takahashi identity.

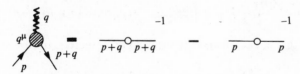

Fig. 7.5. Expansion of $\Gamma_\mu(p,q,p+q)$.

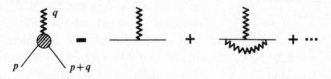

Fig. 7.6. Expansion of $S'_F(p)$.

Let us now calculate $\Gamma_\mu(p, 0, p)$ to lowest order. First of all, from a generalisation of (7.92) to the present case, noting that the $\Gamma(x, y)$ are the relevant inverse propagators and $W = -i \ln Z$, we have

$$\frac{\delta^3 \Gamma}{\delta \bar\psi(x_1) \delta \psi(y_1) \delta A^\mu(x)}$$

$$= -\int du_1 \, dv_1 \, du [iS_F^{-1}(u_1 - x_1) iS_F^{-1}(v_1 - y_1)]$$

$$\times \left\{ -iD_{\mu\nu}^{-1}(u - x)(-i) \frac{\delta^3 Z[0]}{\delta\eta(u_1)\delta\bar\eta(v_1)\delta J^\mu(u)} \right\}, \tag{7.114}$$

where $D_{\mu\nu}$ is the photon propagator function (whose explicit form we do not need). We now need to calculate the third derivative of Z, which itself is given by (7.99). Recalling the general theory in chapter 6, we separate out the interaction term, $e\bar\psi\gamma_\mu\psi A^\mu$, and write

$$Z[\eta, \bar\eta, J_\mu] = N \exp\left[ie \int dz \frac{1}{i} \frac{\delta}{\delta\eta(z)} \gamma^\lambda \frac{1}{i} \frac{\delta}{\delta\bar\eta(z)} \frac{1}{i} \frac{\delta}{\delta J^\lambda(z)} \right] Z_0 \tag{7.115}$$

where the generating functional for free electrons and photons is

$$Z_0 = \exp\left[-i \int dx \, dy \, \bar\eta(x) S_F(x - y)\eta(y) \right]$$

$$\times \exp\left[\frac{i}{2} \int dx \, dy \, J^\mu(x) D_{\mu\nu}(x - y) J^\nu(y). \tag{7.116}$$

This gives, to lowest order,

$$\frac{\delta^3 Z[0]}{\delta\eta(u_1)\delta\bar\eta(v_1)\delta J^\mu(u)} = ie \int dz \, S_F(u_1 - z) S_F(v_1 - z) D_{\mu\nu}(u - z)\gamma^\nu. \tag{7.117}$$

Substituting this in (7.114) and the result in (7.109) gives, to lowest order,

$$\Gamma_\mu(p, q, p + q) = \gamma_\mu \tag{7.118}$$

which, in view of (7.113), satisfies the Ward identity (7.112). In fact, it 'more than' satisfies it, for, in this case of lowest order the vertex function γ_μ is independent of the photon momentum q. We shall see, however, that this feature does not persist in higher orders; to second order (which is order e^3) the Ward identity is satisfied, and $q = 0$ is an essential condition.

Before considering the second order, it is convenient to recast the first order Ward identity in a different form. Differentiating the identity $S_F(p)S_F^{-1}(p) = 1$ with respect to p^μ gives

$$\frac{\partial S_F(p)}{\partial p^\mu} = -S_F(p) \frac{\partial S_F^{-1}(p)}{\partial p^\mu} S_F(p)$$

$$= -S_F(p)\gamma_\mu S_F(p) \tag{7.119}$$

where (7.113) has been used. The right-hand side of this equation has the correct factors to describe the simplest vertex diagram – the first in Fig. 7.5 – in which the external photon has zero momentum. Formally, then, *differentiation of the propagator with respect to p^μ corresponds to insertion of a zero momentum photon line into the internal electron line.*

Now let us consider the next order in Figs. 7.5 and 7.6. We begin by writing out an expansion for the complete electron propagator iS'_F in terms of the bare propagator iS_F. In analogy with (7.73) we write

$$iS'_F = iS_F + iS_F \frac{\Sigma}{i} iS_F + iS_F \frac{\Sigma}{i} iS_F \frac{\Sigma}{i} iS_F + \cdots$$

$$= iS_F \left(1 + \frac{\Sigma}{i} iS'_F \right) \tag{7.120}$$

from which follows

$$S'^{-1}_F = S^{-1}_F - \Sigma. \tag{7.121}$$

Hence, using (7.113),

$$\frac{\partial S'^{-1}_F}{\partial p^\mu} = \frac{\partial S^{-1}_F}{\partial p^\mu} - \frac{\partial \Sigma}{\partial p^\mu}$$

$$= \gamma_\mu - \frac{\partial \Sigma}{\partial p^\mu}. \tag{7.122}$$

Writing the vertex function expansion, shown in Fig. 7.5 as

$$\Gamma_\mu(p, q, p + q) = \gamma_\mu + \Lambda_\mu(p, q, p + q) \tag{7.123}$$

where Λ_μ represents the truly 1PI contribution to Γ_μ, the Ward identity (7.112) now implies that

$$\Lambda_\mu(p, 0, p) = -\frac{\partial \Sigma}{\partial p^\mu}. \tag{7.124}$$

It is now our task to verify this explicitly to lowest order, i.e. when Λ_μ is represented by the second term on the right of Fig. 7.5, and Σ by the second term on the right of Fig. 7.6. Comparing equations (7.120) and (7.121), it is clear that Σ is given by the self-energy bubble in Fig. 7.6, without the external legs. From the Feynman rules this is

$$\frac{\Sigma}{i} = (-ie)^2 \int \frac{d^4k}{(2\pi)^4} \frac{-ig_{\kappa\lambda}}{k^2} \gamma^\kappa \frac{i}{\gamma\cdot(p-k)-m} \gamma^\lambda$$

$$= -e^2 \int \frac{d^4k}{(2\pi)^4} \frac{1}{k^2} \gamma^\lambda S_F(p-k)\gamma_\lambda. \tag{7.125}$$

Here we have taken the photon propagator in the Feynman gauge, i.e. with

$\alpha = 1$ in (7.54). Now using (7.119), we have

$$\frac{\partial \Sigma}{\partial p^\mu} = -ie^2 \int \frac{d^4k}{(2\pi)^4} \frac{1}{k^2} \gamma^\lambda \frac{\partial}{\partial p^\mu} S_F(p-k)\gamma_\lambda$$

$$= ie^2 \int \frac{d^4k}{(2\pi)^4} \frac{1}{k^2} \gamma^\lambda S_F(p-k)\gamma_\mu S_F(p-k)\gamma_\lambda. \tag{7.126}$$

To calculate $\Lambda_\mu(p,q,p+q)$ we simply apply the Feynman rules to the propagators and vertices in Fig. 7.5, and recall from our first order calculation above that these graphs add up to $-ie\Gamma_\mu$ (i.e. that $-ie\Gamma_\mu = -ie\gamma_\mu$ to first order). Hence, using the Feynman gauge for the photon propagator,

$$-ie\Lambda_\mu(p,q,p+q) = (-ie)^3 \int \frac{d^4k}{(2\pi)^4} \frac{-ig_{\kappa\lambda}}{k^2} \gamma^\kappa \frac{i}{\gamma \cdot (p-k) - m} \gamma_\mu$$

$$\times \frac{i}{\gamma \cdot (p-k+q) - m} \gamma^\lambda$$

$$= -e^3 \int \frac{d^4k}{(2\pi)^4} \frac{1}{k^2} \gamma^\lambda S_F(p-k)\gamma_\mu S_F(p-k+q)\gamma_\lambda.$$

Hence

$$\Lambda_\mu(p,q,p+q) = -ie^2 \int \frac{d^4k}{(2\pi)^4} \frac{1}{k^2} \gamma^\lambda S_F(p-k)\gamma_\mu S_F(p-k+q)\gamma_\lambda. \tag{7.127}$$

It is now clear from (7.126) and (7.127) that the Ward identity (7.124) is satisfied.

As mentioned above, the Ward identity comes into its own when we consider renormalisation. Let us briefly anticipate some of the work of chapter 9. The integrals which we have exhibited above are actually divergent, and so are all the other terms which contribute to Figs. 7.5 and 7.6, so the vertex function and complete propagators are highly divergent quantities. In a renormalisable theory, however (and QED is renormalisable), these functions may be represented (at least near $p^2 = m^2$) as infinite constants multiplied by the bare propagators and vertex terms. So we put

$$S_F' \to Z_2 S_F,$$

$$\Gamma_\mu(p,0,p) \to \frac{1}{Z_1} \gamma_\mu. \tag{7.128}$$

The Ward identity then implies that

$$Z_1 = Z_2, \tag{7.129}$$

so the renormalisation of the theory may be done with one constant, not two (actually, there is another constant Z_3, for wave function renormalisation).

The Ward identity holds for the simplest gauge theory, QED, and it is natural to enquire whether analogous identities hold for non-Abelian gauge theories. They indeed do, and were first derived by Slavnov and Taylor, but it turns out that the easiest way of arriving at the Slavnov–Taylor identities is to introduce a rather clever transformation derived by Becchi, Rouet and Stora, under which the effective Lagrangian (7.49) is invariant. This is the subject of the next section.

7.5   Becchi–Rouet–Stora transformation

Our starting point for deriving the Ward identity was to observe that, although the generating functional Z is required to be gauge invariant, the effective Lagrangian is not, because of the gauge-fixing term; in the Abelian case, we could ignore the ghost term. In the non-Abelian case, we have a similar, though more complicated, situation. From equations (7.47–7.50) and (7.52) we have

$$Z = N \int \mathscr{D}A_\mu \mathscr{D}\eta \, \mathscr{D}\bar{\eta} \exp\left(i \int \mathscr{L}_{\text{eff}} \, dx \right)$$

where

$$\mathscr{L}_{\text{eff}} = -\tfrac{1}{4} F^a_{\mu\nu} F^{\mu\nu a} + \mathscr{L}_{\text{GF}} + \mathscr{L}_{\text{FPG}}. \tag{7.130}$$

For the gauge-fixing term we choose the Lorentz gauge

$$\mathscr{L}_{\text{GF}} = -\frac{1}{2\alpha} (\partial^\mu A^a_\mu)^2 \tag{7.131}$$

and the Faddeev–Popov ghost term may be written

$$\begin{aligned}
\mathscr{L}_{\text{FPG}} &= -\bar{\eta}^a (\delta^{ab} \Box - g f^{abc} \partial^\mu A^c_\mu - g f^{abc} A^c_\mu \partial^\mu) \eta^b \\
&= -\bar{\eta}^a \Box \eta^a + g f^{abc} \bar{\eta}^a (\partial^\mu A^c_\mu + A^c_\mu \partial^\mu) \eta^b \\
&= \partial^\mu \bar{\eta}^a \partial_\mu \eta^a - g f^{abc} (\partial^\mu \bar{\eta}^a) A^c_\mu \eta^b + \text{total derivative} \\
&= \partial^\mu \bar{\eta}^a (\partial_\mu \eta^a + g f^{abc} A^b_\mu \eta^c) \\
&\fallingdotseq \partial^\mu \bar{\eta}^a D_\mu \eta^a \\
&= -\bar{\eta}^a \partial^\mu D_\mu \eta^a + \text{total derivative}.
\end{aligned} \tag{7.132}$$

Here we have introduced total derivative terms twice. They may be ignored since they only contribute surface terms to the action. We have also introduced the covariant derivative $D_\mu \eta$ for the general non-Abelian group, as a generalisation of (3.155) for $SU(2)$.

We want now to investigate the behaviour of $\mathscr{L}_{\text{eff}}$ under the gauge transformation (which is the generalisation of (3.124))

$$\delta A_\mu^a = \frac{1}{g} \partial_\mu \Lambda^a + f^{abc} A_\mu^b \Lambda^c$$

$$= \frac{1}{g} (D_\mu \Lambda)^a. \tag{7.133}$$

It was Becchi, Rouet and Stora (BRS) who noticed that if we choose

$$\Lambda^a = -\eta^a \lambda \tag{7.134}$$

where both λ and η^a are Grassmann quantities (so that $\lambda^2 = 0$), and λ is a constant, so that

$$\delta A_\mu^a = -\frac{1}{g} (D_\mu \eta^a) \lambda; \tag{7.135}$$

and if we also demand that

$$\delta \eta^a = -\tfrac{1}{2} f^{abc} \eta^b \eta^c \lambda, \tag{7.136}$$

$$\delta \bar{\eta}^a = -\frac{1}{\alpha g} (\partial^\mu A_\mu^a) \lambda \tag{7.137}$$

(remember that $\bar{\eta}$ and η are independent Grassmann fields, so have independent transformations), then $\mathscr{L}_{\text{eff}}$ is *invariant*. The three equations above constitute the *Becchi–Rouet–Stora transformation*. We shall now prove that $\mathscr{L}_{\text{eff}}$ is indeed invariant under (7.135–7.137).

To begin, it is clear that the gauge field term $\mathscr{L}_0 = -\tfrac{1}{4} F_{\mu\nu}^a F^{\mu\nu a}$ is invariant, since (7.134) is merely a reparametrisation of Λ^a. For the gauge-fixing term we have, under (7.135),

$$\delta \mathscr{L}_{\text{GF}} = \frac{1}{\alpha} (\partial^\mu A_\mu^a) \frac{1}{g} (\partial^\nu D_\nu \eta^a) \lambda, \tag{7.138}$$

and the ghost term gives

$$\delta \mathscr{L}_{\text{FPG}} = -(\delta \bar{\eta}^a) \partial^\mu D_\mu \eta^a - \bar{\eta}^a \partial^\mu (\delta D_\mu \eta^a). \tag{7.139}$$

From (7.137) the first term is

$$-(\delta \bar{\eta}^a) \partial^\mu D_\mu \eta^a = \frac{1}{\alpha g} (\partial^\mu A_\mu^a) \lambda (\partial^\nu D_\nu \eta^a)$$

$$= -\frac{1}{\alpha g} (\partial^\mu A_\mu^a)(\partial^\nu D_\nu \eta^a) \lambda \tag{7.140}$$

since η^a and λ anticommute. (7.138) and (7.140) cancel, so we have finally

$$\delta \mathscr{L}_{\text{eff}} = -\bar{\eta}^a \partial^\mu (\delta D_\mu \eta^a), \tag{7.141}$$

and we must show that, under (7.135) and (7.136), $\delta D_\mu \eta^a = 0$, and hence $\mathscr{L}_{\text{eff}}$ is invariant. This turns out to be a rather long job. We have

$$\delta(D_\mu\eta^a) = \delta[\partial_\mu\eta^a + gf^{abc}A_\mu^b\eta^c]$$
$$= \partial_\mu(\delta\eta^a) + gf^{abc}(\delta A_\mu^b)\eta^c + gf^{abc}A_\mu^b(\delta\eta^c)$$
$$= -\tfrac{1}{2}f^{abc}\partial_\mu(\eta^b\eta^c)\lambda - f^{abc}(\partial_\mu\eta^b + gf^{bmn}A_\mu^b(\delta\eta^c)$$
$$+ gf^{abc}A_\mu^b(-\tfrac{1}{2}\eta^m\eta^n\lambda.$$

Because η is a Grassmann quantity, the derivative in the first term is

$$f^{abc}\partial_\mu(\eta^b\eta^c) = f^{abc}[(\partial_\mu\eta^b)\eta^c + \eta^b(\partial_\mu\eta^c)]$$
$$= f^{abc}[(\partial_\mu\eta^b)\eta^c - (\partial_\mu\eta^c)\eta^b]$$
$$= 2f^{abc}(\partial_\mu\eta^b)\eta^c.$$

Hence

$$\delta(D_\mu\eta^a) = -f^{abc}(\partial_\mu\lambda^b)(\eta^c\lambda + \lambda\eta^c) + gf^{abc}f^{bmn}A_\mu^m\eta^n\eta^c\lambda$$
$$-\tfrac{1}{2}gf^{abc}f^{cmn}A_\mu^b\eta^m\eta^n\lambda.$$

The first term vanishes, since λ and η^c are Grassmann quantities. In the last term, since the structure constants f obey the Jacobi identity (3.148) we have

$$f^{abc}f^{cmn} = -f^{amc}f^{cnb} - f^{anc}f^{cbm},$$

and so

$$\delta(D_\mu\eta^a) = gf^{abc}f^{bmn}A_\mu^m\eta^n\eta^c\lambda$$
$$+ \tfrac{1}{2}gf^{amc}f^{cnb}A_\mu^b\eta^m\eta^n\lambda$$
$$+ \tfrac{1}{2}gf^{anc}f^{cbm}A_\mu^b\eta^m\eta^n\lambda.$$

Interchanging the dummy suffixes $m \leftrightarrow n$ in the last term, this is seen to be equal to the second term. On further relabelling the indices, we have

$$\delta(D_\mu\eta^a) = g(f^{apn}f^{pbm} + f^{amp}f^{pnb})A_\mu^b\eta^m\eta^n\lambda$$
$$= g(-f^{apm}f^{pbn} + f^{amp}f^{pnb})A_\mu^b\eta^m\eta^n\lambda$$
$$= 0. \tag{7.142}$$

Hence $\delta\mathscr{L}_{\text{eff}} = 0$, and the Lagrangian is invariant under the BRS transformations (7.135–7.137). Note in passing that (7.135) and (7.142) imply that

$$\delta^2(A_\mu^a) = 0; \tag{7.143}$$

in highbrow language, the variation in A_μ^a is nilpotent.

We now perform a BRS transformation on Z to yield the Slavnov–Taylor identities, analogous to the Ward identities in electrodynamics.

7.6 Slavnov–Taylor identities

It is convenient to introduce a generating functional dependent on five sources[‡] rather than the expected three:

$$Z[s, x, y; u, v] = \int \mathscr{D}\bar{\eta}\mathscr{D}\eta\mathscr{D}A_\mu \; e^{i\int\mathscr{L}_{\text{tot}}dx}$$

[‡] H. Kluberg-Stern & J.B. Zuber, *Physical Review*, **D12**, 484 (1975).

with

$$\mathcal{L}_{\text{tot}} = \mathcal{L}_{\text{eff}} + s_\mu^a A^{a\mu} + \eta^a x^a + \bar{\eta}^a y^a$$
$$+ u_\mu^a \left(\frac{1}{g} D^\mu \eta\right)^a + v^a (-\tfrac{1}{2} f^{abc} \eta^b \eta^c). \tag{7.144}$$

Of these sources, x, y and u are anticommuting sources. We shall subject Z to a BRS transformation. Our first observation is that the coefficients of the sources u and v are invariant. That the coefficient of u is invariant is the content of (7.142). To prove that the coefficient of v is invariant, we have, under (7.136),

$$\delta(f^{abc} \eta^b \eta^c) = f^{abc}[(\delta\eta^b)\eta^c - \eta^b(\delta\eta^c)]$$
$$= -\tfrac{1}{2} f^{abc}(f^{bmn}\eta^m\eta^n\lambda\eta^c - \eta^b f^{cmn}\eta^m\eta^n\lambda)$$
$$= \tfrac{1}{2}(f^{acb}f^{cmn}\eta^m\eta^n\eta^b + f^{abc}f^{cmn}\eta^m\eta^n\eta^b)\lambda$$
$$= 0 \tag{7.145}$$

where we have relabelled dummy suffixes in the penultimate line. Note, in passing, that from (7.136) this implies that the change in η^a is also nilpotent:

$$\delta^2(\eta^a) = 0. \tag{7.146}$$

Our second observation – or rather task – is to show that the Jacobian of the transformation is unity. The Jacobian is

$$J = \partial\left(\frac{A_\mu^a(x) + \delta A_\mu^a(x), \eta^a(x) + \delta\eta^a(x), \bar{\eta}^a(x) + \delta\bar{\eta}^a(x)}{A_\nu^b(y), \eta^b(y), \bar{\eta}^b(y)}\right). \tag{7.147}$$

The only non-vanishing elements of this determinant are

$$\frac{\delta[A_\mu^a(x) + \delta A_\mu^a(x)]}{\delta A_\mu^b(y)} = \delta_\mu^\nu \delta^4(x - y)(\delta^{ab} - f^{abc}\eta^c\lambda),$$

$$\frac{\delta[\eta^a(x) + \delta\eta^a(x)]}{\delta\eta^b(y)}$$

$$= \delta^4(x - y)\left[\delta^{ab} - \frac{1}{2}\frac{\delta}{\delta\eta^b}(f^{amn}\eta^m\eta^n)\lambda\right]$$
$$= \delta^4(x - y)[\delta^{ab} - \tfrac{1}{2}(f^{amb}\eta^m - f^{abn}\eta^n)\lambda]$$
$$= \delta^4(x - y)(\delta^{ab} + f^{abc}\eta^c\lambda),$$

where we have taken the differentiation as 'right differentiation' (see (6.121b)); and finally

$$\frac{\delta[A_\mu^a(x) + \delta A_\mu^a(x)]}{\delta\eta^b(y)} = \delta^4(x - y) f^{abc} A_\mu^b \lambda.$$

So, in schematic form, the Jacobian is

$$J = \delta_\mu^\nu [\delta^4(x-y)]^3 \begin{vmatrix} 1-f\eta\lambda & fA\lambda & 0 \\ 0 & 1+f\eta\lambda & 0 \\ 0 & 0 & 1 \end{vmatrix}$$

$$= \delta_\mu^\nu [\delta^4(x-y)]^3$$

since $\lambda^2 = 0$. Hence the Jacobian is unity.

As a result of these two observations it now follows that, since $\mathscr{L}_{\text{eff}}$ is invariant under the BRS transformations, the invariance of Z implies that

$$Z = \int \mathscr{D}A_\mu \mathscr{D}\eta \mathscr{D}\bar\eta \exp\left\{ i\left[S + \int dx(s_\mu^a \delta A^{\mu a} + x^a \delta\eta^a + y^a \delta\bar\eta^a) \right] \right\}$$

$$= \int \mathscr{D}A_\mu \mathscr{D}\eta \mathscr{D}\bar\eta e^{iS}\left[1 + \int dx(s_\mu^a \delta A^{\mu a} + x^a \delta\eta^a + y^a \delta\bar\eta^a) \right]$$

(where $S = \int \mathscr{L}_{\text{eff}}\, dx$) hence

$$\int \mathscr{D}A_\mu \mathscr{D}\eta \mathscr{D}\bar\eta e^{iS} \int dx(s_\mu^a \delta A^{\mu a} + x^a \delta\eta^a + y^a \delta\eta^a) = 0, \tag{7.148}$$

where $\delta A^{\mu a}$, $\delta\eta^a$ and $\delta\bar\eta^a$ are given by (7.135–7.137). It will be noticed that these quantities are respectively the coefficients of u and v, and proportional to $\partial^\mu A_\mu^a$, where A_μ^a is the coefficient of s, so (7.148) implies

$$\lambda \int dx \left\{ s_\mu^a(x) \frac{\delta Z}{\delta u_\mu^a(x)} + x^a(x) \frac{\delta Z}{\delta v_\mu^a(x)} - \frac{1}{\alpha} y^a(x) \left[\partial_\mu \frac{\delta Z}{\delta s_\mu^a(x)} \right] \right\} = 0.$$

This equation contains only first order derivatives, which is a consequence of introducing sources u and v for the non-linear terms δA and $\delta\eta$. Putting $Z = e^{iW}$, a similar equation (in fact the same one) holds for W

$$\int dx \left[s^a \frac{\delta W}{\delta u_\mu^a} + x^a \frac{\delta W}{\delta v^a} - \frac{1}{\alpha} y^a \left(\partial_\mu \frac{\delta W}{\delta s_\mu^a} \right) \right] = 0. \tag{7.149}$$

We now convert this into a condition on the generating functional Γ. So we define in the usual way (except that we do not transform the sources u and v)

$$W[s, x, y; u, v] = \Gamma[A, \eta, \bar\eta; u, v] + \int dx(s_\mu^a A^a + x^a \eta^a + y^a \bar\eta^a). \tag{7.150}$$

Then

$$s_\mu^a = -\frac{\delta\Gamma}{\delta A_\mu^a}, \quad x^a = -\frac{\delta\Gamma}{\delta\eta^a}, \quad y^a = -\frac{\delta\Gamma}{\delta\bar\eta^a}. \tag{7.151}$$

In addition,

$$\frac{\delta W}{\delta s_\mu^a} = A^{\mu a}, \quad \frac{\delta W}{\delta u} = \frac{\delta\Gamma}{\delta u}, \quad \frac{\delta W}{\delta v} = \frac{\delta\Gamma}{\delta v}, \tag{7.152}$$

so (7.149) becomes

$$\int dx \left[\frac{\delta\Gamma}{\delta A_\mu^a} \frac{\delta\Gamma}{\delta u^{\mu a}} + \frac{\delta\Gamma}{\delta \eta^a} \frac{\delta\Gamma}{\delta v^a} - \frac{1}{\alpha}(\partial^\mu A_\mu^a) \frac{\delta\Gamma}{\delta \bar{\eta}^a} \right] = 0. \tag{7.153}$$

To obtain a simpler form for this equation, note from (7.132) and (7.144) that the terms in Z involving $\bar{\eta}$ and u are

$$Z = \int \mathcal{D}A_\mu \mathcal{D}\eta \mathcal{D}\bar{\eta} \exp\left\{ i \int dx [-\bar{\eta}^a (\partial^\mu D_\mu \eta)^a + \bar{\eta}^a y^a \right.$$

$$\left. + u_\mu^a (D^\mu \eta)^a + \cdots] \right\},$$

so

$$\frac{\delta Z}{\delta \bar{\eta}^a} = i[y^a - (\partial^\mu D_\mu \eta)^a] Z$$

$$= i y^a Z - \partial_\mu \frac{\delta Z}{\delta u_\mu^a}$$

therefore

$$\frac{\delta W}{\delta \bar{\eta}^a} = y^a - \partial_\mu \left(\frac{\delta W}{\delta u_\mu^a} \right).$$

But $\delta W / \delta \bar{\eta}^a = 0$, and $y^a = -\delta\Gamma/\delta\bar{\eta}^a$, therefore

$$\frac{\delta\Gamma}{\delta\bar{\eta}^a} = -\partial_\mu \left(\frac{\delta\Gamma}{\delta u_\mu^a} \right)$$

and (7.153) becomes

$$\int dx \left[\frac{\delta\Gamma}{\delta u^{a\mu}} \left(\frac{\delta\Gamma}{\delta A_\mu^a} - \frac{1}{\alpha} \partial^\mu (\partial^\nu A_\nu^a) \right) + \frac{\delta\Gamma}{\delta v^a} \frac{\delta\Gamma}{\delta \eta^a} \right] = 0. \tag{7.154}$$

Finally, defining the functional Γ' by

$$\Gamma = \Gamma' - \frac{1}{2\alpha} \int dx (\partial^\nu A_\nu^a)^2, \tag{7.155}$$

(7.154) gives

$$\int dx \left[\frac{\delta\Gamma'}{\delta u^{a\mu}} \frac{\delta\Gamma'}{\delta A_\mu^a} + \frac{\delta\Gamma'}{\delta v^a} \frac{\delta\Gamma'}{\delta \eta^a} \right] = 0. \tag{7.156}$$

This expresses the content of the Slavnov–Taylor identities, though in a different form from that of the original authors. Written as they are here they are in a form which may most easily be shown to imply the renormalisability of Yang–Mills theories.

7.7   A note on ghosts and unitarity

In this chapter we have seen the necessity of introducing ghosts; they were required in order to have a gauge-independent and finite generating functional. As long ago as 1962, however, Feynman saw that ghosts were required in order not to violate unitarity at the '1-loop' level. In this section we shall outline in rather general terms how this argument goes.

We begin by defining unitarity of the S matrix. The operator S is defined by the relation that the probability amplitude that a particular state $|m\rangle$ results, after interaction and scattering, from a particular state $|n\rangle$, is $\langle m|S|n\rangle$. The states $|n\rangle$ are orthonormal and complete:

$$\langle m|n\rangle = \delta_{mn}, \ \sum_m |m\rangle\langle m| = 1. \tag{7.157}$$

The probability that the system ends up in some final state or other is, of course, unity, so

$$\sum_m |\langle m|S|n\rangle|^2 = 1$$

which implies that

$$\sum_m \{\langle m|S|n\rangle *\langle m|S|n\rangle = 1,$$

$$\sum_m \langle n|S^\dagger|m\rangle\langle m|S|n\rangle = 1,$$

$$\langle n|S^\dagger S|n\rangle = 1,$$

where (7.157) has been used. And since the state $|n\rangle$ is arbitrary, we have

$$SS^\dagger = S^\dagger S = 1; \tag{7.158}$$

the S matrix is unitary. Now put

$$S = 1 + iR. \tag{7.159}$$

R is called the reaction matrix. Unitarity then implies that

$$R - R^\dagger = iR^\dagger R$$

or, since $R - R^\dagger = 2i\mathrm{Im}\,R$

$$2\mathrm{Im}\,R = R^\dagger R = RR^\dagger. \tag{7.160}$$

Let us take the matrix elements of this equation between the 2-particle states $\langle p_3 p_4|$ and $|p_1 p_2\rangle$. We have

$$2\,\mathrm{Im}\langle p_3 p_4|R|p_1 p_2\rangle = \sum_n \langle p_3 p_4|R|n\rangle\langle p_1 p_2|R|n\rangle *.$$

Putting $R = (2\pi)^4\delta(p_\mathrm{f} - p_\mathrm{i})T$, and writing the states $|n\rangle$ in terms of the momenta k_i of the particles in it, however many there may be, gives

$$2 \operatorname{Im} \langle p_3 p_4 | T | p_1 p_2 \rangle$$

$$= \frac{1}{(2\pi)^2} \sum_n \int \frac{d^3 k_1}{W_1} \frac{d^3 k_2}{W_2} \cdots \delta^4(p_1 + p_2 - k_1 - k_2 - \cdots - k_n)$$

$$\times \langle p_3 p_4 | T | k_1 k_2 \ldots k_n \rangle \langle p_1 p_2 | T | k_1 k_2 \ldots k_n \rangle^*. \tag{7.161}$$

The summation on the right-hand side is over all *real* (not virtual) intermediate states consistent with energy–momentum conservation. This equation may be represented diagrammatically in the following way

$$\tag{7.162}$$

This is the import of unitarity, with the additional observation that it applies *order by order* in perturbation theory.

The idea is now to apply this fundamental constraint to particular processes, and we begin with the process $e^+ e^- \rightarrow e^+ e^-$ to order e^2 in QED, shown in Fig. 7.7. To this order, only 1-photon exchange is involved, and the amplitude is

$$A = (-ie)^2 \bar{v}(p_2) \gamma_\mu u(p_1) \frac{-ig^{\mu\nu}}{q^2 + i\varepsilon} \bar{u}(p_3) \gamma_\nu v(p_4) \tag{7.163}$$

by a straightforward application of the Feynman rules, when the electron and positron have initial and final momenta $p_1, \ldots, p_4$, and the virtual photon momentum is q. We have chosen the Feynman gauge, where the photon propagator is $-ig^{\mu\nu}/(q^2 + i\varepsilon)$. Here we have deliberately left in the $i\varepsilon$ to emphasise that A is complex, for unitarity is a condition on its *imaginary*

Fig. 7.7. Electron–positron scattering to order e^2.

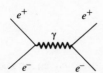

part. It may be shown[‡] that

$$\text{Im} \frac{g^{\mu\nu}}{q^2 + i\varepsilon} = -i\pi g^{\mu\nu}\delta(q^2)\theta(k_0); \tag{7.164}$$

that is, the propagator is replaced by functions making the photon on mass-shell with positive energy – i.e. 'real' except, as we are about to see, it is *not quite real.* For, from the covariant propagator (7.10), we deduce that

$$\sum_{\lambda=0}^{3} \varepsilon_\mu^{(\lambda)}\varepsilon_\nu^{(\lambda)} = -g_{\mu\nu} \tag{7.165}$$

(compare, for example, equation (7.15) for the radiation gauge propagator). Hence

$$\text{Im} \, A = M^\mu \left[\sum_{\lambda=0}^{3} \varepsilon_\mu^{(\lambda)}\varepsilon_\nu^{(\lambda)} \right] M^{\nu\dagger} \tag{7.166}$$

where M^μ is essentially $\bar{v}\gamma^\mu u$, and the other constants have been absorbed. The interesting thing is that *all four* polarisation states occur in (7.166), whereas physical photons are transversely polarised, with $\lambda = 1, 2$ only. So, in the imaginary part of the amplitude, the photons are not quite real, since although they have $q^2 = 0$, they occur in unphysical, as well as physical, polarisation states. Will this make trouble?

The unitarity relation, (7.161) and (7.162), says that $\text{Im} A$ is equal to the sum of the squares of all relevant processes, to order e^2. In fact, there is (in this simple example) only one relevant process, which is electron–positron annihilation, $e^+e^- \to \gamma$, shown in Fig. 7.8. The photon in this process is *entirely* physical, of course. Call the amplitude for the process B; then in the same notation as above

$$B = M^\mu \varepsilon_\mu(q, \lambda)$$

where $\lambda = 1, 2$ only, since the photon must be transversely polarised. Hence the 'unitarity sum' in (7.162) is

$$BB^\dagger = M^\mu \sum_{\lambda=1,2} \varepsilon_\mu(q, \lambda)\varepsilon_\nu(q, \lambda)M^{\nu\dagger} \tag{7.167}$$

Fig. 7.8. Electron–positron annihilation.

[‡] See, for example, R.J. Eden, P.V. Landshoff, D.I. Olive & J.C. Polkinghorne, *The Analytic S-matrix*, section 2.9, Cambridge University Press, 1966.

and this should be equal to (7.166). This will happen if the contribution of the unphysical polarisation states vanishes

$$M^\mu \sum_{\lambda=0,3} \varepsilon_\mu^{(\lambda)}(q)\varepsilon_\nu^{(\lambda)}(q)M^{\nu\dagger} = 0$$

or

$$M^\mu \left[\sum_{\lambda=0}^{3} \varepsilon_\mu^{(\lambda)}\varepsilon_\nu^{(\lambda)} - \sum_{\lambda=1,2} \varepsilon_\mu^{(\lambda)}\varepsilon_\nu^{(\lambda)} \right] M^{\nu\dagger} = 0. \tag{7.168}$$

From (7.165) and (7.64) with $\alpha = 1$ (the axial gauge) (7.168) becomes

$$M^\mu \left[\frac{t^2}{(q\cdot t)^2}q_\mu q_\nu - \frac{q_\mu t_\nu + t_\mu q_\nu}{(q\cdot t)} \right] M^{\nu\dagger} = 0,$$

and this obviously satisfied if

$$q_\mu M^\mu = 0. \tag{7.169}$$

This, however, is a consequence of gauge invariance. The amplitude for Fig. 7.8 is $\varepsilon_\mu M^\mu$, and this must remain unchanged by a gauge transformation $\varepsilon_\mu \to \varepsilon_\mu + \alpha q_\mu$, yielding (7.169). In fact, for a general process, (7.169) is actually a Ward identity, which of course follows from gauge invariance.

We conclude that, in the case of electrodynamics, at least to order e^2, unitarity is satisfied by virtue of gauge invariance. It may in fact be shown that it is satisfied to all orders. Turning to non-Abelian gauge theories, however, the situation becomes more complicated – as, by now, we have learned to expect! In fact, the complications are only those we are by now familiar with – the interaction of the gauge field with itself and the occurrence of ghosts – but the interesting thing is that they are all *required* if unitarity is to be satisfied. Here we shall merely sketch the argument. The reader is referred to ref. (24) for details.

Let us consider the process $q\bar{q} \to q\bar{q}$ to order g^4 in QCD, illustrated in Fig. 7.9. It involves two gluon propagators. We shall call the amplitude for this process A. According to unitarity, $\text{Im}A = \sum B^\dagger B$ where B is the amplitude for all possible processes $q\bar{q} \to \ldots$, of order g^2, which make 'half' the Feynman diagram for $q\bar{q} \to q\bar{q}$ (in the sense that Fig. 7.8 is 'half' of

Fig. 7.9. Quark–antiquark scattering to order g^4 in QCD.

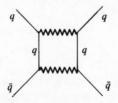

Fig. 7.7). Now the totality of diagrams contributing to $q\bar{q} \to q\bar{q}$ to order g^4 is given in Fig. 7.10. The first two diagrams contribute *distinct* amplitudes, as may be seen when the Feynman rules are applied. The third and fourth diagrams contain *closed loops*, involving a gluon and a ghost. Their contribution to the amplitude is easily evaluated by applying the Feynman rules (7.53), (7.54) and (7.57). Hence B is the sum of the amplitudes for the processes in Fig. 7.11. Here it looks as if the ghost is a real particle, but of course it is not. It only appears in Fig. 7.11 because it contributes *closed loops* to ordinary physical amplitudes. It turns out, by virtue of gauge invariance (Ward identities), that $\text{Im}A = B^\dagger B$, and unitarity is satisfied, but only by virtue of including *all* the diagrams of Fig. 7.11, *including the ghost*, in B.

We see, in conclusion, that when loop diagrams (as distinct from tree diagrams) are considered, unitarity demands the existence of ghosts contributing to closed loops. This was first noticed by Feynman. It is indeed remarkable, and doubtless of profound significance, that ghosts solve, simultaneously, the problems of unitarity and gauge invariance.

Summary
[1]It is demonstrated in the canonical formalism how the usual procedure for finding propagators does not work for the photon, because of

Fig. 7.10. The four diagrams in QCD contributing to $q\bar{q} \to q\bar{q}$ to order g^4. Wavy lines are gluon propagators, dotted lines ghost propagators.

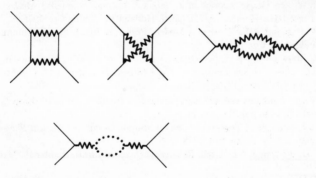

Fig. 7.11. Unitarity and ghosts. The fourth diagram, featuring the ghost, contributes to the unitarity relation.

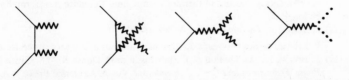

gauge invariance. The impasse is avoided by adding 'gauge-fixing terms' to the Lagrangian. The propagator for transverse photons is found. [2]The analogous problem is considered for non-Abelian fields, but using the path-integral method, rather than the canonical one. It involves finding a well-defined way of taking out the (infinite) factor produced from integration over the group space, and results in the introduction of the Faddeev–Popov ghost field. The Feynman rules are displayed. [3]The self-energy Σ and vertex functions Γ are defined, and the generating functional for Γ defined by a Legendre transformation on W, the generating functional for connected Green's functions. The thermodynamic analogy with field theory is briefly discussed. [4]The Ward–Takahashi identities for QED are derived, in terms of Γ and Σ. Their non-Abelian generalisations, the [6]Slavnov–Taylor identities, are derived by first introducing the [5]Becchi–Rouet–Stora transformations. [7]The connection is exhibited, briefly for QED and even more sketchily for QCD, between the existence of ghosts and the demands of unitarity.

Guide to further reading

The Faddeev–Popov method originated in
(1) L.D. Faddeev & V.N. Popov, *Physics Letters*, **25B**, 29 (1967).

Detailed accounts appear in
(2) L.D. Faddeev & A.A. Slavnov, *Gauge Fields: Introduction to Quantum Theory*, section 3.3, Benjamin/Cummings Publishing Co., 1980.
(3) B.W. Lee 'Gauge theories' in R. Balian & J. Zinn-Justin (eds.), *Methods in Field Theory* (Les Houches, 1975), North-Holland Publishing Co., 1976.
(4) P. Ramond, *Field Theory: A Modern Primer*, p. 303, Benjamin/Cummings Publishing Co., 1981.
(5) C. Itzykson & J.B. Zuber, *Quantum Field Theory*, p. 577, McGraw-Hill, 1980.
(6) J.C. Taylor, *Gauge Theories of Weak Interactions*, chapter 11, Cambridge University Press, 1976.

Vertex functions and self-energy operators are discussed in Itzykson & Zuber (ref. (5), section 6-2-2), and in
(7) D.J. Amit, *Field Theory, the Renormalisation Group, and Critical Phenomena*, chapter 5, McGraw-Hill, 1978.
(8) J.D. Bjorken & S.D. Drell, *Relativistic Quantum Fields*, chapter 19, McGraw-Hill, 1965.
(9) E.M. Lifshitz & L.P. Pitaevskii, *Relativistic Quantum Theory* (Part 2), chapter 11, Pergamon, 1973.
(10) N.N. Bogoliubov & D.V. Shirkov, *Introduction to the Theory of Quantised Fields*, 3rd edition, §28, Wiley, 1980.

An elementary account of thermodynamics and Legendre transformations is to be found in
(11) H.B. Callen, *Thermodynamics*, Wiley, 1961.

For the analogies between statistical mechanics and field theory, see, for example,
(12) E. Brezin, J.C. Le Guillou & J. Zinn-Justin in C. Domb & M.S. Green (eds.), *Phase Transitions and Critical Phenomena*, Vol. 6, Academic Press, 1976.

(13) C. de Dominicis & P.C. Martin, *Journal of Mathematical Physics*, **5**, 14 (1964).

For the Ward–Takahashi identities see

(14) J.C. Ward, *Physical Review*, **78**, 182 (1950); *Proceedings of the Physical Society*, **64** (A), 54 (1951).

(15) Y. Takahashi, *Nuovo Cimento*, **6**, 371 (1957).

Also for example, refs. (8–10) above.

The Becchi–Rouet–Stora transformation is to be found in

(16) C. Becchi, A. Rouet & R. Stora, *Physics Letters*, **52B**, 344 (1974).

For the Slavnov–Taylor identities see

(17) J.C. Taylor, *Nuclear Physics*, **B33**, 436 (1971).

(18) A.A. Slavnov, *Theoretical and Mathematical Physics*, **10**, 99 (1972).

Good reviews appear in ref. (6) above, chapter 12, and

(19) A.A. Slavnov, *Soviet Journal of Particles and Nuclei*, **5**, 303 (1975).

(20) G. 't Hooft & M. Veltman, *Diagrammar*, CERN, Geneva, Report 73–9 (1973) and in ref. (2) above, section 4.6.

The necessity of ghosts in gauge theories was first seen by

(21) R.P. Feynman, *Acta Physica Polonica*, **24**, 697 (1963).

See also

(22) R.P. Feynman in J.R. Klauder (ed.), *Magic without Magic: John Archibald Wheeler*, pp. 355, 377, W.H. Freeman & Co., 1972.

(23) R.P. Feynman in R. Balian & C.H. Llewellyn Smith (eds.), *Interactions électromagnétiques et faibles à haute energie/Weak and electromagnetic interactions at high energy*, p. 121, North-Holland Publishing Co., 1977.

The connection between ghosts and unitarity is very clearly outlined in

(24) I.J.R. Aitchison & A.J.G. Hey, *Gauge Theories in Particle Physics*, section 12.1, Adam Hilger Ltd, Bristol, 1982.

See also (for QED only)

(25) C. Nash, *Relativistic Quantum Fields*, pp. 141–8, Academic Press, 1978.

8

Spontaneous symmetry breaking and the Weinberg–Salam model

The foregoing chapters have dealt with field theories, including gauge theories, and their quantisation. The stage is now almost set for applying this knowledge to particle physics. One curcial bit of scenery, however, is still missing – the idea of 'spontaneous breaking of symmetry'. About 1960 Nambu and Goldstone realised the significance of this notion in condensed matter physics, and Nambu in particular speculated on its application to particle physics. In 1964 Higgs pointed out that the consequences of spontaneous symmetry breaking in gauge theories are very different from those in non-gauge theories. Weinberg and Salam, building on earlier work of Glashow, then applied Higgs' ideas to an $SU(2) \times U(1)$ gauge theory, which they claimed described satisfactorily the weak and electromagnetic interactions together, in other words, in a *unified* way. Serious interest was shown in this theory when 't Hooft proved, in 1971, that it was renormalisable. It has met with notable experimental successes. These matters are the concern of this chapter (with the exception of renormalisation, which we deal with in the next chapter). We begin by explaining spontaneous symmetry breaking, which, when applied to field theory, is a concept that refines our notion of the vacuum.

8.1   What is the vacuum?

We begin by considering two simple physical examples. First, consider the situation illustrated in Fig. 8.1. Place a thin rod of circular cross section vertically on a table, and push down on it along its length, with a force F. If F is small, nothing happens. If F exceeds a critical value, F_{crit}, however, the rod bends, as shown, in a plane which it 'chooses at random'. The symmetric (unbent) configuration becomes unstable when $F > F_{\text{crit}}$, and the new ground state is unsymmetric. Also, there are infinitely many possible new (degenerate) ground states, which are related by a rotational symmetry. The rod can only, of course, choose one of them, but the others are all reached by a rotation. The salient points of this example are

   (i) A parameter (in this case force F) assumes a critical value. Beyond that value,

  (ii) the symmetric configuration becomes unstable, and

 (iii) the ground state is degenerate.

The second example we consider is ferromagnetism. The atoms in the ferromagnet interact through a spin–spin interaction

$$H = -\sum_{i,j} J_{ij} \mathbf{S}_i \cdot \mathbf{S}_j$$

which is a scalar and therefore invariant under rotations. The *ground state*, however, is one in which all the spins (within a domain) are aligned, as in Fig. 8.2, and this is clearly not rotationally invariant. The direction of spontaneous magnetisation is random, and all the degenerate ground states may be reached from a given one by rotation. As Coleman[13] points out, a 'little man' living in such a ferromagnet would have to be very clever to realise that the Hamiltonian H is rotation invariant! The spontaneous magnetisation disappears at high temperature T, when the ground state becomes symmetric (that is, the atoms become randomly oriented).

It is clear that the general situation here is the same as in our first

Fig. 8.1. A rod bent under pressure. The bent position exhibits spontaneous symmetry breaking.

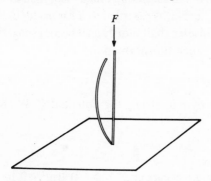

Fig. 8.2. Spin alignment in a ferromagnet.

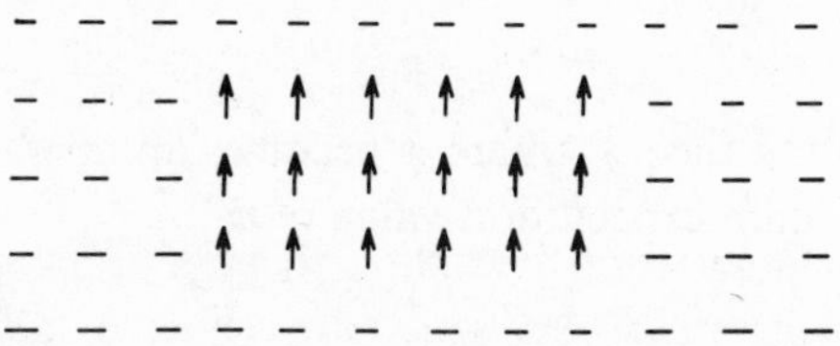

example, the relevant parameter here being T. These two examples exhibit what is known as 'spontaneous braking of symmetry'. In both cases the system possesses a symmetry (rotation symmetry) but the ground state is not invariant under that symmetry; rather, it changes into one of the other (degenerate) ground states.

One subtlety about the ferromagnet is that it must, in principle, be an infinite system. The magnetisation has singled out a particular direction, and a (quantum-mechanical) measurement of direction (angle) will give a sharp answer. But the conjugate variable to angle is angular momentum (recall that $J_z = i\hbar\partial/\partial\phi$), so the angular momentum of the system is *completely undefined*, and must therefore be an infinite sum of all possible values of J. The fact that the ferromagnet is infinite makes it an interesting system to compare with field theory, since a field, of course, is a system with an infinite number of degrees of freedom.

We therefore now look for a similar situation in scalar field theory, in which the symmetry of the Lagrangian is not shared by the ground state solution. In a field theory the ground state is regarded as being the vacuum, so we are in quest of a theory with a new type of vacuum. Since $\mathscr{L}$ must have a symmetry, we choose complex ϕ^4 theory:

$$
\begin{aligned}
\mathscr{L} &= (\partial_\mu\phi)(\partial^\mu\phi^*) - m^2\phi^*\phi - \lambda(\phi^*\phi)^2 \\
&= (\partial_\mu\phi)(\partial^\mu\phi^*) - V(\phi,\phi^*).
\end{aligned}
\tag{8.1}
$$

The λ term is a self-interaction. In the usual scalar field theory, quantisation yields particles of mass m, but here m^2 is regarded as a *parameter* only, and not as a mass term. This is because we shall shortly let it become negative. $\mathscr{L}$ is invariant under the *global* gauge transformation

$$
\phi \to e^{i\Lambda}\phi \quad (\Lambda \text{ const}).
\tag{8.2}
$$

The ground state is obtained by minimising the potential V. We have

$$
\frac{\partial V}{\partial\phi} = m^2\phi^* + 2\lambda\phi^*(\phi^*\phi)
\tag{8.3}
$$

so that when $m^2 > 0$, the minimum occurs at $\phi^* = \phi = 0$. If $m^2 < 0$, however, there is a local maximum at $\phi = 0$, and a minimum at

$$
|\phi|^2 = -\frac{m^2}{2\lambda} = a^2,
\tag{8.4}
$$

i.e. at $|\phi| = a$. In the quantum theory, where ϕ becomes an operator, this condition refers to the vacuum expectation value of ϕ

$$
|\langle 0|\phi|0\rangle|^2 = a^2.
\tag{8.5}
$$

The function V is shown in Fig. 8.3, plotted against ϕ_1 and ϕ_2, where $\phi = \phi_1 + i\phi_2$ (though it should be borne in mind that ϕ is a *field*, not simply a pair of co-ordinates). The minima of V lie along the circle $|\phi| = a$, which form a set of degenerate vacua related to each other by rotation. The physical fields, which are excitations above the vacuum, are then realised by performing perturbations about $|\phi| = a$, not about $\phi = 0$. Let us work in polar co-ordinates, putting

$$\phi(x) = \rho(x)e^{i\theta(x)}, \tag{8.6}$$

so the complex field ϕ is expressed in terms of two real scalar fields ρ and θ. Let us *choose* the vacuum state

$$\langle 0|\phi|0 \rangle = a \tag{8.7}$$

where a is real; then

$$\langle 0|\rho|0 \rangle = a, \quad \langle 0|\theta|0 \rangle = 0. \tag{8.8}$$

We see that this field theoretic example exhibits the same features as the ferromagnet. It has degenerate vacua, which are connected by the symmetry operations of the theory. A particular vacuum involves a particular choice for the values of the field ((8.8) in the field theory, direction of magnetisation for the ferromagnet), and is, of course, not invariant under the symmetry.

Now let us put

$$\phi(x) = [\rho'(x) + a]e^{i\theta(x)}, \tag{8.9}$$

so that ρ' and θ both have vanishing vacuum expectation values. We regard

Fig. 8.3. The potential V has a minimum at $|\phi| = a$, and a local maximum at $\phi = 0$.

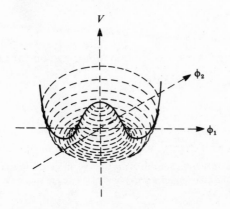

them as the 'physical' fields, and express $\mathscr{L}$ in terms of them. We have, from (8.1),

$$
\begin{aligned}
V &= m^2\rho'^2 + 2m^2 a\rho' + m^2 a^2 \\
&\quad + \lambda(\rho'^4 + 4a\rho'^3 + 6a^2\rho'^2 + 4a^3\rho' + a^4) \\
&= \lambda\rho'^4 + 4a\lambda\rho'^3 + 4\lambda a^2\rho'^2 - \lambda a^4 \\
&= \lambda[(\rho' + a)^2 - a^2]^2 - \lambda a^4 \\
&= \lambda(\phi^*\phi - a^2)^2 - \lambda a^4
\end{aligned}
$$

where (8.4) has been used. In addition,

$$
(\partial_\mu\phi)(\partial^\mu\phi^*) = (\partial_\mu\rho')(\partial^\mu\rho') + (\rho' + a)^2(\partial_\mu\theta)(\partial^\mu\theta)
$$

with $\mathscr{L} = (\partial_\mu\phi)(\partial^\mu\phi^*) - V$. We see that there is a term in ρ'^2, so ρ' has a mass given by

$$
m_{\rho'}^2 = 4\lambda a^2,
$$

but there is no term in θ^2, so θ is a *massless* field. As a result of spontaneous symmetry breaking, what would otherwise be two massive fields (the real parts of ϕ), become one massive and one massless field. We may interpret this with reference to Fig. 8.3. It clearly costs energy to displace ρ' against the restoring forces of the potential, but there are no restoring forces corresponding to displacements along the circular valley $|\phi| = a$, in view of the vacuum degeneracy. Hence for the angular excitations θ, of wavelength λ, we have $\omega \to 0$ as $\lambda \to 0$, so $\omega \propto \lambda$, $E \propto p$, and the relativistic particles are massless. The θ particle is known as a *Goldstone boson*. The important point is that this phenomenon is *general*: spontaneous breaking of a (continuous) symmetry entails the existence of a massless particle, the Goldstone particle[‡]. This statement, known as the Goldstone theorem, will be proved in the next section.

For future reference it is useful to display this result using a 'Cartesian', rather than a polar, decomposition of ϕ. If, instead of (8.9), we have

$$
\phi(x) = a + \frac{\phi_1(x) + i\phi_2(x)}{\sqrt{2}} \tag{8.10}
$$

so that $\langle\phi_1\rangle_0 = \langle\phi_2\rangle_0 = 0$, it is easy to see that (ignoring constant terms)

$$
\begin{aligned}
\mathscr{L} &= \tfrac{1}{2}(\partial_\mu\phi_1)^2 + \tfrac{1}{2}(\partial_\mu\phi_2)^2 - 2\lambda a^2\phi_1^2 \\
&\quad - \sqrt{2}\lambda\phi_1(\phi_1^2 + \phi_2^2) - \frac{\lambda}{4}(\phi_1^2 + \phi_2^2)^2.
\end{aligned} \tag{8.11}
$$

[‡] In this example it has spin zero, but this is not always the case. For example in theories of spontaneous breaking of supersymmetry, there are spin $\tfrac{1}{2}$ Goldstone particles.

Hence the ϕ_2 field is massless, but ϕ_1 has a (mass)^2 of $4\lambda a^2$ – the same result as above.

We conclude by noting the analogy of the above with the ferromagnet. Consider a 'spin wave' of long wavelength λ. It induces a slow variation in the direction of magnetisation, over the specimen, as indicated in Fig. 8.4. Because the forces in a ferromagnet are of *short* range, it requires very little energy to excite this situation, so the frequency of the spin waves approaches zero with increasing λ, i.e. $\omega = ck$. In the relativistic domain, this is equivalent to a massless particle. It is noteworthy, however, that this argument breaks down if there are *long*-range forces, like for example the $1/r$ Coulomb force; in other words if there is a *gauge* field present. In this case, it still costs a finite amount of energy to excite a spin wave of even very long wavelength, since work has to be done against the Coulomb force, so $\omega \rightarrow$ finite as $\lambda \rightarrow \infty$, $k \rightarrow 0$, and the corresponding excitations are *massive*. What is more, so are the photons! It was this situation in solid state physics, discussed in these terms by Anderson,[‡] which led first Higgs[10] and then Weinberg and Salam[20,21] to consider the application of these ideas to the relativistic domain and particle physics. The situation in particle physics, on the face of it, seems to offer no fruitful ground for the application of gauge theories or theories of spontaneous symmetry breakdown. They both predict massless particles; the gauge particles, with spin 1, and the Goldstone bosons, with spin 0 – and, apart from the photon, there are no massless particles in existence. The observation on the ferromagnet above, that the presence of both effects together gets rid of *both* massless particles, is the key to the Weinberg – Salam model of electroweak interactions, described below.

Fig. 8.4. A spin wave inducing a slow spatial variation in the direction of magnetisation in a ferromagnet.

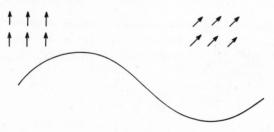

‡ P.W. Anderson, *Physical Review*, **130**, 439 (1963).

8.2   The Goldstone theorem

In the example above, the Lagrangian has a $U(1)$ symmetry, and the two real fields in ϕ form a 2-dimensional representation of $U(1)$. One of these fields has a non-vanishing vacuum expectation value, and there turns out to be one massless particle (the Goldstone boson) and one massive one. It should also be emphasized that the argument above is classical. So two questions arise: first, in the general case where $\mathscr{L}$ is invariant under a symmetry group G, how many Goldstone bosons will there be? And second, what is the status of all this in quantum theory – in particular, how do we prove the existence of massless particles given degenerate vacua? We shall tackle these questions in this order.

The case of a general symmetry group is best approached by considering a specific non-Abelian group, say $SO(3)$. So let us consider an example like the above, except that ϕ_i $(i = 1, 2, 3)$ is an isovector Lorentz-scalar field, and

$$\mathscr{L} = \tfrac{1}{2}\partial_\mu\phi_i\partial^\mu\phi_i - \frac{m^2}{2}\phi_i\phi_i - \lambda(\phi_i\phi_i)^2 \tag{8.12}$$

(summation convention applies).

$\mathscr{L}$ is invariant under isospin rotations, which generate the symmetry group G (in this case $SO(3)$):

$$G: \phi_i \to e^{iQ_k\alpha_k}\phi_i e^{-iQ_k\alpha_k} \; ?$$
$$= (e^{iT_k\alpha_k})_{ij}\phi_j = U_{ij}\phi_j = [U(g)\phi]_i. \tag{8.13}$$

Here α_i are the angles of rotation in isospin space, Q_i are the generators of the group, and T_i a set of matrices obeying the Lie algebra of the group, of the same dimensionality as the representation to which ϕ belongs – in this case 3-dimensional. The matrix $U(g)$, corresponding to the group element g, is a unitary matrix (if T is hermitian), hence we have a unitary representation. Although essential in quantum theory, this is not essential in the classical case, but there is no loss of generality in having one.

We look, as before, for the minimum of the potential $V(\phi_i)$

$$V = \frac{m^2}{2}\phi_i\phi_i + \lambda(\phi_i\phi_i)^2. \tag{8.14}$$

When the parameter $m^2 > 0$, this occurs at $\phi_i = 0$. When $m^2 < 0$, there is a minimum when

$$|\phi_0| = (\phi_1^2 + \phi_2^2 + \phi_3^2)^{\frac{1}{2}} = \left(\frac{-m^2}{4\lambda}\right)^{\frac{1}{2}} \equiv a. \tag{8.15}$$

We again have degenerate vacua and we are free to choose which one is the physical one. We choose

$$\vec{\phi}_0 = a\hat{e}_3. \tag{8.16}$$

The vacuum value of ϕ, ϕ_0, points in the 3 direction in isospin space. This is sketched in Fig. 8.5. It is clear that ϕ_0 is *not* invariant under the full group G, so that there are elements $g \in G$ for which

$$G: \quad \phi_0' = U(g)\phi_0 \neq \phi_0; \tag{8.17}$$

but it *is* invariant under a *subgroup* H of G, in this case rotations about the 3 axis:

$$H: \quad \phi_0' = U(h)\phi_0 = \phi_0$$
$$U(h) = e^{iT_3\alpha_3}. \tag{8.18}$$

On the other hand, of course, V is invariant under the whole of G:

$$V(\phi') = V(\phi), \quad \phi' = U(g)\phi, \tag{8.19}$$

and it is this which gives rise to the Goldstone bosons. How many are there? Putting

$$\phi_3 = \chi + a, \tag{8.20}$$

the physical fields are ϕ_1, ϕ_2 and χ, and it is straightforward to verify (remembering (8.15)) that

$$
\begin{aligned}
V &= \frac{m^2}{2}[\phi_1^2 + \phi_2^2 + (\chi + a)^2] + \lambda[\phi_1^2 + \phi_2^2 + (\chi + a)^2]^2 \\
&= 4a^2\lambda\chi^2 + 4a\lambda\chi(\phi_1^2 + \phi_2^2 + \chi^2) + \lambda(\phi_1^2 + \phi_2^2 + \chi^2)^2 - \lambda a^4 \\
&= \lambda[(\phi_i\phi_i - a^2)^2 - a^4].
\end{aligned} \tag{8.21}
$$

Only the field χ has a quadratic term, and therefore a mass

$$m_\chi^2 = 8a^2\lambda, \quad m_{\phi_1} = m_{\phi_2} = 0, \tag{8.22}$$

so, after spontaneous symmetry breaking, we have *two Goldstone bosons* and one massive scalar field.

Now we can understand this in a very general way. Expanding $V(\phi)$

Fig. 8.5. The vacuum value of $\vec{\phi}$ points in the third direction in isospin space.

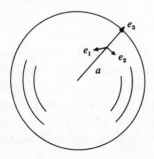

about its minimum, since

$$\frac{\partial V}{\partial \phi_a}\bigg|_{\phi=\phi_0} = 0,$$

we have

$$V(\phi) = V(\phi_0) + \frac{1}{2}\left(\frac{\partial^2 V}{\partial \phi_i \partial \phi_j}\right)_{\phi=\phi_0} \chi_i \chi_j + O(\chi^3) \qquad (8.23)$$

where $\chi(x) = \phi(x) - \phi_0$, and hence the mass matrix is

$$M_{ij} = \left(\frac{\partial^2 V}{\partial \phi_i \partial \phi_j}\right)_{\phi=\phi_0} \geqslant 0. \qquad (8.24)$$

Since $V(\phi_0)$ is the minimum, M_{ij} must be positive or zero. To find out for which fields it is zero, we do a group transformation. The invariance of V, equation (8.19), gives

$$V(\phi_0) = V(U(g)\phi_0) = V(\phi_0) + \frac{1}{2}\left(\frac{\partial^2 V}{\partial \phi_i \partial \phi_j}\right)_{\phi_0} \delta\phi_i \delta\phi_j + \cdots$$

and hence

$$\left(\frac{\partial^2 V}{\partial \phi_i \partial \phi_j}\right)_{\phi_0} \delta\phi_i \delta\phi_j = 0, \qquad (8.25)$$

where $\delta\phi_i$ is the variation in ϕ_i under a group transformation. What is this? From (8.17) and (8.18) it depends on whether the group element g belongs to H or not. If g belongs to H, then $\phi'_0 = \phi_0$ and $\delta\phi_i = 0$, or equivalently

$$\delta\phi\left(\frac{\partial U}{\partial \alpha_3}\right)_{\alpha_3=0} \phi_0 \,\delta\alpha_3 = 0 \qquad (8.26)$$

so (8.25) is satisfied already. If, however, g does not belong to H (if it is, in our example, a rotation about an axis in the 1, 2 plane), then

$$\delta\phi_m = \left[\left(\frac{\partial U}{\partial \alpha_i}\right)_{\alpha_i=0} \phi_0\right]_m \delta\alpha_i \neq 0. \qquad (8.27)$$

(Recall that (8.17), and therefore the above equation, are matrix equations.) In this case, from (8.24) and (8.25),

$$M_{ij}[U'(0)\phi_0]_j = 0$$

and the fields $U'(0)\phi_0$ have *zero mass*. These are the Goldstone bosons. It is now clear that the question of the number of fields with non-zero mass and the number with zero mass is simply a matter of group theory. A field whose mass is not *required* to be zero (though, of course, it may be zero 'by accident') obeys (8.26), and the number of such fields is simply the dimension of the Lie algebra (or order of the Lie group) of H, the subgroup under which the vacuum is invariant. In our case, $H = SO(2) \sim U(1)$ with one generator (T_3), so one field remains massive. The elements of G which do not belong to H, do not form a subgroup (they cannot, since the identity is in H), but form a coset G/H; and the number of Goldstone particles is the dimension of the coset space, which is the number of generators of G that are not also generators of H – in our case $3 - 1 = 2$. These results agree with the explicit calculation above. The interesting finding, however, is that this result does not depend on what representation of G the fields belong to – in our case it was the vector (regular) representation – nor on what form the potential V takes: the number of Goldstone bosons is simply the dimension of G/H. This fact is of great importance when we come to consider the spontaneous breaking of *gauge* symmetries.

Finally, to emphasize the generality of the conclusion, note that it also applies when the symmetry is *not* spontaneously broken. In this case there is a *unique* vacuum (a singlet under G) which is therefore invariant under G itself, so $H = G$, the coset is simply the identity, and there are no Goldstone bosons. At the other extreme, if the vacuum is such that there is *no* subgroup H which leaves one of the vacuum states ϕ_0 invariant, then H is the identity and $G/H = G$, and the number of Goldstone bosons is equal to the order of G. We have now answered the first question posed at the beginning of this section.

We therefore turn to the second question, and ask what the status of the above, classical, argument is in quantum theory. Here the Goldstone theorem states that if there is a field operator $\phi(x)$ with non-vanishing vacuum expectation value $\langle 0 | \phi(x) | 0 \rangle \neq 0$, and which is *not* a singlet under the transformation of some symmetry group, then massless particles must exist in the spectrum of states. There are some rather subtle questions of existence raised by this topic, and for a careful treatment the reader is referred to refs. (5, 6). Here we shall simply outline the proof.

We begin with some preliminary remarks about the symmetry group. If $\mathscr{L}$ is invariant under a group of transformations then (see §3.3) the currents

$$j_\mu^a(x) = \frac{\partial \mathscr{L}}{\partial(\partial^\mu \phi)} \frac{\delta \phi(x)}{\delta \alpha^a}$$

have zero divergence, $\partial^\mu j_\mu^a = 0$, and the corresponding charges

$$Q^a = \int \mathrm{d}^3 x\, j_0^a(x) \tag{8.28}$$

are conserved, $\mathrm{d}Q^a/\mathrm{d}t = 0$, and have the commutation relations of the symmetry group

$$[Q^a, Q^b] = C^{abc} Q^c$$

where C^{abc} are the structure constants of the Lie algebra. The unitary operator corresponding to a group transformation is

$$U = e^{iQ^a \alpha^a}. \tag{8.29}$$

If the vacuum is invariant under the group (i.e. a singlet), $U|0\rangle = |0\rangle$, hence

$$Q^a|0\rangle = 0 \tag{8.30}$$

and the charges annihilate the vacuum. This is the usual case for a symmetry. If this is *not* the case, we may say that we have 'degenerate vacua', and $Q^a|0\rangle = |0\rangle'$, or $Q^a|0\rangle \neq 0$, but strictly speaking we should say that $Q^a|0\rangle$ *does not exist* in Hilbert space–in other words, its norm is infinite.

Returning to the operator $\phi(x)$, since it is not a singlet under the group, there must exist an operator $\phi'(x)$ such that, for some a,

$$[Q^a, \phi'(x)] = \phi(x) \tag{8.31}$$

and, since $\langle 0|\phi(x)|0\rangle \neq 0$,

$$\langle 0|[Q^a, \phi'(x)]|0\rangle = \langle 0|Q^a\phi'(x) - \phi'(x)Q^a|0\rangle \neq 0. \tag{8.32}$$

This means that (8.30) *cannot* apply, so we do not have a symmetry in the usual sense (of degenerate multiplets). (An example of one of the subtleties is that, from above $\langle 0|\phi'(x)Q^a|0\rangle$ exists, whereas $Q^a|0\rangle$ does not.) We now show[6] that (8.32) implies the existence of massless particles. Substituting (8.28) into (8.32) and inserting a complete set of intermediate states we have

$$\sum_n \int d^3y [\langle 0|j_0^a(y)|n\rangle\langle n|\phi'(x)|0\rangle$$
$$- \langle 0|\phi'(x)|n\rangle\langle n|j_0^a(y)|0\rangle]|_{x^0=y^0} \neq 0. \tag{8.33}$$

(The restriction $x^0 = y^0$ arises because this is necessary to prove (8.31).) Now translation invariance implies that

$$j_0^a(y) = e^{-ipy}j_0^a(0)e^{ipy}$$

so (8.33) becomes

$$\sum_n \int d^3y [\langle 0|j_0^a(0)|n\rangle\langle n|\phi'(x)|0\rangle e^{ip_n y}$$
$$- \langle 0|\phi'(x)|n\rangle\langle n|j_0^a(0)|0\rangle e^{-ip_n y}]|_{x^0=y^0}$$
$$= \sum_n \delta^3(\mathbf{p}_n)[\langle 0|j_0^a(0)|n\rangle\langle n|\phi'(x)|0\rangle e^{ip_{n0}y^0}$$
$$- \langle 0|\phi'(x)|n\rangle\langle n|j_0^a(0)|0\rangle e^{-ip_{n0}y^0}]|_{x^0=y^0}$$
$$= \sum_n \delta^3(\mathbf{p}_n)[\langle 0|j_0^a(0)|n\rangle\langle n|\phi'(x)|0\rangle e^{iM_n y^0}$$
$$- \langle 0|\phi'(x)|n\rangle\langle n|j_0^a(0)|0\rangle e^{-iM_n y^0}]|_{x^0=y^0}$$
$$\neq 0 \tag{8.34}$$

‡ E. Fabri & L.E. Picasso, *Physical Review Letters*, **16**, 408 (1966).

where we have performed the spatial integral, and, in view of the delta function in $\mathbf{p}_n$, put $p_{n0} = M_n$, the mass of the intermediate state n. It remains only to show that (8.34) must be *independent of* y_0; if we can show this, we conclude that $M_n = 0$, so all the intermediate states have zero mass, which is the Goldstone theorem. Moreover, these intermediate states *must* exist, in order that (8.34) be non-zero: it will be noticed that the vacuum $(|n\rangle = |0\rangle)$ gives no contribution to the sum. To show that the above expression is independent of y_0, we start from the fact that $j_\mu^a(y)$ is divergenceless:

$$\partial^\mu j_\mu^a(y) = \partial_0 j_0^a(y) + \mathbf{V} \cdot \mathbf{j}^a(y) = 0,$$

which on integration gives

$$\frac{\partial}{\partial y_0} \int d^3y j_0^a(y) = \int d^3y \mathbf{V} \cdot \mathbf{j}^a(y).$$

Hence, since (8.34) is the same as (8.32),

$$\frac{\partial}{\partial y_0} \langle 0|[Q^a, \phi'(x)]|0\rangle$$

$$= \frac{\partial}{\partial y_0} \int d^3y \langle 0|[j_0^a(y), \phi'(x)]|0\rangle$$

$$= -\int d^3y \langle 0|[\mathbf{V} \cdot \mathbf{j}^a(y), \phi'(x)]|0\rangle$$

$$= -\int d\mathbf{S} \cdot \langle 0|[\mathbf{j}(y), \phi'(x)]|0\rangle$$

and under fairly orthodox assumptions[6] this surface integral may be shown to vanish. Hence the Goldstone theorem is proved.

In the 1960s a lot of effort was devoted to searching for a role for Goldstone's theorem in high energy physics. Although there are no zero mass particles, the pion has a seductively low mass and might be *nearly* a Goldstone boson. This would account for the success of the PCAC hypothesis (partially conserved axial current). For accounts of this topic, the reader is referred elsewhere[7,8].

8.3 Spontaneous breaking of gauge symmetries

Now let us see what happens when the symmetry in question is a *gauge* symmetry. Consider the simplest model, that of equation (8.1), but now let us demand invariance under

$$\phi \to e^{i\Lambda(x)}\phi. \tag{8.35}$$

This results (see (3.84)) in the introduction of the electromagnetic field through a covariant derivative, and (8.1) is replaced by

$$\mathcal{L} = (\partial_\mu + ieA_\mu)\phi(\partial^\mu - ieA^\mu)\phi^* - m^2\phi^*\phi$$
$$- \lambda(\phi^*\phi)^2 - \tfrac{1}{4}F_{\mu\nu}F^{\mu\nu}. \tag{8.36}$$

As before, we consider m^2 as a (positive or negative) parameter, so that in the case $m^2 < 0$, and in the absence of a gauge field, the vacuum is at

$$|\phi| = a = \left(\frac{-m^2}{2\lambda}\right)^{\frac{1}{2}} \tag{8.4}$$

Then, as in (8.10), setting

$$\phi(x) = a + \frac{\phi_1(x) + i\phi_2(x)}{\sqrt{2}} \tag{8.10}$$

gives for the Lagrangian, in terms of the physical fields ϕ_1 and ϕ_2:

$$\mathcal{L} = -\tfrac{1}{4}F_{\mu\nu}F^{\mu\nu} + e^2 a^2 A_\mu A^\mu + \tfrac{1}{2}(\partial_\mu \phi_1)^2 + \tfrac{1}{2}(\partial_\mu \phi_2)^2$$
$$- 2\lambda a^2 \phi_1^2 + \sqrt{2}eaA^\mu \partial_\mu \phi_2 + \text{cubic} + \text{quartic terms} \tag{8.37}$$

where we have taken account of (8.4). The interesting term is the second one, proportional to A_μ^2. It indicates that the photon has become *massive*. The scalar field ϕ_1 is a massive field, and ϕ_2 appears to be a massless one, but there is the odd mixed term in $A^\mu \partial_\mu \phi_2$, which would seem to indicate that a propagating photon could turn into a ϕ_2 field, so ϕ_2 does not seem to be a very physical field. In fact, it can be eliminated by a gauge transformation. For infinitesimal Λ in (8.35), (8.10) gives

$$\left.\begin{array}{l} \phi_1' = \phi_1 - \Lambda\phi_2, \\ \phi_2' = \phi_2 + \Lambda\phi_1 + \sqrt{2}\Lambda a. \end{array}\right\} \tag{8.38}$$

This shows that ϕ_2, like A_μ, undergoes an *inhomogeneous* transformation corresponding to a rotation *and translation* in the (ϕ_1, ϕ_2) plane, and so does not have a direct physical interpretation. We may then choose Λ to make $\phi_2 = 0$, and the mixed term above disappears. In this gauge, the Lagrangian (8.37) becomes (putting ϕ_1 rather than ϕ_1')

$$\mathcal{L} = -\tfrac{1}{4}F_{\mu\nu}F^{\mu\nu} + e^2 a^2 A_\mu A^\mu + \tfrac{1}{2}(\partial_\mu \phi_1)^2$$
$$- 2\lambda a^2 \phi_1^2 + \text{coupling terms}. \tag{8.39}$$

This Lagrangian contains two fields only, the photon with spin 1, and ϕ_1 with spin 0, and they are *both massive*. The ϕ_2 field, which in the case of spontaneous breaking of the global symmetry became massless (Goldstone boson), has in this case *disappeared*. And in addition, the gauge field whose presence is due to the fact that we have a *local* symmetry, has now acquired a mass – the '*photon' has become massive*. This phenomenon is called the *Higgs phenomenon*. In this Abelian model it is summarised by saying that spontaneous breaking of a gauge symmetry results, not in the presence of a massless Goldstone boson, but in the disappearance of that field altogether, and the appearance, instead, of a massive, rather than a massless, gauge field. Spontaneous breaking of $U(1)$ symmetry, then, yields the following particle spectrum, depending on whether the symmetry is global or local.

Goldstone mode (SB of global $U(1)$ symmetry):
2 massive scalar fields $\rightarrow$ 1 massive scalar field
+ 1 massless scalar field. (8.40)

Higgs mode (SB of gauge $U(1)$ symmetry):

$$\left. \begin{array}{l} 2 \text{ massive scalar fields} \\ + 1 \text{ photon} \end{array} \right\} \rightarrow \left\{ \begin{array}{l} 1 \text{ massive scalar field} \\ \quad + 1 \text{ massive photon.} \end{array} \right. \qquad (8.41)$$

Note that the number of degrees of freedom is preserved under these transformations. In the Goldstone case, this is trivial, since massless and massive scalar fields each have 1 degree of freedom: $2 = 1 + 1$. In the Higgs case, a massless photon has 2 degrees of freedom, but a massive one 3, since it has a physical transverse polarisation state: $2 + 2 = 1 + 3$. In a manner of speaking, we can say that the photon has eaten a scalar field and acquired a mass; or, more properly, we may compare the situation with the Gupta–Bleuler mechanism (§4.4). In that mechanism, the longitudinal and timelike components of the photon cancel each other, leaving the two transverse components. Here the timelike component of the photon is cancelled by the scalar field, leaving three polarisation states for the photon, rendering it massive.

It should be remarked that although these conclusions about the *particle spectrum* were reached by considering the Lagrangian of (8.39), there is nothing otherwise special about this particular form. In fact, when we come to consider *renormalisation*, it is much more convenient (and, of course, physically equivalent) to consider (8.37). The gauge defined by (8.39) is called the *physical* or *unitary gauge* (or *U gauge*), since in this gauge, only physical particles (i.e. those which would appear in the unitarity condition) appear.

We now turn to the non-Abelian Higgs phenomenon,[14] and for definiteness consider the $O(3)$ model discussed in the previous section. The Lagrangian (8.12), then, needs to be modified by substituting a covariant derivative for the ordinary one, and adding the gauge field term. This gives

$$\mathscr{L} = \tfrac{1}{2}(D_\mu \phi_i)(D^\mu \phi_i) - \frac{m^2}{2} \phi_i \phi_i - \lambda(\phi_i \phi_i)^2 - \tfrac{1}{4} F^i_{\mu\nu} F^{i\mu\nu}. \qquad (8.42)$$

where, from (3.122) and (3.131),

$$D_\mu \phi_i = \partial_\mu \phi_i + g\varepsilon_{ijk} A^j_\mu \phi_k$$
$$F^i_{\mu\nu} = \partial_\mu A^i_\nu - \partial_\nu A^i_\mu + g\varepsilon^{ijk} A^j_\mu A^k_\nu. \qquad (8.43)$$

The potential V has a minimum (for $m^2 < 0$) at

$$|\phi_0| = \left(\frac{-m^2}{4\lambda}\right)^{\frac{1}{2}} = a \tag{8.15}$$

and, as before, we choose the vacuum which points in the 3 direction

$$\vec{\phi}_0 = a\hat{e}_3. \tag{8.16}$$

The physical fields are then ϕ_1, ϕ_2 and $\chi = \phi_3 - a$. After some algebra, it follows easily that

$$\mathcal{L} = \tfrac{1}{2}[(\partial_\mu\phi_1)^2 + (\partial_\mu\phi_2)^2 + (\partial_\mu\chi)^2] + ag[(\partial_\mu\phi_1)A_2^\mu - (\partial_\mu\phi_2)A_1^\mu]$$
$$+ \frac{a^2g^2}{2}[(A_\mu^1)^2 + (A_\mu^2)^2] - \tfrac{1}{4}(\partial_\mu A_\nu^i - \partial_\mu A_\mu^i)^2$$
$$- 4a^2\lambda\chi^2 + \text{cubic} + \text{quartic terms.} \tag{8.44}$$

We have shown explicitly only the terms quadratic in the fields, since for the present purpose they are the only important ones. This Lagrangian is analogous to (8.37); it contains a mixed term in A^μ and ϕ, so is not easy to interpret. To obtain a more 'physical' Lagrangian, we may now make use of the fact that we have a *local* symmetry, so may perform independent gauge transformations at each point in space–time. We therefore select a gauge – the unitary gauge – so that at *every* point in space–time $\vec{\phi}$ lies along the third isospin axis:

$$\vec{\phi}(x) = \hat{e}_3\phi_3 = \hat{e}_3(a + \chi). \tag{8.45}$$

This gets rid of the fields ϕ_1 and ϕ_2, and we have

$$D_\mu\phi_1 = g(a + \chi)A_\mu^2,$$
$$D_\mu\phi_2 = -g(a + \chi)A_\mu^1,$$
$$D_\mu\phi_3 = \partial_\mu\chi,$$

so that

$$(D_\mu\phi_i)^2 = a^2g^2[(A_\mu^1)^2 + (A_\mu^2)^2] + (\partial_\mu\chi)^2$$

and

$$\mathcal{L} = -\tfrac{1}{4}(\partial_\mu A_\nu^i - \partial_\nu A_\mu^i)^2 - \tfrac{1}{2}a^2g^2[(A_\mu^1)^2 + (A_\mu^2)^2]$$
$$+ \tfrac{1}{2}(\partial_\mu\chi)^2 - 4a^2\lambda\chi^2 + \text{cubic} + \text{quartic terms.} \tag{8.46}$$

The remaining particles, then, are 1 massive scalar, 2 massive vectors and 1 massless vector particle. In particular, the Goldstone bosons, present in the spontaneously broken *global* symmetry model, have both disappeared in the *local* symmetry model, and two of the massless gauge fields have become massive. Thus, analogously to (8.40) and (8.41), we may summarise the results for spontaneous breaking of an $O(3)$ symmetric model as follows:

Goldstone mode (global $O(3)$ symmetry):

3 massive scalar fields $\rightarrow$ 1 massive scalar field

$\qquad$ + 2 massless scalar fields. $\hfill$ (8.47)

Higgs mode (local $O(3)$ symmetry):

$$\left.\begin{array}{l}\text{3 massive scalar fields}\\[6pt]\text{+ 3 massless vector fields}\end{array}\right\} \rightarrow \left\{\begin{array}{l}\text{1 massive scalar field}\\[2pt]\text{+ 2 massive vector fields}\qquad(8.48)\\[2pt]\text{+ 1 massless vector field.}\end{array}\right.$$

We may also check that the number of independent modes is preserved: in the Higgs case $3 + 3 \times 2 = 9 = 1 + 2 \times 3 + 2$.

This $O(3)$ model contains all the features of the general non-Abelian case. It should be clear that one massless vector field remains because the subgroup $H \ (= U(1))$ under which the vacuum remains invariant has one generator – it was this circumstance that allowed one scalar field, in the Goldstone case, to remain massive. Thus, the number of massless vector fields is dim H. And on the other hand, the two vector fields which have become massive have done so by absorbing the two Goldstone modes; so the number of massive vector fields is dim G/H. Thus the total number of gauge particles (massive or massless) is dim G, as expected, since the gauge field transforms according to the regular representation of the group. The fact that in the model above, there is also a scalar field remaining is because we chose the scalar fields to belong to an isotriplet. We see again that the outcome of the Higgs mechanism is dictated largely by group theory.

8.4 Superconductivity

Superconductivity provides a nice illustration of the Abelian Higgs model. As everyone knows, superconductivity is the phenomenon, shown by many metals, of having no resistance at very low temperatures. Such metals are therefore capable of carrying persistent currents. These currents effectively screen out magnetic flux, which is therefore zero in a superconductor (the Meissner effect). Another way of stating the Meissner effect is to say that the photons are effectively massive, as in the Higgs phenomenon discussed above. We shall show very briefly how these conclusions follow from the Lagrangian (8.36).

To begin, we consider a *static* situation, so $\partial_0 \phi = 0$, etc., and (8.36) takes the form

$$\mathscr{L} = -\tfrac{1}{2}(\nabla - ie\mathbf{A})\phi \cdot (\nabla + ie\mathbf{A})\phi^* - m^2|\phi|^2 - \lambda|\phi|^4$$
$$\quad - \tfrac{1}{4}(\nabla \times \mathbf{A})^2$$

or

$$-\mathscr{L} = \tfrac{1}{4}(\nabla \times \mathbf{A})^2 + \tfrac{1}{2}|(\nabla - ie\mathbf{A})\phi|^2 + m^2|\phi|^2 + \lambda|\phi|^4. \hfill (8.49)$$

Now $-\mathcal{L}$ is the *Landau–Ginzburg free energy*, where $m^2 = a(T - T_c)$ near the critical temperature $T = T_c$; ϕ is the macroscopic many-particle wave function, and its use is justified by the Bardeen–Cooper–Schrieffer (BCS) theory, according to which, under certain conditions, there is an *attractive* force between electrons, and the field quanta are electron pairs, which, of course, are bosons. At low temperatures, these fall into the same quantum state (Bose–Einstein condensation), and because of this, a many-particle wave function ϕ may be used to describe the macroscopic system. Now, at $T > T_c$, $m^2 > 0$ and the minimum free energy is at $|\phi| = 0$. But when $T < T_c$, $m^2 < 0$ and the minimum free energy is at

$$|\phi|^2 = -\frac{m^2}{2\lambda} > 0; \tag{8.50}$$

which is, of course, an example of spontaneous symmetry breaking. Now $\mathcal{L}$ is invariant under the usual phase transformation

$$\phi \to e^{i\Lambda(x)}\phi, \quad \mathbf{A} \to \mathbf{A} + \frac{1}{e}\nabla\Lambda(x)$$

and the associated conserved current is

$$\mathbf{j} = -\frac{i}{2}(\phi^*\nabla\phi - \phi\nabla\phi^*) - e|\phi|^2\mathbf{A}. \tag{8.51}$$

When $T < T_c$, and ϕ varies only very slightly over the sample, the second of these terms dominates over the first, and

$$\mathbf{j} = \frac{em^2}{2\lambda}\mathbf{A} = -k^2\mathbf{A} \tag{8.52}$$

where k is a positive constant. This is the *London equation*. The electric field $\mathbf{E} = -\partial\mathbf{A}/\partial t = 0$, and Ohm's law defines resistance by $\mathbf{E} = R\mathbf{j}$, so

$$R = 0$$

and we have superconductivity.

The Meissner effect (expulsion of magnetic flux) is easily derived. Ampère's equation is

$$\nabla \times \mathbf{B} = \mathbf{j}.$$

Taking its curl, remembering that $\nabla \cdot \mathbf{B} = 0$, and using (8.52) gives

$$\nabla^2\mathbf{B} = k^2\mathbf{B}. \tag{8.53}$$

Confining ourselves for simplicity to one spatial dimension, (8.53) has the solution

$$B_x = B_0 e^{-kx},$$

so that magnetic field only penetrates the specimen to a characteristic depth $1/k$. When the numerical factors are properly taken into account, we get $1/k \approx 10^{-6}$ cm. Finally, it is clear that (8.53) implies that $\nabla^2 \mathbf{A} = k^2 \mathbf{A}$, or, in a Lorentz covariant form,

$$\Box A_\mu = -k^2 A_\mu$$

which means the 'photons' have a mass k, which is the characteristic feature of the Higgs phenomenon.

Following this application of the Higgs phenomenon to a low energy example, we now describe its application to weak interactions.

8.5   The Weinberg–Salam model

Spontaneous breaking of gauge symmetries was the crucial new ingredient in the model of unified weak and electromagnetic interactions constructed independently by Weinberg and Salam. The general idea was that weak interactions should be mediated by gauge bosons $(W^\pm)$, which are, 'to begin with', massless. The Lagrangian for the theory also contains terms for *massless* electrons, muons and neutrinos, and is invariant under an internal symmetry group, which is a gauge symmetry. A scalar field (the Higgs field) is then introduced with a non-vanishing vacuum-expectation-value. The resulting spontaneous breakdown of symmetry gives masses to e, μ (and τ if desired) and to the gauge bosons, but *not* to the photon and neutrino. It is therefore 'realistic', and has indeed met with a good degree of success in describing weak interactions. The model may also be extended to hadrons, but this will not be described here.

We start with the spinor fields. The Dirac Lagrangian

$$\mathcal{L} = i\bar{\psi}\gamma^\mu \partial_\mu \psi - m\bar{\psi}\psi$$

becomes simply $i\bar{\psi}\gamma \cdot \partial\psi$ if $m = 0$. Writing, as in chapter 2,

$$\psi_L = \left(\frac{1-\gamma_5}{2}\right)\psi, \quad \psi_R = \left(\frac{1+\gamma_5}{2}\right)\psi$$

it follows that

$$i\bar{\psi}\gamma \cdot \partial\psi = i\bar{\psi}_R \gamma \cdot \partial\psi_R + i\bar{\psi}_L \gamma \cdot \partial\psi_L$$

since γ_5 anticommutes with γ_μ. The electron and muon have L and R components, but according to the 2-component neutrino theory, ν_e and ν_μ have L components only, so the lepton Lagrangian is

$$\mathcal{L} = i\bar{e}_R \gamma \cdot \partial e_R + i\bar{e}_L \gamma \cdot \partial e_L + i\bar{\nu}_e \gamma \cdot \partial\nu_e + (e \to \mu). \tag{8.54}$$

It is clear that terms for the τ and its neutrino may also be added if desired; there would then be three 'generations' of leptons. From now on we shall forget about the μ and τ generations, which may trivially be added at any stage.

What internal symmetry does (8.54) possess? The transformations must be between particles whose space–time properties are the same, so the only possibility is a mixing of e_L and v_e. We therefore write the 'isospinor'

$$L = \begin{pmatrix} v_e \\ e_L \end{pmatrix} \tag{8.55}$$

and assign to this doublet a non-Abelian charge $I_w = \frac{1}{2}$ (I_w is 'weak isospin'). v_e has a third component $I_w^3 = \frac{1}{2}$, and e_L has $I_w^3 = -\frac{1}{2}$. This is obviously in straight analogy with (strong) isospin. The remaining particle

$$R = e_R \tag{8.56}$$

is an isosinglet: $I_w = 0$. We have

$$\mathcal{L} = i\bar{R}\gamma\cdot\partial R + i\bar{L}\gamma\cdot\partial L \tag{8.57}$$

and $\mathcal{L}$ is invariant under

$$\left. \begin{array}{l} L \rightarrow e^{-(i/2)\tau\cdot\alpha}L, \\ R \rightarrow R, \end{array} \right\} \tag{8.58}$$

which are rotations in weak isospin space. They generate the group $SU(2)$. More explicitly, the transformations are

$$SU(2): \begin{pmatrix} v_e \\ e_L \\ e_R \end{pmatrix} \rightarrow \left(\begin{array}{c|c} e^{-(i/2)\tau\cdot\alpha} & 0 \\ \hline 0 & 1 \end{array} \right) \begin{pmatrix} v_e \\ e_L \\ e_R \end{pmatrix}. \tag{8.59}$$

The relation between electric charge Q, and I_w^3 is

$$L{:}Q = I_w^3 - \tfrac{1}{2}; \quad R{:}Q = I_w^3 - 1. \tag{8.60}$$

When we come to gauge this symmetry (i.e. make α a space–time function) we will acquire three massless gauge fields. The photon, however, will *not* be one of them, since e_R, being a singlet, will not interact with our gauge fields, but does interact with the photon.

Note, however, that $SU(2)$ is not the maximal symmetry of $\mathcal{L}$. We could also have a simple $U(1)$ transformation on e_R

$$U(1){:}e_R \rightarrow e^{i\beta}e_R. \tag{8.61}$$

How will this affect L? It can be only be by an *overall* phase; in other words, v_e and e_L must pick up the *same* phase (as each other), since otherwise it would be a special case of an $SU(2)$ transformation. This phase, however, is not necessarily the same as that of R. We write, therefore,

$$U(1): \begin{pmatrix} v_e \\ e_L \\ e_R \end{pmatrix} \rightarrow \begin{pmatrix} e^{in\beta} & 0 & 0 \\ 0 & e^{in\beta} & 0 \\ 0 & 0 & e^{i\beta} \end{pmatrix} \begin{pmatrix} v_e \\ e_L \\ e_R \end{pmatrix} \tag{8.62}$$

where n is a number, which we must now find. This $U(1)$ symmetry leads to a conserved charge, of which e_R possesses one value, and v_e and e_L another value. It is clearly not Q, since v_e and e_L have different values of Q. (In other words, the gauge field we get on gauging $U(1)$ is also not the photon field.)

Weinberg suggested that this charge is 'weak hypercharge' Y_w defined by a quasi-Gell-Mann–Nishijima relation

$$Q = I_w^3 + \frac{Y_w}{2}. \tag{8.63}$$

Comparing with (8.60), it is clear that

$$L \text{ has } Y_w = -1,$$
$$R \text{ has } Y_w = -2, \tag{8.64}$$

so, in (8.62), $n = \frac{1}{2}$; the left-handed fields couple, with half the strength of the right-handed field, to the hypercharge gauge field. The $U(1)$ transformation is then

$$U(1): \begin{pmatrix} v_e \\ e_L \\ e_R \end{pmatrix} \rightarrow \begin{pmatrix} e^{i\beta/2} & 0 & 0 \\ 0 & e^{i\beta/2} & 0 \\ 0 & 0 & e^{i\beta} \end{pmatrix} \begin{pmatrix} v_e \\ e_L \\ e_R \end{pmatrix}. \tag{8.65}$$

The Lagrangian (8.54) is then invariant under $SU(2) \otimes U(1)$. (Note that the $U(1)$ could be identified with lepton number, giving $n = 1$. This would then result in a different theory.)

We now gauge the theory. Gauging $SU(2)$ means that we introduce three gauge potentials W_μ^i so that, acting on the isospinor L, the ordinary derivative is replaced by the covariant derivative (3.153):

$$D_\mu L = \partial_\mu L - \frac{i}{2} g \tau \cdot W_\mu L. \tag{8.66}$$

Here g is the $SU(2)$ coupling constant. Gauging $U(1)$ introduces another potential X_μ and coupling constant g', and, from (8.65), since L has half the hypercharge of R, the covariant derivatives are

$$\left.\begin{aligned} D_\mu L &= \partial_\mu L + \frac{i}{2} g' X_\mu L, \\ D_\mu R &= \partial_\mu R + i g' X_\mu R. \end{aligned}\right\} \tag{8.67}$$

Putting (8.66) and (8.67) into (8.57), and adding the gauge-field terms (see (3.131)) gives the Lagrangian

$$\mathcal{L}_1 = i\bar{R}\gamma^\mu(\partial_\mu + ig'X_\mu)R + i\bar{L}\gamma^\mu\left(\partial_\mu + \frac{i}{2}g'X_\mu - \frac{i}{2}g\tau\cdot W_\mu\right)L$$
$$- \tfrac{1}{4}(\partial_\mu W_\nu - \partial_\nu W_\mu + gW_\mu \times W_\nu)^2 - \tfrac{1}{4}(\partial_\mu X_\nu - \partial_\nu X_\mu)^2. \tag{8.68}$$

Next, we introduce an isospinor scalar field (the Higgs field),

$$\phi = \begin{pmatrix} \phi^+ \\ \phi^0 \end{pmatrix}. \tag{8.69}$$

From (8.63), it carries the quantum numbers

$$\phi: I_w = \tfrac{1}{2}, \quad Y_w = 1 \tag{8.70}$$

so that both ϕ^+ and ϕ^0 are *complex* fields (the particle and antiparticle are distinct), and we may put

$$\phi = \begin{pmatrix} \phi^+ \\ \phi^0 \end{pmatrix} = \begin{pmatrix} \dfrac{1}{\sqrt{2}}(\phi_3 + i\phi_4) \\ \dfrac{1}{\sqrt{2}}(\phi_1 + i\phi_2) \end{pmatrix} \tag{8.71}$$

where $\phi_1, \dots, \phi_4$ are real. In addition, by virtue of (8.70), the covariant derivative of ϕ is

$$D_\mu \phi = \left(\partial_\mu - \frac{i}{2} g \boldsymbol{\tau} \cdot \mathbf{W}_\mu - \frac{i}{2} g' X_\mu \right) \phi. \tag{8.72}$$

The Higgs field ϕ also interacts with e^- and v_e with strength G_e, so the overall Lagrangian containing ϕ is

$$\mathscr{L}_2 = (D_\mu \phi)^\dagger (D_\mu \phi) - \frac{m^2}{2} \phi^\dagger \phi - \frac{\lambda}{4}(\phi^\dagger \phi)^2$$

$$- G_e(\bar{L}\phi R + \bar{R}\phi^\dagger L). \tag{8.73}$$

The interaction term in $\mathscr{L}_2$, written out fully, is

$$- G_e(\bar{v}_e e_R \phi^\dagger + \bar{e}_L e_R \phi^0 + \bar{e}_R v_e \phi^- + \bar{e}_R e_L \overline{\phi^0}), \tag{8.74}$$

and

$$\phi^\dagger \phi = (\phi^+)^* \phi^+ + (\phi^0)^* \phi^0 = \tfrac{1}{2}(\phi_1^2 + \phi_2^2 + \phi_3^2 + \phi_4^2). \tag{8.75}$$

As usual, in the case $m^2 > 0$, this describes a scalar field with mass m, and lowest energy state corresponding to $\phi = 0$. But if $m^2 < 0$, the lowest energy state is not at $\phi = 0$, but at

$$(\phi^\dagger \phi)_0 = -\frac{m^2}{\lambda}. \tag{8.76}$$

We choose the isospin frame so that

$$(\phi_1^2)_0 = -\frac{2m^2}{\lambda}, \quad (\phi_2)_0 = (\phi_3)_0 = (\phi_4)_0 = 0,$$

or

$$(\phi_1)_0 = \left(-\frac{2m^2}{\lambda} \right)^{\frac{1}{2}} \equiv \sqrt{2}\eta \tag{8.77}$$

and

$$(\phi)_0 = \begin{pmatrix} 0 \\ \eta \end{pmatrix}, \quad \eta \text{ real.} \tag{8.78}$$

We now have a degenerate vacuum, and spontaneous breaking of the gauge symmetry. For excitations of ϕ above the vacuum, we might expect to have

$$\phi(x) = \begin{pmatrix} \dfrac{\phi_3(x)}{\sqrt{2}} + i\dfrac{\phi_4(x)}{\sqrt{2}} \\[2ex] \eta + \dfrac{\phi_1(x)}{\sqrt{2}} + i\dfrac{\phi_2(x)}{\sqrt{2}} \end{pmatrix}$$

but this is not the case. The fact that the symmetry is *local* means that we may perform a *different* isospin rotation at each point in space, so $\phi(x)$ may be reduced to the form

$$\phi(x) = \begin{pmatrix} 0 \\ \eta + \dfrac{\sigma(x)}{\sqrt{2}} \end{pmatrix} \tag{8.79}$$

at *every* point. (This is the same argument that led to (8.45).) From (8.72), $D_\mu\phi$ is then

$$D_\mu\phi = \begin{pmatrix} 0 \\ \dfrac{1}{\sqrt{2}}\partial_\mu\sigma \end{pmatrix} - \left[\frac{ig}{2}\begin{pmatrix} W_\mu^3 & W_\mu^1 - iW_\mu^2 \\ W_\mu^1 + iW_\mu^2 & -W_\mu^3 \end{pmatrix} + \frac{ig'}{2}X_\mu \right]$$

$$\times \begin{pmatrix} 0 \\ \eta + \dfrac{\sigma}{\sqrt{2}} \end{pmatrix}$$

$$= -\frac{i}{2}\begin{pmatrix} g\eta(W_\mu^1 - iW_\mu^2) + \dfrac{g\sigma}{\sqrt{2}}(W_\mu^1 - iW_\mu^2) \\[2ex] i\sqrt{2}\partial_\mu\sigma + \eta(-gW_\mu^3 + g'X_\mu) + \dfrac{\sigma}{\sqrt{2}}(-gW_\mu^3 + g'X_\mu) \end{pmatrix}.$$

Hence

$$(D_\mu\phi)^\dagger(D_\mu\phi) = \tfrac{1}{2}(\partial_\mu\sigma)^2 + \frac{g^2\eta^2}{4}[(W_\mu^1)^2 + (W_\mu^2)^2]$$

$$+ \frac{\eta^2}{4}(gW_\mu^3 - g'X_\mu)^2 + \text{cubic} + \text{quartic terms.} \tag{8.80}$$

Now we define

$$Z_\mu = \frac{gW_\mu^3 - g'X_\mu}{(g^2 + g'^2)^{\frac{1}{2}}} \equiv \cos\theta_{\rm w} W_\mu^3 - \sin\theta_{\rm w} X_\mu \tag{8.81}$$

and the orthogonal field

$$A_\mu = \frac{g' W_\mu^3 + g X_\mu}{(g^2 + g'^2)^{\frac{1}{2}}} = \sin\theta_W W_\mu^3 + \cos\theta_W X_\mu \qquad (8.82)$$

where the 'Weinberg angle' θ_W is given by

$$\frac{g}{(g^2 + g'^2)^{\frac{1}{2}}} = \cos\theta_W, \quad \frac{g'}{g} = \tan\theta_W. \qquad (8.83)$$

We see, from (8.80), that W_μ^1, W_μ^2 and Z_μ pick up masses

$$M_{W_1}^2 = M_{W_2}^2 = \frac{g^2\eta^2}{2}, \quad M_Z^2 = \frac{g^2\eta^2}{2\cos^2\theta_W} = \frac{M_W^2}{\cos^2\theta_W} \qquad (8.84)$$

and A_μ is massless. A_μ is therefore provisionally identified with the electromagnetic field. The identification may be justified by noting that the lepton gauge-field coupling is, from (8.68) and (8.81–8.83),

$$i\bar{R}\gamma^\mu(\partial_\mu + ig'X_\mu)R + i\bar{L}\gamma^\mu\left(\partial_\mu + \frac{i}{2}g'X_\mu - \frac{i}{2}g\tau\cdot\mathbf{W}_\mu\right)L$$

$$= i\bar{e}\gamma^\mu\partial_\mu e - g\sin\theta_W\bar{e}\gamma^\mu e A_\mu$$

$$+ \frac{g}{\cos\theta_W}(\sin^2\theta_W\bar{e}_R\gamma^\mu e_R - \tfrac{1}{2}\cos 2\theta_W\bar{e}_L\gamma^\mu e_L + \tfrac{1}{2}\bar{v}\gamma^\mu v)Z_\mu$$

$$+ \frac{g}{\sqrt{2}}[(\bar{v}\gamma^\mu e_L W_\mu^\dagger) + \text{h.c.}] \qquad (8.85)$$

where h.c. stands for Hermitian conjugate, and $W_\mu = (W_\mu^1 + iW_\mu^2)/\sqrt{2}$. Note that the A_μ field couples *only* to the electrons, and not to the neutrinos, and that it couples to left and right components equally strongly. We are thus justified in identifying A_μ with the electromagnetic potential, and it follows immediately that the pertinent coupling constant should be e, the proton charge:

■ $$e = g\sin\theta_W. \qquad (8.86)$$

From the last term in (8.85) g is the coupling of the weak field to the electron–neutrino (and muon–neutrino) current. To second order, therefore, this interaction will account for muon decay,

$$\mu^- \to e^- + \bar{v}_e + v_\mu,$$

Fig. 8.6. Muon decay in an intermediate vector boson theory.

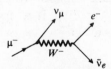

through the diagram of Fig. 8.6, in which the gauge field W is propagated between the two vertices. From (8.85), then, the effective interaction is

$$H_{int} = \frac{g^2}{2} \bar{v}_\mu \gamma^\kappa \mu_L (\text{Prop})_{\kappa\lambda} \bar{e}_L \gamma^\lambda v_e$$

$$= \frac{g^2}{8} \bar{v}_\mu \gamma^\kappa (1 - \gamma_5)_\mu (\text{Prop})_{\kappa\lambda} \bar{e} \gamma^\lambda (1 - \gamma_5) v_e. \tag{8.87}$$

At low q, however, the propagator simply becomes $g_{\kappa\lambda}/M_W^2$, so (8.87) becomes

$$H_{int} = \frac{g^2}{8M_W^2} j_\mu^{\lambda\dagger} j_{e\lambda} \tag{8.88}$$

where

$$j_{l\lambda} = \bar{l} \gamma_\lambda (1 - \gamma_5) v_l$$

is the lepton (l) current. (8.88) is, however, exactly the Fermi current–current interaction

$$H_{int} = \frac{G}{\sqrt{2}} j_\mu^{\lambda\dagger} j_{e\lambda} \tag{8.89}$$

where G is the Fermi constant. Hence we have the equality

$$G = \frac{g^2}{4\sqrt{2} M_W^2}. \tag{8.90}$$

From muon decay[‡], G has the value $G = 1.43 \times 10^{-49}$ erg cm^3, which in 'natural' units $\hbar = c = 1$ is $G \sim 10^{-5} m_p^{-2}$, a 'small' number (if the proton mass sets the scale). It is because G is small that the weak interactions are called weak. In the Weinberg–Salam theory, however, the fundamental coupling constant is g, and, from (8.86), if θ_W is not too small, neither is g. The weak interaction is then not *intrinsically* weak, but, from (8.90), only *appears* weak because M_W is large. In fact, numbers may be put to these quantities. From (8.85) both the neutrinos and charged leptons (e and μ) couple to the *neutral* field Z, so we expect, for example, to observe v_μ–e scattering:

$$v_\mu + e^- \to v_\mu + e^-,$$

through the Feynman diagram of Fig. 8.7. This is an example of a 'neutral current' interaction typically predicted by the Weinberg–Salam theory. This scattering has been observed, with a cross section $\sigma = (1.6 \pm 0.9) \times 10^{-42} (E_v/\text{GeV}) \text{cm}^2$ (see, for example ref. (7), p. 599). It is clear that this cross section gives a value for θ_W (or, equivalently, for g). This is

‡ See, for example, ref. (7), chapter 21.

(ref. (7), p. 686)

$$\sin^2 \theta_W = 0.225 \begin{array}{c} +0.06 \\ -0.05 \end{array}.$$ (8.91)

Equations (8.86) and (8.90) then yield

$$M_W^2 = \frac{e^2}{4\sqrt{2G}\sin^2\theta_W} = (78.6\,\text{GeV}/c^2)^2$$ (8.92)

and (8.84) gives for the Z mass

$$M_Z = 89.3\,\text{GeV}/c^2.$$ (8.93)

The existence of the charged and neutral vector bosons with the above masses is a crucial prediction of the electroweak theory, so the verification is all the more spectacular since these particles have now been found (UA1 Collaboration, *Physics Letters*, **122B**, 103; **126B**, 398 (1983)) at the correct mass. It is worth pointing out that neutral current processes (also observed) follow from any theory whose Lagrangian is invariant under a *global SU*(2) symmetry. It is only when the theory is *gauged* that it becomes renormalisable and that intermediate vector bosons are predicted. For further details of the electroweak theory and its experimental implications, the reader is referred elsewhere.

Finally, an interesting speculation concerns the high temperature behaviour of the electroweak interaction. If the analogy with ferromagnetism holds good, the symmetry, which is spontaneously broken at low temperatures should be restored at high temperatures, and above the critical temperature T_c the W and Z bosons, like the photon, would be massless, and the weak force would become long range like electromagnetism. Kirzhnits and Linde[25] have speculated that this happens, with $T_c \sim 10^{16}$ K, and more formal calculations seem to vindicate this idea. It is based on the prescription that the finite temperature scalar field 2-point function (for example) is obtained by taking the Gibbs average of the 'ordinary' 2-point function:

$$i\Delta_{F\beta}(x-y) = \frac{\text{Tr}\,e^{-\beta H}\langle 0|T(\phi(x)\phi(y))|0\rangle}{\text{Tr}\,e^{-\beta H}}$$ (8.94)

Fig. 8.7. $\nu_\mu e^-$ scattering by Z^0 exchange.

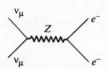

where $\beta = (kT)^{-1}$. If this hypothesis is correct, it will have cosmological implications, since, on the standard 'big bang' model, there was certainly a time when the temperature of the universe was higher than T_c. One interesting question is, when T cools below T_c, how does the spontaneous breaking set in? In a ferromagnet, for example, the direction of spontaneous magnetisation characterises not the whole sample, but only a *domain*, and domain walls separate regions magnetised in different directions. Does this happen in the universe? Another interesting questions is whether, when $T > T_c$, the long-range weak force leads to large-scale repulsive effects between particles with the same Y_w and I_w^3 (just as it does between particles with the same Q in electromagnetism). For these and other related questions we refer the reader to the literature.[28]

Summary

[1]Ferromagnetism is shown to illustrate the phenomenon of spontaneous symmetry breaking (SSB), and a model scalar field theory is exhibited with SSB, which results in the presence of a massless particle, the Goldstone boson. [2]In the general case, where the Lagrangian is invariant under a symmetry group G, but the ground state (vacuum) is only invariant under a subgroup H, the number of Goldstone particles is shown to be $\dim G/H$. It is also shown that the above (classical) arguments also hold in quantum theory: that spontaneous symmetry breaking implies the existence of massless particles. This is the Goldstone theorem. [3]SSB of gauge theories, on the other hand, exhibits quite different phenomena: there are no Goldstone particles, and some (or all) gauge fields become massive. This is shown in Abelian and non-Abelian models. [4]It is shown that superconductivity is a theory with SSB of electromagnetism, an Abelian gauge theory, and the [5]Weinberg–Salam model of unified weak and electromagnetic interactions is described. It exhibits SSB of an $SU(2) \otimes U(1)$ gauge symmetry. Some experimental implications are discussed.

Guide to further reading

The possibility of an analogy between the ground state in many-body physics and the vacuum in quantum field theory was first entertained by

(1) Y. Nambu, *Physical Review Letters*, **4**, 380 (1960); Y. Nambu & G. Jona-Lasinio, *Physical Review*, **122**, 345 (1961).

and is reviewed in

(2) Y. Nambu, in F. Gürsey (ed.), *Group Theoretical Methods and Concepts in Elementary Particle Physics*, Gordon & Breach, 1964

Goldstone's model appeared in

(3) J. Goldstone, *Nuovo Cimento*, **19**, 154 (1961)

and is proved in

(4) J. Goldstone, A. Salam & S. Weinberg, *Physical Review*, **127**, 965 (1962); see also S. Weinberg, in S. Deser, M. Grisaru & H. Pendleton (eds.), *Lectures on Elementary Particles and Quantum Field Theory* (1970 Brandeis Summer Institute) Vol. I, MIT Press, 1970.

Good discussions of the Goldstone theorem appear in
(5) J. Bernstein, *Reviews of Modern Physics*, **46**, 7 (1974).
(6) G.S. Guralnik, C.R. Hagen & T.W.B. Kibble, *Advances in Particle Physics*, Vol. 2 (R.L. Cool & R.E. Marshak (eds.)), Wiley-Interscience, 1968.
(7) T.D. Lee, *Particle Physics and Introduction to Field Theory*, chapters 16, 22 and 24, Harwood Academic Publishers, 1981.
(8) J.C. Taylor, *Gauge Theories of Weak Interactions*, chapter 5, Cambridge University Press, 1966.
(9) L. O'Raigheartaigh, *Reports on Progress in Physics*, **42**, 159 (1979).

For the Higgs phenomenon, see
(10) P.W. Higgs, *Physics Letters*, **12**, 132 (1964); *Physical Review Letters*, **13**, 508 (1964); *Physical Review*, **145**, 1156 (1966).
(11) F. Englert & R. Brout, *Physical Review Letters*, **13**, 321 (1964).
(12) G.S. Guralnik, C.R. Hagen & T.W.B. Kibble, *Physical Review Letters*, **13**, 585 (1964).

Good reviews are to be found in refs. (5–9), and
(13) S. Coleman, in A. Zichichi (ed.), *Laws of Hadronic Matter*, Academic Press, 1975.

For the non-Abelian case, see
(14) T.W.B. Kibble, *Physical Review*, **155**, 1554 (1967).
Also refs. (6), (8), (9) above.

For readable accounts of superconductivity, see, for example,
(15) D.R. Tilley & J. Tilley, *Superfluidity and Superconductivity*, Wiley, 1974.
(16) P.G. de Gennes, *Superconductivity of Metals and Alloys*, Benjamin, 1966.

A good review of the Ginzburg–Landau theory is
(17) M. Cyrot, *Reports on Progress in Physics*, **36**, 103 (1973).

Good accounts of superconductivity for particle physicists are
(18) D.A. Kirzhnits, *Soviet Physics Uspekhi*, **21**, 470 (1978).
(19) I.J.R. Aitchison & A.J.G. Hey, *Gauge Theories in Particle Physics*, chapter 9, Adam Hilger Ltd., Bristol, 1982.

The electroweak unified theory originates in
(20) S. Weinberg, *Physical Review Letters*, **19**, 1264 (1967).
(21) A. Salam in N. Svartholm (ed.), *Proceedings of the Eighth Nobel Symposium*, Almqvist & Wiksell, Stockholm, 1968.
See also
(22) S.L. Glashow, *Nuclear Physics*, **22**, 579 (1961).

For reviews, see refs. (5), (7–9) above, and the 1979 Nobel prize acceptance speeches:
(23) S. Weinberg, *Reviews of Modern Physics*, **52**, 515 (1980); A. Salam, *ibid.*, 525; S.L. Glashow, *ibid.*, 539.

Up-to-date reviews, particularly of experimental data, are to be found in conference proceedings. For example,
(24) L. Durand & L.G. Pondrom (eds.), *Proceedings of the Twentieth International Conference on High Energy Physics* (Madison, Wisconsin, 1980), American Institute of Physics, 1981.

The following papers deal with symmetry restoration at high temperatures.

(25) D.A. Kirzhnits & A.D. Linde, *Physics Letters*, **42B**, 471 (1972).

(26) S. Weinberg, *Physical Review*, **D9**, 3357 (1974).

(27) L. Dolan & R. Jackiw, *Physical Review*, **D9**, 3320 (1974).

For a recent review, see

(28) A.D. Linde, *Reports on Progress in Physics*, **42**, 389 (1979).

For finite temperature Green's functions, see refs. (25), (26) and

(29) E.V. Shuryak. *Physics Reports*, **61**, 71 (1980).

(30) S.W. Hawking, in S.W. Hawking & W. Israel (eds.) *General Relativity: An Einstein Centenary Survey*, Cambridge University Press, 1979, pp. 754–5.

(31) S. Weinberg in A. Zichichi (ed.) *Understanding the Fundamental Constituents of Matter*, Plenum Press, 1978.

9

Renormalisation

If the doors of perception were cleansed, everything would appear to man as it is, infinite.

William Blake

We have seen in previous chapters that integration over internal loops in Feynman diagrams gives divergent results. Since our approach to field theory is based on perturbation theory, however, it is imperative that we make sense of the perturbation series – and that series is one in which higher order terms involve more and more internal integrations, and therefore the possibility of increasing degrees of divergence. It is obvious that, in order for a field theory to be at all sensible or believable, the problems raised by the divergences must be satisfactorily resolved. In this chapter we show how this is done for ϕ^4 theory, electrodynamics (QED) and Yang–Mills theories (QCD). Our general approach is to proceed order by order in perturbation theory (actually in the loop expansion – see below), and show that at each order the quantities of physical interest (masses, coupling constants, Green's functions) can be *renormalised* to finite values. Then (for QED and QCD) we show that this is, in principle, possible to *all orders*; these theories are therefore *renormalisable*. (So is ϕ^4 theory, but we do not prove that.) We begin with ϕ^4 theory.

9.1   Divergences in ϕ^4 theory

We saw in chapter 6 that $\Delta(x - x) = \Delta(0)$ is a divergent quantity, which modifies the free particle propagator and contributes to the self-energy. In momentum space it corresponds to

$$g \int \frac{\mathrm{d}^4 q}{(2\pi)^4} \frac{1}{q^2 - m^2}. \tag{9.1}$$

There are four powers of q in the numerator and two in the denominator, so the integral diverges quadratically at large q (it is *ultra-violet* divergent; in this chapter we do not concern ourselves with the problems of infra-red divergence $(q \to 0)$). This diagram is of order g. Another diagram to diverge is the $O(g^2)$ graph.

$$g^2 \int \frac{d^4q_1}{(2\pi)^4} \frac{d^4q_2}{(2\pi)^4} \frac{\delta(q_1 + q_2 - p_1 - p_2)}{(q_1^2 - m^2)(q_2^2 - m^2)}$$

$$= g^2 \int \frac{d^4q}{(2\pi)^8} \frac{1}{(q^2 - m^2)[(p_1 + p_2 - q)^2 - m^2]}. \qquad (9.2)$$

Here there are four powers of q in both numerator and denominator, so we get a logarithmic divergence.

How may we find the degree of divergence of a particular graph? It is clear that each propagator contributes q^2 to the denominator, and each vertex four powers of q to the numerator, together with a momentum-conservation δ-function. Also, the number of independent momenta (over which integration takes place) is the number of *loops* (1 in the two diagrams above). So consider a diagram of order n, i.e. with n vertices, E external lines, I internal lines and L loops, and, for future purposes, consider that space–time has d dimensions (so that vertices contribute d powers of q to the numerator). We ask for the *superficial*[‡] *degree of divergence* D of this diagram. It is clearly

$$D = dL - 2I. \qquad (9.3)$$

In the above diagrams this formula gives $D = 2$ and $D = 0$, as required. We now want to express D in terms of E and n, so we must eliminate I and L. There are I internal momenta; there is momentum conservation at each vertex (of which there are n), but there is also overall momentum conservation, so there are $n - 1$ relations between the momenta; hence the number of independent momenta is $I - n + 1$. But this is L:

$$L = I - n + 1. \qquad (9.4)$$

In ϕ^4 theory, each vertex gives 4 legs, so there are $4n$ legs, some external and some internal. In doing a leg count, however, the internal ones count twice, because they are connected to two vertices, so

$$4n = E + 2I. \qquad (9.5)$$

Equations (9.3–9.5) give

$$D = d - \left(\frac{d}{2} - 1\right)E + n(d - 4). \qquad (9.6)$$

[‡] See the paragraph below (9.9) for an explanation of 'superficial'.

In the case $d = 4$, we have

$$D = 4 - E, \tag{9.7}$$

which gives the correct result for the diagrams above. It also indicates that diagrams with more external legs than 4 will all converge; for example, if $E = 6$, $D = -2$. Is this true? We shall consider this question below.

For the present, consider the last term in (9.6). It conjures up a horrifying prospect, for if the coefficient of n is greater than 0, D increases with n, so the *complete* theory (summed over all n) will contain an infinite number of terms, each one more divergent than the one before, which is hopeless. In ϕ^4 theory in four space–time dimensions, however, (9.7) shows that D depends on E *only*, not on the order in perturbation theory, so we have only a small number of divergent graphs, and the hope is that the effects of these can be eliminated by (infinite) renormalisation of various physical quantities. If this turns out to be true (which it does) the theory is called *renormalisable*. Of course, we have not yet shown that ϕ^4 theory *is* renormalisable – we do that in §9.3 below. We have shown only that the perturbation series does not give an infinite number of different types of divergent graphs, which is obviously a *sine qua non* for renormalisability. It is instructive, finally, to write the analogous formulae for ϕ^r theory. Equations (9.3) and (9.4) are unchanged. (9.5) becomes $rn = E + 2I$, so (9.6) becomes

$$D = d - \left(\frac{d}{2} - 1\right)E + n\left[\frac{r}{2}(d - 2) - d\right] \tag{9.8}$$

which, when $d = 4$, gives

$$D = 4 - E + n(r - 4). \tag{9.9}$$

This clearly reduces to (9.7) when $r = 4$. But in ϕ^6 theory, for example, $D = 4 - E + 2n$, so the theory is unrenormalisable; ϕ^3 theory, on the other hand gives $D = 4 - E - n$, and is called 'super-renormalisable', since D decreases with increasing n, so there is only a *finite* number of divergent graphs for given E. Equation (9.8) also gives the curious result that in two dimensions $D = 2 - 2n$, so is independent of r.

Fig. 9.1. Diagrams with six external legs in ϕ^4 theory.

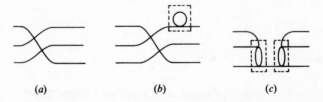

(a)   (b)   (c)

Now let us return to equation (9.7), and enquire whether *all* graphs with $E > 4$ are convergent. In ϕ^4 theory, E is even, so consider the graphs in Fig. 9.1, with $E = 6$, which 'should' be convergent: (*a*) is convergent, as may be seen by writing out the amplitude, but (*b*) contains the 1-loop contribution to the 2-point function, as in (9.1) above, and this is *always* divergent. Similarly (*c*) contains, as marked, two 1-loop contributions to the 4-point function, so is also divergent. This happens with all Feynman diagrams; if they contain hidden 2- or 4-point functions with one loop (or more) they will diverge, despite the formula $D = 4 - E$. This is why D is called the *superficial* degree of divergence. The nice thing is that the converse of this is true. According to *Weinberg's theorem* a Feynman diagram converges if its degree of divergence D, together with the degree of divergence of all its subgraphs, is negative. We shall not prove this.

The two divergent diagrams $G^{(2)}$ and $G^{(4)}$ in (9.1) and (9.2) above are called *primitive divergences*. They are the only primitive divergences in ϕ^4 theory. In the next section we shall consider explicitly how to cope with these divergences.

Dimensional analysis
Dimensional analysis provides some useful insights. The action (in d dimensions)

$$S = \int d^d x \, \mathscr{L}$$

is dimensionless (neglecting the $1/\hbar$ factor). Hence

$$[\mathscr{L}] = L^{-d} \quad (L \text{ is length}),$$

or

$$[\mathscr{L}] = \Lambda^d \quad (\Lambda \text{ is momentum}).$$

The kinetic energy term in $\mathscr{L}$ is $\partial^\mu \phi \partial_\mu \phi$ (ignore the dimensions of $g_{\mu\nu}$), so since $[\partial_\mu] = L^{-1}$, we have

$$[\phi] = L^{1-d/2} \quad \text{or} \quad \Lambda^{d/2-1}. \tag{9.10}$$

Now consider an interaction $g\phi^r$. If $[g] = L^{-\delta}$ (or Λ^δ) then

$$-\delta + r\left(1 - \frac{d}{2}\right) = -d$$

$$\delta = d + r - \frac{rd}{2}. \tag{9.11}$$

So the coupling constant g has the following dimensions in the following

Table 9.1. *Canonical dimensions of various quantities in d-dimensional space–time, and in 4-dimensional space–time. For notes, see the text*

Note	Quantity	Mass dimension in d-dimensional space–time	Mass dimension with $d = 4$
	ϕ	$\dfrac{d}{2} - 1$	1
1	$G^{(n)}(x_1, \ldots, x_n)$	$n\left(\dfrac{d}{2} - 1\right)$	n
2	$G^{(n)}(p_1, \ldots, p_n)$	$-nd + n\left(\dfrac{d}{2} - 1\right)$	$-3n$
		$= -n\left(\dfrac{d}{2} + 1\right)$	
3	$\bar{G}^{(n)}(p_1, \ldots, p_{n-1})$	$d - n\left(\dfrac{d}{2} + 1\right)$	$4 - 3n$
4	$\Gamma^{(2)}(x - y)$	$2 + d$	6
5	$\Gamma^{(n)}(x_1, \ldots, x_n)$	$n\left(\dfrac{d}{2} + 1\right)$	$3n$
6	$\Gamma^{(n)}(p_1, \ldots, p_n)$	$-dn + n\left(\dfrac{d}{2} + 1\right)$	$-n$
		$= n\left(1 - \dfrac{d}{2}\right)$	
7	$\bar{\Gamma}^{(n)}(p_1, \ldots, p_{n-1})$	$d + n\left(1 - \dfrac{d}{2}\right)$	$4 - n$

theories.

$$
\left.
\begin{aligned}
g\phi^4: &\quad \delta = 4 - d \quad [g] = \Lambda^{4-d}, \\
g\phi^3: &\quad \delta = 3 - \frac{d}{2} \quad [g] = \Lambda^{3-d/2}, \\
g\phi^6: &\quad \delta = 6 - 2d \quad [g] = \Lambda^{6-2d}.
\end{aligned}
\right\}
\tag{9.12}
$$

Substituting (9.11) into (9.8) (eliminating r) gives

$$
D = d - \left(\frac{d}{2} - 1\right)E - n\delta.
\tag{9.13}
$$

Hence a renormalisable theory must be one whose coupling constant g has a mass dimension $\delta \geqslant 0$. This is why Fermi's theory of weak interactions, characterised by G_F, with dimensions (mass)$^{-2}$, is unrenormalisable.

It is convenient to know the dimensions (sometimes called 'canonical' or 'engineering' dimensions) of various Green's functions and vertex functions. For ease of reference these are shown in Table 9.1. The entries in the table are explained by the following notes. (1) This follows from the definition. Each $\delta/\delta J(x)$ brings down a $\phi(x)$. For example $G^{(2)}(x, y) \approx \int d^d p \, e^{ip(x-y)}(p^2 - m^2)^{-1}$ clearly has dimension $d - 2$; and $G^{(4)} = \sum G^{(2)} G^{(2)}$. (2) The momentum-space Green's function is obtained by Fourier transform of $G^{(n)}(x_i)$. (3) By translation invariance, overall momentum conservation allows us to define

$$G^{(n)}(p_1, \ldots, p_n) = \bar{G}^{(n)}(p_1, \ldots, p_{n-1}) \delta(P)$$

and $\delta(P)$ has (mass) dimension $-d$. (4) See equation (7.84), remembering that $\delta(x - y)$ has mass dimension d. (5) Follows from the definition, analogous to equation (7.92) for $\Gamma^{(3)}(x_1, x_2, x_3)$. (6) Follows by Fourier transform. (7) Follows from overall momentum conservation, as $\bar{G}^{(n)}(p_i)$ above.

9.2   Dimensional regularisation of ϕ^4 theory

Regularisation is a method of isolating the divergences in Feynman integrals. It makes the task of renormalisation much more explicit and easy to follow. There are several techniques of regularisation. Perhaps the most intuitive one is to introduce a cut-off Λ in the momentum integrals. In electrodynamics a particular example is to modify the free (photon) propagator

$$\frac{1}{k^2} \to \frac{1}{k^2} - \frac{1}{k^2 - \Lambda^2} = -\frac{\Lambda^2}{k^2(k^2 - \Lambda^2)}.$$

Similar to this is the Pauli–Villars regularisation in which a fictitious field of mass M is introduced. In both these cases the limit $\Lambda \to \infty$ $(M \to \infty)$ is taken, and the renormalised quantities are independent of Λ (M). These methods become problematic, however, particularly when non-Abelian gauge theories are considered. A trouble-free and elegant method is that of dimensional regularisation, which has therefore become popular, and will be outlined in this section. The idea is to treat the loop integrals (which cause the divergences) as integrals over d-dimensional momenta, and then take the limit $d \to 4$. It turns out that the singularities of 1-loop graphs are simple poles in $d - 4$. In this section we show how to apply this technique to ϕ^4 theory, and in §9.3 we show how renormalisation is performed, to 'get rid of' the unwanted infinite expressions.

We first need to generalise the 4-dimensional Lagrangian

$$\mathscr{L} = \tfrac{1}{2} \partial_\mu \phi \partial^\mu \phi - \frac{m^2}{2} \phi^2 - \frac{g}{4!} \phi^4$$

to d dimensions. Since ϕ has dimension $\frac{1}{2}d - 1$ and $\mathcal{L}$ dimension d (see (9.10)) g is dimensionless in four dimensions, but to keep it dimensionless in d dimensions, it must be multiplied by μ^{4-d} where μ is an *arbitrary mass parameter*. So

$$\mathcal{L} = \tfrac{1}{2}\partial_\mu\phi\partial^\mu\phi - \frac{m^2}{2}\phi^2 - \frac{\mu^{4-d}g}{4!}\phi^4. \tag{9.14}$$

Now we use the corresponding Feynman rules to calculate the order g correction to the free propagator. The loop integral (cf. (9.1)) is

$$\underset{\textstyle \bigcirc}{\quad} \tfrac{1}{2}g\mu^{4-d}\int\frac{d^dp}{(2\pi)^d}\frac{1}{p^2 - m^2}. \tag{9.15}$$

The $\frac{1}{2}$ is the symmetry factor, and we are working in d-dimensional 'Minkowski' (not Euclidean) space. Throughout this chapter we shall be dealing with integrals like the one above, and they are derived in the Appendix to the chapter. From equation (9A.5) the above expression is

$$-\frac{ig}{32\pi^2}m^2\left(\frac{4\pi\mu^2}{m^2}\right)^{2-d/2}\Gamma\left(1 - \frac{d}{2}\right). \tag{9.16}$$

The gamma function has poles at zero and the negative integers, so we see that the divergence of the integral manifests itself as a simple pole as $d \to 4$. From Appendix 9B, we have

$$\Gamma(-n + \varepsilon) = \frac{(-1)^n}{n!}\left[\frac{1}{\varepsilon} + \psi_1(n+1) + O(\varepsilon)\right] \tag{9.17}$$

where

$$\psi_1(n+1) = 1 + \frac{1}{2} + \cdots + \frac{1}{n} - \gamma$$

and $\gamma = -\psi_1(1) = 0.577$ is the Euler–Mascheroni constant. Putting

$$\varepsilon = 4 - d \tag{9.18}$$

gives

$$\Gamma(1 - d/2) = \Gamma(-1 + \varepsilon/2) = -\frac{2}{\varepsilon} - 1 + \gamma + O(\varepsilon). \tag{9.19}$$

So, expanding (9.16) about $d = 4$, gives (using $a^\varepsilon = 1 + \varepsilon\ln a + \cdots$)

$$\frac{-igm^2}{32\pi^2}\left[-\frac{2}{\varepsilon} - 1 + \gamma + O(\varepsilon)\right]\left[1 + \frac{\varepsilon}{2}\ln\left(\frac{4\pi\mu^2}{m^2}\right)\right]$$

$$= \frac{igm^2}{16\pi^2\varepsilon} + \frac{igm^2}{32\pi^2}\left[1 - \gamma + \ln\left(\frac{4\pi\mu^2}{m^2}\right)\right] + O(\varepsilon)$$

$$= \frac{igm^2}{16\pi^2\varepsilon} + \text{finite}. \tag{9.20}$$

The finite part of the correction to the propagator is not very important, but note that it depends on the arbitrary mass μ.

Now let us calculate the 4-point function to order g^2. The Feynman rules give (for the s-channel contribution – see below)

$= \frac{1}{2}g^2(\mu^2)^{4-d}\int \frac{d^d p}{(2\pi)^d}\frac{1}{(p^2-m^2)}\frac{1}{(p-q)^2-m^2}.$

$$(9.21)$$

The denominators in the integrand are combined by using the Feynman formula

$$\frac{1}{ab}=\int_0^1 \frac{dz}{[az+b(1-z)]^2} \tag{9.22}$$

(which is obtained by observing that

$$\frac{1}{ab}=\frac{1}{b-a}\left(\frac{1}{a}-\frac{1}{b}\right)=\frac{1}{b-a}\int_a^b \frac{dx}{x^2}$$

and putting $x=az+b(1-z)$; a and b are to be considered as complex variables, to avoid the singularity at $a=b$). Then

$$\frac{1}{p^2-m^2}\frac{1}{(p-q)^2-m^2}=\int_0^1 \frac{dz}{[p^2-m^2-2pq(1-z)+q^2(1-z)]^2}$$

By changing variables to

$$p'=p+q(1-z)$$

we see that the denominator in the integrand is the square of $p'^2-m^2+q^2z(1-z)$. But $d^d p'=d^d p$, so relabelling $p'\to p$, (9.21) becomes

$$\frac{1}{2}g^2(\mu^2)^{4-d}\int_0^1 dz\int \frac{d^d p}{(2\pi)^d}\frac{1}{[p^2-m^2+q^2z(1-z)]^2}.$$

The last integral is now in the form of equation (9A.4), so we get

$$\frac{ig^2}{2}(\mu^2)^{4-d}\left(\frac{1}{4\pi}\right)^{d/2}\frac{\Gamma(2-d/2)}{\Gamma(2)}\int_0^1 dz\,[q^2z(1-z)-m^2]^{d/2-2}$$

$$=\frac{ig^2}{32\pi^2}(\mu^2)^{2-d/2}\Gamma(2-d/2)\int_0^1 dz\left[\frac{q^2z(1-z)-m^2}{4\pi\mu^2}\right]^{d/2-2}.$$

$$(9.23)$$

In the limit $d\to 4$, equation (9.17) gives

$$\Gamma(2-d/2)=\Gamma(\varepsilon/2)=\frac{2}{\varepsilon}-\gamma+O(\varepsilon) \tag{9.24}$$

so (9.23) becomes

$$\frac{ig^2\mu^\varepsilon}{32\pi^2}\left(\frac{2}{\varepsilon}-\gamma+O(\varepsilon)\right)\left\{1-\frac{\varepsilon}{2}\int_0^1 dz\ln\left[\frac{q^2z(1-z)-m^2}{4\pi\mu^2}\right]\right\}$$

$$=\frac{ig^2\mu^\varepsilon}{16\pi^2\varepsilon}-\frac{ig^2\mu^\varepsilon}{32\pi^2}\left\{\gamma+\int_0^1 dz\ln\left[\frac{q^2z(1-z)-m^2}{4\pi\mu^2}\right]\right\}. \qquad (9.25)$$

Here, the leading term depends on μ, and the finite term depends on the total momentum $(p_1+p_2)^2=q^2=s$. Putting

$$F(s,m,u)=\int_0^1 dz\ln\left[\frac{sz(1-z)-m^2}{4\pi\mu^2}\right] \qquad (9.26)$$

we have, finally, for (9.21),

$$\frac{ig^2\mu^\varepsilon}{16\pi^2\varepsilon}-\frac{ig^2\mu^\varepsilon}{32\pi^2}[\gamma+F(s,m,\mu)]=\frac{ig^2\mu^\varepsilon}{16\pi^2\varepsilon}+\text{finite}. \qquad (9.27)$$

We have now obtained in explicit form the lowest order corrections to the 2- and 4-point functions of ϕ^4 theory. We conclude by writing out the corresponding vertex (1PI) functions $\Gamma^{(2)}(p)$ and $\Gamma^{(4)}(p_i)$. In the notation of (7.72), equation (9.20) is Σ/i, so to order g

$$\Sigma=-\frac{gm^2}{16\pi^2\varepsilon}+\text{finite}$$

and $\Gamma^{(2)}$ is given by equation (7.79) (ignoring the finite part of Σ):

$$\Gamma^{(2)}(p)=p^2-m^2\left(1-\frac{g}{16\pi^2\varepsilon}\right). \qquad (9.28)$$

It is clearly infinite in the limit $\varepsilon\to 0$. The 4-point function $\Gamma^{(4)}(p_1,\ldots,p_4)$ is given by the analogue of equation (7.92), but in momentum space $\Gamma^{(4)}(p_1,\ldots,p_4)=G^{(2)}(p_1)^{-1}\ldots G^{(2)}(p_4)^{-1}G^{(4)}(p_1,\ldots,p_4)$ where the Green's function $G^{(4)}$ is the sum of the order g term (equation (6.111)), equation (9.27) above, and the 'crossed' terms, obtained by substituting the Mandelstam variables t and u, for s, where

$$s=(p_1+p_2)^2, \quad t=(p_1+p_3)^2, \quad u=(p_1+p_4)^2. \qquad (9.29)$$

The four functions $G^{(2)-1}$ have the effect of amputating the external legs in the usual way, so we have (the amputated legs, as in chapter 7, being denoted by dashed lines)

$$= -ig\mu^\varepsilon + \frac{3ig^2\mu^\varepsilon}{16\pi^2\varepsilon} - \frac{ig^2\mu^\varepsilon}{32\pi^2}$$

$$\times [3\gamma + F(s, m, \mu) + F(t, m, \mu) + F(u, m, \mu)]$$

$$= -ig\mu^\varepsilon\left(1 - \frac{3g}{16\pi^2\varepsilon}\right) + \text{finite}. \tag{9.30}$$

This, again, is infinite. To be physically sensible, the vertex functions $\Gamma^{(2)}$ and $\Gamma^{(4)}$ should be finite. We shall see in the next section how this renormalisation is achieved. Meanwhile, it is worth remarking that the corrections we have made are not to the same order in the coupling g. $\Gamma^{(2)}$ is to order g, and $\Gamma^{(4)}$ to order g^2. The parameter which is the same in our calculation of $\Gamma^{(2)}$ and $\Gamma^{(4)}$ is the *number of loops* – in this case one. There are reasons for believing that an expansion in the number of loops has more physical relevance than an expansion in powers of g. We shall conclude this section by reviewing this argument.

Loop expansion

What we wish to show is that an expansion in L is equivalent to an expansion in $\hbar$. To see this, recall that $\mathscr{L}$ has the dimensions of Planck's constant, so the formula (6.66) for the generating functional (of a scalar field theory) should properly be written

$$Z[J(x)] = \int \mathscr{D}\phi \exp\left\{\frac{i}{\hbar}\int[\mathscr{L} + \hbar J(x)\phi(x)]\,dx\right\}. \tag{9.31}$$

Splitting up $\mathscr{L}$ into 'free' and 'interacting' parts $\mathscr{L} = \mathscr{L}_0 + \mathscr{L}_{\text{int}}$ (though actually, as we shall see below, the distinction is somewhat artificial), we then obtain, in place of (6.76),

$$Z[J] = \exp\left\{\frac{i}{\hbar}\mathscr{L}_{\text{int}}\left[\frac{1}{i}\frac{\delta}{\delta J}\right]\right\}Z_0[J] \tag{9.32}$$

where

$$Z_0[J] = \int \mathscr{D}\phi \exp\left[\frac{i}{\hbar}\int dx(\mathscr{L}_0 + \hbar J\phi)\right]. \tag{9.33}$$

Finally, in place of (6.13), we have

$$Z_0[J] = N\exp\left[-\tfrac{1}{2}i\hbar\int dx\,dy\,J(x)\Delta_F(x - y)J(y)\right]. \tag{9.34}$$

It is now clear from (9.32) that each vertex contributes a factor $\hbar^{-1}$, and from (9.34) that each propagator contributes a factor $\hbar$, to a general graph in nth order perturbation theory. Hence such a graph contributes a factor $\hbar^{I-n} = \hbar^{L-1}$ using (9.4), and an expansion in the number of loops is an expansion in powers of $\hbar$ around the classical theory. Note that we have assumed that our general graph has no propagators associated with the external legs. What we are expanding therefore is the vertex function $\Gamma^{(n)}$ rather than the Green's function $G^{(n)}$. We shall calculate the loop expansion of Γ below, when we consider the case of spontaneous symmetry breaking.

9.3   Renormalisation of ϕ^4 theory

In this section I describe two approaches to renormalisation. The first is more intuitive, but the second, the method of counterterms, is the one normally used in high energy physics. The approaches, are, of course, completely equivalent, and to some extent illuminate each other. The purpose of the endeavour is to render physical quantities finite.

We begin by considering the vertex functions $\Gamma^{(2)}$ and $\Gamma^{(4)}$, equations (9.28) and (9.30) above. To the (1-loop) approximation we are considering, they should be finite, so we put

$$\Gamma^{(2)}(p) = p^2 - m_1^2 \tag{9.35}$$

where m_1 is a new parameter, defined by this equation. It is taken to be *finite*, and to represent the *physical* mass (at least in this approximation). The original mass m is taken to be *infinite*, and to have no direct physical significance. It is the mass the particle would have if no interactions were present – and since they always are present, it is an unobservable quantity. It is related to m_1 by

$$m^2 = m_1^2 + \frac{m^2 g}{16\pi^2 \varepsilon}$$

$$= m_1^2 \left(1 + \frac{g}{16\pi^2 \varepsilon} \right) \tag{9.36}$$

where we have substituted $m \to m_1$ in the loop correction. This is valid since the error involved is of order g^2. The physical mass, m_1, is also called the *renormalised mass* and is given by

$$m_1^2 = - \Gamma^{(2)}(0). \tag{9.37}$$

We now apply a similar treatment to $\Gamma^{(4)}$. First let us write (9.30) as

$$i\Gamma^{(4)}(p_i) = g\mu^\varepsilon + \frac{g^2 \mu^\varepsilon}{32\pi^2} \left[\frac{6}{\varepsilon} - 3\gamma - F(s, m, \mu) - F(t, m, \mu) - F(u, m, \mu) \right]. \tag{9.38}$$

We now *define* a new parameter g_1, the *renormalised coupling constant*, by

$$g_1 = g\mu^\varepsilon + \frac{g^2 \mu^\varepsilon}{32\pi^2} \left[\frac{6}{\varepsilon} - 3\gamma - 3F(0, m, \mu) \right]. \tag{9.39}$$

Rearranging this expression by substituting g_1 for g and m_1 for m where appropriate (the error involved being, as before, a 2-loop term) gives

$$g = g_1 \mu^{-\varepsilon} - \frac{3g_1^2 \mu^{-2\varepsilon}}{32\pi^2}\left[\frac{2}{\varepsilon} - \gamma - F(0, m_1, \mu)\right]. \tag{9.40}$$

Here g_1 is regarded as *finite*, and g, the original coupling constant, as infinite. Now g_1 would be the value of the measured coupling if ϕ^4 theory were realistic (and the 1-loop approximation was a good enough one to compare with experiment; otherwise a higher approximation would have to be taken) and if measurements were taken at the (unphysical) point $s = t = u = 0$. The variable g is the 'bare' coupling, which would be the measured one only if higher order interactions did not contribute, which they inevitably do. So g has no direct physical significance. Expression $\Gamma^{(4)}$ in terms of g_1 then gives

$$i\Gamma^{(4)}(p_i) = g_1 - \frac{g_1^2 \mu^{-\varepsilon}}{32\pi^2}$$

$$\times [F(s, m_1, \mu) + F(t, m_1, \mu) + F(u, m_1, \mu) - 3F(0, m, \mu)]. \tag{9.41}$$

It follows immediately that

$$i\Gamma^{(4)}(p_i = 0) = g_1 \tag{9.42}$$

since, when $p_1 = p_2 = p_3 = p_4 = 0$, $s = t = u = 0$. So the coupling constant g_1, by virtue of its definition (9.39), is the value of $i\Gamma^{(4)}$ when all external momenta vanish. It should be clear, however, that if our objective is only to make the theory finite, there are other ways of defining the renormalised mass and coupling constant. For example, a common alternative to (9.42) is to define g_1 by the value of $\Gamma^{(4)}$ at the symmetrical point $p_i^2 = m^2$, $p_i p_j = -m^2/3$ ($i \neq j$), so that $s = t = u = 4m^2/3$. Note also that $\Gamma^{(4)}(p_i)$ depends on μ, the arbitrary mass it was necessary to introduce when we embarked on dimensional regularisation. It is obvious that physical quantities must also be independent of μ; this is what underlies the renormalisation group, to be studied below.

We have now renormalised ϕ^4 theory to one loop. What happens in the 2-loop approximation? The relevant diagrams are shown in Fig. 9.2. By analysing the integrals involved, it is found that (see, for example, Amit[24]), for $\Gamma^{(2)}$, the addition of the 2-loop diagram changes the renormalised mass

Fig. 9.2. Two loop terms in $\Gamma^{(2)}$ and $\Gamma^{(4)}$.

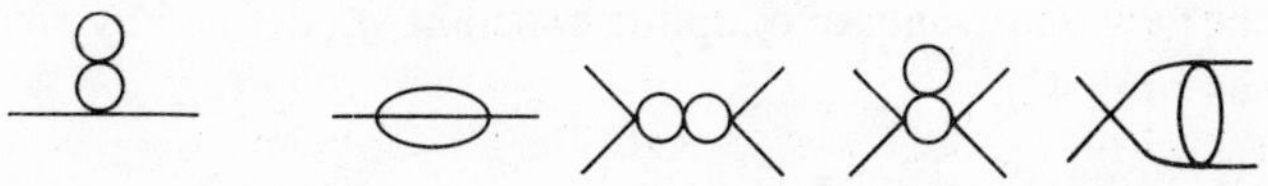

m_1 (this is hardly surprising), but $\Gamma^{(2)}$ contains an additional divergence, stemming from the second diagram in Fig. 9.2. The natural question then is, will this divergence disappear on coupling constant renormalisation? Turning to $\Gamma^{(4)}$ it is seen that the 2-loop graphs do indeed give extra divergent contributions. Some of these disappear when the mass renormalisation is taken into account, and the rest can be made finite by redefining the renormalised coupling constant g_1. So $\Gamma^{(4)}$ is *finite* to two loops, with mass and coupling constant renormalisation. $\Gamma^{(2)}$, however, remains *divergent* – the coupling constant renormalisation does *not* remove the divergence in the second diagram of Fig. 9.2. It is removed by absorption in a multiplicative factor, and we define a renormalised 2-point function $\Gamma_r^{(2)}$ by

$$\Gamma_r^{(2)} = Z_\phi(g_1, m_1, \mu)\Gamma^{(2)}(p, m_1, \mu). \tag{9.43}$$

$\Gamma_r^{(2)}$ is now finite, and Z_ϕ infinite; $Z_\phi^{\frac{1}{2}}$ is called the *wave function* (or *field*) *renormalisation constant*. Z_ϕ may be expanded in the number of loops, giving

$$\begin{aligned} Z_\phi &= 1 + g_1 Z_1 + g_1^2 Z_2 + \cdots \\ &= 1 + g_1^2 Z_2 + \cdots, \end{aligned} \tag{9.44}$$

since there is no 1-loop contribution. Equation (9.43) amounts to renormalisation of the field amplitude, but the value of this cannot be entirely arbitrary. Analogously to (9.37) and (9.42), we demand that at some point – say $p^2 = 0$ – the field amplitude is unity, so that

$$\left. \frac{\partial}{\partial p^2} \Gamma_r^{(2)} \right|_{p^2 = 0} = 1. \tag{9.45}$$

Clearly, the choice of the point $p^2 = 0$ is arbitrary. The fact that $\Gamma^{(2)}(p, m_1, \mu)$ was divergent means, of course, that, in the 2-loop approximation, m_1 is infinite (in the limit $\varepsilon \to 0$). The renormalised vertex function $\Gamma_r^{(2)}$, however, now gives a *finite* mass m_r (r for 'renormalised')

$$m_r^2 = Z_\phi m_1^2;$$

that is to say, the divergences between Z_ϕ and m_1^2 cancel. And, just as the introduction of Z_ϕ has changed the renormalised mass from m_1 to m_r, it also changes the renormalised coupling constant. It is clear that the analogous equation to (9.43) is

$$\Gamma_r^{(4)} = Z_\phi^2 \Gamma^{(4)}(p, m_1, \mu), \tag{9.46}$$

so that the new renormalised coupling constant, g_r, defined by an equation analogous to (9.42), is

$$\begin{aligned} i\Gamma_r^{(4)}(p_i = 0) &= g_r \mu^\varepsilon = Z_\phi^2 g_1 \mu^\varepsilon, \\ g_r &= Z_\phi^2 g_1. \end{aligned} \tag{9.47}$$

Z_ϕ is a function of $g_1 = g\mu^{-\varepsilon}$, so, writing in this dependence explicitly, the renormalised n-particle vertex function is

$$\Gamma_r^{(n)}(p_i, g_r, m_r, \mu) = Z_\phi^{n/2}(g\mu^{-\varepsilon})\Gamma^{(n)}(p_i, g, m) \tag{9.48}$$

or

$$\Gamma^{(n)}(p_i, g, m) = Z_\phi^{-n/2}(g\mu)^{-\varepsilon}\Gamma_r^{(n)}(p_i, g_r, m_r, \mu). \tag{9.49}$$

We now have the result that, to two loops, the vertex functions (or, equivalently, Green's functions), mass and coupling constant in ϕ^4 theory may be made finite by suitable renormalisation. The question is, does this desirable property hold to all orders? This is the question of renormalisability, but it is difficult to answer. We refer the reader to the literature. It may be worth mentioning that the greatest trouble in proving renormalisability is caused by the phenomenon of 'overlapping divergences'. These are present in the second graph of Fig. 9.2, which (putting $m = 0$ for convenience) is clearly given by

$$i\int \frac{d^4k_1\, dk_2}{k_1^2 k_2^2 (p - k_1 - k_2)^2}.$$

There are two loops, so two integration variables. The overall degree of divergence is $8 - 6 = 2$. If, however, k_1 (say) is held constant and we integrate over k_2, the degree of divergence is $4 - 4 = 0$, so the graph still gives a logarithmic divergence. This is the phenomenon of overlapping divergences – it is impossible to separate out the divergencies in k_1 and k_2. It turns out that overlapping divergences give no trouble in QED because of gauge invariance, as we shall see below. In this respect ϕ^4 theory is more difficult than QED.

Let us briefly survey what we have done. Starting from a Lagrangian (6.65), we find that the Green's functions which contain one or more internal loops are divergent. They can, however, be made finite, by a renormalisation of the mass, coupling constant and field amplitude, from their 'bare' to their 'physical' values. This view of renormalisation has the merit of a clear physical interpretation, and has analogies in solid state physics, where, for instance, electrons are ascribed an 'effective mass' to take into account their interaction with the lattice. This corresponds exactly with the above view of mass renormalisation.

An alternative view of renormalisation may be taken, however, and is, in fact, the one more commonly adopted in high energy physics. This is to regard the parameters m and g in the original Lagrangian as the *physical* mass and coupling constant. The fact that this Lagrangian does not give finite Green's functions then requires that *extra terms* be added to the Lagrangian, to cancel the divergences. These terms are called *counter-terms*. They may be constructed at each order in perturbation theory. We now outline how this is done.

Counter-terms

Consider mass renormalisation at the 1-loop level, given by equations (9.28), (9.35) and (9.36). We may describe this as follows. The 1-loop modification to the free propagator is

$$\underline{\quad O \quad} = \frac{igm^2}{16\pi^2\varepsilon} + \text{finite},$$

which diverges as $\varepsilon \to 0$. We therefore add to $\mathscr{L}$ a term

$$\delta\mathscr{L}_1 = -\frac{gm^2}{32\pi^2\varepsilon}\phi^2 \equiv -\frac{\delta m^2}{2}\phi^2. \tag{9.50}$$

This is treated as an *interaction*, and gives rise to the additional Feynman rule

$$\underline{\quad \times \quad} = -\frac{igm^2}{16\pi^2\varepsilon} = -i\delta m^2. \tag{9.51}$$

Then, to order g, the complete inverse propagator is (cf. equations (7.77) and (7.78))

$$\underline{\quad \oslash \quad}^{-1} = \left[\underline{\quad\quad} + \underline{\quad O \quad} + \underline{\quad \times \quad}\right]^{-1} \tag{9.52}$$

or

$$\Gamma^{(2)}(p) = i[G^{(2)}p]^{-1}$$
$$= i\left[\frac{p^2 - m^2}{i} - \left(\frac{igm^2}{16\pi^2}\frac{1}{\varepsilon} + \text{finite}\right) + \frac{igm^2}{16\pi^2\varepsilon}\right]$$
$$= p^2 - m^2, \tag{9.53}$$

ignoring, as before, the finite term (or including it in m^2). Here m^2 is taken to be a *finite* quantity, the physical particle mass, and is equal, in an appropriate order of perturbation theory, to $-\Gamma^{(2)}(0)$. The Lagrangian is now $\mathscr{L} + \delta\mathscr{L}_1, \delta\mathscr{L}_1$ being a counter-term, which is divergent. It may appear strange to introduce what is clearly a 'mass' term (9.50) into the Lagrangian and call it an 'interaction', but, in fact, there is no contradiction. To see this, consider for example the free theory given by

$$\mathscr{L} = \tfrac{1}{2}(\partial_\mu\phi)(\partial^\mu\phi) - \tfrac{1}{2}m^2\phi^2,$$

and regard it as describing a *massless* field ϕ, given by the first term in $\mathscr{L}$, with an *interaction* given by the second term. The Feynman rules are then

$$\underline{\quad\quad} \quad \frac{i}{p^2},$$

$$\underline{\quad \times \quad} \quad -im^2,$$

and the complete propagator is

$$= \frac{i}{p^2} + \frac{i}{p^2}(-im^2)\frac{i}{p^2} + \frac{i}{p^2}(-im^2)\frac{i}{p^2}(-im^2)\frac{i}{p^2} + \ldots$$

$$= \frac{i}{p^2 - m^2}$$

which is the usual propagator for a *massive* field.

A similar treatment may be made of $\Gamma^{(4)}$. As seen in (9.30), $\Gamma^{(4)}$, calculated to order g^2 from the Lagrangian $\mathscr{L}$, diverges as $\varepsilon \to 0$:

$$= \quad + \quad + \text{(2 crossed)}$$

$$= -ig\mu^\varepsilon\left(1 - \frac{3g}{16\pi^2\varepsilon}\right) + \text{finite.}$$

So by adding a counter-term

$$\delta\mathscr{L}_2 = -\frac{1}{4!}\frac{3g^2\mu^\varepsilon}{16\pi^2\varepsilon}\phi^4 = -\frac{Bg\mu^\varepsilon}{4!}\phi^4$$

to the Lagrangian, we get an additional interaction with Feynman rule

$$-\frac{3ig^2\mu^\varepsilon}{16\pi^4\varepsilon}$$

which will render $\Gamma^{(4)}$ finite

$$= \quad + \quad + \text{(2 crossed)} +$$

$$= -ig\mu^\varepsilon + \text{finite.} \tag{9.54}$$

Finally, the divergence of $\Gamma^{(2)}$ at the 2-loop level, described (but not proved) above, necessited multiplying $\Gamma^{(n)}$ by a factor $Z_\phi^{n/2}$. This is clearly equivalent to adding a counter-term

$$\delta\mathscr{L}_3 = -\frac{A}{2}(\partial_\mu\phi)^2$$

to the Lagrangian, with $1 + A = Z_\phi$. Finite Green's functions may then be obtained by adding to the original Lagrangian $\mathscr{L}$ counter-terms $\mathscr{L}_{CT}$:

$$\mathscr{L} = \tfrac{1}{2}(\partial_\mu\phi)^2 - \frac{m^2}{2}\phi^2 - \frac{g\mu^\varepsilon}{4!}\phi^4, \tag{9.55}$$

$$\mathscr{L}_{CT} = \frac{A}{2}(\partial_\mu\phi)^2 - \frac{\delta m^2}{2}\phi^2 - \frac{Bg\mu^\varepsilon}{4!}\phi^4. \tag{9.56}$$

The total Lagrangian, called the 'bare Lagrangian' $\mathscr{L}_B$, is

$$\mathscr{L}_B = \mathscr{L} + \mathscr{L}_{CT} = \left(\frac{1+A}{2}\right)(\partial_\mu\phi)^2 - \frac{(m^2+\delta m^2)}{2}\phi^2$$

$$- (1+B)\frac{g\mu^\varepsilon}{4!}\phi^4 \tag{9.57}$$

so we see that the effect of the counter-terms is clearly equivalent to multiplying ϕ, m and g by 'renormalisation factors' Z: that is, defining the bare quantities

$$\phi_B = \sqrt{Z_\phi}\,\phi, \quad Z_\phi = 1 + A,$$

$$m_B = Z_m m, \quad Z_m^2 = \frac{m^2+\delta m^2}{1+A},$$

$$g_B = \mu^\varepsilon Z_g g, \quad Z_g = \frac{1+B}{(1+A)^2}, \tag{9.58}$$

the bare Lagrangian is

$$\mathscr{L}_B = \tfrac{1}{2}(\partial_\mu\phi_B)^2 - \frac{m_B^2}{2}\phi_B^2 - \frac{g_B}{4!}\phi_B^4. \tag{9.59}$$

In the language of counter-terms, a theory is renormalisable if the counter-terms, required to cancel the divergences at each order in perturbation theory, are of the *same form* as those appearing in the original Lagrangian. If this is the case, 'bare' quantities may be defined by (infinite) multiplicative renormalisation factors, as above, and the bare Lagrangian is of the same form as the original Lagrangian. The bare Lagrangian, moreover, is the 'true' Lagrangian of the theory, in that it yields finite physical quantities to any desired order.

9.4   Renormalisation group

We have seen how to renormalise ϕ^4 theory. In the technique of dimensional regularisation, we saw that it was necessary to introduce a new parameter μ, with the dimensions of mass. The renormalised 1PI function $\Gamma_r^{(n)}$ depends on μ, as shown in equation (9.48), through the dependence of Z_ϕ on μ. In other words, the unrenormalised function $\Gamma^{(n)}$ given by (9.49) is independent of μ, so is *invariant* under the *group* of transformations

$$\mu \to e^s \mu. \tag{9.60}$$

These transformations form the *renormalisation group*. Introducing the dimensionless differential operator $\mu(\partial/\partial\mu)$ we have

$$\mu \frac{\partial}{\partial\mu} \Gamma^{(n)} = 0 \tag{9.61}$$

or, from (9.49),

$$\mu \frac{d}{d\mu} [Z_\phi^{-n/2}(g\mu^{-\varepsilon})\Gamma_r^{(n)}(p_i, g_r, m_r, \mu)] = 0 \tag{9.62}$$

where g_r and m_r depend on μ. Performing the differentiation in (9.62) and multiplying by $Z_\phi^{n/2}$ gives

$$\left[-n\mu \frac{\partial}{\partial\mu} \ln \sqrt{Z_\phi} + \mu \frac{\partial}{\partial\mu} + \mu \frac{\partial g_r}{\partial\mu} \frac{\partial}{\partial g_r} + \mu \frac{\partial m_r}{\partial\mu} \frac{\partial}{\partial m_r} \right] \Gamma_r^{(n)} = 0. \tag{9.63}$$

For simplicity, let us from now on write g for g_r, m for m_r and $\Gamma^{(n)}$ for $\Gamma_r^{(n)}$, so that, despite appearances, we are always dealing with renormalised quantities. Then, defining the quantities

$$\left. \begin{aligned} \gamma(g) &= \mu \frac{\partial}{\partial\mu} \ln \sqrt{Z_\phi}, \\[2mm] \beta(g) &= \mu \frac{\partial g}{\partial\mu}, \\[2mm] m\gamma_m(g) &= \mu \frac{\partial m}{\partial\mu}, \end{aligned} \right\} \tag{9.64}$$

equation (9.63) becomes

$$\left[\mu \frac{\partial}{\partial\mu} + \beta(g) \frac{\partial}{\partial g} - n\gamma(g) + m\gamma_m(g) \frac{\partial}{\partial m} \right] \Gamma^{(n)} = 0. \tag{9.65}$$

This is the *renormalisation group equation* (RG equation). It expresses the invariance of the renormalised $\Gamma^{(n)}$ under a change of regularisation parameter μ.

Let us now write down a similar equation expressing the invariance of $\Gamma^{(n)}$ under a change of *scale*. Let $p \to tp$, $m \to tm$, $\mu \to tm$. $\Gamma^{(n)}$ has a mass dimension D given by (see Table 9.1 – where the 1 PI function is denoted $\bar{\Gamma}^{(n)}$)

$$D = d + n\left(1 - \frac{d}{2}\right) = 4 - n + \varepsilon\left(\frac{n}{2} - 1\right) \tag{9.66}$$

where $d = 4 - \varepsilon$. Then

$$\Gamma^{(n)}(tp_i, g, m, \mu) = t^D \Gamma^{(n)}(p_i, g, t^{-1}m, t^{-1}\mu)$$

$$= \mu^D F\left(g, \frac{t^2 p_i^2}{m\mu}\right), \tag{9.67}$$

so

$$\left(t\frac{\partial}{\partial t} + m\frac{\partial}{\partial m} + \mu\frac{\partial}{\partial \mu} - D\right)\Gamma^{(n)} = 0. \tag{9.68}$$

Now eliminate $\mu(\partial\Gamma/\partial\mu)$ between (9.65) and (9.68):

$$\left[-t\frac{\partial}{\partial t} + \beta\frac{\partial}{\partial g} - n\gamma(g) + m(\gamma_m(g) - 1)\frac{\partial}{\partial m} + D\right]$$

$$\times \Gamma^{(n)}(tp, g, m, \mu) = 0. \tag{9.69}$$

This equation expresses directly the effect on $\Gamma^{(n)}$ of scaling up momenta by a factor t. Note that if $\beta = \gamma(g) = \gamma_m(g) = 0$, the effect is given simply by the canonical dimension D, as would be expected from a naive scaling argument. It is the effects of *interaction* which give rise to the need for renormalisation and therefore to non-vanishing functions β, $\gamma(g)$ and $\gamma_m(g)$, and thence to a departure from pure scaling behaviour of the Green's functions. Note particularly that if we start with a *massless* theory, the Lagrangian is *scale invariant*, but the Green's functions are *not*, because β and $\gamma(g)$ are non-vanishing. They contribute to so-called 'anomalous' dimensions. The origin of anomalous dimensions is therefore profound: renormalisation inevitably introduces a *scale*, in the shape of a mass μ in dimensional regularisation, or in the shape of a momentum cut-off Λ in a cut-off regularisation, so even a scale invariant classical theory does not give rise to a scale invariant quantum theory.

We now wish to find a solution of (9.69). The equation expresses the fact that a change in t may be compensated by a change in m and g, and an overall factor. So we expect that there should be functions $g(t)$, $m(t)$ and $f(t)$ such that

$$\Gamma^{(n)}(tp, m, g, \mu) = f(t)\Gamma^{(n)}(p, m(t), g(t), \mu). \tag{9.70}$$

Differentiating this with respect to t:

$$\frac{\partial}{\partial t}\Gamma^{(n)}(tp, m, g, \mu)$$

$$= \frac{df}{dt}\Gamma^{(n)}(p, m(t), g(t), \mu) + f(t)\left(\frac{\partial m}{\partial t}\frac{\partial\Gamma^{(n)}}{\partial t} + \frac{\partial g}{\partial t}\frac{\partial\Gamma^{(n)}}{\partial g}\right),$$

or

$$t\frac{\partial}{\partial t}\Gamma^{(n)}(tp,m,g,\mu)$$

$$= \left(t\frac{\mathrm{d}f}{\mathrm{d}t} + f(t)t\frac{\partial m}{\partial t}\frac{\partial}{\partial m} + f(t)t\frac{\partial g}{\partial t}\frac{\partial}{\partial g} \right)\Gamma^{(n)}(p,m(t),g(t),\mu)$$

$$= \left(t\frac{\mathrm{d}f}{\mathrm{d}t} + tf(t)\frac{\partial m}{\partial t}\frac{\partial}{\partial m} + tf(t)\frac{\partial g}{\partial t}\frac{\partial}{\partial g} \right)\frac{1}{f(t)}\Gamma^{(n)}(tp,m,g,\mu),$$

hence

$$\left(-t\frac{\partial}{\partial t} + \frac{t}{f}\frac{\mathrm{d}f}{\mathrm{d}t} + t\frac{\partial m}{\partial t}\frac{\partial}{\partial m} + t\frac{\partial g}{\partial t}\frac{\partial}{\partial g} \right)\Gamma^{(n)}(tp,m,g,\mu) = 0. \qquad (9.71)$$

Now compare (9.71) with (9.69). The coefficients of $\partial/\partial g$ give

$$t\frac{\partial g(t)}{\partial t} = \beta(g); \qquad (9.72)$$

$g(t)$ is called a *running coupling constant*. Knowledge of the function $\beta(g)$ enables us to find $g(t)$; and of particular interest is the asymptotic limit of $g(t)$ as $t \to \infty$. Along with (9.72) we have $g(1) = g$. Comparing coefficients of $\partial/\partial m$ in the two equations gives

$$t\frac{\partial m}{\partial t} = m[\gamma_m(g) - 1] \qquad (9.73)$$

and the remaining terms yield

$$\frac{t}{f}\frac{\mathrm{d}f}{\mathrm{d}t} = D - n\gamma(g).$$

This equation can be integrated to give

$$f(t) = t^D \exp\left[-\int_0^t \frac{n\gamma(g(t))\mathrm{d}t}{t} \right],$$

which, on substitution into (9.70) gives (using (9.66) and taking the limit $\varepsilon \to 0$)

$$\Gamma^{(n)}(tp,m,g,\mu) = t^{4-n}\exp\left[-n\int_0^t \frac{\gamma(g(t))\mathrm{d}t}{t} \right]\Gamma^{(n)}(p,m(t),g(t),\mu). \qquad (9.74)$$

This is the solution to the RG equation (9.69), in terms of the running coupling constant $g(t)$ and running mass $m(t)$. The exponential term is the 'anomalous dimension'. The physics at large momentum is governed by $m(t)$

and $g(t)$, and a particular use of the renormalisation group is to study the large (or even the small) momentum behaviour of quantum field theory. An equation like (9.74) allows us to speculate outside the domain of perturbation theory. Let us examine some possible behaviours of $g(t)$ as $t \to \infty$, i.e. at large momentum, and assume that (9.72) is still valid there:

$$t\frac{\partial g(t)}{\partial t} = \beta(g(t)). \tag{9.72}$$

First, suppose that $\beta(g)$ has the form shown in Fig. 9.3. The zeros of β, at $g = 0$ and $g = g_0$, are called *fixed points*. Then we may see that as $t \to \infty$, a value of g near to g_0 tends towards g_0. For if $g < g_0, \beta > 0$, so g increases with increasing t, and is driven towards g_0. If, on the other hand, $g > g_0$, then $\beta < 0$, so g decreases with increasing t, so is driven *back* towards g_0. Hence $g(\infty) = g_0$; g_0 is called an *ultra-violet stable* fixed point. By an analogous argument, if g is small, then as $t \to 0, g \to 0$, and $g(0) = 0$ is called an *infra-red stable* fixed point.

Second, suppose that $\beta(g)$ has the form shown in Fig. 9.4. Again there are two fixed points, but the sign of β is reversed, so $g = g_0$ is an infra-red stable fixed point, and $g = 0$ an ultra-violet stable fixed point. For this latter behaviour, perturbation theory gets better and better at higher energies, and in the infinite momentum limit, the coupling constant vanishes. This is known as *asymptotic freedom*.

As an example, let us investigate the asymptotic behaviour of ϕ^4 theory, assuming that the 1-loop expression (9.39) for the renormalised coupling

Fig. 9.3. A possible form of the β function; g_0 is an ultra-violet stable fixed point, and $g = 0$ an infra-red stable fixed point.

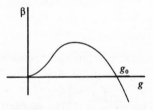

Fig. 9.4. Another form of the β function; g_0 is infra-red stable, and $g = 0$ is ultra-violet stable, giving asymptotic freedom.

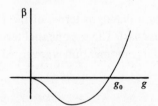

constant is a reliable pointer to the asymptotic regime. Ignoring the finite corrections, we then have

$$g_1 = g\mu^\varepsilon\left(1 + \frac{3g}{16\pi^2\varepsilon}\right) \tag{9.75}$$

so that

$$\mu\frac{\partial g_1}{\partial \mu} = \varepsilon g\mu^\varepsilon + \frac{3g^2}{16\pi^2}\mu^\varepsilon.$$

The β function is given by (9.64), where g in that equation refers to the *renormalised* coupling constant, and the limit $\varepsilon \to 0$ has been taken; so we obtain

$$\beta(g) = \lim_{\varepsilon \to 0}\mu\frac{\partial g_1}{\partial \mu} = \frac{3g^2}{16\pi^2} > 0. \tag{9.76}$$

From (9.72), or perhaps more directly by noting that if $s = \ln t$, so that $t\partial/\partial t = \partial/\partial s$, equation (9.72) becomes

$$\frac{\partial}{\partial s}g(s) = \beta(g(s)). \tag{9.77}$$

We see that the effective (running) coupling constant increases with s, i.e. with increasing momentum, so ϕ^4 theory is *not* asymptotically free. In fact, it is easily seen that the solution to (9.76) is

$$g = \frac{g_0}{1 - ag_0\ln(\mu/\mu_0)}$$

where $a = 3/16\pi^2$, so g increases with increasing μ.

This chapter has so far been entirely devoted to renormalisation of ϕ^4 theory, in the hope that the ideas and techniques necessary for renormalisation may be learned on an example unencumbered by other complications which the real world has to offer. Nevertheless, we have by no means exhausted the subject of renormalisation and the renormalisation group. In particular, the important topics of mass-dependent and -independent renormalisation prescriptions, and the Callan–Symanzik equation, similar in form to the renormalisation group equation, have not been touched on. Because of the introductory nature of this book, I make no apology for this, and press on towards the real world first of all in the shape of quantum electrodynamics.

9.5   Divergences and dimensional regularisation of QED

The only particles in QED are photons and electrons. Divergences occur in several types of Feynman diagram, for example the electron and photon self-energy diagrams of Figs. 9.5 and 9.6. We shall treat these diagrams in the same systematic way in which we analysed ϕ^4 theory, and shall show that there is only a finite number of primitively divergent diagrams, and hence that QED is, in principle, renormalisable. We shall then regularise the Feynman integrals, using dimensional regularisation again (Pauli–Villars regularisation does not preserve gauge invariance in non-Abelian gauge theories, so it is simplest to use dimensional regularisation throughout). In the following sections we shall see how to renormalise QED explicitly to order e^2 (one loop), and how the Ward identity guarantees that QED is renormalisable to all orders. We begin, then, by analysing the divergences of Feynman integrals.

The general formula for the superficial degree of divergence D of a Feynman graph in d-dimensional space–time is analogous to equation (9.3):

$$D = dL - 2P_i - E_i \qquad (9.78)$$

where

$$\left.\begin{array}{l} L = \text{number of loops}, \\ P_i = \text{number of internal photon lines}, \\ E_i = \text{number of internal electron lines}, \\ d = \text{dimension of space–time}. \end{array}\right\} \qquad (9.79)$$

In addition, let

$$\left.\begin{array}{l} n = \text{number of vertices}, \\ P_e = \text{number of external photon lines}, \\ E_e = \text{number of external electron lines}. \end{array}\right\} \qquad (9.80)$$

As before, L = number of independent momenta for integration = number of internal lines $- n$ (because of momentum conservation at each vertex) $+ 1$

Fig. 9.5. Electron self-energy diagram.

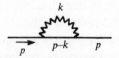

Fig. 9.6. Photon self-energy diagram.

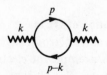

(because overall momentum conservation holds in any case):

$$L = E_i + P_i - n + 1. \tag{9.81}$$

Now each vertex gives two electron legs. If they are external, they are counted once, and if internal, twice, so

$$2n = E_e + 2E_i. \tag{9.82}$$

The analogous relation for photons is clearly

$$n = P_e + 2P_i. \tag{9.83}$$

Equations (9.78) and (9.81) give

$$D = (d - 1)E_i + (d - 2)P_i - d(n - 1)$$

which, on substituting for E_i and P_i from (9.82) and (9.83), gives

$$D = d + n\left(\frac{d}{2} - 2\right) - \left(\frac{d-1}{2}\right)E_e - \left(\frac{d-2}{2}\right)P_e. \tag{9.84}$$

When $d = 4$ this yields

$$D = 4 - \frac{3E_e}{2} - P_e, \tag{9.85}$$

showing that D is independent of n, the *sine qua non* for renormalisability.

Let us check (9.85) for the two self-energy diagrams. The electron self-energy diagram, Fig. 9.5, has $E_e = 2$, $P_e = 0$, so $D = 1$. The Feynman rules give

$$-i\Sigma(p) = (-ie)^2 \int \frac{d^4k}{(2\pi)^4} \gamma^\mu \frac{i}{p\!\!\!/ - k\!\!\!/ - m} \frac{-ig_{\mu\nu}}{k^2}\gamma^\nu, \tag{9.86}$$

which has four powers of k in the numerator and three in the denominator, so $D = 1$, as predicted. The photon self-energy diagram, Fig. 9.6, has $E_e = 0$, $P_e = 2$, so $D = 2$. The photon self-energy is denoted $\Pi_{\mu\nu}$, and is also called *vacuum polarisation*. Unlike electron self-energy, it has no classical counterpart. The Feynman rules give

$$i\Pi^{\mu\nu}(k) = -(-ie)^2 \int \frac{d^4p}{(2\pi)^4} \mathrm{Tr}\left(\gamma^\mu \frac{i}{p\!\!\!/ - m} \gamma^\nu \frac{i}{p\!\!\!/ - k\!\!\!/ - m}\right). \tag{9.87}$$

The overall minus sign on the right-hand side comes from the closed fermion loop. It is clear that this integral is quadratically divergent, as anticipated. Note that this graph gives a modified photon propagator, so that, in the Feynman gauge, to one loop

$$iD'_{\mu\nu}(k) = -i\frac{g_{\mu\nu}}{k^2} + \left(\frac{-ig_{\mu\alpha}}{k^2}\right)i\Pi^{\alpha\beta}(k)\left(\frac{-ig_{\beta\nu}}{k^2}\right)$$

$$= \text{〰〰〰} + \text{〰〰◯〰〰}. \tag{9.88}$$

Actually, although the electron and photon self-energy graphs are superficially linearly and quadratically divergent (respectively), they both turn out to be logarithmically divergent only, as will be seen below.

These self-energy graphs are *primitively divergent*, and the question we must answer is what other primitively divergent graphs there are in QED. There are in fact three more. The first one of them is the vertex graph shown in Fig. 9.7. It has $E_e = 2$, $P_e = 1$, so (9.85) indicates $D = 0$, a logarithmic divergence. The Feynman rules give (cf. equation (7.127))

$$-ie\Lambda_\mu(p, q, p + q) = (-ie)^3 \int \frac{d^4k}{(2\pi)^4} \frac{-ig_{\rho\sigma}}{(k + p)^2} \gamma^\rho \frac{i}{\not{k} - \not{q} - m} \gamma_\mu \frac{i}{\not{k} - m} \gamma^\sigma,$$

$$(9.89)$$

which is indeed logarithmically divergent. This vertex graph, and the two self-energy graphs above, all have the property that the removal of their infinities results in a redefinition of various physical quantities, viz. electron mass and wave function normalisation, and electric charge. In other words, no extra terms in the Lagrangian are required of a type which are not there already. But the two remaining primitively divergent graphs could pose a serious threat to the renormalisability of QED. The first of these is the 3-photon coupling shown in Fig. 9.8. It has $E_e = 0$, $P_e = 3$, so $D = 1$. It turns out, however, that this graph is cancelled by a similar graph with the electron arrows reversed, and so may be ignored.[‡] This is the content of

Fig. 9.7. Vertex graph.

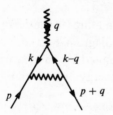

Fig. 9.8. A 3-photon vertex.

[‡] Arrows in a Feynman diagram correspond to the direction of charge flow.

Furry's theorem, which actually follows from the C invariance of the Lagrangian. $\mathscr{L}$ is invariant under charge conjugation:

$$\psi \to \psi^c = C\bar{\psi}^T, \quad A_\mu \to -A_\mu,$$

and it then follows immediately that Green's functions, with an odd number of external photon lines, are zero. (In an analogous way, in ϕ^4 theory, $\mathscr{L}$ is invariant under $\phi \to -\phi$, so Green's functions with an odd number of ϕ legs are zero.) The second diagram which could cause trouble for renormalisability is Fig. 9.9, the scattering of light by light. Here $E_e = 0$, $P_e = 4$, so $D = 0$ and the graph is superficially logarithmically divergent. It gives a non-zero cross section for photon–photon scattering, a process for which there is no classical analogue. It turns out, however, that due to gauge invariance, this graph is actually convergent, so causes no trouble.

We have isolated the three primitive divergences of QED, so our task now is to calculate them using dimensional regularisation. We must first generalise the Lagrangian (7.100) (with $\alpha = 1$ and neglecting source terms, which are unnecessary here),

$$\mathscr{L} = i\bar{\psi}\gamma^\mu\partial_\mu\psi - m\bar{\psi}\psi - eA^\mu\bar{\psi}\gamma_\mu\psi$$
$$- \tfrac{1}{4}(\partial_\mu A_\nu - \partial_\nu A_\mu)^2 - \tfrac{1}{2}(\partial_\mu A^\mu)^2,$$

to d dimensions. Recall from above that the mass dimension of $\mathscr{L}$ is d, $[\mathscr{L}] = d$, so if

$$[\psi] = \frac{d-1}{2}, \quad [A_\mu] = \frac{d}{2} - 1,$$

then all the terms in $\mathscr{L}$ above, except the third one, have the correct dimension. To get the third one right, we must multiply e by $\mu^{2-d/2}$, where μ is an arbitrary mass. Hence we have

$$\mathscr{L} = i\bar{\psi}\gamma^\mu\partial_\mu\psi - m\bar{\psi}\psi - e\mu^{2-d/2}A^\mu\bar{\psi}\gamma_\mu\psi$$
$$- \tfrac{1}{4}(\partial_\mu A_\nu - \partial_\nu A_\mu)^2 - \tfrac{1}{2}(\partial_\mu A^\mu)^2. \tag{9.90}$$

Recall that the extension to d dimensions is to be made only for internal loops, so external wave functions like $u(p)$, for example, may be left in their 4-dimensional form. All we have to consider, then, is the definition and

Fig. 9.9. The scattering of light by light.

algebra of Dirac matrices in d dimensions. We have

$$\{\gamma_\mu, \gamma_\nu\} = 2g_{\mu\nu} \tag{9.91}$$

where $g_{\mu\nu}$ is the metric tensor in d-dimensional Minkowski space (with signature $+ - - - \cdots$), so that $\delta_\mu^\mu = d$, hence

$$\gamma^\mu \gamma_\mu = d, \quad \gamma_\mu \gamma_\nu \gamma^\mu = (2 - d)\gamma_\nu. \tag{9.92}$$

In addition,

$$\text{Tr(odd no. of } \gamma \text{ matrices)} = 0, \tag{9.93}$$

$$\text{Tr } I = f(d), \quad \text{Tr } \gamma_\mu \gamma_\nu = f(d)g_{\mu\nu}, \tag{9.94}$$

$$\text{Tr } \gamma_\mu \gamma_\kappa \gamma_\nu \gamma_\lambda = f(d)(g_{\mu\kappa}g_{\nu\lambda} - g_{\mu\nu}g_{\kappa\lambda} + g_{\mu\lambda}g_{\kappa\nu}), \tag{9.95}$$

where $f(d)$ is an arbitrary well-behaved function with $f(4) = 4$ – its explicit form is irrelevant. What we *cannot* do in d dimensions is define the analogue of γ_5. In four dimensions, we have

$$\gamma_5 = \frac{i}{4!} \varepsilon^{\mu\nu\rho\sigma} \gamma_\mu \gamma_\nu \gamma_\rho \gamma_\sigma, \tag{9.96}$$

but the Levi–Civita symbol $\varepsilon^{\mu\nu\rho\sigma}$ is specific to $d = 4$. This difficulty has a bearing on the problem of chiral anomalies, mentioned in §9.10 below.

Now let us consider the primitively divergent diagrams, starting with the fermion self-energy graph of Fig. 9.5. Expression (9.86), duly generalised to d dimensions, gives

$$\begin{aligned}
\text{----} \quad = \Sigma(p) &= -ie^2\mu^{4-d}\int \frac{d^dk}{(2\pi)^d}\gamma_\mu \frac{1}{\not{p} - \not{k} - m}\gamma_\nu \frac{g^{\mu\nu}}{k^2} \\
&= -ie^2\mu^{4-d}\int \frac{d^dk}{(2\pi)^d} \frac{\gamma_\mu(\not{p} - \not{k} + m)\gamma^\mu}{[(p-k)^2 - m^2]k^2}.
\end{aligned}$$

Introducing the Feynman parameter z (see (9.22)) yields

$$\Sigma(p) = -i\mu^{4-d}e^2\int_0^1 dz \int \frac{d^dk}{(2\pi)^d} \frac{\gamma_\mu(\not{p} - \not{k} + m)\gamma^\mu}{[(p-k)^2 z - m^2 z + k^2(1-z)]^2}.$$

Defining $k' = k - pz$ gives

$$\Sigma(p) = -i\mu^{4-d}e^2\int_0^1 dz \int_0^1 \frac{d^dk'}{(2\pi)^d} \frac{\gamma_\mu(\not{p} - \not{p}z - \not{k}' + m)\gamma^\mu}{[k'^2 - m^2 z + p^2 z(1-z)]^2}.$$

The term linear in k' integrates to zero, so

$$\Sigma(p) = -i\mu^{4-d}e^2\int_0^1 dz \gamma_\mu(\not{p} - \not{p}z + m)\gamma^\mu$$

$$\times \int \frac{d^dk'}{(2\pi)^d} \frac{1}{[k'^2 - m^2 z + p^2 z(1-z)]^2}.$$

Note that in four dimensions the momentum integral is, as claimed above, logarithmically, and not linearly, divergent. The integral is performed with the help of equation (9A.5), giving

$$\Sigma(p) = \mu^{4-d}e^2 \frac{\Gamma(2-d/2)}{(4\pi)^{d/2}} \int_0^1 dz \gamma_\mu[\not{p}(1-z)+m]\gamma^\mu$$

$$\times [m^2 z - p^2 z(1-z)]^{d/2-2}$$

As $d \to 4$, $\Gamma(2-d/2)$ develops a pole. Putting $\varepsilon = 4 - d$, and using equations (9.24) and (9.92) then gives

$$\Sigma(p) = -\frac{e^2}{16\pi^2}\Gamma(\varepsilon/2) \int_0^1 dz\{2\not{p}(1-z) - 4m - \varepsilon[\not{p}(1-z)+m]\}$$

$$\times \left(\frac{m^2 z - p^2 z(1-z)}{4\pi\mu^2}\right)^{-\varepsilon/2}$$

$$= \frac{e^2}{8\pi^2\varepsilon}(-\not{p}+4m) + \frac{e^2}{16\pi^2}\left\{\not{p}(1+\gamma) - 2m(1+2\gamma)\right.$$

$$\left. + 2\int_0^1 dz[\not{p}(1-z) - 2m]\ln\left(\frac{m^2 z - p^2 z(1-z)}{4\pi\mu^2}\right)\right\}$$

$$= \frac{e^2}{8\pi^2\varepsilon}(-\not{p}+4m) + \text{finite}. \tag{9.97}$$

Next, we calculate the vacuum polarisation graph. From (9.87) we get, in d dimensions,

$$\Pi_{\mu\nu}(k) = i\mu^{4-d}e^2 \int \frac{d^d p}{(2\pi)^d} \text{Tr}\left[\gamma_\mu \frac{1}{\not{p}-m}\gamma_\nu \frac{1}{\not{p}-\not{k}-m}\right]$$

$$= ie^2\mu^{4-d} \int \frac{d^d p}{(2\pi)^d} \frac{\text{Tr}[\gamma_\mu(\not{p}+m)\gamma_\nu(\not{p}-\not{k}+m)]}{(p^2-m^2)[(p-k)^2-m^2]}.$$

Introducing the Feynman parameter, and putting $p' = p - kz$ then gives

$$\Pi_{\mu\nu} = ie^2\mu^{4-d} \int_0^1 dz \int \frac{d^d p'}{(2\pi)^d}$$

$$\times \frac{\text{Tr}[\gamma_\mu(\not{p}'+\not{k}z+m)\gamma_\nu(\not{p}'-\not{k}(1-z)+m)]}{[p'^2 - m^2 + k^2 z(1-z)]^2}.$$

Because of (9.93), and because the terms odd in p' give no contribution to the integral (from (9A.6)), the numerator N in the momentum integral is

$$N = [p'^\kappa p'^\lambda - k^\kappa k^\lambda z(1-z)]\, \text{Tr}(\gamma_\mu \gamma_\kappa \gamma_\nu \gamma_\lambda) + m^2 \text{Tr}(\gamma_\mu \gamma_\nu)$$

which gives, from (9.94) and (9.95),

$$N = f(d)\{2p'_\mu p'_\nu - 2z(1-z)(k_\mu k_\nu - k^2 g_{\mu\nu})$$

$$- g_{\mu\nu}[p'^2 - m^2 - k^2 z(1-z)]\}.$$

Hence (putting $p' \to p$)

$$
\Pi_{\mu\nu}(k) = ie^2 \mu^{4-d} f(d) \int_0^1 dz \int \frac{d^d p}{(2\pi)^d} \left\{ \frac{2p_\mu p_\nu}{[p^2 - m^2 + k^2 z(1-z)]^2} \right.
$$
$$
\left. - \frac{2z(1-z)[k_\mu k_\nu - g_{\mu\nu}k^2]}{[p^2 - m^2 + k^2 z(1-z)]^2} - \frac{g_{\mu\nu}}{[p^2 - m^2 + k^2 z(1-z)]} \right\}.
$$

From equations (9A.5) and (9A.7) the contributions of the first and third terms in the integrand cancel. The middle term leads to a logarithmically (not quadratically – see above) divergent integral, whose value is obtained from (9A.5) (using (9.24)):

$$
\Pi_{\mu\nu}(k) = \frac{e^2}{2\pi^2}(k_\mu k_\nu - g_{\mu\nu}k^2)
$$
$$
\times \left\{ \frac{1}{3\varepsilon} - \frac{\gamma}{6} - \int_0^1 dz \, z(1-z) \ln\left[\frac{m^2 - k^2 z(1-z)}{4\pi\mu^2} \right] + O(\varepsilon) \right\}.
$$

As expected by now, the divergent part is a pole in ε. The finite part contains terms depending on k^2, so for small k^2 we have

$$
\Pi_{\mu\nu}(k) = \frac{e^2}{6\pi^2}(k_\mu k_\nu - g_{\mu\nu}k^2)\left(\frac{1}{\varepsilon} + \frac{k^2}{10m^2} + \cdots \right)
$$
$$
= \frac{e^2}{6\pi^2\varepsilon}(k_\mu k_\nu - g_{\mu\nu}k^2) + \text{finite}. \tag{9.98}
$$

Finally, we evaluate the vertex graph of Fig. 9.7. Equation (9.89) gives (note the change of notation)

$$
-ie\mu^{2-d/2}\Lambda_\mu(p, q, p')
$$
$$
= (-ie\mu^{2-d/2})^3 \int \frac{d^d k}{(2\pi)^d} \, \gamma_\nu \frac{i}{\not{p'} - \not{k} - m} \gamma_\mu \frac{i}{\not{p} - \not{k} - m} \gamma_\rho \frac{-ig^{\nu\rho}}{k^2}
$$
$$
= -(e\mu^{2-d/2})^3 \int \frac{d^d k}{(2\pi)^d}
$$
$$
\times \frac{\gamma_\nu(\not{p'} - \not{k} + m)\gamma_\mu(\not{p} - \not{k} + m)\gamma^\nu}{k^2[(p-k)^2 - m^2][(p'-k)^2 - m^2]}
$$

Now we introduce the 2-parameter Feynman formula, analogous to (and easily derived from) (9.22)

$$
\frac{1}{abc} = 2 \int_0^1 dx \int_0^{1-x} dy \frac{1}{[a(1-x-y) + bx + cy]^3}, \tag{9.99}
$$

which gives

$$
\Lambda_\mu = \frac{2ie^2\mu^{4-d}}{(2\pi)^d} \int_0^1 dx \int_0^{1-x} dy \int d^d k \times
$$

$$\times \frac{\gamma_\nu(\not{p}' - \not{k} + m)\gamma_\mu(\not{p} - \not{k} + m)\gamma^\nu}{[k^2 - m^2(x+y) - 2k(px + p'y) + p^2x + p'^2y]^3}.$$

Defining $k' = k - px - p'y$ then gives (with $k' \to k$)

$$\Lambda_\mu(p, q, p') = \frac{2ie^2\mu^{4-d}}{(2\pi)^d} \int_0^1 dx \int_0^{1-x} dy \int d^dk$$

$$\times \frac{\gamma_\nu[\not{p}'(1-y) - \not{p}x - \not{k} + m]\gamma_\mu[\not{p}(1-x) - \not{p}'y - \not{k} + m]\gamma^\nu}{[k^2 - m^2(x+y) + p^2x(1-x) + p'^2y(1-y) - 2p{\cdot}p'xy]^3}.$$

(9.100)

This integral contains convergent and divergent pieces. The part of the numerator quadratic in k is divergent, the rest convergent, so we put

$$\Lambda_\mu = \Lambda_\mu^{(1)} + \Lambda_\mu^{(2)}.$$

(9.101)

The divergent part $\Lambda_\mu^{(1)}$ may be written, from (9A.7), as

$$\Lambda_\mu^{(1)}(p, q, p')$$

$$= \frac{e^2}{2}\mu^{4-d}\left(\frac{1}{4\pi}\right)^{d/2}\Gamma(2 - d/2)\int_0^1 dx \int_0^{1-x} dy$$

$$\times \frac{\gamma_\nu\gamma_\rho\gamma_\mu\gamma^\rho\gamma^\nu}{[m^2(x+y) + p^2x(1-x) + p'^2y(1-y) + 2p{\cdot}p'xy]^{2-d/2}}.$$

By some 'diracology' similar to (9.92),

$$\gamma_\nu\gamma_\rho\gamma_\mu\gamma_\sigma\gamma^\nu = (2-d)\gamma_\rho\gamma_\mu\gamma_\sigma + 2(\gamma_\mu\gamma_\sigma\gamma_\rho - \gamma_\rho\gamma_\sigma\gamma_\mu)$$

so

$$\gamma_\nu\gamma_\rho\gamma_\mu\gamma^\rho\gamma^\nu = (2-d)^2\gamma_\mu.$$

Putting, as usual, $\varepsilon = 4 - d$, so that $(2 - d)^2 = 4 - 2\varepsilon$, we get

$$\Lambda_\mu^{(1)}(p, q, p') = \frac{e^2}{8\pi^2\varepsilon}\gamma_\mu + \text{finite}.$$

(9.102)

The convergent part of Λ_μ is that part with no k in the numerator of the integrand. Since it is convergent, we may put $d = 4$ and perform the integration over k using (9A.5), giving

$$\Lambda_\mu^{(2)}(p, q, p') = \frac{e^2}{16\pi^2}\int_0^1 dx \int_0^{1-x} dy$$

$$\times \frac{\gamma_\nu[\not{p}'(1-y) - \not{p}x + m]\gamma_\mu[\not{p}(1-x) - \not{p}'y + m]\gamma^\nu}{m^2(x+y) - p^2x(1-x) - p'^2y(1-y) + 2p{\cdot}p'xy}.$$

(9.103)

We now have explicit expressions for the three primitively divergent Feynman diagrams in QED, and have found three divergent terms and one convergent term. We shall see in the next section that the divergent terms may be cancelled by counter-terms in the Lagrangian, and subsequent renormalisation; and the convergent term gives an anomalous magnetic moment to the electron.

9.6   1-loop renormalisation of QED

We summarise, for convenience, the divergent quantities found above:

$$\Sigma(p) = \frac{e^2}{8\pi^2\varepsilon}(-\not{p} + 4m) + \text{finite}, \tag{9.97}$$

$$\Pi_{\mu\nu}(k) = \frac{e^2}{6\pi^2\varepsilon}(k_\mu k_\nu - g_{\mu\nu}k^2) + \text{finite}, \tag{9.98}$$

$$\Lambda_\mu^{(1)}(p,q,p') = \frac{e^2}{8\pi^2\varepsilon}\gamma_\mu + \text{finite}. \tag{9.102}$$

It is worth remarking, incidentally, that the divergent parts of Σ and Λ above satisfy the Ward identity (7.124). We saw in §7.4, by explicit calculation, that the Ward identity is satisfied to order e^2 (that is, to the 1-loop approximation). Here we see that dimensional regularisation preserves the Ward identity.

Now let us consider what counter-terms it is necessary to add to $\mathscr{L}$ to make the above quantities finite. We begin with the electron self-energy, which, as we saw in (7.121), modifies the electron inverse propagator to

$$\begin{aligned}
\Gamma^{(2)}(p) = S_F'(p)^{-1} &= S_F(p)^{-1} - \Sigma(p) \\
&= \not{p} - m - \frac{e^2}{8\pi^2\varepsilon}(-\not{p} + 4m) \\
&= \not{p}\left(1 + \frac{e^2}{8\pi^2\varepsilon}\right) - m\left(1 + \frac{e^2}{2\pi^2\varepsilon}\right).
\end{aligned} \tag{9.104}$$

(Note that we are neglecting the finite corrections. This corresponds to 'minimal subtraction'.) Because the coefficients of $\not{p}$ and m are not equal, we need two counter-terms; one for the overall magnitude of the propagator, giving a contribution to the electron wave function normalisation, and one for the electron mass. So to

$$\mathscr{L}_1 = i\bar{\psi}\not{\partial}\psi - m\bar{\psi}\psi$$

we add

$$(\mathscr{L}_1)_{CT} = iB\bar{\psi}\not{\partial}\psi - A\bar{\psi}\psi \tag{9.105}$$

giving the total (bare) Lagrangian

$$(\mathscr{L}_1)_B = i(1+B)\bar{\psi}\not{\partial}\psi - (m+A)\bar{\psi}\psi \tag{9.106}$$

where A and B are chosen to give a finite electron propagator to order e^2.

$$-i\Sigma(p) \qquad -iA \qquad iB\not{p}$$

(9.107)

(It is easy to check that the 'interaction' $iB\not{p}$ modifies the propagator from $i/\not{p}$ to $i/(1 + B)\not{p}$.) Hence

$$\frac{e^2}{8\pi^2\varepsilon}(-\not{p} + 4m) + A - B\not{p} = \text{finite},$$ (9.108)

whence, ignoring the finite terms,

$$\left.\begin{array}{l} A = -\dfrac{me^2}{2\pi^2\varepsilon}, \\[2mm] B = -\dfrac{e^2}{8\pi^2\varepsilon} \end{array}\right\}$$ (9.109)

and

$$Z_2 = 1 + B = 1 - \frac{e^2}{8\pi^2\varepsilon}.$$ (9.110)

Defining the 'bare' wave function by

$$\psi_{\mathrm{B}} = \sqrt{Z_2}\,\psi$$ (9.111)

enables us to write the bare Lagrangian (9.106) as

$$(\mathscr{L}_1)_{\mathrm{B}} = i\bar{\psi}_{\mathrm{B}}\not{\partial}\psi_{\mathrm{B}} - m_{\mathrm{B}}\bar{\psi}_{\mathrm{B}}\psi_{\mathrm{B}}$$ (9.112)

where the bare mass m_{B} is given by

$$m_{\mathrm{B}} = Z_2^{-1}(m + A)$$

$$= m\left(1 - \frac{e^2}{2\pi^2\varepsilon}\right)\left(1 + \frac{e^2}{8\pi^2\varepsilon}\right)$$

$$= m\left(1 - \frac{3e^2}{8\pi^2\varepsilon}\right) = m + \delta m.$$ (9.113)

Since the propagator is the Fourier transform of the vacuum expectation value of the time-ordered product of fields, the 'bare' propagator is the Fourier transform of

$$\langle 0|\,T(\bar{\psi}_{\mathrm{B}}\psi_{\mathrm{B}})|0\rangle = Z_2\langle 0|\,T(\bar{\psi}\psi)|0\rangle,$$

so, recalling that in this 'counter-term' treatment of renormalisation, ψ refers to the *physical* electron (i.e. with interactions included) and ψ_{B} to the bare electron, we see that Z_2^{-1} is the probability of finding a bare electron propagating, in the propagation of a physical electron. In our not entirely rigorous treatment, the *renormalised* electron wave function and the renormalised electron mass may be taken to be essentially ψ and m, the 'original' ones; but this is only true if we ignore the finite correction terms, as we did above. Indeed, it follows trivially that, ignoring these terms, the renormalised 2-point vertex is

$$\underset{-i\Sigma(p)}{\text{diagram}} = \underset{}{\text{diagram}} - \underset{-i\Sigma(p)}{\text{diagram}} - \underset{-iA}{\text{diagram}} - \underset{iB\not{p}}{\text{diagram}}$$

or

$$\Gamma^{(2)}(p) = i(S'_F(p))^{-1}$$
$$= \not{p} - m - (\Sigma(p) + A - B\not{p})$$
$$= \not{p} - m + \text{finite terms} \qquad (9.114)$$

where we have used (9.108). The finite terms will, in general, give a renormalised wave function and mass, different from ψ and m. They are determined by deciding the momenta at which the various vertex functions vanish, as explained in §9.3. Apart from this consideration, however, the philosophy of renormalisation by adding counter-terms to the Lagrangian is, in essence, the reverse of the philosophy of renormalisation first discussed in §9.3, by redefining mass, etc., to be finite when interactions are taken into account. This necessitates the bare mass (etc.) in the original Lagrangian having an infinite value. In the counter-term procedure, we start off with a physical (finite) mass, and this is kept finite by adding (infinite) counter-terms to $\mathscr{L}$. The resulting bare mass m_B (defined, for example, by (9.113)) is what we 'started out' with in the first renormalisation procedure.

Let us turn now to the vacuum polarisation tensor (9.98). From (9.88) it gives rise to a modified photon propagator

$$D'_{\mu\nu}(k) = D_{\mu\nu}(k) - D_{\mu\alpha}(k)\Pi^{\alpha\beta}(k)D_{\beta\nu}^{(k)} + \cdots$$
$$= -\frac{g_{\mu\nu}}{k^2} - \frac{g_{\mu\alpha}}{k^2}\frac{e^2}{6\pi^2}\left[(k^\alpha k^\beta - g^{\alpha\beta}k^2)\left(\frac{1}{\varepsilon} + \frac{k^2}{10m^2}\right)\right]\frac{g_{\beta\nu}}{k^2} + \cdots$$
$$= -\frac{g_{\mu\nu}}{k^2}\left(1 - \frac{e^2}{6\pi^2\varepsilon} - \frac{e^2}{60\pi^2}\frac{k^2}{m^2}\right) - \frac{e^2}{6\pi^2\varepsilon}\frac{1}{k^2}\frac{k_\mu k_\nu}{k^2} + \cdots. \qquad (9.115)$$

Here we have expressed $D_{\mu\nu}$ in the Feynman gauge. Note (1) that the coefficient of the Feynman gauge part of the new propagator contains an infinite (as $\varepsilon \to 0$) part, and also a finite (ε-independent) part, proportional to k^2; (2) that the resultant propagator is *not* in the Feynman gauge, because of the term in $k_\mu k_\nu$. Physical quantities are, of course, gauge invariant, so are not affected by this. Nonetheless, the infinite terms in $D'_{\mu\nu}$ must be removed by adding counter-terms to the original Lagrangian. The Lagrangian giving a Feynman gauge propagator was

$$\mathscr{L}_2 = -\tfrac{1}{4}F_{\mu\nu}F^{\mu\nu} - \tfrac{1}{2}(\partial_\mu A^\mu)^2 = \tfrac{1}{2}A^\mu g_{\mu\nu}\Box A^\nu, \tag{9.116}$$

so the required counter-term is

$$(\mathscr{L}_2)_{\text{CT}} = -\frac{C}{4}F_{\mu\nu}F^{\mu\nu} - \frac{E}{2}(\partial_\mu A^\mu)^2. \tag{9.117}$$

C and E will be different because of observation (2) above. Then the bare Lagrangian is

$$(\mathscr{L}_2)_{\text{B}} = -\left(\frac{1+C}{4}\right)F_{\mu\nu}F^{\mu\nu} + \text{gauge terms}$$

$$= -\frac{Z_3}{4}F_{\mu\nu}F^{\mu\nu} + \text{gauge terms} \tag{9.118}$$

with

$$Z_3 = 1 - \frac{e^2}{6\pi^2\varepsilon}. \tag{9.119}$$

$(\mathscr{L}_2)_{\text{B}}$ then gives a finite photon propagator to order e^2.

The electron self-energy graph gave rise to a bare electron mass different from the physical one. Will the same thing happen here? In other words, will the photon acquire a finite mass by virtue of the self-energy effects of vacuum polarisation? Needless to say, this would be a disaster, but, in fact, it does not happen, and the reason stems from the fact that $\Pi_{\mu\nu}(k)$ is of the form (cf. (9.98))

$$\Pi^{\alpha\beta}(k) = (k^\alpha k^\beta - g^{\alpha\beta}k^2)\Pi(k^2). \tag{9.120}$$

This, in turn, follows because gauge invariance demands

$$k_\alpha \Pi^{\alpha\beta}(k) = 0 \tag{9.121}$$

and (9.120) is the most general Lorentz covariant structure consistent with this. Substituting (9.120) into (9.115) gives, to one loop,

$$D'_{\mu\nu} = D_{\mu\nu} - D_{\mu\alpha}(k^\alpha k^\beta - g^{\alpha\beta}k^2)\Pi(k^2)D_{\beta\nu},$$

and hence, putting $D_{\mu\nu} = -g_{\mu\nu}/k^2$,

$$D'_{\mu\nu} = \frac{1}{k^2[1+\Pi(k^2)]}\left(-g_{\mu\nu} - \frac{k_\mu k_\nu}{k^2}\Pi(k^2)\right) \tag{9.122}$$

$\Pi(k^2)$ contains divergences. In dimensional regularisation

$$\Pi(k^2) = \frac{e^2}{6\pi^2}\left(\frac{1}{\varepsilon} + \frac{k^2}{10m^2}\right) = \frac{e^2}{6\pi^2\varepsilon} + \Pi_{\text{f}}(k^2) \tag{9.123}$$

where $\Pi_{\text{f}}(k^2)$ is finite and tends to zero as $k^2 \to 0$. The complete propagator may then be written

$$D'_{\mu\nu} = \frac{-g_{\mu\nu}}{k^2[1 + \Pi(k^2)]} + \text{gauge terms}$$

$$= \frac{-g_{\mu\nu}}{k^2\left[1 + \dfrac{e^2}{6\pi^2\varepsilon} + \Pi_{\mathrm{f}}(k^2)\right]} + \text{gauge terms}$$

$$= \frac{-Z_3 g_{\mu\nu}}{k^2[1 + \Pi_{\mathrm{f}}(k^2)]} + \text{gauge terms} \qquad (9.124)$$

where we have used (9.119). The Lagrangian (9.118), however, which contains the counter-terms, suggests the definition of the *bare* field

$$A_{\mathrm{B}}^\mu = Z_3^{\frac{1}{2}} A^\mu \qquad (9.125)$$

whose propagator $D'_{\mu\nu} \sim \langle 0 | T(A_{\mu\mathrm{B}} A_{\nu\mathrm{B}}) | 0 \rangle \sim Z_3 \langle 0 | T(A_\mu A_\nu) | 0 \rangle = Z_3 \tilde{D}'_{\mu\nu}$, where $\tilde{D}'_{\mu\nu}$ is the *renormalised* complete propagator; from (9.124) it is

$$\tilde{D}'_{\mu\nu} = \frac{-g_{\mu\nu}}{k^2[1 + \Pi_{\mathrm{f}}(k^2)]} + \text{gauge terms.} \qquad (9.126)$$

(We are using the notations $D'_{\mu\nu}$ and $\tilde{D}'_{\mu\nu}$ of Bjorken and Drell.[3])

We see, then, that the photon mass remains zero after renormalisation. In addition, as desired and anticipated, renormalisation gets rid of the infinite terms in the propagator $D'_{\mu\nu}$; but it does *not* get rid of the finite correction term in k^2, which will give rise to physical effects. Ignoring the gauge terms, the renormalised propagator is

$$\tilde{D}'_{\mu\nu} = \frac{-g_{\mu\nu}}{k^2}\left(1 - \frac{e^2}{60\pi^2}\frac{k^2}{m^2} + O(k^4)\right). \qquad (9.127)$$

The correction term (known as the Uehling term) gives rise to a modification of the Coulomb potential in co-ordinate space. The potential between charges e a distance r apart is now

$$\frac{e^2}{4\pi r} + \frac{e^2}{60\pi^2 m^2}\delta^3(r).$$

The extra term modifies the hydrogen atom energy levels, and gives a significant contribution to the Lamb shift, which splits the degeneracy of the $2S_{\frac{1}{2}}$ and $2P_{\frac{1}{2}}$ levels. QED predicts that the $2S_{\frac{1}{2}}$ level is 1057.9 MHz above $2P_{\frac{1}{2}}$, of which -27.1 MHz comes from the vacuum polarisation graph above. Since agreement between theory and experiment is better than 0.1 MHz, we have an excellent confirmation of the above theory.

We turn finally to the vertex function, and its divergent part $\Lambda_\mu^{(1)}$ given by (9.102). It is clear that it may be eliminated by adding to the Lagrangian a counter-term

$$(\mathscr{L}_3)_{\mathrm{CT}} = -De\mu^{2-d/2}\bar{\psi}A\!\!\!/\psi \tag{9.128}$$

with

$$D = -\frac{e^2}{8\pi^2\varepsilon} \tag{9.129}$$

so that

$$(\mathscr{L}_3)_{\mathrm{B}} = -(1+D)e\mu^{\varepsilon/2}A^\mu\bar{\psi}\gamma_\mu\psi$$
$$\equiv -Z_1 e\mu^{\varepsilon/2}A^\mu\bar{\psi}\gamma_\mu\psi \tag{9.130}$$

with

$$Z_1 = 1 - \frac{e^2}{8\pi^2\varepsilon}. \tag{9.131}$$

The total bare Lagrangian (to one loop) for QED is, from (9.106), (9.118) and (9.130),

$$\mathscr{L}_{\mathrm{B}} = iZ_2\bar{\psi}\gamma^\mu\partial_\mu\psi - (m+A)\bar{\psi}\psi$$
$$- Z_1 e\mu^{\varepsilon/2}A^\mu\bar{\psi}\gamma_\mu\psi - \frac{Z_3}{4}(\partial_\mu A_\nu - \partial_\nu A_\mu)^2 + \text{gauge terms} \tag{9.132}$$

with

$$Z_1 = Z_2 = 1 - \frac{e^2}{8\pi^2\varepsilon},$$
$$Z_3 = 1 - \frac{e^2}{6\pi^2\varepsilon}, \quad A = -\frac{me^2}{2\pi^2\varepsilon}. \tag{9.133}$$

$\mathscr{L}_{\mathrm{B}}$, as distinct from the original Lagrangian (9.90), gives (to one loop) finite self-energy and vertex, and e and m stand for the experimental charge and mass. We may, alternatively, write $\mathscr{L}_{\mathrm{B}}$ in terms of 'bare' quantities. From (9.111), (9.113), (9.125) and the following definition of bare charge e_{B}:

$$e_{\mathrm{B}} = e\mu^{\varepsilon/2}\frac{Z_1}{Z_2 Z_3^{\frac{1}{2}}} = e\mu^{\varepsilon/2}Z_3^{-\frac{1}{2}}, \tag{9.134}$$

we have

$$\mathscr{L}_{\mathrm{B}} = i\bar{\psi}_{\mathrm{B}}\gamma^\mu\psi_{\mathrm{B}} - m_{\mathrm{B}}\bar{\psi}_{\mathrm{B}}\psi_{\mathrm{B}} - e_{\mathrm{B}}A_{\mathrm{B}}^\mu\bar{\psi}_{\mathrm{B}}\gamma_\mu\psi_{\mathrm{B}}$$
$$- \tfrac{1}{4}(\partial_\mu A_{\mathrm{B}\nu} - \partial_\nu A_{\mathrm{B}\mu})^2. \tag{9.135}$$

We have now absorbed all the infinite quantities into the definitions of bare quantities, and the fact that we are able to do this, keeping a Lagrangian of the same form as the original ((9.135) is of the same form as (9.90)) means that, to this order, QED is renormalisable. The proof that it is renormalisable to all orders is given in the next section. We close this section by making two points of less formal, more physical, interest. First, the *finite* contribution $\Lambda_\mu^{(2)}$ to the vertex function has the physical consequence that a Dirac particle has an 'anomalous' magnetic moment, which we can calculate. Second, an application of the renormalisation group argument to (9.134) will give a prediction about the asymptotic behaviour of QED.

Anomalous magnetic moment of the electron

Having removed the infinite part of the vertex function by renormalisation, we now turn our attention to the finite contribution $\Lambda_\mu^{(2)}$ shown in (9.103). We recall that the complete vertex function is given by (7.123) as $\gamma_\mu + \Lambda_\mu$, and this expression must be sandwiched between spinors $\bar{u}(p')\ldots u(p)$. To warm up, let us show that an electromagnetic current $\bar{u}(p')\gamma_\mu u(p)$ describes a particle with the 'Dirac' magnetic moment with $g_s = 2$ (see §2.6). We have the definitions (2.99) and (2.170)

$$\gamma_\mu\gamma_\nu + \gamma_\nu\gamma_\mu = 2g_{\mu\nu},$$

$$\gamma_\mu\gamma_\nu - \gamma_\nu\gamma_\mu = -2i\sigma_{\mu\nu},$$

and the Dirac equations (2.140) and (2.141)

$$\gamma_\mu p^\mu u(p) = mu(p),$$
$$\bar{u}(p')\gamma_\mu p'^\mu = m\bar{u}(p').$$

From these it is simple to deduce that

$$\gamma_\mu u(p) = \frac{1}{m}(p_\mu - i\sigma_{\mu\nu}p^\nu)u(p)$$

$$\bar{u}(p')\gamma_\mu = \frac{1}{m}\bar{u}(p')(p'_\mu + i\sigma_{\mu\nu}p'^\nu).$$

Hence

$$\bar{u}(p')\gamma_\mu u(p) = \tfrac{1}{2}\bar{u}(p')[\gamma_\mu u(p)] + \tfrac{1}{2}[\bar{u}(p')\gamma_\mu]u(p)$$

$$= \frac{1}{2m}\bar{u}(p')[(p_\mu + p'_\mu) + i\sigma_{\mu\nu}q^\nu]u(p) \qquad (9.136)$$

where $q = p' - p$. By comparison with §2.6 it may be seen that it is the second term, in $\sigma_{\mu\nu}q^\nu$, that yields the magnetic moment $g_s = 2$. We must now include the effect of $\Lambda_\mu^{(2)}$ to calculate the total vertex

$$\bar{u}(p')\Gamma_\mu u(p) = \bar{u}(p')(\gamma_\mu + \Lambda_\mu^{(2)})u(p), \qquad (9.137)$$

with $\Lambda_\mu^{(2)}$ given by (9.103). Sandwiched between the spinors, we may replace, in the numerator of that expression, p' by m on the left, and p by m on the right. Moreover, since (2.99) gives, for example $\not{p}\gamma_\mu = 2p_\mu - \gamma_\mu\not{p}$, the numerator of the integrand of (9.103) becomes

$$-4m(y - xy - x^2)p_\mu - 4m(x - xy - y^2)p'_\mu + D\gamma_\mu$$

where the term in γ_μ is not exhibited explicitly, because it does not contribute to the magnetic moment, so is not of interest here. (It is actually infra-red divergent, which is a problem we do not go into.) In the denominator of the integrand, putting $p^2 = p'^2 = m^2$, $(p - p')^2 = q^2 = 0$ gives an expression $m^2(x + y)^2$, so, sandwiched between spinors, and ignoring the term in γ_μ we have

$$
\begin{aligned}
\Lambda_\mu^{(2)} &= \frac{-e^2}{4\pi^2 m} \int_0^1 dx \int_0^{1-x} dy \frac{1}{(x + y)^2} \\
&\quad \times [(y - xy - x^2)p_\mu + (x - xy - y^2)p'_\mu] \\
&= \frac{-e^2}{16\pi^2 m}(p_\mu + p'_\mu).
\end{aligned}
$$

Substituting this in (9.136):

$$\bar{u}(p')(p + p')_\mu u(p) = \bar{u}(p')(2m\gamma_\mu - i\sigma_{\mu\nu}q^\nu)u(p),$$

it turns out that the term in γ_μ cancels the one we neglected above, so (9.137) gives the total vertex as

$$\bar{u}(p')\Gamma_\mu u(p) = \bar{u}(p')\left[\frac{(p + p')_\mu}{2m} + \left(1 + \frac{\alpha}{2\pi}\right)\frac{i\sigma_{\mu\nu}q^\nu}{2m}\right]u(p) \qquad (9.138)$$

where $\alpha = e^2/4\pi$; $\alpha/2\pi$ gives the lowest order correction to the magnetic moment of the electron, which therefore has a gyromagnetic ratio

$$\frac{g}{2} = 1 + \frac{\alpha}{2\pi} + O(\alpha^2). \qquad (9.139)$$

This was first calculated by Schwinger in 1948, and agreed with the contemporary experimental results. Since then g has been calculated to order α^3, and a recent comparison of theory and experiment gives, for the electron[29]

$$
\begin{aligned}
a_{\text{th}} &= \tfrac{1}{2}(g - 2) \\
&= 0.5\left(\frac{\alpha}{\pi}\right) - 0.32848\left(\frac{\alpha}{\pi}\right)^2 + 1.49\left(\frac{\alpha}{\pi}\right)^3 \\
&= (1\ 159\ 652.4 \pm 0.4) \times 10^{-9} \\
a_{\text{exp}} &= (1\ 159\ 652.4 \pm 0.2) \times 10^{-9}. \qquad (9.140)
\end{aligned}
$$

For the muon, agreement between experiment and this purely electromagnetic calculation is not so good, but neither is it expected to be, since there are contributions from hadrons, W and Z bosons, and the Higgs boson.

Asymptotic behaviour

The asymptotic behaviour of QED (in so far as the 1-loop approximation gives a reliable prediction) may be inferred from equation (9.134), which, with (9.133) gives

$$e_B = e\mu^{\varepsilon/2}\left(1 + \frac{e^2}{12\pi^2\varepsilon}\right).$$

In the limit $\varepsilon \to 0$, the bare charge e_B is independent of μ, so we may deduce how e scales with μ. Differentiating the above equation gives

$$\mu\frac{\partial e}{\partial \mu} = -\frac{\varepsilon}{2}e + \frac{e^3}{12\pi^2} + \frac{e^5}{96\pi^4\varepsilon}.$$

Ignoring the last term, and letting $\varepsilon \to 0$ then gives

$$\mu\frac{\partial e}{\partial \mu} = \frac{e^3}{12\pi^2}. \tag{9.141}$$

But from (9.64), this is $\beta(e)$. So we see that, like ϕ^4 theory, $\beta > 0$ and the running coupling constant e increases with increasing scale (i.e. with μ), so asymptotically gets stronger. Despite the smallness of the fine structure constant $\alpha = e^2/4\pi = 1/137$, then, perturbation theory in QED is not good at large momenta.

The solution to (9.141) is

$$e^2(\mu) = \frac{e^2(\mu_0)}{1 - \frac{e^2(\mu_0)}{6\pi^2}\ln\frac{\mu}{\mu_0}}, \tag{9.142}$$

from which we see explicitly the increase of e with μ. We also see the so-called *Landau singularity*,

$$\mu = \mu_0\exp\left(\frac{6\pi^2}{e^2(\mu_0)}\right).$$

Fig. 9.10. Polarisation of a dielectric medium by a charge, or of the vacuum in QED.

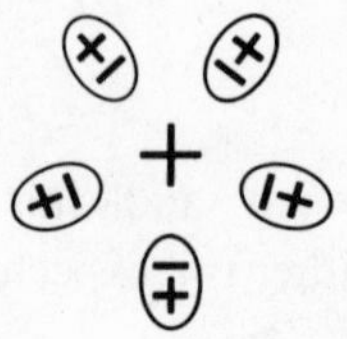

The increase of e with μ, i.e. with *decreasing distance*, has an analogy in macroscopic electrostatics in a dielectric medium. Here, the presence of an electric charge polarises the medium, as shown in Fig. 9.10. If the free charge density is ρ_f and the polarisation charge density is ρ_p, we have the equations (in Heaviside–Lorentz units)

$$\mathbf{\nabla}\cdot\mathbf{E} = \rho_f \quad \text{in vacuo,}$$

$$\mathbf{\nabla}\cdot\mathbf{E} = \rho_f + \rho_p \quad \text{in medium.}$$

Putting

$$\mathbf{\nabla}\cdot\mathbf{P} = -\rho_p$$

gives

$$\mathbf{\nabla}\cdot(\mathbf{E} + \mathbf{P}) = \rho_f$$

or

$$\mathbf{\nabla}\cdot\mathbf{D} = \rho_f$$

where $\mathbf{D} = \mathbf{E} + \mathbf{P}$. If $\mathbf{P} = \alpha\mathbf{E}$, then $\mathbf{D} = (1 + \alpha)\mathbf{E}$, so finally

$$\mathbf{\nabla}\cdot\mathbf{E} = \left(\frac{1}{1 + \alpha}\right)\rho_f, \tag{9.143}$$

and we see that the effect of the medium is to 'screen' the original charge. From a distance, the charge (as measured by $\mathbf{E}$) is measured to be less than it is nearby. At a distance much less than molecular or atomic dimensions, the 'bare', unscreened charge is seen.

The same situation holds in QED. Here the photon self-energy graph (Fig. 9.6) may be interpreted by saying the vacuum is 'filled' with virtual e^+e^- pairs which screen a bare charge, in the manner drawn in Fig. 9.10. So the charge e increases with decreasing distance, i.e. with an increasing momentum (or mass) scale. We shall see below that the situation in this regard is very different in QCD.

9.7 Renormalisability of QED

In this section we shall prove that quantum electrodynamics is renormalisable to all orders of perturbation theory. The proof follows closely that of Jauch and Rohrlich,[31] which in turn is based on the key paper of Ward.[34]

We have seen above that the complete propagators and vertex functions S_F', D' and Γ_μ are all divergent (to second, and to all orders). We have the equations (cf. (7.121), (7.123), (7.124))

$$S_F'(p)^{-1} = S_F(p)^{-1} - \Sigma(p) \tag{9.144}$$

$$D'(k)^{-1} = D(k)^{-1} - \Pi(k), \tag{9.145}$$

$$\Gamma_\mu(p, q, p + q) = \gamma_\mu + \Lambda_\mu(p, q, p + q), \tag{9.146}$$

$$-\frac{\partial \Sigma(p)}{\partial p^\mu} = \Lambda_\mu(p, 0, p),\tag{9.147}$$

where

$$D_{\mu\nu}(k) = g_{\mu\nu}D(k), \quad D'_{\mu\nu}(k) = g_{\mu\nu}D'(k),\tag{9.148}$$

$$\Pi_{\mu\nu}(k) = -g_{\mu\nu}\Pi(k).\tag{9.149}$$

(We are working in the Feynman gauge.) In (9.144) $\Sigma(p)$ is the *proper* self-energy, and similarly $\Pi(k)$ in (9.145) represents the proper vacuum polarisation diagrams. (9.147) is the Ward identity. Σ, Π and Λ are given by relevant Feynman diagrams. The idea is to analyse the divergences of these to show that the corresponding divergences in the propagators and vertex function can all be removed by a multiplicative renormalisation, i.e. that we can define *finite* propagators and vertex functions

$$\tilde S_F = \frac{1}{Z_2}S'_F, \quad \tilde D_F = \frac{1}{Z_3}D'_F, \quad \tilde\Gamma_\mu = Z_1\Gamma_\mu\tag{9.150}$$

which obey the correct functional equations, so that all the infinities of the theory are absorbed into Z_1, Z_2, Z_3 and mass renormalisation.

We are concerned with divergencies of Feynman diagrams. In chapter 7 we defined the class of *proper* (1-particle irreducible) Feynman diagrams. We now define a subclass of them, called *irreducible*. The lowest order graphs for Σ and Π (⌇⌇⌇ , ⌇⌇◯⌇⌇) are examples of *self-energy* graphs, and the lowest order graph for Λ_μ (—⌇⌇—) is an example of a *vertex* graph. Consider, now, an arbitrary Feynman diagram, say the one in Fig. 9.11(*a*). It is proper, and contains self-energy and vertex parts. To this diagram corresponds another diagram, called its *skeleton*, which is obtained by replacing every self-energy part by a line, and every vertex part by a corner (i.e. a 'bare' vertex). The skeleton of Fig. 9.11(*a*) is drawn in Fig. 9.11(*b*). A graph which is its own skeleton is called *irreducible*. Otherwise it is reducible. Actually, the irreducible graph in Fig. 9.11(*b*) is convergent (see equation (9.85)), and, by the same formula Fig. 9.11(*a*) is *superficially* convergent, but it is actually divergent, because of the self-energy and vertex insertions.

To analyse these divergences, consider first irreducible graphs. The only

Fig. 9.11. (*a*) An arbitrary Feynman diagram. (*b*) Its skeleton.

(*a*)   (*b*)

irreducible self-energy graphs are ⟿ and ᗐ, which have already been analysed. There is, however, an infinite number of irreducible vertex diagrams, three of which are shown in Fig. 9.12. From (9.85) $D = 0$ for all of them, so they are logarithmically divergent, and only one infinite constant appears in the expression for Λ_μ, so, as in the second order case we have

$$\Lambda_\mu = L\gamma_\mu + \Lambda_v^{(f)} \tag{9.151}$$

where L is infinite and $\Lambda_\mu^{(f)}$ is finite and has the definition

$$\bar{u}(p)\Lambda_\mu^{(f)}(p, 0, p)u(p) = 0 \tag{9.152}$$

which is true to second order, as readily seen in the last section. We conclude that for *all* irreducible graphs, the divergences may be isolated.

Now consider reducible graphs. Any reducible graph is obtained from its skeleton by inserting the relevant self-energy and vertex parts in place of the propagators (lines) and bare vertices. First, consider the vertex part V, with skeleton V_s. Writing Λ_μ as a functional of S, D, γ (the Dirac matrix) and e, as well as a function of p' and p, we have (subscript s stands for skeleton)

$$\Lambda_\mu(p, p'; S_F, D_F, \gamma, e) = \Lambda_{\mu s}(p, p'; S_F', D_F', \Gamma, e). \tag{9.153}$$

Hence

$$\Gamma_\mu(p, p') = \gamma_\mu + \Lambda_{\mu s}(p, p'; S_F', D_F', \Gamma, e). \tag{9.154}$$

With regard to self-energy parts the situation is more complicated, because of *overlapping divergences*, mentioned in §9.3 above, which do not occur in vertex functions. The graph of Fig. 9.13(a), for example, gives a

Fig. 9.12. Examples of irreducible vertex diagrams.

Fig. 9.13. Examples of overlapping divergences in QED.

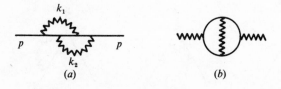

$$(a) \qquad\qquad\qquad (b)$$

contribution to $\Sigma(p)$ of

$$e^4 \int d^4k_1 d^4k_2 \frac{1}{k^2} \gamma^\mu \frac{1}{\not p - \not k_1 - m} \gamma^\rho \frac{1}{\not p - \not k_1 - \not k_2 - m} \gamma_\mu \frac{1}{k^2} \frac{1}{\not p - \not k_2 - m} \gamma_\rho.$$

The integral has a superficial degree of divergence $D = 1$, so is linearly divergent. But if k_1 is held fixed, $D = 0$ and it is still logarithmically divergent over k_2, so the divergences from each integration cannot be separated out. Fig. 9.13(b) also has overlapping divergences. They are a major obstacle in the separation of divergences and therefore in the proof of renormalisability, but were overcome in a paper by Ward,[34] by using the Ward identity. The general idea is, instead of separating out the finite part in $\Sigma(p)$, separate it out in $\partial\Sigma/\partial p^\mu = -\Lambda_\mu$, a vertex part – this has been done already. For the fourth order self-energy diagram in Fig. 9.13(a), since differentiation with respect to p is equivalent to insertion of a zero momentum photon, we get three diagrams for $\partial\Sigma/\partial p$, shown in Fig. 9.14. These diagrams must be added. Mathematically, this is summed up in the Ward–Takahashi identity (7.111):

$$S_F'(p)^{-1} - S_F'(p_0)^{-1} = (p - p_0)^\mu \Gamma_\mu(p, p_0). \tag{9.155}$$

Equations (9.154) and (9.155) are coupled equations for the electron propagators and self-energy diagrams. All the terms are infinite, but the problem of overlapping divergences in self-energy diagrams has been obviated by an appeal to the Ward–Takahashi identity.

Now the situation is similar with regard to photon self-energy diagrams Π. They also possess overlapping divergences – for example Fig. 9.13(b). So we proceed in an analogous way. Define the operator Δ_μ (analogous to Λ_μ) by

$$\Delta_\mu(k) = -\frac{\partial \Pi(k)}{\partial k^\mu} \tag{9.156}$$

(analogous to $\Lambda_\mu = -\partial\Sigma/\partial p^\mu$). This results in 3-photon vertex graphs; on differentiation, Fig. 9.13(b) gives the two graphs in Fig. 9.15. (Note, by the way, that these are not zero, despite Furry's theorem. This states that for

Fig. 9.14. The three diagrams obtained by differentiating Fig. 9.13(a) with respect to p.

$(q = 0)$

every diagram with an odd number of external photon lines, there exists another diagram, with the electron momenta reversed, such that the sum vanishes. Here we do not take the sum.) Like Λ_μ, Δ_μ is logarithmically divergent. If Δ_μ is analogous to Λ_μ, what is analogous to Γ_μ? It is denoted W_μ, and defined by

$$W_\mu(k) = -2k_\mu + \Delta_\mu(k). \tag{9.157}$$

We can now find a relation between the complete photon propagator D_F' and W_μ, analogous to the Ward–Takahashi identity (9.155). From (9.115), (9.148) and (9.149) we first verify (9.145):

$$D' = D + D\Pi D + D\Pi D\Pi D + \cdots$$

$$= \frac{D}{1 - \Pi D}$$

$$(D')^{-1} = D^{-1} - \Pi,$$

and then, putting $D = -1/k^2$,

$$\frac{\partial D^{-1}}{\partial k^\mu} = -2k^\mu$$

hence

$$\frac{\partial (D')^{-1}}{\partial k^\mu} = -2k_\mu - \frac{\partial \Pi}{\partial k^\mu}$$

$$= -2k_\mu + \Delta_\mu(k)$$

$$= W_\mu, \tag{9.158}$$

where we have used (9.156, 9.157). This is the relation we wanted between D' and W_μ. Finally, we have, analogous to equation (9.153) ($\Delta_{\mu s}$ = skeleton of Δ_μ),

$$\Delta_\mu(k; S_F, D_F, k, e) = \Delta_{\mu s}(k; S_F', D_F', W, e) \tag{9.159}$$

and (9.157) then becomes

$$W_\mu(k) = -2k_\mu + \Delta_{\mu s}(k; S_F', D_F', W, e). \tag{9.160}$$

We are now in possession of a set of coupled functional equations (9.154), (9.155), (9.158) and (9.160). They all refer to divergent quantities. We want the corresponding set of equations for *finite* functions. First, we may find the

Fig. 9.15. The two diagrams obtained by differentiating Fig. 9.13(*b*) with respect to k.

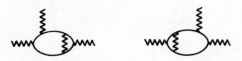

finite vertex functions $\tilde{\Lambda}_\mu(p, p')$ and $\tilde{\Delta}_\mu(k)$. Since Λ and Δ are logarithmically divergent, only one subtraction is necessary to define convergent functions:

$$\tilde{\Lambda}_\mu(p, p') = \Lambda_{\mu s}(p, p') - \Lambda_{\mu s}(p_0, p_0)|_{\not{p}_0 = m}, \tag{9.161}$$

$$\tilde{\Delta}_\mu(k^2) = \Delta_{\mu s}(k^2) - \Delta_{\mu s}(\mu^2), \tag{9.162}$$

where μ is an invariant photon mass and p_0 is the momentum of an on-shell electron, so that $p_0^2 = m^2$. Remember that $\Lambda_{\mu s}$ is to be sandwiched between spinors $\bar{u}(p) \dots u(p')$; the instruction $\not{p}_0 = m$ means that $\not{p}_0$ is to be commuted through to the right, and then put equal to m. From (9.151) and (9.152),

$$\Lambda_{\mu s}(p_0, p_0)|_{\not{p}_0 = m} = L\gamma_\mu \tag{9.163}$$

where L is an infinite constant. We now *define finite* propagators and vertex functions, denoted $\tilde{\ }$, by equations analogous to (9.154), (9.155), (9.160), (9.158):

$$\tilde{\Gamma}_\mu(p, p') = \gamma_\mu + \tilde{\Lambda}_{\mu s}(p, p'; \tilde{S}_F, \tilde{D}_F, \tilde{\Gamma}, e_{\text{ren}}), \tag{9.164}$$

$$\tilde{S}_F(p)^{-1} - \tilde{S}_F(p_0)^{-1} = (p - p_0)^\mu \tilde{\Gamma}_\mu(p, p_0), \tag{9.165}$$

$$\tilde{W}_\mu(k) = -2k_\mu + \tilde{\Delta}_{\mu s}(k; \tilde{S}_F, \tilde{D}_F, \tilde{W}, e_{\text{ren}}), \tag{9.166}$$

$$\frac{\partial \tilde{D}_F(k)^{-1}}{\partial k^\mu} = \tilde{W}_\mu(k). \tag{9.167}$$

The electron and photon propagators are normalised to

$$\tilde{S}_F(p_0)^{-1} = \not{p}_0 - m, \tag{9.168}$$

$$k^2 \tilde{D}_F(k^2)|_{k^2 = \mu^2} = 1. \tag{9.169}$$

We shall now show that $\tilde{\Gamma}_\mu$, $\tilde{S}_F$, $\tilde{D}_F$ and $\tilde{W}_\mu$ are related to Γ_μ, S'_F, D'_F and W_μ by the relations (9.150), *provided* that the charge e is also renormalised to

$$e_r^2 = Z_3 e^2. \tag{9.170}$$

To show this, first consider Γ_μ – more particularly Λ_μ. Suppose that, apart from the original (bare) vertex, it is of order $(e^2)^n$. Then it contains $2n$ propagators S_F, n propagators D_F and $(2n + 1)$ factors γ_μ. So, introducing the transformations (9.150) and (9.170),

$$\left.\begin{array}{ll} S_F \to \dfrac{1}{Z_2} S_F = \tilde{S}_F, & D_F \to \dfrac{1}{Z_3} D_F = \tilde{D}_F, \\[2mm] \Gamma_\mu \to Z_1 \Gamma_\mu = \tilde{\Gamma}_\mu, & e^2 \to Z_3 e^2 = e_{\text{ren}}^2, \end{array}\right\} \tag{9.171}$$

we have

$$\Lambda_{\mu s} \to Z_1 \Lambda_{\mu s}$$

(where we have used the Ward identity $Z_1 = Z_2$), or

$$\Lambda_{\mu s}\left[\frac{1}{Z_2} S'_F, \frac{1}{Z_3} D'_F, Z_1 \Gamma_\nu, Z_3 e^2\right]$$

$$= Z_1 \Lambda_{\mu s}[S'_F, D'_F, \Gamma_v, e^2]. \tag{9.172}$$

Now, from (9.164), (9.161) and (9.163),

$$\tilde{\Gamma}_\mu = \gamma_\mu + \tilde{\Lambda}_{\mu s}$$
$$= \gamma_\mu + \Lambda_{\mu s} - L\gamma_\mu$$
$$= (1 - L)\left(\gamma_\mu + \frac{1}{1 - L}\Lambda_{\mu s}\right)$$
$$= Z_1\left\{\gamma_\mu + \frac{1}{Z_1}\Lambda_{\mu s}[\tilde{S}_F, \tilde{D}_F, \tilde{\Gamma}, e_r^2]\right\}$$

which becomes, using (9.172),

$$= Z_1\{\gamma_\mu + \Lambda_{\mu s}[S'_F, D'_F, \Gamma_v, e^2]\}$$
$$= Z_1\Gamma_\mu \tag{9.173}$$

with $Z_1 = 1 - L$. So the renormalisation (9.171) is the correct one to yield (9.164). We see that a *subtraction* (equation (9.161)) is equivalent to *multiplicative renormalisation*, as first expressed by Dyson.

Similarly we may show that (9.166) holds if

$$\tilde{W}_\mu = Z_3 W_\mu. \tag{9.174}$$

This is shown by considering a typical graph for Δ_μ. A term of order $(e^2)^n$ contains $(n - 1) D_F$ functions, and $(2n + 1) \Gamma$ and S_F functions (consider, for

example, , of order $(e^2)^3$), so under (9.171)

$$\Delta_{\mu s} \to Z_3 \Delta_{\mu s}$$

or

$$\Delta_{\mu s}\left[\frac{1}{Z_2}S'_F, \frac{1}{Z_3}D'_F, Z_1\Gamma, Z_3 e^2\right] = Z_3 \Delta_{\mu s}[S'_F, D'_F, \Gamma, e^2]. \tag{9.175}$$

To show that equations (9.166) and (9.174) are consistent, recall the definition (9.162), which, together with (9.166), implies

$$\tilde{W}_\mu(k) = -2k_\mu + \Delta_{\mu s}(k^2) - \Delta_{\mu s}(\mu^2)$$
$$= -2k_\mu[1 + \tfrac{1}{2}\Delta_s(\mu^2)] + \Delta_{\mu s}(k^2)$$
$$= Z_3\left[-2k_\mu + \frac{1}{Z_3}\Delta_{\mu s}(k^2)\right]$$
$$= Z_3 W_\mu(k) \tag{9.176}$$

where $Z_3 = 1 + \tfrac{1}{2}\Delta_s(\mu^2)$ (we have put $\Delta_{\mu s} = k_\mu\Delta_s$, which is reasonable since a photon vertex $\propto k_\mu$). This is (9.174), so we have again shown the equivalence of subtraction and multiplicative renormalisation. This concludes the proof that QED is renormalisable to all orders.

9.8   Asymptotic freedom of Yang–Mills theories

In this section we shall perform calculations analogous to those in §9.6, with the aim of showing that at high energies the running coupling in Yang–Mills theories approaches zero, a property known as asymptotic freedom. It turns out that asymptotic freedom is a property possessed by all non-Abelian gauge theories, and, so far as is known, *only* by non-Abelian gauge theories.[36] We shall not study the general case, but confine ourselves, for definiteness and because of the physical relevance of QCD, to $SU(3)$ gauge symmetry. We shall work, as in §9.6, in the 1-loop approximation.

The key to asymptotic behaviour in the quantity $\beta(g)$ defined by (9.64), and g is the Yang–Mills coupling constant (charge), whose physical and bare values are related by an equation analogous to (9.134):

$$\left.\begin{aligned}
\beta(g) &= \mu\frac{\partial g}{\partial \mu}, \\
g_{\mathrm{B}} &= g\mu^{\varepsilon/2}Z_1 Z_2^{-1} Z_3^{-\frac{1}{2}}.
\end{aligned}\right\} \tag{9.177}$$

Here Z_1 is the renormalisation constant for the quark–gluon–gluon vertex, Z_2 that for the quark wave function, and Z_3 that for the gluon wave function (self-energy). We could perform the calculation without referring to quarks, for example finding the renormalisation of the 3-gluon vertex, but it is a bit simpler, and also instructive, to follow as closely as possible the calculations we performed in QED.

We begin, then, by calculating the quark self-energy diagram, shown in Fig. 9.16; a, b, c and d are $SU(3)$ labels. Applying the Feynman rules (7.54) and (7.60) and putting $\alpha = 1$ (the Feynman gauge) gives

$$-\mathrm{i}\Sigma^{ab}(p) = -g^2\mu^{4-d}\int\frac{\mathrm{d}^d k}{(2\pi)^d}\gamma_\mu\frac{1}{p\!\!\!/ - k\!\!\!/ - m}$$

$$\times \gamma_\nu\frac{g^{\mu\nu}}{k^2}(T^c)_{ad}(T^c)_{db}.$$

It is seen immediately that this is simply $(T^c T^c)_{ab}$ multiplied by the

Fig. 9.16. Quark self-energy diagram.

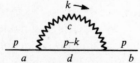

corresponding self-energy expression in QED:

$$\Sigma^{ab}(p) = (T^c T^c)_{ab} \Sigma(\text{QED})$$

$$= (T^c T^c)_{ab} \frac{g^2}{8\pi^2 \varepsilon}(-\not{p} + 4m) \tag{9.178}$$

where we have used (9.97). It remains to calculate the purely group theoretic factor $(T^c T^c)_{ab}$ (c is summed over), where $T^c = \lambda^c/2$, and the λ matrices are displayed in (3.180). It is simple to see that

$$T^c T^c = \tfrac{1}{4}(\lambda_1^2 + \lambda_2^2 + \cdots + \lambda_8^2)$$

$$= \tfrac{4}{3}\mathbf{1} \tag{9.179}$$

where $\mathbf{1}$ is the 3×3 unit matrix, so

$$(T^c T^c)_{ab} = \tfrac{4}{3}\delta_{ab}. \tag{9.180}$$

It is common to denote this quantity $C_2(F)$, and by some group theory it can be shown that for the group $SU(N)$ it has the value

$$(T^c T^c)_{ab} = C_2(F)\delta_{ab}, \tag{9.181}$$

$$C_2(F) = \frac{N^2 - 1}{2N} \quad \text{for } SU(N)$$

$$= \tfrac{4}{3} \quad \text{for } SU(3). \tag{9.182}$$

From (9.178) we then have

$$\Sigma^{ab}(p) = \frac{g^2}{6\pi^2 \varepsilon}(-\not{p} + 4m)\delta^{ab}. \tag{9.183}$$

The fermion (quark) wave function is then renormalised by $\sqrt{Z_2}$ (cf. (9.111)), where Z_2 is given by (cf. (9.110))

$$Z_2 = 1 - \frac{g^2}{6\pi^2 \varepsilon}. \tag{9.184}$$

Next, we turn to vacuum polarisation in QCD, or gluon self-energy. The complete gluon propagator is

$$\tag{9.185}$$

the internal loops corresponding to gluons, ghosts and quarks. The last term is actually zero, since the loop gives a contribution proportional to

$$\int \frac{d^d k}{k^2},$$

but in the method of dimensional regularisation we have

$$\int d^d k (k^2)^{a-1} = 0 \quad (a = 0, 1, \ldots n).$$

We have, then, three loops to calculate. The first is displayed in Fig. 9.17, and gives a contribution

$$i\Pi^{ab}_{\mu\nu}(1) = \text{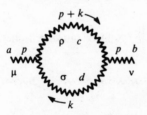}$$

$$= -\tfrac{1}{2} g^2 \mu^{4-d} f^{acd} f^{bdc} \int \frac{d^d k}{(2\pi)^d} \frac{E_{\mu\nu}}{(p+k)^2 k^2} \tag{9.186}$$

where

$$E_{\mu\nu} = [(-2p-k)_\sigma g_{\mu\rho} + (p+2k)_\mu g_{\sigma\rho} + (p-k)_\rho g_{\mu\sigma}]$$
$$\times [(p-k)^\rho g^\sigma_\nu + (p+2k)_\nu g^{\sigma\rho} + (-k-2p)^\sigma g^\rho_\nu]. \tag{9.187}$$

The factor $\tfrac{1}{2}$ in (9.186) is the symmetry factor. Using $g^\sigma_\sigma = d$, (9.187) becomes

$$E_{\mu\nu} = p_\mu p_\nu (d-6) + (p_\mu k_\nu + k_\mu p_\nu)(2d-3) + k_\mu k_\nu (4d-6)$$
$$+ g_{\mu\nu}[(2p+k)^2 + (p-k)^2]. \tag{9.188}$$

The procedure is now standard. Introduce the Feynman parameters, put $k' = k + pz$, and use the formulae (9A.4–9A.6) (actually, the terms linear in k' integrate to zero). Finally, the pole terms are extracted using (9.19) and (9.24) and we finish up with

$$\Pi^{ab}_{\mu\nu}(1) = \frac{-g^2}{16\pi^2 \varepsilon} f^{acd} f^{bcd} [\tfrac{11}{3} p_\mu p_\nu - \tfrac{19}{6} g_{\mu\nu} p^2]. \tag{9.189}$$

The ghost contribution is shown in Fig. 9.18. Applying the Feynman

Fig. 9.17. Gluon loop contribution to vacuum polarisation in QCD.

Fig. 9.18. Ghost loop contribution to vacuum polarisation in QCD.

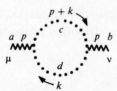

rules (not forgetting the minus sign for the ghost loop) we have

$$i\Pi^{ab}_{\mu\nu}(2) = \text{(diagram)}$$

$$= g^2 f^{cad} f^{dbc} \mu^{4-d} \int \frac{d^d k}{(2\pi)^d} \frac{(p+k)_\mu k_\nu}{(p+k)^2 k^2}$$

and by straightforward application of the techniques used above, we get

$$\Pi^{ab}_{\mu\nu}(2) = \frac{g^2}{16\pi^2 \varepsilon} f^{acd} f^{bcd} [\tfrac{1}{3} p_\mu p_\nu + \tfrac{1}{6} g_{\mu\nu} p^2]. \tag{9.190}$$

Adding the ghost and gluon contributions together we have

$$\Pi^{ab}_{\mu\nu}(1+2) = \frac{g^2}{8\pi^2 \varepsilon} f^{acd} f^{bcd} \tfrac{5}{3} (g_{\mu\nu} p^2 - p_\mu p_\nu). \tag{9.191}$$

Finally, we come to the quark contribution, shown in Fig. 9.19. This is

$$i\Pi^{ab}_{\mu\nu}(3) = \text{(diagram)}$$

$$= -g^2 \mu^{4-d} \int \frac{d^d k}{(2\pi)^d} (T^a)_{dc} (T^b)_{cd}$$

$$\times \text{Tr}\left[\gamma_\mu \frac{1}{\not{p} + \not{k} - m} \gamma_\nu \frac{1}{\not{k} - m} \right]. \tag{9.192}$$

This expression is simply $(T^a)_{dc}(T^b)_{cd} = \text{Tr}(T^a T^b)$ times the corresponding expression for QED – see below equation (9.97). Hence, from (9.98), we get

$$\Pi^{ab}_{\mu\nu}(3) = \text{Tr}(T^a T^b) \frac{g^2}{6\pi^2 \varepsilon} (p_\mu p_\nu - g_{\mu\nu} p^2). \tag{9.193}$$

We now calculate the group theoretic factors appearing in (9.191) and (9.193). The λ matrices (3.180) are normalised such that

$$\text{Tr } \lambda^a \lambda^b = 2\delta^{ab}.$$

Hence, with $T^a = \lambda^a/2$,

$$\text{Tr}(T^a T^b) = \tfrac{1}{2}\delta^{ab}.$$

This is what would be substituted in (9.193) if only quarks belonging to *one* representation of $SU(3)$ contributed to the vacuum polarisation; or, in other words, if there were only one 'flavour' of quark carrying the $SU(3)$ colour

Fig. 9.19. Quark loop contribution to vacuum polarisation in QCD.

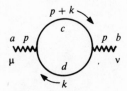

label. But we know this is not true; there are at least five flavours of quark, probably at least six, and possibly more. So if n_F is the number of quark flavours, we have

$$\text{Tr}(T^a T^b) = \frac{n_F}{2}\delta^{ab}. \tag{9.194}$$

Finally, turning to (9.191), it may easily be checked from (3.182) that

$$f^{acd} f^{bcd} = 3\delta^{ab}. \tag{9.195}$$

This number 3 is a Casimir operator of the group, commonly denoted $C_2(G)$, and the above equation is more generally written

$$f^{acd} f^{bcd} = \delta^{ab} C_2(G), \tag{9.196}$$

$$C_2(G) = N \quad \text{for } SU(N)$$
$$= 3 \quad \text{for } SU(3). \tag{9.197}$$

Gathering our results together, the vacuum polarisation tensor becomes

$$\Pi^{ab}_{\mu\nu}(1 + 2 + 3) = \frac{g^2}{8\pi^2\varepsilon}(g_{\mu\nu}p^2 - p_\mu p_\nu)\left(\tfrac{5}{3}C_2(G) - \frac{2n_F}{3}\right)\delta^{ab}$$

$$= \frac{g^2}{8\pi^2\varepsilon}(g_{\mu\nu}p^2 - p_\mu p_\nu)\left(5 - \frac{2n_F}{3}\right)\delta^{ab}. \tag{9.198}$$

The renormalisation constant Z_3 required to cancel this divergence in a counter-term follows immediately by comparing equations (9.98) and (9.119):

$$Z_3 = 1 + \frac{g^2}{8\pi^2\varepsilon}\left(5 - \frac{2n_F}{3}\right). \tag{9.199}$$

An interesting feature of (9.198) and (9.199) is that the contributions of the 'pure Yang–Mills' terms (gluons and ghosts) and of the quarks, to $\Pi_{\mu\nu}$, are opposite in sign. We shall see below that a consequence of this is that asymptotic freedom depends on the number of quark flavours.

We must now calculate the quark–gluon vertex function. Two distinct Feynman diagrams contribute to this, and are shown in Figs. 9.20 and 9.21. Figure 9.20 is analogous to Fig. 9.7 in QED, but Fig. 9.21 is definitely non-

Fig. 9.20. Correction to the quark–gluon vertex.

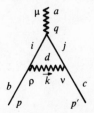

Abelian in character, with its 3-gluon coupling. The contribution of Fig. 9.20 to the vertex function is

$$-ig\mu^{2-d/2}(\Lambda_\mu^a)_{cd}(p,q,p')(1)$$

$$=(-ig\mu^{2-d/2})^3\int\frac{d^dk}{(2\pi)^d}\gamma_\nu(T^d)_{cj}\frac{i}{p\!\!\!/''-k\!\!\!/-m}$$

$$\times\gamma_\mu(T^a)_{ji}\frac{i}{p\!\!\!/-k\!\!\!/-m}(T^d)_{ib}\gamma_\rho\frac{-ig^{\nu\rho}}{k^2}.$$

On comparing this with the corresponding equation for QED (below equation (9.98)), we see (hardly surprisingly) that it differs only by a group theoretic factor

$$\Lambda_\mu^a(p,q,p')(1)=(T^dT^aT^d)\Lambda_\mu(p,q,p')(\text{QED}). \tag{9.200}$$

The group theoretic factor is easily evaluated:

$$T^dT^aT^d=T^d[T^a,T^d]+T^dT^dT^a=if^{adc}T^dT^c+C_2(F)T^a$$

where we have used $T^a=\lambda^a/2$, the commutation relations (3.181), and (9.181). Using the commutation relations again, and (9.196), gives finally

$$T^dT^aT^d=-\tfrac{1}{2}f^{adc}f^{dcb}T^b+C_2(F)T^a$$
$$=[-\tfrac{1}{2}C_2(G)+C_2(F)]T^a. \tag{9.201}$$

Substituting (9.201) and (9.102) into (9.200), we get

$$\Lambda_\mu^a(1)=\frac{g^2}{8\pi^2\varepsilon}\left[-\frac{C_2(G)}{2}+C_2(F)\right]\gamma_\mu T^a. \tag{9.202}$$

We now evaluate the vertex contribution of Fig. 9.21. It is

$$-ig\mu^{2-d/2}\Lambda_\rho^a(2)$$

$$=(-ig)^2(-g)(\mu^{4-d})^{\frac{3}{2}}\int\frac{d^dk}{(2\pi)^d}\gamma_\mu(T^b)_{dm}\frac{-i}{(k-p)^2}f_{abc}$$

$$\times[(p-q-k+p)_\nu g_{\mu\rho}+(k-p-q+k)_\rho g_{\mu\nu}$$

$$+(q-k-p+q)_\mu g_{\nu\rho}]\frac{-i}{(q-k)^2}\frac{i}{k\!\!\!/-m}\gamma_\nu(T^c)_{mn}$$

Fig. 9.21. Correction to the quark–gluon vertex involving the 3-gluon vertex.

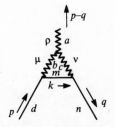

or

$$\Lambda_\rho^a(2) = \frac{g^2\mu^{4-d}}{(2\pi)^d} f^{abc} T^b T^c I_\rho \tag{9.203}$$

where

$$I_\rho = \int d^d k$$

$$\times \frac{\gamma_\mu(\not k + m)\gamma_\nu[(2p - q - k)_\nu g_{\mu\rho} + (2k - p - q)_\rho g_{\mu\nu} + (2q - p - k)_\mu g_{\nu\rho}]}{(k^2 - m^2)(k - p)^2(q - k)^2}. \tag{9.204}$$

I_ρ is evaluated using (9.99) and putting $k' = k - px - qy$, which gives

$$I_\rho = 2\int dx \int dy \int d^d k' \gamma^\mu(\not k' + \not p x + \not q y + m)\gamma^\nu$$

$$\times \{[(2 - x)p - (1 + y)q - k']_\nu g_{\mu\rho}$$

$$+ [2k' + (2x - 1)p + (2y - 1)q]_\rho g_{\mu\nu}$$

$$+ [(2 - y)q - (1 + x)p - k']_\mu g_{\nu\rho}\}$$

$$\times [k'^2 - m^2(1 - x - y) + p^2 x + q^2 y - (px + zy)^2]^{-3}. \tag{9.205}$$

In this integral, the terms linear in k' integrate to zero. Those with no k' in the numerator are finite (convergent) in the limit $d \to 4$, so may be ignored. The divergent part comes from terms in the numerator quadratic in k', which are

$$N_\rho = \gamma^\mu \not k' \gamma^\nu(-k'_\nu g_{\mu\rho} + 2k'_\rho g_{\mu\nu} - k'_\mu g_{\nu\rho})$$

and by a simple bit of 'Diracology' this becomes

$$N_\rho = -2\gamma_\rho k'^2 + (4 - 2d)k'_\rho \not k'.$$

Hence the divergent term is

$$I_\rho = 2\int dx \int dy \int d^d k$$

$$\times \frac{(4 - 2d)k_\rho \not k - 2\gamma_\rho k^2}{[k^2 - m^2(1 - x - y) + p^2 x + q^2 y - (px + qy)^2]^3}.$$

which, by applying the relevant formulae (9A.7), (9A.8) and (9B.1) from the appendices and putting $\varepsilon = 4 - d$, gives a pole part

$$I_\rho = -\frac{6i\pi^2}{\varepsilon}\gamma_\rho. \tag{9.206}$$

Now the group theory factor in (9.203) is, from (9.196),

$$f^{abc} T^b T^c = \frac{i}{2} f^{abc} f^{bcd} T^d$$

$$= \frac{i}{2} C_2(G) T^a, \tag{9.207}$$

so putting (9.207) and (9.206) into (9.203) gives finally (putting $\rho \to \mu$, and $d = 4$)

$$\Lambda_\mu^a(2) = \frac{g^2}{8\pi^2\varepsilon}\left[\frac{3C_2(G)}{2}\right]\gamma_\mu T^a. \tag{9.208}$$

Adding the two vertex contributions (9.202) and (9.208) gives

$$\Lambda_\mu^a = \frac{g^2}{8\pi^2\varepsilon}[C_2(G) + C_2(F)]\gamma_\mu T^a$$

$$= \frac{g^2}{8\pi^2\varepsilon}\frac{13}{3}\gamma_\mu T^a, \tag{9.209}$$

using (9.197) and (9.182). The corresponding renormalisation constant Z_1 is

$$Z_1 = 1 - \frac{g}{8\pi^2\varepsilon}\frac{13}{3}. \tag{9.210}$$

We now bring together equations (9.177), (9.184), (9.199) and (9.210) to give

$$g_B = g\mu^{\varepsilon/2}\left(1 - \frac{g^2}{8\pi^2\varepsilon}\frac{13}{3}\right)\left(1 + \frac{g^2}{8\pi^2\varepsilon}\frac{4}{3}\right)\left[1 - \frac{g^2}{8\pi^2\varepsilon}\left(\frac{5}{2} - \frac{n_F}{3}\right)\right]$$

$$= g\mu^{\varepsilon/2}\left[1 + \frac{g^2}{16\pi^2\varepsilon}\left(-11 + \frac{2n_F}{3}\right)\right], \tag{9.211}$$

from which follows (in the limit $\varepsilon \to 0$)

$$\beta(g) = \mu\frac{\partial g}{\partial\mu} = \frac{g^3}{16\pi^2}\left(-11 + \frac{2n_F}{3}\right). \tag{9.212}$$

If the number of quark flavours is $n_F \leqslant 16$, then $\beta < 0$ and g decreases with increasing mass (momentum) scale μ, so the theory is *asymptotically free*. It seems likely that in nature $n_F < 16$, so asymptotic freedom is a property possessed by QCD, and is the justification of the parton model (see §1.10) according to which partons behave almost like free particles when interacting at high momentum transfer with photons, inside a hadron.

Finally, we may deduce the form of the running coupling constant $\alpha_s' = g^2/4\pi$. Writing

$$\beta(\bar{g}) = \frac{d\bar{g}}{dt}$$

where $t = \ln\mu$, equation (9.212) may be written as

$$\frac{d\bar{g}}{dt} = -b\bar{g}^3$$

$$b = \frac{11 - 2n_F/3}{16\pi^2}. \tag{9.213}$$

Writing this as

$$\frac{d}{dt}(\bar{g}^{-2}) = 2b$$

the solution is clearly

$$\frac{1}{\bar{g}^2} = \frac{1}{g^2} + 2bt,$$

$$\bar{g}^2 = \frac{g^2}{1 + 2btg^2},$$

or

$$\alpha_s(t) = \frac{\alpha_0}{1 + 8\pi bt\alpha_0}$$

where $\alpha_0 = g^2/4\pi$. Now $t = \ln \mu$, which in deep inelastic scattering experiments we may represent as $\frac{1}{2}\ln(Q^2/\Lambda^2)$, so we write

$$\alpha_s(Q^2) = \frac{1}{1/\alpha_0 + 4\pi b \ln(Q^2/\Lambda^2)} \tag{9.214}$$

and we see that $\alpha_s(Q^2)$ goes to zero like $(\ln Q^2)^{-1}$. Ignoring the $1/\alpha_0$ in the denominator, and using (9.213), (9.214) may be written

$$\alpha_s(Q^2) = \frac{4\pi}{(11 - 2n_F/3)\ln Q^2/\Lambda^2} \tag{9.215}$$

where Λ is a scale 'chosen' by the world in which we live.

9.9 Renormalisation of pure Yang–Mills theories

In this section we give a brief outline of the proof of renormalisability of pure Yang–Mills theories; that is, non-Abelian gauge theories with no spontaneous symmetry breaking. The case of spontaneous symmetry breaking is postponed to §9.11. The treatment in this section is based on the lectures of Lee[6] and the book by Taylor,[30] to which the reader is referred for further details.

Before embarking on the proof proper it is useful to check that the power counting argument, presented above for ϕ^4 and QED, indicates renormalisability. For simplicity let us deal with the case when no spinor fields are present. Then there are three types of vertex – 3-vector particles, 4-vector particles, and vector-ghost-ghost. Let

$$\left.\begin{array}{l} n_3 = \text{number of 3-vector vertices,} \\ n_4 = \text{number of 4-vector vertices,} \\ n_g = \text{number of vector-ghost-vector vertices,} \\ V_e = \text{number of external (vector) lines,} \\ V_i = \text{number of internal vector lines,} \\ G_i = \text{number of internal ghost lines.} \end{array}\right\} \tag{9.216}$$

The equation for the superficial degree of divergence is analogous to (9.78)

$$D = 4L - 2V_i - 2G_i + n_3 + n_g. \tag{9.217}$$

Here we have put $d = 4$, and n_3 and n_g are present because these vertices contain one power of momentum (see the Feynman rules). Analogous to (9.81) we have, for the number of loops

$$L = V_i + G_i - n_3 - n_4 - n_g + 1$$
$$= V_i - n_3 - n_4 + 1 \tag{9.218}$$

since $n_g = G_i$. Finally, analogous to (9.82), we have

$$4n_4 + 3n_3 + 3n_g = 2V_i + 2G_i + V_e. \tag{9.219}$$

Substituting (9.219) and (9.218) into (9.217) easily gives

$$D = 4 - V_e, \tag{9.220}$$

showing that the number of primitively divergent graphs is finite. Hence Yang–Mills theory should be renormalisable. (We shall see, when we consider spontaneously broken theories, that the above argument does not hold because the vector propagators are different, and the possibility of disaster looms.)

We return now to the task of showing that the theory is indeed renormalisable. The crucial physical objects are the 1PI functions Γ. What has to be shown is that we can rescale the gauge, ghost and source fields (see equation (7.144)) and coupling constant

$$A_{a\mu} = Z^{\frac{1}{2}}(\varepsilon)A^r_{a\mu}, \quad \eta_a = \tilde{Z}^{\frac{1}{2}}(\varepsilon)\eta^r_a, \quad \tilde{\eta}_a = \tilde{Z}^{\frac{1}{2}}(\varepsilon)\eta^r_a,$$

$$u_{a\mu} = \tilde{Z}^{\frac{1}{2}}(\varepsilon)u^r_a, \quad v_a = Z^{\frac{1}{2}}(\varepsilon)v^r_a, \quad g = \frac{X(\varepsilon)}{\tilde{Z}(\varepsilon)Z^{\frac{1}{2}}(\varepsilon)}g^r, \tag{9.221}$$

so that the generating function Γ is finite as $\varepsilon \to 0$, and gauge invariance is preserved *at each order*. This requirement is contained in the Slavnov–Taylor identity:

$$\int dx \left(\frac{\delta \Gamma'}{\delta u^{a\mu}} \frac{\delta \Gamma'}{\delta A^a_\mu} + \frac{\delta \Gamma'}{\delta v^a} \frac{\delta \Gamma'}{\delta \eta^a} \right) = 0. \tag{7.156}$$

Defining

$$\Gamma_1 * \Gamma_2 = \int dx \left(\frac{\delta \Gamma_1}{\delta u} \frac{\delta \Gamma_2}{\delta A} + \frac{\delta \Gamma_1}{\delta v} \frac{\delta \Gamma_2}{\delta \eta} \right), \tag{9.222}$$

we write (7.156) in the symbolic form

$$\Gamma * \Gamma = 0. \tag{9.223}$$

We now perform a loop expansion ($=$ expansion in powers of $\hbar$) on Γ

$$\Gamma = \sum_{n=0}^{\infty} \Gamma_n \tag{9.224}$$

and also on the renormalisation constants, so that $Z(\varepsilon) = 1 + \sum_{n=1} Z_n(\varepsilon)$, etc. Substituting (9.224) into (9.223) gives, up to nth order,

$$\sum_{p+q=n} \Gamma_p * \Gamma_q = 0. \tag{9.225}$$

The problem which faces us is that Γ turns out to be divergent. It is made finite by adding counter-terms to the action. This suggests the following notation. The action with no counter-terms (CT) is (apart from finite additions) the renormalised action, so we put

$$
\begin{aligned}
S_0 &= S^r, \\
S_1 &= S^r + (S_1)_{CT}, \\
S_2 &= S^r + (S_2)_{CT}, \\
&\vdots \\
S_\infty &= S^B,
\end{aligned}
\tag{9.226}
$$

where S^B is the bare action.

Now $\Gamma_0(S_0)$ is finite, because it is the zero-loop functional, which contains no divergences. Actually, as we shall see in §9.11, $\Gamma_0(S_0) = S_0$. $\Gamma_1(S_0)$, however, diverges, and may be split into finite and divergent parts:

$$\Gamma_1(S_0) = \Gamma_1^f(S_0) + \Gamma^{div}(S_0). \tag{9.227}$$

To zeroth order (9.225) reads

$$S_0 * S_0 = 0 \tag{9.228}$$

and to first order, with (9.227),

$$S_0 * \Gamma_1^f(S_0) + \Gamma_1^f(S_0) * S_0 = 0 \tag{9.229}$$

$$S_0 * \Gamma_1^{div}(S_0) + \Gamma_1^{div}(S_0) * S_0 = 0. \tag{9.230}$$

We now have to find a counter-term $(S_1)_{CT}$ such that $\Gamma_1(S_1)$ is finite, and

$$S_1 * S_1 = 0 \tag{9.231}$$

in parallel with (9.228). (Our technique is clearly a recursive one, and the action must always satisfy this identity.) To order $\hbar$, the new action is

$$S_1 = S_0 - \Gamma_1^{div}(S_0) \tag{9.232}$$

and the renormalised generating functional is

$$\Gamma_1(S_1) = \Gamma_1(S_0) + S_1 - S_0 \tag{9.233}$$

which, by construction, is finite. To show that S_1, defined by (9.232), obeys (9.231), we observe that

$$
\begin{aligned}
S_1 * S_1 &= S_0 * S_0 - S_0 * \Gamma_1^{div} - \Gamma_1^{div} * S_0 + \Gamma_1^{div} * \Gamma_1^{div} \\
&= \Gamma_1^{div} * \Gamma_1^{div} \sim O(\hbar^2)
\end{aligned}
\tag{9.234}
$$

where we have used (9.228) and (9.230). So, to order $\hbar$, (9.231) is satisfied; it is clear that it can be made to be satisfied *exactly* by adding an appropriate finite term of order $\hbar^2$ to (9.232).

We have thus made the zero- and one-loop functionals finite, by counter-terms which respect the Slavnov–Taylor identities. The proof that this is possible for Γ_n is made by induction; Γ_{n-1} is assumed to be finite, and one shows how to construct a finite Γ_n. To nth order, (9.225) reads

$$\Gamma_n * \Gamma_0 + \Gamma_0 * \Gamma_n = -\Gamma_{n-1} * \Gamma_1 - \Gamma_1 * \Gamma_{n-1} - \cdots.$$

Since all terms on the right-hand side are finite (by hypothesis) the divergent part of Γ_n satisfies

$$\Gamma_n^{\text{div}}(S_{n-1}) * S_0 + S_0 * \Gamma_n^{\text{div}}(S_{n-1}) = 0 \tag{9.235}$$

which is, from (9.222),

$$\left[\left(\frac{\delta S_0}{\delta u} \frac{\delta}{\delta A} + \frac{\delta S_0}{\delta v} \frac{\delta}{\delta \eta} \right) + \left(\frac{\delta S_0}{\delta A} \frac{\delta}{\delta u} + \frac{\delta S_0}{\delta \eta} \frac{\delta}{\delta v} \right) \right] \Gamma_n^{\text{div}} = 0. \tag{9.236}$$

We write this as

$$\mathscr{G} \Gamma_n^{\text{div}} = 0 \tag{9.237}$$

with

$$\mathscr{G} = \mathscr{G}_0 + \mathscr{G}_1,$$

$$\mathscr{G}_0 = \frac{\delta S_0}{\delta u} \frac{\delta}{\delta A} + \frac{\delta S_0}{\delta v} \frac{\delta}{\delta \eta},$$

$$\mathscr{G}_1 = \frac{\delta S_0}{\delta A} \frac{\delta}{\delta u} + \frac{\delta S_0}{\delta \eta} \frac{\delta}{\delta v}. \tag{9.238}$$

Now $\mathscr{G}$ has the important property, which can be proved, that it is nilpotent:

$$\mathscr{G}^2 = 0 \tag{9.239}$$

which means that a special solution to (9.237) is $\Gamma_n^{\text{div}} = \mathscr{G} F(A, \eta, \bar{\eta}, u, v)$ where F is some function. In fact, the general solution (9.237), first conjectured by Kluberg-Stern and Zuber[‡], and later proved by Joglekar and Lee,[‡‡] is

$$\Gamma_n^{\text{div}} = G(A) + \mathscr{G} F(A, \eta, \bar{\eta}, u, v) \tag{9.240}$$

where $G(A)$ is a gauge-invariant functional

$$D^\mu(A) \frac{\delta}{\delta A_\mu} G(A) = 0. \tag{9.241}$$

[‡] H. Kluberg-Stern & J.-B. Zuber, *Physical Review* **D12**, 427, 482, 3159 (1975).
[‡‡] S.D. Joglekar & B.W. Lee, *Annals of Physics*, **97**, 160 (1976).

Explicit results may be obtained from (9.240) and (9.241), and these yield explicit expressions, in the loop expansion, for the renormalisation constants $Z, \tilde{Z}$ and X, of equation (9.221).

9.10   Chiral anomalies

In QED and QCD, the only type of interaction between matter and gauge fields is a vector interaction, of the form $g_V J_\mu W^\mu$ where W^μ is the gauge field and $J_\mu \sim \bar{\psi}\gamma_\mu\psi$ is the vector current (internal indices being suppressed) of the Fermi matter field. The vector current is conserved:

$$\partial^\mu J_\mu = 0, \tag{9.242}$$

and this leads to a Ward identity for the vertex function: for the graph of Fig. 9.22, for example, whose amplitude is

$$W^\mu \langle p'|J_\mu|p \rangle, \tag{9.243}$$

current conservation implies

$$(p' - p)^\mu J_\mu = q^\mu J_\mu = 0 \tag{9.244}$$

which is the Ward identity.

In the Weinberg–Salam theory, however – more generally, in a gauge theory of weak interactions, commonly called *quantum flavour-dynamics* or QFD – there is also an *axial vector* coupling between matter and gauge fields, $g_A J_\mu^5 W^\mu$, with

$$J_\mu^5 = \bar{\psi}\gamma_\mu\gamma_5\psi. \tag{9.245}$$

Simple use of the Dirac equation shows that

$$\partial^\mu J_\mu^5 = 2im\bar{\psi}\gamma_5\psi \equiv 2mJ_5. \tag{9.246}$$

J_5 may be called the chiral density. The axial current is not conserved unless $m = 0$, but (9.246) does give rise to an *axial Ward identity*, even in the case $m \neq 0$. To keep matters simple, we shall consider below the situation when $m = 0$, so the axial current is exactly conserved. This simplification is amply justified because the complications which arise in the massive case remain unaffected by putting $m = 0$.

Fig. 9.22. Expansion of the vector coupling between gauge and matter fields to order e^3.

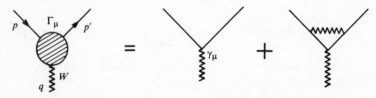

The statements above are of course purely formal. What we now do is to set about verifying them in *perturbation theory*. For the vector coupling, this was done in chapter 7; an expansion of the vertex function Γ_μ to order e^3 gives the two diagrams of Fig. 9.22, and it is found that the Ward identity

$$(p' - p)^\mu \Gamma_\mu = 0$$

is indeed satisfied.

An analogous expansion of the axial vertex is shown in Fig. 9.23, up to order e^5. In each diagram in the expansion, the lowest vertex is $\gamma_\mu \gamma_5$, an axial coupling, and the others are vector couplings. What we find is that the last graph, containing a 'triangle' closed loop of Fermi fields, fails to satisfy the axial Ward identities, giving rise to the so-called axial, or chiral, or triangle anomaly. The serious nature of this anomaly lies in the fact that, as has been emphasised already, the Ward identity (and its appropriate generalisation in the non-Abelian case) is essential to proving the renormalisability of gauge theories; so the triangle anomaly threatens renormalisability of the Weinberg–Salam model, which would be a disaster! The only way of saving renormalisability is to ensure that the *total* contribution of the triangle graphs is zero, so the anomalies cancel. This is a condition on the fermion content of the theory, which, remarkably, turns out to be satisfied in the Weinberg–Salam model if there exist quarks as well as leptons, and if the quarks carry an additional $SU(3)$ (colour) label.

This is the sequence of arguments on which we shall now embark, but before doing so it is worth noting that the chiral anomaly is actually rather an old problem, and was first encountered in an analysis of $\pi^0 \to 2\gamma$ decay by Steinberger in 1949 and Schwinger in 1951. It was ignored for some time but

Fig. 9.23. Expansion of the axial vector vertex between gauge and matter fields to order e^5.

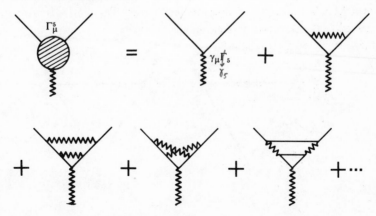

resurfaced in the work of Sutherland in 1966 who showed that application of current algebra and PCAC to π^0 decay gave a very small rate (actually a vanishing rate in the unphysical limit $m_\pi \to 0$), in contradiction with experiment. It was Adler, and Bell and Jackiw, who later realised the important role of the axial anomaly in the Sutherland paradox, and, in fact, the anomaly is sometimes called the Adler–Bell–Jackiw or ABJ anomaly. The Sutherland paradox disappears when the anomaly cancellation condition is applied.

Let us now consider the fermion triangle in Fig. 9.23. (Actually, it is not only the AVV triangle ($A = \gamma_\mu \gamma_5$ vertex, $V = \gamma_\mu$ vertex) which contributes to Γ_μ^5, but also the AAA triangle, and also square and pentagon configurations, which occur in higher orders. It turns out, however, that when the anomaly is cancelled in the VVA graph, it disappears in all of them, so it is sufficient to consider the VVA triangle. It also turns out that the anomaly is unaffected by radiative corrections. So, again, we need only consider the simplest VVA graph. For details of these matters, the student is referred to the literature.) There are two contributions to it, shown in Fig. 9.24. The fermion contribution to the amplitude (i.e. we ignore the gauge field propagators) is

$$T_{\kappa\lambda\mu}(p_1,p_2) = S_{\kappa\lambda\mu}(p_1,p_2) + S_{\lambda\kappa\mu}(p_2,p_1), \tag{9.247}$$

since the second graph in Fig. 9.24 is obtained from the first by interchanging $\kappa \leftrightarrow \lambda$, $p_1 \leftrightarrow p_2$. The Feynman rules give (ignoring the coupling constants, which in a real process will depend on θ_W)

$$S_{\kappa\lambda\mu}(p_1,p_2) = -(-\mathrm{i})^3 \int \frac{\mathrm{d}^4k}{(2\pi)^4}$$

$$\times \mathrm{Tr}\left(\gamma_\kappa \frac{\mathrm{i}}{k\!\!\!/ - p\!\!\!/_1 - m} \gamma_\mu \gamma_5 \frac{\mathrm{i}}{k\!\!\!/ + p\!\!\!/_2 - m} \gamma_\lambda \frac{\mathrm{i}}{k\!\!\!/ - m} \right).$$

Fig. 9.24. The triangle graphs.

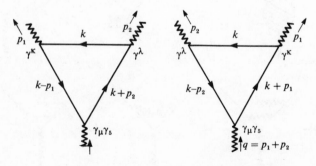

Putting $m = 0$ (see above) gives

$$S_{\kappa\lambda\mu} = -(2\pi)^{-4} \int d^4k \frac{\text{Tr}[\gamma_\kappa(\not{k}-\not{p}_1)\gamma_\mu\gamma_5(\not{k}+\not{p}_2)\gamma_\lambda\not{k}]}{(k-p_1)^2(k+p_2)^2k^2}. \tag{9.248}$$

The expected Ward identities, following from conservation of the vector and axial currents at the three vertices, are

$$\left.\begin{array}{l} (p_1 + p_2)^\mu T_{\kappa\lambda\mu} = 0 \quad (A), \\ p_1^\kappa T_{\kappa\lambda\mu} = 0 \quad (V), \\ p_2^\lambda T_{\kappa\lambda\mu} = 0 \quad (V). \end{array}\right\} \tag{9.249a,b,c}$$

The (V) identities follow from conservation of charge, and what we shall show is that, if these two hold, then the (A) identity (9.249a) does not hold, unless we impose extra conditions.

Note, incidently, that S is a *linearly* divergent integral. Also, it is symmetric under the intercharge $(p_1, \kappa) \leftrightarrow (p_2, \lambda)$, so that the presence of the crossed graph in (9.247) simply results in a factor 2. Hence the identities (9.249) should apply to $S_{\kappa\lambda\mu}$ alone.

We shall proceed to show, following a *naive* and incorrect argument, that equations (9.249) *all* appear to hold[‡]. We shall then have to investigate the fallacy. First, consider (9.249a). We have

$$(p_1 + p_2)^\mu S_{\kappa\lambda\mu}$$
$$= -(2\pi)^{-4} \int d^4k \frac{\text{Tr}[\gamma_\kappa(\not{k}-\not{p}_1)(\not{p}_1+\not{p}_2)\gamma_5(\not{k}+\not{p}_2)\gamma_\lambda\not{k}]}{k^2(k-p_1)^2(k+p_2)^2}. \tag{9.250}$$

Using the identity

$$(\not{p}_1 + \not{p}_2)\gamma_5 = -(\not{k}-\not{p}_1)\gamma_5 - \gamma_5(\not{k}+\not{p}_2),$$

(9.250) becomes

$$(p_1 + p_2)^\mu S_{\kappa\lambda\mu} = (2\pi)^{-4} \int d^4k \frac{\text{Tr}[\gamma_\kappa\gamma_5(\not{k}+\not{p}_2)\gamma_\lambda\not{k}]}{k^2(k+p_2)^2}$$
$$+ (2\pi)^{-4} \int d^4k \frac{\text{Tr}[\gamma_\kappa(\not{k}-\not{p}_1)\gamma_5\gamma_\lambda\not{k}]}{k^2(k-p_1)^2}. \tag{9.251}$$

By a symmetry argument, this must vanish. For $S_{\kappa\lambda\mu}$ is a rank 3 pseudotensor (i.e. with abnormal parity), so the left-hand side of (9.251) is a rank 2 pseudotensor. Each integral on the right-hand side, however, depends only on *one* momentum, so the equation is impossible, and the right-hand side must be zero – verifying (9.249a).

[‡] I follow here the argument of P.H. Frampton, Ohio State University lectures, 1978 (unpublished).

Now consider (9.249*b*). We have

$$p_1^\kappa S_{\kappa\lambda\mu} = -(2\pi)^{-4} \int d^4k \frac{\text{Tr}[\not{p}_1(\not{k}-\not{p}_1)\gamma_\mu\gamma_5(\not{k}+\not{p}_2)\gamma_\lambda\not{k}]}{k^2(k-p_1)^2(k+p_2)^2}. \tag{9.252}$$

Introducing the change of variable

$$k'_\mu = (k+p_2)_\mu, \tag{9.253}$$

and using the cyclic property of traces (and being careless) gives

$$p_1^\kappa S_{\kappa\lambda\mu} = -(2\pi)^{-4} \int d^4k'$$

$$\times \frac{\text{Tr}[(\not{k}'-\not{p}_2)\not{p}_1(\not{k}'-\not{p}_1-\not{p}_2)\gamma_\mu\gamma_5\not{k}'\gamma_\lambda]}{(k'-p_2)^2(k'-p_1-p_2)^2 k'^2}. \tag{9.254}$$

Now observing that

$$\not{p}_1 = (\not{k}'-\not{p}_2) - (\not{k}'-\not{p}_1-\not{p}_2)$$

gives

$$p_1^\kappa S_{\kappa\lambda\mu} = -\int \frac{d^4k'}{(2\pi)^4}$$

$$\times \left\{ \frac{\text{Tr}[(\not{k}'-\not{p}_1-\not{p}_2)\gamma_\mu\gamma_5\not{k}'\gamma_\lambda]}{k'^2(k'-p_1-p_2)^2} - \frac{\text{Tr}[(\not{k}'-\not{p}_2)\gamma_\mu\gamma_5\not{k}'\gamma_\lambda]}{k'^2(k'-p_2)^2} \right\},$$

which must vanish, by the same argument as above. This verifies (9.249*b*). Similarly, (9.249*c*) may apparently be verified by making the change of variable $k'' = k - p_1$, and putting $\not{p}_2 = (\not{k}'' + \not{p}_1 + \not{p}_2) - (\not{k}'' + \not{p}_1)$. We now appear to have demonstrated the truth of the three Ward identities (9.249), but our proof contained a fallacy.

The source of the fallacy is that changing the variable of integration (as in (9.253)) in a *linearly divergent* integral changes the value of the integral by a finite amount. The divergent part of $S_{\kappa\lambda\mu}$ is

$$S_{\kappa\lambda\mu} = -(2\pi)^{-4} \int d^4k \frac{\text{Tr}(\gamma_\kappa \not{k}\gamma_\mu\gamma_5\not{k}\gamma_\lambda\not{k})}{k^6}. \tag{9.255}$$

If k is shifted to $k' = k + a$, the change in $S_{\kappa\lambda\mu}$ is calculated using the following considerations

$$\int d^4k F(k) = \int d^4k' F(k'-a) = \int d^4k' \left[F(k') - a\frac{\partial F}{\partial k} + \cdots \right].$$

The last term is changed into a surface integral by Gauss's theorem. In our calculation, $F \sim k^{-3}$, so since the hypersurface $\sim k^3$, the surface integral is non-vanishing. To find its value, putting $k'_\mu = (k+a)_\mu$ in (9.255), gives

$$S'_{\kappa\lambda\mu} = S_{\kappa\lambda\mu} + U_{\kappa\lambda\mu\nu}a^\nu \tag{9.256}$$

where

$$U_{\kappa\lambda\mu\nu} = -(2\pi)^{-4}\int d^4k\, \frac{\partial}{\partial k^\nu}\left[\frac{\mathrm{Tr}(\gamma_\kappa \slashed{k}\gamma_\mu\gamma_5\slashed{k}\gamma_\lambda\slashed{k})}{k^6}\right]$$

$$= -(2\pi)^{-4}\int d^4k\, \frac{\partial}{\partial k^\nu}\left[\frac{\mathrm{Tr}(\gamma_5\slashed{k}\gamma_\lambda\slashed{k}\gamma_\kappa\slashed{k}\gamma_\mu)}{k^6}\right]. \tag{9.257}$$

Now we use the trace formula

$$\mathrm{Tr}(\gamma_5\gamma_\rho\gamma_\lambda\gamma_\sigma\gamma_\kappa\gamma_\tau\gamma_\mu) = 4i\varepsilon_{\kappa\tau\mu\alpha}(\delta^\alpha_\rho g_{\lambda\sigma} - \delta^\alpha_\lambda g_{\rho\sigma} + \delta^\alpha_\sigma g_{\rho\lambda})$$
$$- 4i\varepsilon_{\rho\lambda\sigma\alpha}(\delta^\alpha_\kappa g_{\tau\mu} - \delta^\alpha_\tau g_{\kappa\mu} + \delta^\alpha_\mu g_{\kappa\tau}) \tag{9.258}$$

which gives

$$\mathrm{Tr}(\gamma_5\slashed{k}\gamma_\lambda\slashed{k}\gamma_\kappa\slashed{k}\gamma_\mu) = 4i\varepsilon_{\kappa\lambda\mu\alpha}k^\alpha k^2.$$

Hence

$$U_{\kappa\lambda\mu\nu} = -\frac{4i}{(2\pi)^4}\varepsilon_{\kappa\lambda\mu\alpha}\int d^4k\, \frac{\partial}{\partial k^\nu}\left(\frac{k^\alpha}{k^4}\right). \tag{9.259}$$

We transform this integral to a Euclidean space, so that $k_4 = ik_0$. Then we observe that if $\alpha \neq \nu$ the integral vanishes, since it is odd in k; and if $\alpha = \nu$ we have $(k^\alpha)^2 = \frac{1}{4}k^2$ ($\alpha = 1,\ldots,4$). So, transforming to a 3-dimensional surface integral yields (using $\oint d\Omega = 2\pi^2$ for S^3)

$$U_{\kappa\lambda\mu\nu} = \frac{4}{(2\pi)^4}\varepsilon_{\kappa\lambda\mu\alpha}\int d^4k_E\, \frac{\partial}{\partial k^\nu}\left(\frac{k^\alpha}{k^4}\right)$$

$$= \frac{1}{(2\pi)^4}\varepsilon_{\kappa\lambda\mu\nu}\int d^4k_E\, \frac{\partial}{\partial k^\alpha}\left(\frac{k^\alpha}{k^4}\right)$$

$$= \frac{1}{(2\pi)^4}\varepsilon_{\kappa\lambda\mu\nu}\oint (d^3 S_E)\alpha\frac{k^\alpha}{k^4}$$

$$= \frac{1}{(2\pi)^4}\varepsilon_{\kappa\lambda\mu\nu}\oint \frac{k_\alpha}{k}(k^3\, d\Omega)\frac{k^\alpha}{k^4}$$

$$= \frac{1}{8\pi^2}\varepsilon_{\kappa\lambda\mu\nu}. \tag{9.260}$$

Referring back to the change of variable introduced in (9.253), we see from (9.256) and (9.260) that $S_{\kappa\lambda\mu}$ should be replaced by

$$S'_{\kappa\lambda\mu}(p_1, p_2) = \frac{1}{8\pi^2}\varepsilon_{\kappa\lambda\mu\nu}p_2^\nu$$

and therefore the Ward identity corresponding to (9.249b) should now read

$$p_1^\kappa S_{\kappa\lambda\mu} = \frac{1}{8\pi^2}\varepsilon_{\kappa\lambda\mu\nu}p_1^\kappa p_2^\nu. \tag{9.261}$$

Similarly, (9.249c) becomes

$$p_2^\lambda S_{\kappa\lambda\mu} = -\frac{1}{8\pi^2}\varepsilon_{\kappa\lambda\mu\nu}p_2^\lambda p_1^\nu, \tag{9.262}$$

whereas the axial Ward identity (9.249a) remains satisfied. We see that it is impossible to satisfy both the vector and the axial vector Ward identities, but of the two the V identity has more right to be regarded as sacrosanct because it corresponds to conservation of charge. In order to *retain* (9.249b, c), then, we redefine the amplitude for the triangle graph to be (cf. (9.247))

$$T'_{\kappa\lambda\mu}(p_1, p_2) = S_{\kappa\lambda\mu}(p_1, p_2) + S_{\lambda\kappa\mu}(p_2, p_1)$$

$$+ \frac{1}{4\pi^2}\varepsilon_{\kappa\lambda\mu\nu}(p_1 - p_2)^\nu \tag{9.263}$$

which satisfies the Ward identities (cf. (9.249))

$$\left.\begin{array}{ll} p_1^\kappa T'_{\kappa\lambda\mu} = 0 & (V), \\ p_2^\lambda T'_{\kappa\lambda\mu} = 0 & (V), \\ (p_1 + p_2)^\mu T'_{\kappa\lambda\mu} = \dfrac{1}{2\pi^2}\varepsilon_{\kappa\lambda\mu\nu}p_2^\mu p_1^\nu & (A). \end{array}\right\} \tag{9.264a, b, c}$$

The vector Ward identities are now satisfied, whereas the axial Ward identity contains an anomaly, which no method of regularisation can avoid. Indeed, in dimensional regularisation the existence of the anomaly is already hinted at by the impossibility of defining a suitable generalisation of γ_5 in d dimensions, as mentioned above.

On reworking the above calculation in the case of massive fermions ($m \neq 0$), it is seen that the axial anomaly is unchanged. On the other hand, in the massless case, the fact that the Ward identity (9.264c) contains an anomaly indicates that the axial current J_μ^5 is not conserved, i.e. there should be an additional term in (9.246). In fact, it is not difficult to see that amending (9.246) to

$$\partial^\mu J_\mu^5 = 2m J_5 + \frac{e^2}{8\pi^2}F_{\mu\nu}\tilde{F}^{\mu\nu} \tag{9.265}$$

where $\tilde{F}^{\mu\nu}$ (the dual of $F^{\mu\nu}$) is defined by (2.234), yields (9.264c), since

$$\begin{aligned} F_{\mu\nu}\tilde{F}^{\mu\nu} &= \tfrac{1}{2}\varepsilon_{\mu\nu\rho\sigma}(\partial^\mu A^\nu - \partial^\nu A^\mu)(\partial^\rho A^\sigma - \partial^\sigma A^\rho) \\ &= 2\varepsilon_{\mu\nu\rho\sigma}\partial^\mu A^\nu \partial^\rho A^\sigma \\ &= \partial^\mu(2\varepsilon_{\mu\nu\rho\sigma}A^\nu\partial^\rho A^\sigma). \end{aligned} \tag{9.266}$$

Further, since this term is a total divergence the 'new' axial current

$$\tilde{J}_\mu^5 = J_\mu^5 - \frac{e^2}{4\pi^2}\varepsilon_{\mu\nu\rho\sigma}A^\nu\partial^\rho A^\sigma \tag{9.267}$$

is divergenceless in the limit $m = 0$:

$$\partial^\mu J^5_\mu = 0 \quad (m = 0),$$

but is not gauge invariant, so cannot be considered as a physical current.

Cancellation of anomalies

The discovery of anomalies in gauge theories is particularly unpleasant because the fact that the Ward identities are not all satisfied means that gauge theories involving axial, as well as vector currents are not, after all, renormalisable if there is only one type of fermion contributing to the triangle diagram. The only hope for retaining renormalisability lies in the possibility that if there are several fermions their separate contributions may cancel. The Weinberg–Salam model involves axial vector currents, and so is an obvious testing-ground for this idea.

There are many processes in this model which would involve the triangle graph, and it turns out that the same cancellation condition emerges from every example, so, to keep matters as simple as possible, consider the diagram in Fig. 9.25, which shows a contribution to neutrino–neutrino scattering, with W^+, W^- and Z^0 coupling to the legs of the triangle, round which propagate the fermions of the theory. In practice, of course, since this process is of such a high order it gives an extremely small contribution to neutrino scattering, but it is crucial in principle that its contribution vanishes *exactly*, otherwise renormalisability is lost. Let M_a, M_b and M_c be the coupling matrices at the three vertices of the triangle for the fermion-gauge boson interaction. It is characteristic of the Weinberg–Salam model that left-handed fermions couple differently from right-handed fermions. In fact, the right-handed particles have no weak isospin, so do not couple to W^+ and W^- (which is what makes this example simple). In general, however, the triangle contribution of the left-handed fermions is proportional to

$$\begin{aligned}(1 - \gamma_5)&(\text{Tr } M^L_a M^L_b M^L_c + \text{Tr } M^L_a M^L_c M^L_b) \\ &= (1 - \gamma_5)\text{Tr } M^L_a \{M^L_b, M^L_c\},\end{aligned} \tag{9.268}$$

Fig. 9.25. A contribution to ν–ν scattering in the Weinberg–Salam model, involving the triangle diagram.

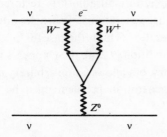

the two terms coming, of course, from the two diagrams of Fig. 9.24. Similarly, the right-handed fermions contribute a term proportional to

$$(1 + \gamma_5)\text{Tr}\, M_a^R\{M_b^R, M_c^R\}. \tag{9.269}$$

The chiral anomaly vanishes if the terms in γ_5 cancel

$$\text{Tr}\, M_a^L\{M_b^L, M_c^L\} = \text{Tr}\, M_a^R\{M_b^R, M_c^R\}. \tag{9.270}$$

Turning to Fig. 9.25 and labelling the W^-, W^+ and Z vertices a, b and c respectively, we have, from the Lagrangian (8.85),

$$M_a^L = gI_+^W, \quad M_b^L = gI_-^W, \tag{9.271}$$

where $I_\pm^W = \frac{1}{2}(I_1^W \pm iI_2^W)$, and I_1^W and I_2^W are the components of weak isospin. To find M_c^L, we rewrite the coupling of the Z boson in (8.85) (with $\theta = \theta_W$) as

$$\frac{g}{\cos\theta}[\tfrac{1}{2}(\bar{v}\gamma^\mu v - \bar{e}_L\gamma^\mu e_L) + \sin^2\theta(\bar{e}_L\gamma^\mu e_L + \bar{e}_R\gamma^\mu e_R)]Z_\mu$$

$$= g\sec\theta(J_3^\mu + \sin^2\theta J_{em}^\mu)Z_\mu. \tag{9.272}$$

Hence

$$M_c^L = g\sec\theta(I_3^W + \sin^2\theta Q). \tag{9.273}$$

Since the right-handed fermions do not couple to $W^\pm$, the right-hand side of (9.270) is zero, and the equation becomes

$$\text{Tr}\,(I_{3L}^W + \sin^2\theta Q_L)\{I_{+L}^W, I_{-L}^W\} = 0.$$

Since $\{I_+, I_-\} = 1$, and $\text{Tr}\, I_3 = \frac{1}{2}\text{Tr}\,\tau_3 = 0$, this condition becomes

$$\sum_i Q_L^i = 0; \tag{9.274}$$

that is, the sum of the electric charges of the left-handed fermions should be zero. In the Weinberg–Salam model with leptons only, this is not satisfied. When hadrons – quarks – are included as well, however, we have, for the case of one generation, and quarks with three colours

$$Q_e + Q_{v_e} + 3(Q_u + Q_d)$$
$$= -1 + 0 + 3(\tfrac{2}{3} - \tfrac{1}{3}) = 0; \tag{9.275}$$

the chiral anomaly disappears, and renormalisability is restored. This condition clearly remains intact if other generations of particles are added with the same charges as the first generation. This is the case for (μ^-, v_μ, c, s) and (τ^-, v_τ, t, b). The anomaly-free condition (9.274), then, appears to shed some light on lepton–hadron symmetry, but allows for an arbitrary number of generations. A solution to the generation problem must be sought elsewhere.

9.11   Renormalisation of Yang–Mills theories
with spontaneous symmetry breakdown

We come now to the final topic of importance for the application of gauge theories to particle physics, which is the proof that Yang–Mills theories with spontaneous symmetry breaking are renormalisable. Formally, what this amounts to showing is that the divergence structure of non-Abelian gauge theories is unaffected by spontaneous symmetry breakdown. On the face of it, the presence of massive vector mesons, which is, of course, a consequence of these theories, seems to destroy renormalisability, but a clever choice of gauge (the 't Hooft gauge) may be made, which restores it. We shall first describe the 't Hooft gauges, and then show more formally, by using the effective potential, that spontaneous symmetry breakdown does not affect the renormalisability of gauge theories. Finally, some consequences of the effective potential will be mentioned.

't Hooft's gauges

We saw in chapter 8 that spontaneous breaking of a gauge theory makes the gauge field quantum massive. It is sufficient for our purposes to consider the Abelian case, where this conclusion follows from the Lagrangian (8.39). By comparison with the Proca Lagrangian (4.95), it describes a vector meson ('photon') of mass

$$M = \sqrt{2}ea. \tag{9.276}$$

We shall now show why we expect theories containing massive vector mesons to be unrenormalisable. The proof follows simply from finding the propagator. We do this, as usual, by finding the inverse of the quadratic (in A_μ) part of $\mathcal{L}$. Dropping surface terms, then, we write (4.95) as

$$\mathcal{L} = \tfrac{1}{2}A^\mu[g_{\mu\nu}(\Box + M^2) - \partial_\mu\partial_\nu]A^\nu. \tag{9.277}$$

In momentum space, the operator in square brackets is

$$g_{\mu\nu}(-k^2 + M^2) + k_\mu k_\nu.$$

Writing its inverse as $Ag^{\nu\rho} + Bk^\nu k^\rho$, we have

$$[g_{\mu\nu}(-k^2 + M^2) + k_\mu k_\nu][Ag^{\nu\rho} + Bk^\nu k^\rho] = \delta_\mu^\rho$$

$$A(-k^2 + M^2)\delta_\mu^\rho + k_\mu k^\rho(A + BM^2) = \delta_\mu^\rho$$

$$A = \frac{-1}{k^2 - M^2}, \quad B = -\frac{A}{M^2}$$

so the vector meson propagator is

$$D_{\mu\nu} = \frac{1}{k^2 - M^2}\left(-g^{\mu\nu} + \frac{k^\mu k^\nu}{M^2}\right). \tag{9.278}$$

The disastrous property of this propagator is that as $k \to \infty$, $D_{\mu\nu} \to$ const, whereas, for example, scalar particles and photons possess propagators scaling as k^{-2} at large k. It is easy to see, by the power counting argument, that we have apparently lost renormalisability, since if the vector particles have propagators, given by (9.278), the superficial degree of divergence of an arbitrary Feynman diagram will be, instead of (9.217),

$$D = 4L - 2G_i + n_3 + n_g. \tag{9.279}$$

Equations (9.218, 9.219) are unchanged, so (9.220) becomes replaced by

$$D = 4 - 4n_4 + 3n_3 + n_g - 2V_e. \tag{9.280}$$

Since D increases with n_3 and n_g, the theory is not renormalisable.

We must recall, however, that (8.39) was obtained from the original Lagrangian (8.37) by performing a particular gauge transformation to get rid of the unwanted mixed term $A^\mu \partial_\mu \phi_2$ in (8.39). This is equivalent to adding a gauge-fixing term. To be more general, however, this unwanted term disappears if we add to (8.37) a gauge-fixing term of the form

$$\mathscr{L}_{GF} = -\frac{1}{2\xi}(\partial_\mu A^\mu + \xi M \phi_2)^2 \tag{9.281}$$

where ξ is a parameter determining the gauge. Depending on the value of ξ, we get a class of gauges called the *'t Hooft gauges*. Adding (9.281) to (8.37) gives

$$\mathscr{L} = -\tfrac{1}{4}F_{\mu\nu}F^{\mu\nu} + \frac{M^2}{2}A_\mu A^\mu - \frac{1}{2\xi}\partial_\mu A^\mu \partial_\nu A^\nu$$

$$+ \tfrac{1}{2}(\partial_\mu \phi_1)^2 - 2\lambda a^2 \phi_1^2 + \tfrac{1}{2}(\partial_\mu \phi_2)^2 - \frac{\xi}{2}M^2 \phi_2^2$$

$$+ \text{coupling terms}. \tag{9.282}$$

The term quadratic in A^μ is

$$\tfrac{1}{2}A^\mu \left[g_{\mu\nu}(\Box + M^2) - \partial_\mu \partial_\nu \left(1 - \frac{1}{\xi}\right) \right] A^\nu.$$

The vector meson propagator follows by repeating the argument which leads from (9.277) to (9.278). We get

$$D_{\mu\nu} = \frac{1}{k^2 - M^2} \left[-g_{\mu\nu} + (1 - \xi)\frac{k_\mu k_\nu}{k^2 - \xi M^2} \right]. \tag{9.283}$$

For finite ξ, as $k_\mu \to \infty$, $D_{\mu\nu} \to k^{-2}$, so renormalisability is *restored*. We have a *renormalisable gauge* (or *R gauge*). On the other hand, the ϕ_2 field has reappeared in (9.282), and will have a propagator with a pole at $k^2 = \xi M^2$. We know that the ϕ_2 field is unphysical (there are only four degrees of freedom – see (8.41) – so as well as the massive spin 1 field we are only allowed one more scalar field, which is ϕ_1) and it may be shown that this pole cancels the other pole in the vector propagator (9.283), also at $k^2 = \xi M^2$; they are both unphysical.

Particular choices of ξ are, for example, $\xi = 1$, which we call the Feynman gauge, since it gives a propagator analogous to that in QED, and $\xi = 0$, which for the same reason we call the Landau gauge. The choice $\xi \to \infty$ gives back the propagator (9.278), which suggests non-renormalisability, but the ϕ_2 field becomes *manifestly* unphysical.

To summarise, we have a class of gauges. Gauges with ξ finite are manifestly renormalisable (R gauges) but are complicated by the presence of five field components, rather than the desired four. Gauges with $\xi \to \infty$ are not manifestly renormalisable, but are physical in the sense that only four fields contribute to the unitarity condition (U gauge). Since S-matrix elements are gauge-independent, they are independent of ξ, and so will be physically sensible, and give rise to renormalisable theory.

Finally, we should mention that the above analysis, carried out for an *Abelian* gauge theory, is also applicable to the non-Abelian case. For details, the reader should consult the literature.

The effective potential

The notion of the effective potential has been found to be a very useful one in connection with theories exhibiting spontaneously broken symmetry. It enables them to be viewed in essentially the same way as theories with unbroken symmetry, and allows one to calculate quantum corrections to the classical picture of spontaneous symmetry breaking which was given in the previous chapter. This has led to interesting results concerning the connections between radiative corrections and spontaneous symmetry breaking, which we will touch on at the end of the chapter.

For simplicity, consider the case of a real scalar field, with Lagrangian and classical action

$$\left. \begin{array}{l} \mathscr{L} = \tfrac{1}{2}(\partial_\mu \phi)^2 - V(\phi); \quad V(\phi) = \dfrac{m^2}{2}\phi^2 - \dfrac{g}{4!}\phi^4, \\[2mm] S[\phi] = \displaystyle\int \mathrm{d}^4x \mathscr{L}(\phi(x)). \end{array} \right\} \qquad (9.284)$$

The Lagrangian is invariant under $\phi \to -\phi$, but this symmetry is not shared by the solution

$$\left. \frac{\mathrm{d}V}{\mathrm{d}\phi} \right|_{\phi=\phi_0} = 0 \qquad (9.285)$$

where $\phi_0 \neq 0$. The symmetry has been spontaneously broken. These statements are classical. Quantum considerations enter with loops (remember the loop expansion is an expansion in $\hbar$), and the divergences which they introduce necessitate renormalisation. The renormalisation conditions – see, for example, (9.37) and (9.42) – are stated in terms of $\Gamma^{(n)}$ the 1PI functions. The generating functional for the $\Gamma^{(n)}(x_1, \ldots, x_n)$ is $\Gamma[\phi]$, so, in short, to take full account of quantum effects we must work in terms of $\Gamma[\phi]$. It is called the *effective action*, a name which will be justified below.

The connected generating functional W is defined by (see (5.68), (6.66) and (6.105))

$$e^{iW[J]} = \langle 0^+ | 0^- \rangle_J. \tag{9.286}$$

The *classical field* ϕ_c (see (7.81)) is defined by

$$\phi_c(x) = \frac{\delta W[J]}{\delta \phi(x)}_J = \frac{\langle 0^+ | \phi_{0N}(x) | 0^- \rangle_J}{\langle 0^+ | 0^- \rangle_J} \tag{9.287}$$

and is seen to depend on the source $J(x)$. The vacuum expectation value $\langle \phi \rangle$ is defined as

$$\langle \phi \rangle = \lim_{J \to 0} \phi_c. \tag{9.288}$$

From (7.80), $\Gamma[\phi_c]$ is defined as

$$\Gamma[\phi_c] = W[J] - \int dx J(x) \phi_c(x), \tag{9.289}$$

and from (7.81) it obeys

$$\frac{\delta \Gamma[\phi_c]}{\delta \phi_c(x)} = -J(x). \tag{9.290}$$

When $J(x) \to 0$, $\phi_c(x)$ becomes a constant, which, from above, is $\langle \phi \rangle$, so $\langle \phi \rangle$ is a solution of the equation

$$\left. \frac{d\Gamma[\phi_c]}{d\phi_c} \right|_{\langle \phi \rangle} = 0. \tag{9.291}$$

The usual expansion of $\Gamma[\phi_c]$, in powers of ϕ_c, is

$$\Gamma[\phi_c] = \sum_{n=0}^{\infty} \frac{1}{n!} \int dx_1 \ldots dx_n \Gamma^{(n)}(x_1, \ldots, x_n) \phi_c(x_1) \ldots \phi_c(x_n), \tag{9.292a}$$

or, equivalently, by Fourier transform

$$\Gamma[\phi_c] = \sum_{n=0}^{\infty} \frac{1}{n!} \int dp_1 \ldots dp_n \delta^4(p_1 + \cdots + p_n)$$
$$\times \Gamma^{(n)}(p_1, \ldots, p_n) \tilde{\phi}_c(p_1) \ldots \tilde{\phi}_c(p_n). \tag{9.292b}$$

An alternative expansion is one in terms of ϕ_c and its derivatives, as follows:

$$\Gamma[\phi_c] = \int dx [-U(\phi_c(x)) - \tfrac{1}{2}(\partial_\mu \phi_c)^2 Z(\phi_c(x)) + \ldots]. \tag{9.293}$$

The function (not functional) $U(\phi_c)$ is called the *effective potential*. We shall see below that in the classical limit it is equal to the potential V. In the case where $\phi_c(x) = \langle \phi \rangle = a$, a constant, all terms in the above expansion vanish, except the first one, so

$$\Gamma[a] = -\Omega U(a) \tag{9.294}$$

where Ω is the total volume of space–time.

Comparing the expansions (9.292*b*) and (9.293) we have

$$U(a) = -\sum_{n=0}^{\infty} \frac{1}{n!} a^n \Gamma^{(n)}(p_i = 0). \tag{9.295}$$

Renormalisation conditions, such as (9.37) and (9.42), on $\Gamma^{(2)}(p_i = 0)$ and $\Gamma^{(4)}(p_i = 0)$ may then be stated in terms of U:

$$\frac{\delta S[\phi, J]}{\delta\phi(x)}\bigg|_{\langle\phi\rangle} = m^2 \tag{9.296}$$

$$-i\frac{d^4 U(\phi_c)}{d\phi_c^4}\bigg|_{\langle\phi\rangle} = g. \tag{9.297}$$

In addition, the condition (9.291) on the vacuum expectation value becomes

$$\frac{dU(\phi_c)}{d\phi_c}\bigg|_{\langle\phi\rangle} = 0. \tag{9.298}$$

To investigate the properties of the spontaneously broken theory, we now define a new quantum field ϕ':

$$\phi' = \phi - \langle \phi \rangle \tag{9.299}$$

which has vanishing vacuum expectation value. So, in the words of Coleman,[40] 'We have recreated the entire structure of our study of spontaneous symmetry breakdown in classical field theory. The only difference is that, instead of working with the classical potential V, we work with the effective potential U.'[‡] Moreover, since all the divergences of the theory have been taken up by the counter-terms *before* the renormalisation conditions (9.296) and (9.297) are applied, it is clear that no new divergences occur in a theory with spontaneous symmetry breaking, over and above those in an unbroken theory. So, again in Coleman's words, 'The divergence structure of a renormalisable field theory is not affected by the occurrence of spontaneous symmetry breakdown.' The physical implication of this important result is that the Weinberg–Salam model is renormalisable, since, although there are massive vector particles, which in the ordinary way would destroy renormalisability, the existence of these is due to the spontaneous symmetry breaking of a non-Abelian gauge theory.

[‡] Coleman denotes the classical potential by U, and the effective potential by V.

'Pure' non-Abelian gauge theories are renormalisable, as we saw in §9.9, and we have just seen that spontaneous breaking of the symmetry does not affect renormalisation. These results are the climax of a beautiful and important chapter in twentieth century physics.

Loop expansion of the effective potential

We come to final task, which is to calculate the effective potential. We employ equation (9.289), and begin by calculating $W[J]$ by the method of saddle-point evaluation of the path integral. Suppose we have an integral of the form

$$I = \int e^{-f(x)} \, dx \tag{9.300}$$

and $f(x)$ is stationary at some point x_0, then by expanding $f(x)$ about x_0:

$$f(x) = f(x_0) + \tfrac{1}{2}(x - x_0)^2 f''(x_0) + \cdots, \tag{9.301}$$

the integral is seen to be

$$I \approx e^{-f(x_0)} \int e^{-\frac{1}{2}(x-x_0)^2 f''(x_0)}. \tag{9.302}$$

Hence the integral becomes a Gaussian, and is easily evaluated.

Restoring Planck's constant to the definition of $W[J]$, we have (ignoring overall normalisation)

$$e^{(i/\hbar)W[J]} = \int \mathscr{D}\phi \, e^{(i/\hbar)S[\phi, J]} \tag{9.303}$$

where

$$S[\phi, J] = \int d^4x [\mathscr{L}(\phi) + \hbar\phi(x)J(x)]. \tag{9.304}$$

From (9.284, 9.285) we have

$$\left. \frac{\delta S[\phi, J]}{\delta \phi(x)} \right|_{\phi_0} = \hbar J(x). \tag{9.305}$$

Now we expand the action around ϕ_0:

$$\begin{aligned}
S[\phi, J] &= S[\phi_0, J] + \int dx [\phi(x) - \phi_0] \left. \frac{\delta S}{\delta \phi(x)} \right|_{\phi_0} \\
&\quad + \frac{1}{2} \int dx \, dy [\phi(x) - \phi_0][\phi(y) - \phi_0] \left. \frac{\delta^2 S}{\delta \phi(x)\delta \phi(y)} \right|_{\phi_0} + \cdots \\
&= S[\phi_0] + \hbar \int dx \, \phi(x)J(x) \\
&\quad + \frac{1}{2} \int dx \, dy [\phi(x) - \phi_0] \left. \frac{\delta^2 S}{\delta \phi(x)\delta \phi(y)} \right|_{\phi_0} [\phi(y) - \phi_0] + \cdots,
\end{aligned} \tag{9.306}$$

where (9.305) has been used. By functional differentiation of the action (9.284), it is easily seen that

$$\frac{\delta^2 S}{\delta\phi(x)\delta\phi(y)}\Big|_{\phi_0} = -[\Box + V''(\phi_0)]\delta(x - y). \tag{9.307}$$

By putting $\phi' = \phi - \phi_0$, (9.306) then becomes

$$S[\phi, J] = S[\phi_0, J] + \hbar \int dx\phi'(x)J(x)$$

$$-\frac{1}{2}\int dx\phi'(x)[\Box + V''(\phi_0)]\phi'(x) + \cdots.$$

Substituting this into (9.303) and appealing to saddle-point integration (9.302) gives (relabelling ϕ' as ϕ)

$$\exp\left(\frac{i}{\hbar}W\right) = \exp\left\{\frac{i}{\hbar}S[\phi_0, J]\right\}$$

$$\times \int \mathcal{D}\phi \exp\left\{\frac{-i}{\hbar}\frac{1}{2}\int dx\phi[\Box + V''(\phi_0)]\phi\right\}. \tag{9.308}$$

To obtain the loop expansion (equal to the expansion in $\hbar$) we rescale the field ϕ into $\hbar^{\frac{1}{2}}\phi$, so the $\hbar$ in the final exponent above vanishes. To evaluate the integral in this exponent, we go over to Euclidean space (which eliminates the i), and use the functional formula (6.30), giving

$$\exp\left(\frac{i}{\hbar}W\right) = \exp\left\{\frac{i}{\hbar}S[\phi_0, J]\right\}\{\det[\Box + V''(\phi_0)]\}^{-\frac{1}{2}}. \tag{9.309}$$

Using the general formula

$$\det A = e^{\text{Tr}\ln A} \tag{9.310}$$

gives

$$W[J] = S[\phi_0] + \hbar \int dx\phi_0(x)J(x) + \frac{i\hbar}{2}\text{Tr}\ln[\Box + V''(\phi_0)]. \tag{9.311}$$

This is the loop expansion of $W[J]$; corrections terms (of order $\hbar^2$) have been ignored.

To find the loop expansion of $\Gamma[\phi_c]$, we substitute (9.311) into (9.289), but first we need to find $S[\phi_c]$ in terms of $S[\phi_0]$, which is easily done. Putting $\phi_0 = \phi_c - \phi_1$,

$$S[\phi_0] = S[\phi_c - \phi_1]$$

$$= S[\phi_c] - \int dx\phi_1(x)\frac{\delta S}{\delta\phi(x)}\Big|_{\phi_0} + O(\hbar^2)$$

$$= S[\phi_c] - \hbar \int dx\phi_1(x)J(x) + O(\hbar^2). \tag{9.312}$$

Substituting (9.311), (9.312) into (9.289) then gives, on putting the source $J \to 0$,

$$\Gamma[\phi_c] = S[\phi_c] + \frac{i\hbar}{2} \operatorname{Tr} \ln[\Box + V''(\phi_0)]. \tag{9.313}$$

Now we put $\phi_c(x) = a$ (constant). Then $\Gamma[a]$ is given by (9.294), and it follows from (9.284) that $S[a] = -\Omega V(a)$, so (9.313) implies that

$$U(a) = V(a) - \frac{i\hbar}{2} \Omega^{-1} \operatorname{Tr} \ln[\Box + V''(a)]. \tag{9.314}$$

This shows that in the classical limit, which is the tree approximation, the effective potential $U(a)$ becomes the same as the classical potential $V(a)$ – as indeed (9.313) shows that the effective action becomes the same as the classical action.

Since the trace of an operator is the sum (integral) over its eigenvalues, equation (9.314), on going into (Euclidean) momentum space, may be written,

$$\begin{aligned} U(a) &= V(a) + \frac{\hbar}{2} \int \frac{dk_E}{(2\pi)^4} \ln[k_E^2 + V''(a)] \\ &= V(a) + \frac{\hbar}{2} \int \frac{dk_E}{(2\pi)^4} \ln\left(k_E^2 + m^2 - \frac{ga^2}{2}\right). \end{aligned} \tag{9.315}$$

Now consider the following question. When $m^2 > 0$ (in ϕ^4 theory) the vacuum is non-degenerate, and when $m^2 < 0$ it is degenerate; what sort of vacuum do we have when $m^2 = 0$? The answer is provided by appealing to the effective potential. Imposing the condition that the renormalised mass is still zero (equation (9.296)) it turns out that the minimum of the effective potential occurs for $\langle \phi \rangle \neq 0$, so there *is* spontaneous symmetry breaking, induced by radiative corrections.[‡]

Appendix A: integration in d dimensions

We work in d-dimensional 'Minkowski' space, with one timelike and $(d-1)$ spacelike dimensions. We are interested in integrals of the type

$$I_d(q) = \int \frac{d^d q}{(p^2 + 2pq - m^2)^\alpha} \tag{9A.1}$$

where $p = (p_0, \mathbf{r})$. We introduce polar co-ordinates

$$(p_0, r, \phi, \theta_1, \theta_2, \dots, \theta_{d-3})$$

[‡] Actually, as pointed out by Coleman and Weinberg,[56] this is slightly misleading, since in ϕ^4 theory 'the new minimum lies very far outside the expected range of validity of the 1-loop approximation'. This criticism does not apply, however, to massless scalar electrodynamics.

so that

$$d^d p = dp_0 r^{d-2} \, dr \, d\phi \sin\theta_1 \, d\theta_1 \sin^2\theta_2 \, d\theta_2 \ldots \sin^{d-3}\theta_{d-3} \, d\theta_{d-3}$$

$$= dp_0 r^{d-2} \, dr \, d\phi \prod_{k=1}^{d-3} \sin^k\theta_k \, d\theta_k$$

$$(-\infty < p_0 < \infty, \quad 0 < r < \infty, \quad 0 < \phi < 2\pi, \quad 0 < \theta_i < \pi).$$

Then

$$I_d(q) = 2\pi \int_{-\infty}^{\infty} dp_0 \int_0^{\infty} r^{d-2} \, dr \int_0^{\pi} \frac{\prod_1^{d-3} \sin^k\theta_k \, d\theta_k}{(p^2 + 2pq - m^2)^\alpha}.$$

Now we use the formula

$$\int_0^{\pi/2} (\sin\theta)^{2n-1}(\cos\theta)^{2m-1} \, d\theta = \frac{1}{2} \frac{\Gamma(n)\Gamma(m)}{\Gamma(n+m)}^\ddagger$$

and put $m = \frac{1}{2}$ (remembering that $\Gamma(\frac{1}{2}) = \sqrt{\pi}$), giving

$$\int_0^{\pi} (\sin\theta)^k \, d\theta = \sqrt{\pi} \frac{\Gamma\left(\dfrac{k+1}{2}\right)}{\Gamma\left(\dfrac{k+2}{2}\right)}$$

and hence

$$I_d(q) = \frac{2\pi^{(d-1)/2}}{\Gamma\left(\dfrac{d-1}{2}\right)} \int_{-\infty}^{\infty} dp_0 \int_0^{\infty} \frac{r^{d-2} \, dr}{(p_0^2 - r^2 + 2pq - m^2)^\alpha}.$$

This integral is 'Lorentz' invariant, so we evaluate it in the frame $q_\mu = (\mu, \mathbf{0})$. Then $2pq = 2\mu p_0$. Changing variables to $p'_\mu = p_\mu + q_\mu$, which implies that $p_0'^2 - q^2 = p_0^2 + 2\mu p_0$, we have

$$I_d(q) = \frac{2\pi^{(d-1)/2}}{\Gamma\left(\dfrac{d-1}{2}\right)} \int_{-\infty}^{\infty} dp'_0 \int_0^{\infty} \frac{r^{d-2} \, dr}{[p_0'^2 - r^2 - (q^2 + m^2)]^\alpha}. \tag{9A.2}$$

The Euler beta function is

$$B(x, y) = \frac{\Gamma(x)\Gamma(y)}{\Gamma(x+y)} = 2 \int_0^{\infty} dt \, t^{2x-1}(1 + t^2)^{-x-y\ddagger}$$

(valid for $\mathrm{Re}\,x > 0, \mathrm{Re}\,y > 0$), so putting

$$x = \frac{1+\beta}{2}, \quad y = \alpha - \frac{1+\beta}{2}, \quad t = \frac{s}{M}$$

---

‡ See, for example, M. Abramowitz & I.A. Stegun, *Handbook of Mathematical Functions*, section 6.2, Dover Publications, 1965.

we have

$$\int_0^\infty ds \frac{s^\beta}{(s^2 + M^2)^\alpha} = \frac{\Gamma\left(\frac{1+\beta}{2}\right)\Gamma\left(\alpha - \frac{1+\beta}{2}\right)}{2(M^2)^{\alpha - (1+\beta)/2}\Gamma(\alpha)}. \tag{9A.3}$$

Substituting this into (9A.2), with $\beta = d - 2$, $M^2 = -p_0'^2 + q^2 + m^2$, gives

$$I_d(q) = (-1)^{-\alpha}\pi^{(d-1)/2}\frac{\Gamma\left(\alpha - \frac{d-1}{2}\right)}{\Gamma(\alpha)}$$

$$\times \int_{-\infty}^\infty \frac{dp_0'}{(q^2 + m^2 - p_0'^2)^{\alpha - (d-1)/2}}$$

$$= (-1)^{2\alpha + (d-1)/2}\pi^{(d-1)/2}\frac{\Gamma\left(\alpha - \frac{d-1}{2}\right)}{\Gamma(\alpha)}$$

$$\times \int_{-\infty}^\infty \frac{dp_0'}{[p_0'^2 - (q^2 + m^2)]^{\alpha - (d-1)/2}}.$$

To evaluate this integral, we invoke (9A.3) again to obtain

$$I_d(q) = i\pi^{d/2}\frac{\Gamma\left(\alpha - \frac{d}{2}\right)}{\Gamma(\alpha)}\frac{1}{[-(q^2 + m^2)]^{\alpha - d/2}}$$

$$= (-1)^\alpha i\pi^{d/2}\frac{\Gamma\left(\alpha - \frac{d}{2}\right)}{\Gamma(\alpha)}\frac{1}{(q^2 + m^2)^{\alpha - d/2}}. \tag{9A.4}$$

Hence (from (9A.1))

$$\int \frac{d^d p}{(p^2 + 2pq - m^2)^\alpha} = i\pi^{d/2}\frac{\Gamma\left(\alpha - \frac{d}{2}\right)}{\Gamma(\alpha)}\frac{1}{[-q^2 - m^2]^{\alpha - d/2}}. \tag{9A.5}$$

Differentiating this with respect to q^μ:

$$-\alpha\int d^d p \frac{2p_\mu}{(p^2 + 2pq - m^2)^{\alpha + 1}}$$

$$= i\pi^{d/2}\frac{\Gamma\left(\alpha - \frac{d}{2}\right)}{\Gamma(\alpha)}\left(-\alpha + \frac{d}{2}\right)\frac{2q_\mu}{[-q^2 - m^2]^{\alpha - d/2 + 1}}.$$

Using $\beta\Gamma(\beta) = \Gamma(\beta + 1)$ and putting $\alpha + 1 \to \alpha$ gives

$$\int d^d p \frac{p_\mu}{(p^2 + 2pq - m^2)^\alpha} = -i\pi^{d/2}\frac{\Gamma\left(\alpha - \frac{d}{2}\right)}{\Gamma(\alpha)}\frac{q_\mu}{(-q^2 - m^2)^{\alpha - d/2}}. \tag{9A.6}$$

Differentiating again with respect to q^ν yields

$$\int d^d p \frac{p_\mu p_\nu}{(p^2 + 2pq - m^2)^\alpha} = \frac{i\pi^{d/2}}{\Gamma(\alpha)} \frac{1}{(-q^2 - m^2)^{\alpha - d/2}}$$

$$\times \left[q_\mu q_\nu \Gamma\left(\alpha - \frac{d}{2}\right) + \tfrac{1}{2} g_{\mu\nu}(-q^2 - m^2)\Gamma\left(\alpha - 1 - \frac{d}{2}\right) \right]. \quad (9A.7)$$

Hence, contracting, we have

$$\int d^d p \frac{p^2}{(p^2 + 2pq - m^2)^\alpha} = \frac{i\pi^{d/2}}{\Gamma(\alpha)} \frac{1}{(-q^2 - m^2)^{\alpha - d/2}}$$

$$\times \left[q^2 \Gamma\left(\alpha - \frac{d}{2}\right) + \frac{d}{2}(-q^2 - m^2)\Gamma\left(\alpha - 1 - \frac{d}{2}\right) \right]. \quad (9A.8)$$

Note that the naive use of these formulae requires some mathematical justification. Equation (9A.1) only converges if $d < 2\alpha$. It has to be shown that this integral may be analytically continued into one which is convergent for any d (actually $d < 6$ will do). Details are given in 't Hooft and Veltman.[9]

Appendix B: the gamma function[‡]
The purpose of this appendix is to prove that

$$\Gamma(-n + \varepsilon) = \frac{(-1)^n}{n!}\left[\frac{1}{\varepsilon} + \psi_1(n + 1) + O(\varepsilon) \right] \quad (9B.1)$$

where

$$\psi_1(n + 1) = 1 + \tfrac{1}{2} + \cdots + \frac{1}{n} - \gamma \quad (9B.2)$$

and γ is the Euler–Mascheroni constant. We need the formulae

$$z\Gamma(z) = (z + 1), \quad (I)$$

$$\psi_1(z) = \frac{d \ln \Gamma(z)}{dz} = \frac{\Gamma'(z)}{\Gamma(z)}, \quad (II)$$

$$\frac{1}{\Gamma(z)} = z e^{\gamma z} \prod_{n=1}^{\infty} \left(1 + \frac{z}{n}\right) e^{-z/n}. \quad (III)$$

II is a definition of $\psi_1(z)$, and III is the Weierstrass representation of $\Gamma(z)$; γ, the Euler–Mascheroni constant, is

$$\gamma = \lim_{n \to \infty} \left(1 + \tfrac{1}{2} + \cdots + \frac{1}{n} - \ln n\right)$$

$$= 0.5\,772\,157.$$

[‡] I am grateful to Dr W.A.B. Evans for his help with this appendix.

Formulae II and III appear in standard texts, for example, P.M. Morse & H. Feshbach, *Methods of Theoretical Physics*, Vol. I, p. 422, McGraw-Hill, 1953. From III

$$-\ln\Gamma(z) = \ln z + \gamma z + \sum_{r=1}^{\infty}\left[\ln\left(1+\frac{z}{r}\right) - \frac{z}{r}\right]$$

therefore

$$-\frac{d}{dz}\ln\Gamma(z) = \frac{1}{z} + \gamma + \sum_{r=1}^{\infty}\left[\frac{1}{1+\dfrac{z}{r}}\frac{1}{r} - \frac{1}{r}\right]$$

$$= \frac{1}{z} + \gamma + \sum_{r=1}^{\infty}\left(\frac{1}{z+r} - \frac{1}{r}\right).$$

Hence from II

$$\psi_1(z) = -\gamma - \frac{1}{z} + \sum_{r=1}^{\infty}\left(\frac{1}{r} - \frac{1}{r+z}\right).$$

When $z = n = $ integer

$$\psi_1(n) = -\gamma + \sum_{r=1}^{n-1}\frac{1}{r}; \quad \psi_1(1) = -\gamma.$$

By a Taylor expansion ($\varepsilon \ll 1$)

$$\Gamma(1+\varepsilon) = \Gamma(1) + \varepsilon\Gamma'(1) + O(\varepsilon^2)$$
$$= 1 + \varepsilon\Gamma(1)\psi_1(1) + O(\varepsilon^2)$$
$$= 1 - \varepsilon\gamma + O(\varepsilon^2).$$

Hence from I

$$\Gamma(\varepsilon) = \frac{1}{\varepsilon}\Gamma(1+\varepsilon)$$

$$= \frac{1}{\varepsilon} - \gamma + O(\varepsilon).$$

Using I again

$$\Gamma(-1+\varepsilon) = \frac{-1}{1-\varepsilon}\Gamma(\varepsilon)$$

$$= -(1 + \varepsilon + \varepsilon^2 + \cdots)\left(\frac{1}{\varepsilon} - \gamma + O(\varepsilon)\right)$$

$$= -\left[\frac{1}{\varepsilon} + 1 - \gamma + O(\varepsilon)\right].$$

Similarly

$$\Gamma(-2+\varepsilon) = \frac{-1}{2-\varepsilon}\Gamma(-1+\varepsilon)$$

$$= \frac{(-1)^2}{2}\left[\frac{1}{\varepsilon} + (1 + \tfrac{1}{2} - \gamma) + O(\varepsilon)\right],$$

and in the general case

$$\Gamma(-n+\varepsilon) = \frac{(-1)^n}{n!}\left[\frac{1}{\varepsilon} + \left(1 + \tfrac{1}{2} + \cdots + \frac{1}{n} - \gamma\right) + O(\varepsilon)\right]$$

which is (9B.1).

Summary

[1]The formula for the superficial degree of divergence shows that the only primitive divergences in ϕ^4 theory containing 1 loop are the 2-point and 4-point functions. [2]These divergences are made explicit by dimensional regularisation, and are seen to be poles in $\varepsilon = 4 - d$. [3]It is then outlined to one and two loops how ϕ^4 theory is renormalised; first, by regarding the original parameters in the Lagrangian as 'bare' (and infinite), and the (finite) renormalised ones as 'physical'; and second, by regarding the original parameters as physical, and therefore finite, counter-terms being added to $\mathscr{L}$ at each order to keep the physical quantities finite. [4]The renormalisation group is introduced and is shown to give some indication of the asymptotic behaviour of field theories at high energy. Asymptotic freedom is explained, and ϕ^4 theory is seen not to be asymptotically free. [5]The power counting formula for the superficial degree of divergence of Feynman diagrams in QED is derived, and the primitively divergent diagrams isolated. The electron self-energy, vacuum polarisation and vertex graphs are evaluated using dimensional regularisation. [6]QED is renormalised to one loop, and the anomalous magnetic moment of the electron is calculated. It is shown that QED is not asymptotically free, but that at smaller distances (larger momenta) the effective charge increases, as it does in macroscopic dielectrics ('screening'). [7]It is shown that QED is renormalisable to arbitrary order. [8]By explicit calculation of the relevant Feynman diagrams, QCD is shown to be asymptotically free if the number of quark flavours is less than 16. [9]It is briefly shown that non-Abelian gauge theories are renormalisable. [10]The chiral anomaly, which threatens renormalisability, is explained and the condition for cancellation of the anomaly derived. The cancellation condition holds if quarks, carrying colour, are included in the Weinberg–Salam model. [11]In general, theories

398   *Renormalisation*

involving massive vector particles are not renormalisable, but it is shown how choosing a particular gauge, the 't Hooft gauge, spontaneously broken gauge symmetries (with massive vector particles) may be made manifestly renormalisable. The formalism of the effective potential is introduced to show how the divergence structure of gauge theories is unaffected by the occurrence of spontaneous symmetry breaking. The effective potential, evaluated in the loop expansion, indicates interesting connections between spontaneous symmetry breaking and radiative corrections.

Guide to further reading

Renormalisation is treated in many standard textbooks, for example

(1) N.N. Bogoliubov & D.V. Shirkov, *Introduction to the Theory of Quantised Fields*, (3rd. edition), Wiley-Interscience, 1980.
(2) S.S. Schweber, *An Introduction to Relativistic Quantum Field Theory*, Harper & Row, 1961.
(3) J.D. Bjorken & S.D. Drell, *Relativistic Quantum Fields*, McGraw-Hill, 1965.
(4) C. Itzykson & J.B. Zuber, *Quantum Field Theory*, McGraw-Hill, 1980.
(5) C. Nash, *Relativistic Quantum Fields*, Academic Press, 1978.

Good modern surveys are provided by the lectures of C.G. Callan, B.W. Lee and D.J. Gross in

(6) R. Balian & J. Zinn-Justin (eds.), *Méthodes en Théorie des Champs/Methods in Field Theory* (Les Houches Summer School 1975), North-Holland Publishing Co., 1976. Reprinted by World Scientific, Singapore, 1981.

An excellent introdution to renormalisation is

(7) S. Coleman 'Renormalisation and symmetry: a review for non-specialists', in A. Zichichi (ed.), *Problems of the Fundamental Interactions*, Editrice Compositori, Bologna, 1973.

Weinberg's theorem on the convergence of Feynman diagrams is in

(8) S. Weinberg, *Physical Review*, **118**, 838 (1960).
See also ref. (4) p. 405, and ref. (3) p. 322.

The original papers on dimensional regularisation are

(9) C.G. Bollini, *Nuovo Cimento*, **12B**, 20 (1972); J.F. Ashmore, *Nuovo Cimento Lettere*, **4**, 289 (1972); G. 't Hooft & M. Veltman, *Nuclear Physics*, **44B**, 189 (1972).

For a review see

(10) G. Leibbrandt, *Reviews of Modern Physics*, **47**, 849 (1975).

Renormalisation of φ^4 theory to all orders is treated in ref. (1), by Callan and Gross in ref. (6), and in

(11) W. Zimmermann, in S. Deser, M. Grisaru & H. Pendleton (eds.), *Lectures on Elementary Particles and Quantum Field Theory* (Brandeis Summer Institute, 1970), Vol. I, MIT Press, 1970.
(12) A. Rouet, *Rivista del Nuovo Cimento*, **3**, no. 7 (1980).

The renormalisation group is discussed in ref. (1) and in

(13) M. Gell-Mann & F.E. Low, *Physical Review* **95**, 1300 (1954).

For an early review, see

(14) L.D. Landau 'On the Quantum Theory of Field', in W. Pauli (ed.), *Niels Bohr and the Development of Physics*, Pergamon Press, 1955.

For the Callan–Symanzik equation, see

(15) C.G. Callan, *Physical Review*, **D2**, 154 (1970).

(16) K. Symanzik, *Communication in Mathematical Physics*, **18**, 227 (1970).

Good reviews of the renormalisation group and the Callan–Symanzik equation are to be found in

(17) S. Coleman '*Dilatations*', in ref. (7) above.

(18) D. Politzer, *Physics Reports*, **C14**, 129 (1974).

(19) A. Peterman, *Physics Reports*, **53**, 157 (1979).

(20) A.A. Vladimirov & D.V. Shirkov, *Soviet Physics Uspekhi*, **22**, 860 (1980).

(21) J. Iliopoulos, C. Itzykson & A. Martin, *Reviews of Modern Physics*, **47**, 165 (1975).

(22) K. Higashijima & K. Nishijima, *Progress of Theoretical Physics*, **64**, 2179 (1980).

(23) P. Ramond, *Field Theory: A Modern Primer*, Benjamin/Cummings, 1981.

An excellent account, though written more with condensed matter than fields in mind, is

(24) D.J. Amit, *Field Theory, the Renormalisation Group, and Critical Phenomena*, McGraw-Hill, 1978.

Dimensional regularisation of QED is outlined in ref. (7) (particularly by 't Hooft and Veltman). For recent reviews of this, as well as of renormalisation and the renormalisation group in QED, see

(25) R. Coquereaux, *Annals of Physics*, **125**, 401 (1980).

(26) S. Narisan, *Physics Reports*, **84**, 263 (1982).

The anomalous magnetic moment of the electron was calculated by

(27) J. Schwinger, *Physical Review*, **73**, 416 (1948),

and measured by

(28) P. Kusch & H.M. Foley, *Physical Review*, **72**, 1256 (1947); *ibid.* **73**, 412 (1948).

For a recent account of (g-2) see

(29) F.H. Combley, *Reports on Progress in Physics*, **42**, 1889 (1979).

The muon magnetic moment is discussed in ref. (29) and

(30) J.C. Taylor, *Gauge Theories of Weak Interactions*, section 16.2, Cambridge University Press, 1976.

Good textbook accounts of the renormalisibility of QED are to be found in

(31) J.M. Jauch & F. Rohrlich, *The Theory of Photons and Electrons*, 2nd. edition, chapter 10, Springer, 1976.

(32) S.S. Schweber, H.A. Bethe & F. de Hoffmann, *Mesons and Fields, vol. 1 (Fields)*, Row, Peterson & Co., 1956.

(33) K. Huang, *Quarks, Leptons and Gauge Fields*, chapter 9, World Scientific, Singapore, 1982.

The paper by Ward on which our approach is based is

(34) J.C. Ward, *Physical Review*, **78**, 182 (1950).

The discovery of asymptotic freedom in Yang–Mills theories was made by

(35) H.D. Politzer, *Physical Review Letters*, **30**, 1346 (1973); D.J. Gross & F. Wilczek, *ibid.* **30**, 1343 (1973), *Physical Review* **D8**, 3633 (1973), *ibid.* **D9**, 980 (1974).

See also

(36) S. Coleman & D.J. Gross, *Physical Review Letters*, **31**, 851 (1973).

For a review, see Politzer, ref. (18).

1-loop renormalisation of non-Abelian gauge theories is partially reviewed in

(37) S. Narison, *Physics Reports*, **84**, 263 (1982).

Good accounts appear in refs. (4), (6) and (30) above, and in

(38) L.D. Faddeev & A.A. Slavnov, *Gauge Fields: Introduction to Quantum Theory*, Benjamin-Cummings, 1980.

(39) E. Leader & E. Predazzi, *An Introduction to Gauge Theories and the New Physics*, Cambridge University Press, 1982.

(40) S. Coleman, in A. Zichichi (ed.), *Laws of Hadronic Matter*, Academic Press, 1975.

Important landmarks in the study of the renormalisability of pure Yang–Mills theories are the following papers.

(41) G. 't Hooft, *Nuclear Physics*, **B33**, 173 (1971); *ibid.*, **B35**, 167 (1971).

(42) A.A. Slavnov, *Theoretical and Mathematical Physics*, **10**, 99 (1972); *Soviet Journal of Particles and Nuclei*, **5**, 303 (1975).

(43) J.C. Taylor, *Nuclear Physics*, **B33**, 436 (1971).

(44) B.W. Lee & J. Zinn-Justin, *Physical Review*, **D5**, 3121, 3137, 3155; **D7**, 1049 (1972).

(45) G. 't Hooft & M.T. Veltman, *Nuclear Physics*, **B50**, 318 (1972).

Reviews appear in the lectures by Lee, ref. (6) above, in chapter 14 of Taylor's book (30), in section 12-4 of ref. (4), and in

(46) E.S. Abers & B.W. Lee, *Physics Reports*, **9C**, 1 (1973).

(47) W. Marciano & H. Pagels, *Physics Reports*, **36**, 137 (1978).

(48) G. Costa & M. Tonin, *Rivista del Nuovo Cimento*, **5** (no. 1), 29 (1975).

The chiral anomaly first appeared in work on $\pi^0 \to 2\gamma$ decay by

(49) J. Steinberger, *Physical Review*, **76**, 1180 (1949); J. Schwinger, *Physical Review*, **82**, 664 (1951).

The Sutherland paradox on $\pi^0 \to 2\gamma$ is reported in

(50) D.G. Sutherland, *Physics Letters*, **23**, 384 (1966); *Nuclear Physics*, **B2**, 433 (1967).

This stimulated further work on the chiral anomaly:

(51) S.L. Adler, *Physical Review*, **177**, 2426 (1969); J.S. Bell & R. Jackiw, *Nuovo Cimento*, **60A**, 47 (1967).

For reviews, see S.L. Adler in S. Deser *et al.*, ref. (11) above, and K. Huang, ref. (33), chapter 11, and

(52) R. Jackiw in S.B. Treiman, R. Jackiw & D.J. Gross, *Lectures on Current Algebra and its Applications*, Princeton University Press, 1972.

(53) J. Wess, *Acta Physica Austriaca Supplementum*, **9**, 494 (1972).

't Hooft's gauges originate in

(54) G. 't Hooft, *Nuclear Physics*, **B35**, 167 (1971).

The effective potential was introduced by

(55) J. Goldstone, A. Salam & S. Weinberg, *Physical Review*, **127**, 965 (1962); G. Jona-Lasinio, *Nuovo Cimento*, **34**, 1790 (1964).

Its significance was further explored by

(56) S. Coleman & S. Weinberg, *Physical Review*, **D7**, 1883 (1973).

See also Coleman (40). Good textbook accounts appear in Itzykson and Zuber (4), p. 448, Ramond (23), p. 99, and Huang (33), p. 200.

10

Topological objects in field theory

Un fourmi de dix-huit mètres
Avec un chapeau sur la tête,
Ça n'existe pas;
Pourquoi pas?

French song

No-one can deny the success which quantum field theory, in the perturbative approximation, has enjoyed over the last half century. One need only mention the interpretation of quantised fields as particles, the description of scattering processes, the precise numerical agreements in quantum electrodynamics, the successful prediction of the W particle, and the beginnings of an understanding of the strong interaction through quantum chromodynamics. Yet despite these successes, the question of how to describe *the basic matter fields of nature* has remained unanswered – except, of course, through the introduction of quantum numbers and symmetry groups. As far as field theory goes, the matter fields are treated as point objects. Even in classical field theory these present us with unpleasant problems, in the shape of the infinite self-energy of a point charge. In the quantum theory, these divergences do not disappear; on the contrary, they appear to get worse, and despite the comparative success of renormalisation theory the feeling remains that there ought to be a more satisfactory way of doing things.

Now it turns out that non-linear classical field theories possess extended solutions, commonly known as solitons, which represent stable configurations with a well-defined energy which is nowhere singular. May this be of relevance to particle physics? Since non-Abelian gauge theories are non-linear, it may well be, and the last ten years have seen the discovery of vortices, magnetic monopoles and 'instantons', which are soliton solutions to the gauge-field equations in two space dimensions (i.e. a 'string' in

401

3-dimensional space), three space dimensions (localised in space but not in time) and 4-dimensional space–time (localised in space and time). If gauge theories are taken seriously then so must these solutions be. It will be seen that they do give rise to new physics and there is even the hope that they may solve the problem of quark confinement.

Not the least interesting feature of this subject is the branch of mathematics which it involves; for the stability of these solitons arises from the fact that the boundary conditions fall into distinct classes, of which the vacuum belongs only to one. These boundary conditions are characterised by a particular correspondence (mapping) between the group space and co-ordinate space, and because these mappings are not continuously deformable into one another they are *topologically* distinct. The relevant notions in topology will be developed as we go along. We begin our survey with the 'sine–Gordon' equation which has no relevance to particle physics but whose soliton solutions are quite well understood, and therefore form a good introduction to the subject.

10.1   The sine–Gordon kink

The sine–Gordon equation

$$\frac{\partial^2 \phi}{\partial t^2} - \frac{\partial^2 \phi}{\partial x^2} + a \sin b\phi = 0 \tag{10.1}$$

describes a scalar field in one space and one time dimension. It possesses moving, as well as stationary, solutions. To find moving solutions, we want a field of the form

$$\phi(x, t) = f(x - vt) = f(\xi).$$

It is easy to check that

$$f(\xi) = \frac{4}{b} \arctan e^{\pm \gamma \xi} \tag{10.2}$$

is a solution, where $\gamma = (1 - v^2)^{-\frac{1}{2}}$. The appearance of this wave is shown in Fig. 10.1. It is *solitary wave*, which moves without changing shape or size, and therefore without dissipation, in strong contrast to the waves set up when, for instance, a stone is thrown into a pond. These waves spread out and the energy is dissipated. Solitary waves (solitons) have been observed, for example, moving along canals. In this case they are solutions of the Korteweg de Vries equation.

Since solitons are solutions of non-linear wave equations the superposition principle is not obeyed. This means that when two solitons meet the resultant wave form is a complicated one, but the surprising thing is that, asymptotically, the solitons separate out again – they 'pass through' one another. This property is, of course, of interest to particle physicists, though we shall not develop it any further here. Another consequence of the fact that the superposition principle does not hold is that the quantisation of solitons becomes non-trivial. We shall not follow this matter any further either. Instead, we turn to the stationary solutions of the sine–Gordon equation, which possess an interest of a different type.

It is clear that (10.1) possesses an infinite number of constant solutions (which, as we shall see in a moment, have zero energy):

$$\phi = \frac{2\pi n}{b}, \quad n = 0, \pm 1, \pm 2, \ldots; \tag{10.3}$$

that is, the sine–Gordon equation possesses a degenerate vacuum.[‡] ('Vacuum' here does not, of course, mean the state in Hilbert space, but simply a classical field configuration of zero energy.) The Lagrangian for the sine–Gordon equation is

$$\mathscr{L} = \frac{1}{2}\left(\frac{\partial\phi}{\partial t}\right)^2 - \frac{1}{2}\left(\frac{\partial\phi}{\partial x}\right)^2 - V(\phi) \tag{10.4}$$

with

$$V(\phi) = \frac{a}{b}(1 - \cos b\phi),$$

where the constant has been chosen so that the solutions (10.3) have $V = 0$. They therefore have zero energy since the energy density of the field configuration is

$$\mathscr{H} = \frac{1}{2}\left(\frac{\partial\phi}{\partial t}\right)^2 + \frac{1}{2}\left(\frac{\partial\phi}{\partial x}\right)^2 + V(\phi). \tag{10.5}$$

Fig. 10.1. A solitary wave (soliton).

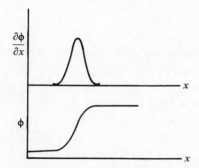

[‡] It is this property that is crucial in what follows, and so a corresponding analysis could be made for other field theories with degenerate vacua, for example the ϕ^4 model with $m^2 < 0$ considered in chapter 8.

Note that we may write

$$V(\phi) = \frac{ab}{2}\phi^2 - \frac{ab^3}{4!}\phi^4 + \cdots, \tag{10.6}$$

so putting $m^2 = ab, \lambda = ab^3$, we have

$$V(\phi) = \frac{m^2}{2}\phi^2 - \frac{\lambda}{4!}\phi^4 + \cdots, \tag{10.7}$$

and m stands for the 'particle' mass and λ for the self-interaction coupling.

The potential V in (10.4) is shown in Fig. 10.2 with the (zero energy) ground state given by (10.3). Now construct the following configuration. Let ϕ approach one of the zeros of V (say $n = 0$) as $x \to -\infty$, but a *different* zero (say $n = 1$) as $x \to \infty$. Between these two there is clearly a region where

$$\phi \neq \frac{2\pi n}{b}, \quad \frac{\partial\phi}{\partial x} \neq 0,$$

and therefore, from (10.5), where there is a positive energy density. We assume the configuration is static, so $\partial\phi/\partial t = 0$. Because of the boundary conditions on ϕ, we expect the total energy to be finite. Let us find what it is. For a stationary solution to the sine–Gordon equation we have

$$\frac{\partial^2\phi}{\partial x^2} = \frac{\partial V}{\partial\phi}$$

which gives on integration

$$\frac{1}{2}\left(\frac{\partial\phi}{\partial x}\right)^2 = V(\phi), \tag{10.8}$$

the integration constant being zero. From (10.5) and (10.8), the energy of the stationary soliton is

$$\begin{aligned}
E &= \int \mathscr{H}\,dx \\
&= \int\left[\frac{1}{2}\left(\frac{\partial\phi}{\partial x}\right)^2 + V(\phi)\right]dx \\
&= \int 2V(\phi)\,dx \\
&= \int_0^{2\pi/b}[2V(\phi)]^{\frac{1}{2}}\,d\phi
\end{aligned}$$

Fig. 10.2. The sine–Gordon potential $V(\phi)$.

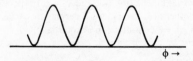

$\phi \to$

where we have put in the integration limits given by (10.3) between $n = 0$ and $n = 1$. This integral is now easily performed on substituting (10.4). We have

$$E = \left(\frac{2a}{b}\right)^{\frac{1}{2}} \int_0^{2\pi/b} (1 - \cos \phi)^{\frac{1}{2}} \, d\phi$$

$$= \left(\frac{2a}{b^3}\right)^{\frac{1}{2}} \int_0^{2\pi} (1 - \cos \alpha)^{\frac{1}{2}} \, d\alpha$$

$$= 8\left(\frac{a}{b^3}\right)^{\frac{1}{2}}$$

$$E = \frac{8m^3}{\lambda} \tag{10.9}$$

where in the last step we have used the substitution in (10.7). So this soliton has a finite energy, with the interesting property that the energy is *inversely* proportional to the coupling constant. This may indeed be a useful property for particle physics.

There is a simple model which makes this soliton easy to visualise. Consider an infinite horizontal string with pegs attached to it at equally spaced intervals, and connect each peg to its neighbour with a small spring (the 'coupling'). Each peg is also acted on by gravity. The ground state corresponds to every peg hanging vertically. The soliton we have found, with $n = 0 \to 1$, corresponds to the situation in Fig. 10.3. This soliton – and others of this type (see below) – is called a *kink*. It should be clear from the peg model that the kink is stable, and cannot decay into the ground state with $E = 0$. This would involve a (semi-) infinite number of pegs turning over, which would need a (semi-) infinite amount of energy. But what is the *mathematical* reason for the stability of the kink? It is to be found in the boundary conditions. 'Space' in this model is an infinite line, whose boundary is two points (the end-points). At these two points the 1-kink solution has $n = 0$ and $n = 1$, and this is *not* continuously deformable into $n = 0$ and $n = 0$ (the ground state). The kink, then, is a 'topological' object. Its existence depends on the topological properties of the space (in particular, its boundary, which in this case is a discrete set). This conclusion is a general one; that is to say, the stability of soliton solutions in non-linear field theories is a consequence of topology.

Finally, the stability of the soliton (kink) obviously signals a *conservation law*: there must be a conserved charge Q, equal to an integer N (the

Fig. 10.3. Pegs on a line representing the kink (soliton) solution to the sine–Gordon equation.

difference between the two integers in (10.3)), and a corresponding divergenceless current J^μ ($\mu = 0, 1$). They are easy to construct. With

$$J^\mu = \frac{b}{2\pi}\varepsilon^{\mu\nu}\partial_\nu\phi \tag{10.10}$$

($\varepsilon^{\mu\nu}$ is antisymmetric, with $\varepsilon^{01} = 1$), we have the *identity* $\partial_\mu J^\mu = 0$, and the charge is

$$\begin{aligned}
Q &= \int_{-\infty}^{\infty} J^0 \, \mathrm{d}x \\
&= \frac{b}{2\pi}\int_{-\infty}^{\infty}\frac{\partial\phi}{\partial x}\,\mathrm{d}x \\
&= \frac{b}{2\pi}[\phi(\infty) - \phi(-\infty)] = N.
\end{aligned} \tag{10.11}$$

The interesting thing is that the current J^μ does *not* follow from the invariance of $\mathscr{L}$ under any symmetry transformation. It is therefore *not* a Noether current. Its divergencelessness follows independently of the equations of motion.

We consider, in the following sections, examples of solitons in gauge theories, beginning with one in two space dimensions – the vortex.

10.2   Vortex lines

Now consider a scalar field in 2-dimensional space. The 'boundary' of this space is the circle at infinity, denoted S^1. We construct a field whose value on the boundary is

$$\phi = a\mathrm{e}^{in\theta} \quad (r \to \infty) \tag{10.12}$$

where r and θ are polar co-ordinates in the plane, a is a constant, and, to make ϕ single-valued, n is an integer. We propose this form, rather than simply $\phi = a$, because it is a generalisation to two dimensions of (10.3). ((10.3) is a solution of the sine–Gordon equation, whereas it is yet to be seen what equation describes the 2-dimensional solitons we are in the process of developing.) From (10.12), we have

$$\nabla\phi = \frac{1}{r}(ina\,\mathrm{e}^{in\theta})\hat{\boldsymbol{\theta}}. \tag{10.13}$$

The Lagrangian and Hamiltonian functions are

$$\mathscr{L} = \frac{1}{2}\left(\frac{\partial\phi}{\partial t}\right)^2 - \tfrac{1}{2}|\nabla\phi|^2 - V(\phi), \tag{10.14}$$

$$\mathscr{H} = \frac{1}{2}\left(\frac{\partial\phi}{\partial t}\right)^2 + \tfrac{1}{2}|\nabla\phi|^2 + V(\phi). \tag{10.15}$$

Now let us consider a static configuration with, for example,

$$V(\phi) = [a^2 - \phi^*\phi]^2 \tag{10.16}$$

so that $V = 0$ on the boundary. Then as $r \to \infty$

$$\mathcal{H} = \tfrac{1}{2}|\nabla\phi|^2 = \frac{n^2 a^2}{2r^2}$$

and the energy (mass) of the static configuration is

$$E \approx \int^{\infty} \mathcal{H} r \, dr \, d\theta = \pi n^2 a^2 \int^{\infty} \frac{1}{r} dr.$$

This is *logarithmically divergent*; the kink, as it stands, cannot be generalised to two dimensions – nor to more that two, for it turns out that in all these cases the energy is divergent.

To proceed, we add a gauge field, so that what counts is the covariant derivative

$$D_\mu \phi = \partial_\mu \phi + ieA_\mu \phi. \tag{10.17}$$

By choosing A_μ of the form

$$\mathbf{A} = \frac{-1}{e}\nabla(n\theta) \quad (r \to \infty),$$

i.e.

$$A_r \to 0, \quad A_\theta \to -\frac{n}{er} \quad (r \to \infty), \tag{10.18}$$

we find that at $r = \infty$

$$D_\theta \phi = \frac{1}{r}\left(\frac{\partial\phi}{\partial\theta}\right) + ieA_\theta\phi = 0, \quad D_r\phi = 0 \tag{10.19}$$

so $D_\mu\phi \to 0$ on the boundary at infinity. The Lagrangian is now

$$\mathcal{L} = -\tfrac{1}{4}F_{\mu\nu}^2 + |D_\mu\phi|^2 - V(\phi). \tag{10.20}$$

Since (10.18) is a *pure gauge*,

$$A_\mu \to \partial_\mu\chi \quad (r \to \infty), \tag{10.21}$$

then $F_{\mu\nu} \to 0$. For a static configuration $\mathcal{H} = -\mathcal{L}$, and with $V(\phi)$ given by (10.16) we have $\mathcal{H} \to 0$ as $r \to \infty$, making possible a field configuration of *finite* energy. We shall now see that the effect of adding the gauge field is to give the soliton *magnetic flux*. Consider the integral $\oint \mathbf{A} \cdot d\mathbf{l}$ round the circle S^1 at infinity. By Stokes' theorem, this is $\int \mathbf{B} \cdot d\mathbf{S} = \Phi$, the flux enclosed, hence

$$\Phi = \oint \mathbf{A} \cdot d\mathbf{l} = \oint A_\theta r \, d\theta = -\frac{2\pi n}{e}, \tag{10.22}$$

and the flux is *quantised*. So we have, after all, constructed a 2-dimensional field configuration, consisting of a charged scalar field and a gauge field (the electromagnetic field!). It carries magnetic flux, and since $D_\mu \phi \to 0$ and $F_{\mu\nu} \to 0$ on the boundary at infinity, it appears to have finite energy. It is clear that by adding a third dimension (the z axis) on which the fields have no dependence, this configuration becomes a vortex line. Apart from the presence of the scalar field, it is the same as the solenoid discussed in §3.4 under the Bohm–Aharonov effect; and just as that effect was attributable to the topology of the gauge group $U(1)$, so here also we shall see that it is this same topology which ensures stability of the vortex.

It will not have escaped the reader's notice that the Lagrangian (10.20) with $V(\phi)$ given by (10.16) is that of the Higgs model – see (8.36) and (8.4) – that is, scalar electrodynamics with spontaneous symmetry breaking. Actually, we saw in §8.4 that this Lagrangian is the relativistic version of the Landau–Ginzburg free energy, which describes superconductivity. It is known that on the occasions when magnetic flux *does* penetrate super-conductors (that is, in type II superconductors), it does so in quantised flux lines, called Abrikosov flux lines. It is these that the present solutions are describing, the field ϕ in superconductivity being the BCS condensate.

To be a little more systematic, let us start from the Higgs Lagrangian (8.36):

$$\mathscr{L} = -\tfrac{1}{4}F_{\mu\nu}F^{\mu\nu} + |(\partial_\mu + ieA_\mu)\phi|^2 - m^2\phi^*\phi - \lambda(\phi^*\phi)^2. \qquad (8.36)$$

Spontaneous symmetry breaking is signalled by $m^2 < 0$, and the vacuum is then given by (8.4)

$$|\phi|_{\mathrm{vac}} = a = \left(\frac{-m^2}{2\lambda}\right)^{\frac{1}{2}}. \qquad (8.4)$$

The equations of motion obtained from (8.36) are

$$D^\mu(D_\mu\phi) = -m^2\phi - 2\lambda\phi|\phi|^2, \qquad (10.23)$$

$$ie(\phi\partial_\mu\phi^* - \phi^*\partial_\mu\phi) + 2e^2 A_\mu|\phi|^2 = \partial^\nu F_{\mu\nu}. \qquad (10.24)$$

We must first check that these equations allow the solutions (10.12) and (10.18) at infinity. Since by construction (see (10.19)) $D_\mu\phi = 0$ as $r \to \infty$, the left-hand side of (10.23) vanishes; and so does the right-hand side if ϕ takes on its vacuum value (8.4). Since A_μ is a pure gauge (see (10.21)) $F_{\mu\nu} = 0$ as $r \to \infty$, so the right-hand side of (10.24) vanishes. In view of (10.12) and (10.18) the left-hand side vanishes identically when $\mu = r$, and when $\mu = \theta$ it vanishes when ϕ assumes the vacuum value (8.4). Hence our particular choices for A_μ and ϕ are allowed by the equations of motion.

As r becomes finite, and particularly as $r \to 0$, of course, the values of A_μ and ϕ change. Let us now treat the problem as one in three dimensions, with cylindrical symmetry about the z axis. Then, since there is magnetic flux, the magnetic field component B_z must be non-zero, which means that A cannot be a pure gauge everywhere. Also, continuity requires that $\phi \to 0$ as $r \to 0$; since this is not the vacuum value, the 2-dimensional soliton will have an energy, and the vortex will have a corresponding mass per unit length. The forms of A and ϕ are found from the equations of motion. Taking B with a z component only, and A with a θ component only, we have

$$B = B_z = \frac{1}{r}\frac{d}{dr}[rA(r)], \quad A(r) = A_\theta = A. \tag{10.25}$$

In addition, ϕ is of the form

$$\phi = \chi(r)e^{in\theta} \tag{10.26}$$

with

$$\chi(r)\underset{r\to 0}{\longrightarrow} 0, \quad \chi(r)\underset{r\to\infty}{\longrightarrow} a. \tag{10.27}$$

In the static case, the equation of motion (10.23) then becomes

$$(\partial_i + ieA_i)^2\phi - (m^2 + 2\lambda|\phi|^2)\phi = 0$$

which, on summing over the r and θ components, gives

$$\frac{1}{r}\frac{d}{dr}\left(r\frac{d\chi}{dr}\right) - \left[\left(\frac{n}{r} - eA\right)^2 + m^2 + 2\lambda\chi^2\right]\chi = 0. \tag{10.28}$$

On the other hand, taking the θ component of (10.24) gives (recall equations (2.217–2.221))

$$\frac{ie}{r}(2in)\chi^2 + 2e^2A\chi^2 = -\partial_i F_{\theta i} = -\partial_i(\varepsilon_{\theta ij}B_j)$$

and hence

$$\frac{d}{dr}\left(\frac{1}{r}\frac{d}{dr}(rA)\right) - 2e\left(\frac{n}{r} + eA\right)\chi^2 = 0. \tag{10.29}$$

One should now solve the coupled non-linear equations of motion (10.28) and (10.29). No exact analytic solution, however, has yet been found. In the approximation where $\chi \simeq a$ a constant (i.e. for $r \to \infty$), Nielsen and Olesen[8] found (with c a constant of integration and K_1 and K_0 modified Bessel functions)

$$A = -\frac{n}{er} - \frac{c}{e}K_1(|e|ar)\underset{r\to\infty}{\longrightarrow} -\frac{n}{er} - \frac{c}{e}\left(\frac{\pi}{2|e|ar}\right)^{\frac{1}{2}} e^{-|e|ar} + \cdots$$

with magnetic field

$$B_z = c\chi K_0(|e|ar) \to \frac{c}{e}\left(\frac{\pi a}{2|e|r}\right)^{\frac{1}{2}} e^{-|e|ar} + \cdots . \tag{10.30}$$

To obtain the variation of the scalar field, we put

$$\chi(r) = a + \rho(r);$$

then[8]

$$\rho(r) \simeq e^{-\sqrt{-m^2}\, r} \cdot (-a) \tag{10.31}$$

(recall that $-m^2 > 0$). These solutions are sketched in Fig. 10.4.

Why are these solutions stable? As with the kink, the reason is topological. The Lagrangian is invariant under a symmetry group – in this case $U(1)$, the electromagnetic gauge group. The field ϕ (with boundary value given by (10.12)) is a representation of $U(1)$. The group space of $U(1)$ is a circle S^1, since an element of $U(1)$ may be written $\exp(i\theta) = \exp[i(\theta + 2\pi)]$, so the space of all values of θ is a line with $\theta = 0$ identified with $\theta = 2\pi$, and the line becomes a circle S^1. The field ϕ in (10.12) is a representation basis of $U(1)$, but it is the boundary value of the field in a 2-dimensional space. This boundary is clearly a circle S^1 (the circle $r \to \infty$, $\theta = (0 \to 2\pi)$). Hence ϕ defines a mapping of the boundary S^1 in physical space onto the group space S^1:

$$\phi: S^1 \to S^1, \tag{10.32}$$

the mapping being specified by the integer n. Now a solution characterised by one value of n is stable since it cannot be continuously deformed into a solution with a different value of n (a rubber band which fits twice round a circle cannot be continously deformed into one which goes once round the circle). This is to say (see §3.4) that the *first homotopy group* of S^1, the group space of $U(1)$, is not trivial:

$$\pi_1(S^1) = \mathbb{Z}. \tag{10.33}$$

$\mathbb{Z}$ is the additive group of integers.

The status of a topological argument like this is that it provides a very general condition which *must* be fulfilled in order that solitons exist in a particular model. If, as in the model above, the topological argument indicates that soliton solutions are possible in principle then one goes to the

Fig. 10.4. The variations of the scalar and magnetic fields in the Nielsen–Olesen solution.

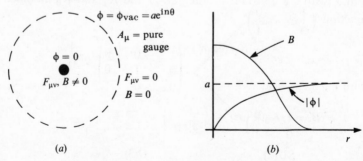

(a)   (b)

equations of motion to find them. Topology therefore provides *existence arguments*. As an example, let us enquire whether stringlike solutions to (spontaneously broken) gauge theories exist when the gauge group is $SU(2)$. This is the group of 2×2 matrices

$$U = u_0 + i \sum_{j=1}^{3} u_j \sigma_j$$

where σ_j are the Pauli matrices, and the condition that U is unitary and has unit determinant is

$$u_0^2 + u_1^2 + u_2^2 + u_3^2 = 1. \tag{10.34}$$

Now this is the equation for the unit sphere S^3 in 4-dimensional Euclidean space E^4; that is, the group space of $SU(2)$ is S^3. There will exist stable vortices in an $SU(2)$ gauge theory if the mappings of the group onto the S^1 boundary fall into distinct classes; that is, if $\pi_1(S^3)$ is non-trivial. But $\pi_1(S^3)$ is, in fact, trivial, for S^3 is a simply connected space; every closed curve S^1 on S^3 may be shrunk to a point, so the boundary conditions may all be shrunk to the trivial constant condition $\phi = \text{const}$, and no vortices exist.

The group $O(3)$, on the other hand, is not simply connected but doubly connected. For example, corresponding to the $O(3)$ matrix

$$\begin{vmatrix} \cos\alpha & -\sin\alpha & 0 \\ \sin\alpha & \cos\alpha & 0 \\ 0 & 0 & 1 \end{vmatrix}, \tag{10.35}$$

which corresponds to rotation about the z axis through an angle $-\alpha$, is the $SU(2)$ matrix

$$\begin{pmatrix} e^{i\alpha/2} & 0 \\ 0 & e^{-i\alpha/2} \end{pmatrix}. \tag{10.36}$$

Now $\alpha = 0$ clearly gives the identity matrix in both cases, but $\alpha = 2\pi$ gives the identity again in $O(3)$ and minus the identity in $SU(2)$. This is the origin of the well-known statement that vectors do not change sign on rotation through 2π but spinors do. In other words, corresponding to two elements of $SU(2)$ (the identity, and minus the identity) there is only one element of $O(3)$:

$$\begin{pmatrix} 1 & 0 \\ 0 & 1 \end{pmatrix}$$
$$\begin{pmatrix} -1 & 0 \\ 0 & -1 \end{pmatrix} \Bigg\rangle \begin{vmatrix} 1 & 0 & 0 \\ 0 & 1 & 0 \\ 0 & 0 & 1 \end{vmatrix}.$$

$$SU(2) \qquad\qquad O(3)$$

There is a 2–1 mapping of $SU(2)$ onto $O(3)$. The group space of $O(3)$, accordingly, is obtained from that of $SU(2)$ by *identifying opposite points* on the 3-space S^3, since they correspond to the same $O(3)$ transformation. This space is *doubly connected*, as we shall now show. We consider *closed curves* S^1 in the group space of $O(3)$. Each curve corresponds to a continuous set of rotations, starting (say) from the identity 0, and returning there. One possible type of closed curve is the path c_1 in Fig. 10.5. This corresponds to a series of rotations, the angle of which nowhere exceeds π. If the angle does exceed π, then the path in group space becomes like c_2. On reaching the angle π at the point A, the path reappears at the opposite point A', and eventually returns to the origin O. It is clear that c_1 is homotopic (may be shrunk) to a point, where c_2 is homotopic to a line. The reader may convince himself that a closed path in which the angle of rotation exceeds 2π reappears at opposite points on the surface of S^3 *twice*, and is therefore homotopic to a point. Similarly, one in which the angle exceeds 3π is homotopic to a straight line. Consequently, there are only two types of closed path S^1 in the group space of $O(3)$: those homotopic to a point and those homotopic to a line. This means there is one non-trivial vortex in an $O(3)$ gauge theory. The vortices may have 'charges' (flux) 1 or 0, with the algebra $0 + 0 = 0, 1 + 0 = 1, 1 + 1 = 0$, so two non-trivial vortices will annihilate each other. (It may be parenthetically remarked that whether the gauge group is $SU(2)$ or $O(3)$ depends on *what particles exist*: if there are particles with 'isospin' of $\frac{1}{2}, \frac{3}{2}, \frac{5}{2}$, etc. then the gauge group is $SU(2)$, but if all the particles have integral isospin the group is $O(3)$.)

One way of making lines of magnetic flux is to place two opposite magnetic charges close together. So an obvious question is: if gauge theories allow flux lines, do they also allow magnetic charges? In fact they do, and these are called 't Hooft–Polyakov magnetic monopoles, after their discoverers. Like vortices, these monopoles owe their stability (and therefore existence) to the non-trivial topological properties of the gauge group. In this respect they are completely different from the 'ordinary' magnetic monopoles, which are *point* magnetic charges, and which may be introduced into Maxwell's equations to make them symmetric between

Fig. 10.5. Two types of closed path in the group space of $O(3)$.

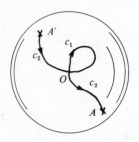

electricity and magnetism. Dirac showed that the prescriptions of quantum theory imply a remarkable quantisation condition for point magnetic charges, and for this reason they are sometimes referred to as Dirac monopoles. We shall study Dirac monopoles in the next section, and 't Hooft–Polyakov monopoles after that. This will serve to acquaint the reader with the idea of magnetic charge, as well as to demonstrate the difference between the two types of monopole.

10.3   The Dirac monopole

Consider a magnetic monopole of strength g at the origin. The magnetic field is radial and is given by a Coulomb-type law

$$\mathbf{B} = \frac{g}{r^3}\mathbf{r} = -g\nabla\left(\frac{1}{r}\right) \tag{10.37}$$

(we are using Gaussian units). Since $\nabla^2 1/r = -4\pi\delta^3 r$, we have

$$\nabla\cdot\mathbf{B} = 4\pi g\delta^3 r \tag{10.38}$$

corresponding to a point magnetic charge, as desired. Since $\mathbf{B}$ is radial, the total flux through a sphere surrounding the origin is

$$\Phi = 4\pi\mathbf{r}^2 B = 4\pi g. \tag{10.39}$$

Consider a particle with electric charge e in the field of this monopole. Its wave function is

$$\psi = |\psi|\exp\left[\frac{i}{\hbar}(\mathbf{p}\cdot\mathbf{r} - Et)\right].$$

In the presence of an electromagnetic field, $\mathbf{p} \to \mathbf{p} - e/c\mathbf{A}$, so

$$\psi \to \psi\exp\left(-\frac{ie}{\hbar c}\mathbf{A}\cdot\mathbf{r}\right);$$

or the phase α changes by

$$\alpha \to \alpha - \frac{e}{\hbar c}\mathbf{A}\cdot\mathbf{r}.$$

Consider a closed path at fixed r, θ, with ϕ ranging from 0 to 2π. The total change in phase is

$$\Delta\alpha = \frac{e}{\hbar c}\oint\mathbf{A}\cdot d\mathbf{l}$$

$$= \frac{e}{\hbar c}\int\text{curl}\,\mathbf{A}\cdot d\mathbf{S}$$

$$= \frac{e}{\hbar c}\int\mathbf{B}\cdot d\mathbf{S}$$

$$= \frac{e}{\hbar c}(\text{Flux through cap}) = \frac{e}{\hbar c}\Phi(r, \theta); \tag{10.40}$$

$\Phi(r, \theta)$ is the flux through the cap defined by a particular r and θ, as shown by the shaded area in Fig. 10.6. As θ is varied the flux through the cap varies. As $\theta \to 0$ the loop shrinks to a point and the flux passing through the cap approaches zero:

$$\Phi(r, 0) = 0.$$

As the loop is lowered over the sphere the cap encloses more and more flux until, eventually, at $\theta \to \pi$ we should have, from (10.39),

$$\Phi(r, \pi) = 4\pi g. \tag{10.41}$$

However, as $\theta \to \pi$ *the loop has again shrunk to a point* so the requirement that $\Phi(r, \pi)$ is finite entails, from (10.40), that *A is singular at $\theta = \pi$*. This argument holds for all spheres of all possible radii, so it follows that **A** is singular along *the entire negative z axis*. This is known as the *Dirac string*. It is clear that by a suitable choice of coordinates the string may be chosen to be along any direction, and, in fact, need not be straight, but must be continuous.

The singularity in **A** gives rise to the so-called Dirac veto – that the wave function vanish along the negative z axis. Its phase is therefore indeterminate there and referring to (10.40) there is *no necessity* that as $\theta \to \pi$, $\Delta\alpha \to 0$. We must have $\Delta\alpha = 2\pi n$, however, in order for ψ to be single-valued. From (10.40) and (10.41) we then have

$$2\pi n = \frac{e}{\hbar c} 4\pi g,$$

■ $$eg = \tfrac{1}{2} n\hbar c. \tag{10.42}$$

This is the Dirac quantisation condition. It implies that the product of *any* electric with *any* magnetic charge is given by the above. Then, in principle, if there exists a magnetic charge anywhere in the universe all electric charges will be quantised:

$$e = n\frac{\hbar c}{2g}.$$

This is a possible explanation for the observed 'quantisation' of electric

Fig. 10.6.

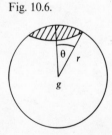

charge (see the footnote on p. 87), though nowadays this is more commonly ascribed to the existence of quarks and non-Abelian symmetry groups. Note, however, that the quantisation condition has an explicit dependence on Planck's constant, and therefore on the quantum theory. In units $\hbar = c = 1$ (10.42) becomes

■ $\qquad eg = \tfrac{1}{2}n.$ (10.43)

Let us now derive an expression for the vector potential A_μ. As seen above, it is singular. This much is clear from (10.38), for if $\mathbf{B} = \text{curl}\,\mathbf{A}$ and $\mathbf{A}$ is regular div $\mathbf{B} = 0$, and no magnetic charges may exist. From the argument above, $\mathbf{A}$ is constructed by considering the pole as the end-point of a string of magnetic dipoles whose other end is at infinity. This gives

$$A_x = g\frac{-y}{r(r+z)}, \quad A_y = g\frac{x}{r(r+z)}, \quad A_z = 0 \tag{10.44}$$

or

$$A_r = A_\theta = 0, \quad A_\phi = \frac{g}{r}\frac{1-\cos\theta}{\sin\theta}. \tag{10.45}$$

$\mathbf{A}$ is clearly singular along $r = -z$. If, on the other hand, the Dirac string were chosen to be along $r = z$, we should have

$$A_r = A_\theta = 0, \quad A_\phi = -\frac{g}{r}\frac{1+\cos\theta}{\sin\theta}. \tag{10.46}$$

The rationale for writing the alternative expressions (10.45) and (10.46) is that the Dirac string singularity is clearly unphysical, and in these expressions it is in different places. The only *physical* singularity in $\mathbf{A}$ is at the origin, where, from (10.38), div $\mathbf{B} = \text{div}\,(\text{curl}\,\mathbf{A})$ is singular. Since it is obviously desirable to get rid of unphysical singularities, this suggests the following construction. Divide the space surrounding the monopole – the sphere, essentially – into two overlapping regions R_a and R_b, as shown in Fig. 10.7. R_a excludes the negative z axis (S pole) and R_b excludes the

Fig. 10.7. R_a and R_b are overlapping domains on the sphere. R_a excludes the S pole, R_b the N pole.

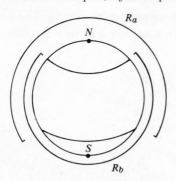

positive z axis (N pole). In each region $\mathbf{A}$ is defined differently:

$$A_r^a = A_\theta^a = 0, \quad A_\phi^a = \frac{g}{r}\frac{1 - \cos\theta}{\sin\theta}, \tag{10.47}$$

$$A_r^b = A_\theta^b = 0, \quad A_\phi^b = -\frac{g}{r}\frac{1 + \cos\theta}{\sin\theta}. \tag{10.48}$$

Referring to (10.45) and (10.46), it is clear that $\mathbf{A}^a$ and $\mathbf{A}^b$ are *both finite in their own domain*. In the region of overlap, however, they are not the same, but are related by a *gauge transformation* ($\hbar = c = 1$):

$$A_\phi^b = A_\phi^a - \frac{2g}{r\sin\theta} = A_\phi^a - \frac{i}{e}S\nabla_\phi S^{-1} \tag{10.49}$$

with

$$S = \exp(2ige\phi) \tag{10.50}$$

The covariant form of (10.49) is

$$A_\mu^b = A_\mu^a - \frac{i}{e}S\partial_\mu S^{-1}. \tag{10.51}$$

The requirement that the gauge transform function S be single-valued as $\phi \to \phi + 2\pi$ is clearly the Dirac quantisation condition (10.43). To check that (10.47) and (10.48) really do represent a monopole, we calculate the total magnetic flux through a sphere surrounding the origin.

$$\Phi = \int F_{\mu\nu}\,dx^{\mu\nu}$$

$$= \oint \operatorname{curl}\mathbf{A}\cdot d\mathbf{S}$$

$$= \int_{R_a} \operatorname{curl}\mathbf{A}\cdot d\mathbf{S} + \int_{R_b} \operatorname{curl}\mathbf{A}\cdot d\mathbf{S}.$$

Here we take R_a and R_b as not actually overlapping, but having a common boundary, which for convenience is taken to be the equator $\theta = \pi/2$. Since R_a and R_b have boundaries Stokes' theorem is applicable, and since the equator bounds R_a in a positive orientation and R_b in a negative one we have

$$\Phi = \oint_{\theta=\pi/2} \mathbf{A}^a\cdot d\mathbf{l}^a - \oint_{\theta=\pi/2} \mathbf{A}^b\cdot d\mathbf{l}^b$$

$$= \frac{i}{e}\oint\frac{d}{d\phi}(\ln S^{-1})\,d\phi$$

$$= 4\pi g$$

from (10.50), and using (10.47) and (10.48). This agrees with (10.41).

This construction is due to Wu and Yang, and is, in essence, a fibre bundle formulation of the magnetic monopole. The base space (3-dimensional space R^3 minus the origin $\approx R^3 - (\text{point}) \approx S^2 \times R^1$) is parameterised in two independent ways, corresponding to two overlapping but not identical regions. In each region the vector potential is given by a different expression. Readers familiar with the Möbius strip will recognize a similarity here. There is no unique parameterisation of the Möbius strip; *locally* it is the direct product of an interval $(0, 1)$ and a circle, but globally the circle has to be divided into two distinct overlapping regions, with a different parameterisation of the strip in each region.

There is thus a fibre-bundle formulation of the Dirac monopole. The base space is essentially S^2 (the sphere surrounding the monopole) and the group space in S^1 (since the gauge group is $U(1)$). The fibre bundle is not $S^2 \times S^1$ but S^3, which is *locally* the same as $S^2 \times S^1$ but is globally distinct. For further details on the fibre-bundle formulation, the reader is referred to the literature.

10.4   The 't Hooft–Polyakov monopole

In the context of Maxwell's electrodynamics, with Abelian gauge group $U(1)$, it is clear that although magnetic charges may be 'added' to the theory, there is no necessity for doing this. A theory with monopoles is more symmetric between electricity and magnetism than one without, but this does not amount to a *requirement* that monopoles exist. They may or may not; the above considerations do not allow us to decide. When the gauge symmetry is enlarged to a non-Abelian group, however, and spontaneous symmetry breaking is introduced, the field equations yield a solution which corresponds to a magnetic charge. If such theories are correct, then, magnetic monopoles *must* exist, and should therefore be looked for. It is a matter of natural curiosity to enquire where the magnetic charge in this model comes from, since the matter and gauge fields in the theory carry electric charge *only*. It will not surprise the reader to hear that the origin of the magnetic charge is topological. The theoretical possibility of monopoles of this type was discovered in 1974 by 't Hooft and Polyakov.

We consider a theory with an $O(3)$ symmetry group, containing the gauge field $F_{\mu\nu}^a$ (a is the group index) and an isovector Higgs field ϕ^a. The Lagrangian is (cf. (8.42))

$$\mathcal{L} = -\tfrac{1}{4}F_{\mu\nu}^a F^{\mu\nu a} + \tfrac{1}{2}(D_\mu \phi^a)(D^\mu \phi^a)$$

$$-\frac{m^2}{2}\phi^a\phi^a - \lambda(\phi^a\phi^a)^2 \tag{10.52}$$

where

$$F^a_{\mu\nu} = \partial_\mu A^a_\nu - \partial_\nu A^a_\mu + e\varepsilon^{abc} A^b_\mu A^c_\nu, \Big\}$$
$$D_\mu \phi^a = \partial_\mu \phi^a + e\varepsilon^{abc} A^b_\mu \phi^c. \quad \Big\}$$

$$(10.53)$$

We are interested in static solutions in which the gauge potentials have the non-trivial form

$$A^a_i = -\varepsilon_{iab} \frac{r^b}{er^2} \quad (r \to \infty), \Big\}$$
$$A^a_0 = 0. \qquad\qquad\qquad \Big\}$$

$$(10.54)$$

and the scalar field is

$$\phi^a = F \frac{r^a}{r} \quad (r \to \infty)$$

$$(10.55)$$

with $F^2 = -m^2/4\lambda$. These expressions have a remarkable form because of the mixing they employ between space and isospace indices. For example, (10.55) describes a field which, in the x direction in space, has only an isospin '1' component, in the y direction, only a '2' component, and in the z direction, only a '3' component. In a manner of speaking , it is 'radial' – Polyakov calls it a 'hedgehog' solution. It can be shown[19] that there exist regular solutions to the field equations derived from (10.52), which have the asymptotic form (10.54), (10.55). For example, the equation of motion of ϕ is

$$-(m^2 + 4\lambda\phi^b\phi^b)\phi^a = D_\mu(D^\mu\phi^a).$$

Equation (10.55) implies $|\phi| = F$, so the left-hand side of the above equation vanishes at infinity. It is easy to see that $D_\mu\phi^a$ also vanishes; for with $i = x, y, z$ we have

$$D_i\phi^a = F\partial_i\left(\frac{r^a}{r}\right) + e\varepsilon^{abc}A^b_i F\frac{r^c}{r}$$

$$= F\left(\frac{\delta^{ia}}{r} - \frac{r^i r^a}{r^3}\right) - \varepsilon^{abc}\varepsilon_{ibm}\frac{Fr^m r^c}{r^3}$$

$$= 0.$$

Hence, at infinity, ϕ takes on its vacuum value and is covariantly constant, but has the non-trivial boundary condition (10.55), rather than the more usual ('Abelian') condition $\phi^{1,2} = 0, \phi^3 \neq 0$. On the other hand, $F^a_{\mu\nu}$ is not zero at infinity. We shall see below that there is a radial magnetic field. This solution is sketched in Fig. 10.8.

Now let us generalise the definition of the electromagnetic field $F_{\mu\nu}$ so that it reduces to the usual one when the scalar field ϕ has only a third component. We put

$$F_{\mu\nu} = \frac{1}{|\phi|}\phi^a F^a_{\mu\nu} - \frac{1}{e|\phi|^3}\varepsilon_{abc}\phi^a(D_\mu\phi^b)(D_\nu\phi^c).$$

$$(10.56)$$

It is quite clear that when

$$A_\mu^{1,2} = 0, \quad A_\mu^3 \equiv A_\mu \neq 0, \atop \phi^{1,2} = 0, \quad \phi^3 = F \neq 0 \Bigg\} \tag{10.57}$$

this gives the usual $F_{\mu\nu}$, so long as $A_\mu^3 = A_\mu$, the Maxwell vector potential. Now, defining

$$A_\mu = \frac{1}{|\phi|}\phi^a A_\mu^a \tag{10.58}$$

a straightforward calculation gives

$$F_{\mu\nu} = \partial_\mu A_\nu - \partial_\nu A_\mu - \frac{1}{e|\phi|^3}\varepsilon_{abc}\phi^a(\partial_\mu\phi^b)(\partial_\nu\phi^c). \tag{10.59}$$

This is similar to, but more complicated than, the usual definition of the electromagnetic field, but it reduces to it when ϕ becomes fixed in isospace. Inserting the asymptotic conditions (10.54) and (10.55), it is easily seen that $A_\mu = 0$, so all the electromagnetic field is contributed by the Higgs field; and we find

$$F_{0i} = 0, \quad F_{ij} = -\frac{1}{er^3}\varepsilon_{ijk}r^k. \tag{10.60}$$

This corresponds to a radial magnetic field (see (2.221))

$$B_k = \frac{r^k}{er^3}. \tag{10.61}$$

The magnetic flux is, from (10.39),

$$\Phi = \frac{4\pi}{e},$$

Fig. 10.8. The asymptotic forms of the gauge and scalar fields constituting a 't Hooft–Polyakov monopole. Polyakov calls it a 'hedgehog' solution.

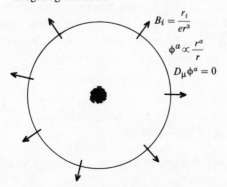

so by comparison with (10.41) the magnetic charge g is such that

$$eg = 1. \tag{10.62}$$

From (10.43), this is twice the Dirac unit. We conclude that the configuration of gauge and scalar fields with asymptotic form (10.54–10.55) carries a magnetic charge – that is, when viewed from infinity, there is a radial magnetic field. It has been shown by 't Hooft that this configuration is everywhere non-singular, and therefore has a finite energy. He estimates the monopole mass to be of the order $137M_W$, where M_W is a typical vector boson mass, so the monopoles are extremely heavy. The mass is inversely proportional to e^2 (cf. (10.9)).

What is the origin of this magnetic charge? How does it come about that a configuration of fields carrying electric charge only can arrange itself in such a way as to simulate a magnetic charge? To answer this, we write the magnetic current K_μ as

$$\begin{aligned} K^\mu &= \partial_\nu \tilde{F}^{\mu\nu} \\ &= \tfrac{1}{2}\varepsilon^{\mu\nu\rho\sigma}\partial_\nu F_{\rho\sigma} \end{aligned} \tag{10.63}$$

where $\tilde{F}_{\mu\nu}$ is the dual of $F_{\mu\nu}$ – cf. equation (2.236) which holds when no magnetic sources are present. From (10.59) we then have

$$K^\mu = -\frac{1}{2e}\varepsilon^{\mu\nu\rho\sigma}\varepsilon_{abc}\partial_\nu\hat{\phi}^a\partial_\rho\hat{\phi}^b\partial_\sigma\hat{\phi}^c \tag{10.64}$$

where

$$\hat{\phi}^a = \frac{1}{|\phi|}\phi^a.$$

We see that the magnetic current *depends on the Higgs field only*, as noticed above in (10.60). Moreover, this current is *identically conserved*:

$$\partial_\mu K^\mu = 0. \tag{10.65}$$

This property is reminiscent of the current (10.10) for the sine–Gordon kink. The conservation of both these currents does not follow from a symmetry of the Lagrangian, so they are not Noether currents. It will be recalled that the sine–Gordon charge (10.11) – the 'kink number' – depends simply on the non-trivial boundary conditions. We anticipate the same phenomenon here. The conserved magnetic charge is

$$\begin{aligned} M &= \frac{1}{4\pi}\int K^0 \mathrm{d}^3x \\ &= -\frac{1}{8\pi e}\oint_{S^2} \varepsilon_{ijk}\varepsilon_{abc}\hat{\phi}^a\partial_j\hat{\phi}^b\partial_k\hat{\phi}^c(\mathrm{d}^2S)_i. \end{aligned} \tag{10.66}$$

Here the integral is taken over the sphere S^2 at infinity, which, of course, is the *boundary* of the static field configuration ϕ. Since ϕ must be *single-valued*, as $(dS)_i$ covers the sphere once, the vector ϕ will be covered an *integral number* of times, say d. It then follows [21] that the integral in (10.66) is $8d$, hence

$$M = \frac{d}{e}, \quad d \text{ integer.} \tag{10.67}$$

Since ϕ^a is an isovector, the unit vector $\hat{\phi}$ describes a sphere S^2 in field space (isospace), so the boundary describes a mapping of the sphere S^2 in coordinate space onto the $\hat{\phi}$ manifold, which is S^2.[‡]

$$\hat{\phi}: S^2 \text{ in field space} \to S^2 \text{ in coordinate space;} \tag{10.68}$$

d is called the *Brouwer degree* of this mapping. It is necessarily integral. So equation (10.67) displays explicitly the topological nature of the 't Hooft–Polyakov monopole.

In the model considered by 't Hooft, the non-Abelian group is $SO(3)$, electromagnetism being represented by the Abelian subgoup $U(1)$. An interesting question is: how does the existence of magnetic charge in non-Abelian gauge theories depend on the gauge group? To answer this we begin by reflecting on equation (10.68). It is obvious that in general terms what is important is the $\hat{\phi}$ manifold (so the gauge theory *must* be spontaneously broken). What is the space of the $\hat{\phi}$ manifold in general? In chapter 8 we learned that it is the vacuum manifold. If the symmetry group of the theory is G (in this case $SO(3)$), and the unbroken subgroup is H (in this case $U(1)$), then transformations belonging to H leave the vacuum manifold invariant. So the space of $\hat{\phi}$ is the set of transformations in G which are *not* related by a transformation belonging to H. This is the definition of a coset space. In schematic terms the elements of the gauge G may be written

$$G = H + HM_1 + HM_2 + \cdots \tag{10.69}$$

where H denotes the elements of the subgroup H, and $M_1, M_2, \ldots$ belong to G but not to H, and are all different. The vacuum manifold is essentially the space of the elements M_i, and that is the coset space G/H. Consulting (10.68) again, the existence of magnetic monopoles requires a non-trivial mapping of G/H onto S^2, the boundary in coordinate space. As we saw in chapter 3 (equation (3.114)) these mappings form a group, in this case the *second homotopy group of* G/H, $\pi_2(G/H)$. Magnetic monopoles will exist if this group is non-trivial. We now invoke a mathematical theorem involving homotopy groups.[3,22]

[‡] To be precise, the space of $\hat{\phi}$, an isovector, is S^2 with opposite points identified. I am grateful to Mr S. Vokos for pointing this out to me.

Theorem: $\pi_2(G/H)$ is isomorphic to the kernel of the
natural homomorphism of $\pi_1(H)$ into $\pi_1(G)$.   (10.70)

To explain the terms in this theorem: $\pi_1(H)$ and $\pi_1(G)$ are the first homotopy groups of H and G. They are trivial if the groups are simply connected, isomorphic to $\mathbb{Z}_2(C_2)$ if the groups are doubly connected, etc. Since every closed path in H is also a closed path in G, there is a natural mapping of $\pi_1(H)$ into $\pi_2(G)$; this is called a homomorphism. The kernel of the homomorphism is the set of elements of $\pi_1(H)$ which are mapped onto the *identity* of $\pi_1(G)$.

Let us watch this theorem in action by applying it to the 't Hooft case, where $G = SO(3), H = U(1)$. Since $SO(3)$ is doubly connected (see above), $\pi_1(G) = \mathbb{Z}_2$. On the other hand, $U(1)$ is infinitely connected (the group space is a circle, and a closed curve going n times round a circle cannot be continuously deformed into one going $m (\neq n)$ times round), so $\pi_1(H) = \mathbb{Z}$, the additive group of integers. So the kernel of the mapping of $\pi_1(H)$ into $\pi_1(G)$ is the additive group of *even* integers, hence

$$\pi_2(SO(3)/U(1)) = \text{additive group of even integers.} \quad (10.71)$$

This is consistent with what we found; the monopole charge was *twice* the Dirac quantum.

The trouble is that the true non-Abelian electroweak group is not $SO(3)$, but $SU(2) \times U(1)$, given by the Weinberg–Salam model. (The $SO(3)$ model is the Georgi–Glashow model.[‡] Its salient characteristic is that the only neutral current in it is the electromagnetic current. It was therefore rendered obsolete by the discovery of weak neutral current events, such as $v_e + p \rightarrow v_e + p + \pi^0$.) Moreover, the electromagnetic subgroup, although given by $U(1)$, is irregularly embedded in $SU(2) \times U(1)$, and so is *non-compact*, with the consequence that magnetic monopoles do not exist in the Weinberg–Salam model. To see this argument, note that in this model there are two $U(1)$ subgroups, so that a particle with a third component of weak isospin I_3^w and weak hypercharge Y^w, will transform under these $U(1)$ groups by

$$\exp(i\alpha I_3^w)\exp(i\beta Y^w). \quad (10.72)$$

The group space of $U(1)$ is a circle, or, equivalently, a line with the points 0 and 2π identified. Hence the group space of $U(1) \times U(1)$ may be represented, as in Fig. 10.9, by a square $ABCD$, with the edges AC and AB identified with BD and CD respectively. This is a *torus* T^2. (In general the group space of the direct product of n groups $U(1)$ is a toroid T^n.) The group element (10.72) will correspond to a point (α, β) in the group space T^2 of Fig. 10.9. Electric charge Q in the Weinberg–Salam model is given by

‡ H. Georgi & S.L. Glashow, *Physical Review Letters*, **28**, 1494 (1972).

$$Q = \sin\theta_{\rm W} I_3^{\rm w} + \cos\theta_{\rm W} Y^{\rm w}$$

where $\theta_{\rm W}$ is the Weinberg angle (cf. equation (8.82)). Under an electromagnetic gauge transformation through an angle γ, the state vector for a particle with charge Q is multiplied by

$$\exp(i\gamma Q) = \exp\left[i(\gamma \sin\theta_{\rm W} I_3^{\rm w} + \gamma \cos\theta_{\rm W} Y^{\rm w})\right],$$

and to this transformation corresponds a point in group space given by

$$\alpha = \gamma \sin\theta_{\rm W}, \quad \beta = \gamma \cos\theta_{\rm W}$$

hence

$$\alpha/\beta = \tan\theta_{\rm W} = \text{irrational}. \tag{10.73}$$

The above condition corresponds to a line in group space $Aaa'bb'cc'dd'ee'\dots$ (see Fig. 10.9), which, since α/β is an irrational number, is a line of *infinite length*. It winds round the torus without ever meeting itself again. Hence the electromagnetic gauge group in the Weinberg–Salam model has infinite volume, and is *non-compact*. It follows that $\pi_1(H)$ does not exist (or is trivial) so $\pi_2(G/H)$ is also trivial and no monopoles exist. If nature is 'grand-unified', however, and the electroweak group $SU(2) \times U(1)$ is a subgroup of a grand-unified semisimple group, say $SU(5)$, then this argument no longer holds, and monopoles may exist. These questions have recently come to life following a claim that a monopole has been discovered.[‡]

Comparing the 't Hooft–Polyakov monopole with the Dirac monopole

Fig. 10.9. The group space of $U(1) \otimes U(1)$ is a square with opposite edges identified, hence a torus. An electromagnetic gauge transformation traces out the line $Aaa'bb'cc'dd'ee'\dots$.

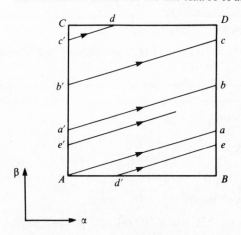

[‡] B. Cabrera, *Physical Review Letters*, **48**, 1378 (1982). But see also B. Cabrera *et al., ibid.*, **51**, 1933 (1983).

of §10.3, it will seem as if they have almost nothing in common; to be more precise, nothing at all except that they both possess magnetic charge. This is not quite true, however, and it may be helpful to conclude this section by showing how the two may be related. We start with a Dirac monopole with a string singularity along the negative z axis. The vector potential is therefore given by (10.45). Now we submerge this in an $SU(2)$ theory, with the vector potential aligned in the third direction in isospin space. Using the matrix potential $A_\mu = A_\mu^a \tau^a$ then gives

$$A_0 = A_r = A_\theta = 0, \quad A_\phi = \tau_3 \left(-\frac{g}{r} \right) \left(\frac{1 - \cos\theta}{\sin\theta} \right). \tag{10.74}$$

In addition, we introduce a scalar field ϕ, with vacuum expectation value F, also aligned along the third direction in isospin space:

$$\phi = \tau_3 F. \tag{10.75}$$

Now we transform A_μ and ϕ by a space-dependent isospin gauge transformation. A general $SU(2)$ gauge transformation may be characterised by the Euler angles (α, β, γ) and written

$$S = e^{(i/2)\alpha\tau_3} e^{(i/2)\beta\tau_2} e^{(i/2)\gamma\tau_3}$$

$$= \left(\begin{matrix} \cos\beta/2\, e^{i(\alpha + \gamma/2)} & \sin\beta/2\, e^{i(-\gamma + \alpha/2)} \\ -\sin\beta/2\, e^{i(\gamma - \alpha/2)} & \cos\beta/2\, e^{-i(\alpha + \gamma/2)} \end{matrix} \right).$$

Now we put $\gamma = -\alpha = \phi$, $\beta = -\theta$, giving

$$S = \left(\begin{matrix} \cos\theta/2 & -e^{-i\phi}\sin\theta/2 \\ e^{i\phi}\sin\theta/2 & \cos\theta/2 \end{matrix} \right) \tag{10.76a}$$

hence

$$S^{-1} = \left(\begin{matrix} \cos\theta/2 & e^{-i\phi}\sin\theta/2 \\ -e^{i\phi}\sin\theta/2 & \cos\theta/2 \end{matrix} \right). \tag{10.76b}$$

The transformation law for A_μ is (as in (3.162), but with $A_\mu = A_\mu^a \tau^a$ and $g \to e$)

$$A'_\mu = S A_\mu S^{-1} + \frac{2i}{e} S \partial_\mu S^{-1}. \tag{10.77}$$

From (10.76b) it follows that

$$\partial_r S^{-1} = 0, \quad \partial_\theta S^{-1} = \frac{1}{2r} \left(\begin{matrix} -\sin\theta/2 & e^{-i\phi}\cos\theta/2 \\ -e^{i\phi}\cos\theta/2 & -\sin\theta/2 \end{matrix} \right),$$

$$\partial_\phi S^{-1} = \frac{-i}{r\sin\theta} \left(\begin{matrix} 0 & e^{-i\phi}\sin\theta/2 \\ e^{i\phi}\sin\theta/2 & 0 \end{matrix} \right).$$

Substituting these into (10.77), using (10.74) and putting $g = 1/e$ (from (10.62)) gives, after straightforward manipulations,

$$A_0' = A_r' = 0,$$

$$A_\theta' = \frac{1}{er}(\tau_1 \sin \phi - \tau_2 \cos \phi),$$

$$A_\phi' = \frac{1}{er}(\tau_1 \cos \theta \cos \phi + \tau_2 \cos \theta \sin \phi - \tau_3 \sin \theta).$$

The Cartesian components of A may then be found; for example,

$$A_x' = A_r' \cos \phi \sin \theta + A_\theta' \cos \theta \cos \phi - A_\phi' \sin \phi$$

$$= \frac{1}{er}\left[\tau_2 \left(\frac{-z}{r} \right) + \tau_3 \left(\frac{y}{r} \right) \right]. \tag{10.78}$$

This is the 'hedgehog' form (10.54). Under the same transformation (10.76) the Higgs field (10.75) becomes

$$\phi' = S\phi S^{-1}$$

$$= F \begin{pmatrix} \cos \theta & e^{-i\phi} \sin \theta \\ -e^{i\phi} \sin \theta & -\cos \theta \end{pmatrix}$$

$$= F (\sin \theta \cos \phi \tau_1 + \sin \theta \cos \phi \tau_2 + \cos \theta \, \tau_3), \tag{10.79}$$

i.e.

$$\phi'^a = F\frac{r^a}{r} \tag{10.80}$$

as in (10.55). As a result of this transformation, the string singularity of the Dirac potential *disappears*, and the source of the monopole resides, as it were, in the Higgs field. From equation (10.59) we may say that the gauge transformation transfers responsibility for the monopole from the first (Dirac) term, to the second (topological, Higgs) one. Thus the Dirac and 't Hooft–Polyakov monopoles are not so unconnected as they at first appear.

10.5 Instantons

Our final example of soliton solutions is concerned with those which are localised in time as well as in space, and which 't Hooft has therefore christened 'instantons'. (An alternative name, suggested by Polyakov, is 'pseudo particles'.) It is not surprising that such solutions exist, since the gauge-field equations are fully relativistic, so allow a topological non-triviality in time as well as in space. Moreover, the gauge group $SU(2)$ plays a rather special role as may be seen from the following consideration. To begin with, space–time is considered to be 'Euclideanised' so that it becomes E^4. Its boundary is then S^3, the 3-sphere. On the other hand, it was seen in (10.34) above that the group space of $SU(2)$ is also S^3. Hence topologically non-trivial solutions to the $SU(2)$ gauge-field equations are possible if there exist non-trivial (non-homotopic) mappings of S^3 onto S^3, that is if $\pi_3(S^3)$ is non-trivial (see (3.114)). And indeed it is:

$$\pi_3(S^3) = \mathbb{Z}. \tag{10.81}$$

It follows that instantons are therefore possible in the *pure* gauge theory; spontaneous symmetry breaking is unnecessary. This distinguishes instantons from monopoles. The plan in this section will be to write down the instanton solution, exhibiting its topological nature, and then to mention briefly the physical consequences that follow from the existence of instantons. There is a considerable amount of literature on this topic, so our treatment will be very introductory and the reader is referred to the many excellent reviews to broaden his knowledge. In addition, instantons have aroused the interest of a number of pure mathematicians, and many papers explore their connections with topology and algebraic geometry. But these matters are beyond the scope of this book, and the reader is again referred elsewhere.

We begin with some mathematical preliminaries. Euclidean space has coordinates (x_1, x_2, x_3, x_4) with (see (6.16))

$$x_0 = -ix_4 \tag{10.82}$$

(and $x_0 = ct$). The Euclidean field tensor $F_{\mu\nu}^a$ is defined[31] in the same way as the Minkowski tensor (see (3.169)):

$$F_{\mu\nu}^a = \partial_\mu A_\nu^a - \partial_\nu A_\mu^a + g\varepsilon^{abc} A_\mu^b A_\nu^c \tag{10.83}$$

with

$$A_\mu = \tfrac{1}{2}\sigma^a A_\mu^a, \quad F_{\mu\nu} = \tfrac{1}{2}\sigma^a F_{\mu\nu}^a. \tag{10.84}$$

This takes the form (see (3.166))

$$F_{\mu\nu} = \partial_\mu A_\nu - \partial_\nu A_\mu - ig[A_\mu, A_\nu]. \tag{10.85}$$

Defining

$$\partial_{[\mu} A_{\nu]} \equiv \partial_\mu A_\nu - \partial_\nu A_\mu \tag{10.86}$$

this becomes

$$F_{\mu\nu} = \partial_{[\mu} A_{\nu]} - ig[A_\mu, A_\nu]. \tag{10.87}$$

The *dual* of $F_{\mu\nu}$, denoted $\tilde{F}_{\mu\nu}$, is defined by

$$\tilde{F}_{\mu\nu} = \tfrac{1}{2}\varepsilon_{\mu\nu\rho\sigma} F_{\rho\sigma} \tag{10.88}$$

(remembering that in Euclidean space there is no need to distinguish upper and lower indices). With $\varepsilon_{1234} = 1$ this yields

$$\tilde{\tilde{F}}_{\mu\nu} = F_{\mu\nu}, \tag{10.89}$$

whereas in Minkowski space, since when $\varepsilon^{0123} = 1$, then $\varepsilon_{0123} = -1$, so

$$\tilde{\tilde{F}}_{\mu\nu} = -F_{\mu\nu} \quad \text{(in Minkowski space)}. \tag{10.90}$$

Under gauge transformations

$$A'_\mu = SA_\mu S^{-1} - \frac{i}{g}(\partial_\mu S)S^{-1},$$ (10.91)

$$F'_{\mu\nu} = SF_{\mu\nu}S^{-1}.$$ (10.92)

Now we define

$$K_\mu = \tfrac{1}{4}\varepsilon_{\mu\nu\kappa\lambda}\left(A^a_\nu\partial_\kappa A^a_\lambda + \frac{g}{3}\varepsilon_{abc}A^a_\nu A^b_\kappa A^c_\lambda\right)$$

$$= \varepsilon_{\mu\nu\kappa\lambda}\operatorname{Tr}\left(\tfrac{1}{2}A_\nu\partial_\kappa A_\lambda - \frac{ig}{3}A_\nu A_\kappa A_\lambda\right).$$ (10.93)

Then

$$\partial_\mu K_\mu = \tfrac{1}{4}\operatorname{Tr}\tilde{F}_{\mu\nu}F_{\mu\nu} = \tfrac{1}{8}\tilde{F}^a_{\mu\nu}F^a_{\mu\nu}$$ (10.94)

so that $\operatorname{Tr}\tilde{F}F$ is a total divergence.

Proof. Because of the cyclic property of the trace

$$\partial_\mu K_\mu = \varepsilon_{\mu\nu\kappa\lambda}\operatorname{Tr}[\tfrac{1}{2}(\partial_\mu A_\nu)(\partial_\kappa A_\lambda) - ig(\partial_\mu A_\nu)A_\kappa A_\lambda]$$

On the other hand,

$$\operatorname{Tr}F_{\mu\nu}\tilde{F}_{\mu\nu} = \tfrac{1}{2}\varepsilon_{\mu\nu\kappa\lambda}\operatorname{Tr}\{\partial_{[\mu}A_{\nu]} - ig[A_\mu, A_\nu]\}\{\partial_{[\kappa}A_{\lambda]} - ig[A_\kappa, A_\nu]\}$$

$$= 2\varepsilon_{\mu\nu\kappa\lambda}\operatorname{Tr}(\partial_\mu A_\nu)(\partial_\kappa A_\lambda)$$

$$\quad - 2ig\varepsilon_{\mu\nu\kappa\lambda}\operatorname{Tr}A_\mu A_\nu(\partial_\kappa A_\lambda)$$

$$\quad - 2ig\varepsilon_{\mu\nu\kappa\lambda}\operatorname{Tr}(\partial_\mu A_\nu)A_\kappa A_\lambda$$

$$\quad - 2g^2\varepsilon_{\mu\nu\kappa\lambda}\operatorname{Tr}A_\mu A_\nu A_\kappa A_\lambda.$$

Because of the cyclic trace property, the second two terms above are equal and the last one vanishes. Hence (10.94) is proved.

Now consider a 4-dimensional volume V^4 in E^4, with boundary $\partial V^4 \sim S^3$. Suppose it is a pure vacuum, $A_\mu = 0, F_{\mu\nu} = 0$. Then $K_\mu = 0$. The field equations (in the absence of matter)

$$D_\mu F_{\mu\nu} = 0$$ (10.95)

are clearly satisfied over the whole region V^4, as is the Bianchi identity

$$D_\mu \tilde{F}_{\mu\nu} = 0$$ (10.96)

which, of course, *must* be satisfied. Applying Gauss's theorem to (10.94) gives

$$\int_{V^4}\operatorname{Tr}F_{\mu\nu}\tilde{F}_{\mu\nu}\,d^4x = 4\int_{V^4}\partial_\mu K_\mu\,d^4x$$

$$= 4\oint_{\partial V^4}K_\perp\,d^3x.$$ (10.97)

This is trivially satisfied if V^4 is a pure vacuum.

Now we perform a (space–time-dependent) gauge transformation at the boundary S^3

$$A_\mu \to -\frac{i}{g}(\partial_\mu S)S^{-1} \quad \text{(on } S^3\text{)}, \tag{10.98}$$

i.e.

$$F_{\mu\nu} = 0$$

so the boundary becomes a 'pure gauge' vacuum; and take

$$S = \frac{x_4 + i\mathbf{x}\cdot\boldsymbol{\sigma}}{\sqrt{\tau^2}} \tag{10.99}$$

where

$$\tau^2 = x_4^2 + \mathbf{x}^2. \tag{10.100}$$

Then after some straightforward (but lengthy) algebra we find

$$\left.\begin{aligned}
A_i &= \frac{i}{g\tau^2}[x_i - \sigma_i(\boldsymbol{\sigma}\cdot\mathbf{x} + ix_4)], \\
A_4 &= -\frac{1}{g\tau^2}\boldsymbol{\sigma}\cdot\mathbf{x},
\end{aligned}\right\} \tag{10.101}$$

and

$$K_\mu = \frac{2x_\mu}{g^2\tau^4}. \tag{10.102}$$

Equation (10.97) then yields

$$\begin{aligned}
\int \text{Tr}\, F_{\mu\nu}\tilde{F}_{\mu\nu}\, d^4x &= 4\oint_{S^3} K_\perp\, d^3\sigma \\
&= \frac{8\tau}{g^2\tau^4}\oint_{S^3} d(\text{area}) \\
&= \frac{16\pi^2}{g^2},
\end{aligned} \tag{10.103}$$

using the fact that the area of the 3-sphere of radius τ is $2\pi^2\tau^3$. (The area of the unit sphere S^n is $\pi^{n/2}2^{n+1}(n/2)!/(n!)$.) We see immediately that $F_{\mu\nu}$ *cannot be zero over the whole volume V^4*, although it does vanish on the boundary. It will be appreciated that this is a consequence of the fact that K_μ is not gauge invariant.

The above situation is sketched in Fig. 10.10. The field strength $F_{\mu\nu}$ is non-zero inside the volume V^4, but vanishes on the boundary S^3, where A_μ becomes a pure gauge. It is clear that (10.98) is *not* a solution to the gauge-field equations over the whole space, but is simply the asymptotic form as $\tau^2 \to \infty$. How are we to understand this? We first show that the integral (10.103) above defines a *topological index*. It is called the *Pontryagin index* (or Pontryagin class), and denoted q:

$$q = \frac{g^2}{16\pi^2} \operatorname{Tr} \int F_{\mu\nu} \tilde{F}_{\mu\nu} \, d^4x. \tag{10.104}$$

Then in the case we are considering we have

$$q = \frac{g^2}{4\pi^2} \int \partial_\mu K_\mu \, d^4x = 1. \tag{10.105}$$

We shall show that q is the degree of the mapping of the group space, S^3, of $SU(2)$ onto the co-ordinate space boundary S^3. Putting (10.98) into (10.93) gives

$$K_\mu = \frac{1}{6g^2} \varepsilon_{\mu\nu\kappa\lambda} \operatorname{Tr}(S^{-1}\partial_\nu S)(S^{-1}\partial_\lambda S)(S^{-1}\partial_\kappa S),$$

hence

$$\begin{aligned} q &= \frac{1}{24\pi^2} \oint_{S^3} \varepsilon_{\mu\nu\kappa\lambda} \hat{n}_\mu \operatorname{Tr}(S^{-1}\partial_\nu S)(S^{-1}\partial_\lambda S)(S^{-1}\partial_\kappa S) \, d^3\sigma \\ &= \frac{1}{24\pi^2} \oint_{S^3} \frac{\partial(g)}{\partial(\sigma)} d^3\sigma \\ &= \frac{1}{24\pi^2} \int_G d^3g \end{aligned} \tag{10.106}$$

where d^3g is the invariant element of volume in group space. Hence q gives the (Brouwer) degree of the mapping $S^3 \to S^3$.

This solution, then, is like the soliton solution, except that E^4 has one

Fig. 10.10. The instanton. Inside the volume V^4 the field strength $F_{\mu\nu}$ is non-vanishing, but $F_{\mu\nu}$ vanishes on the boundary S^3.

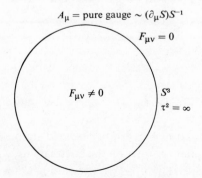

time and three space dimensions. The similarity is that as one of these co-ordinates passes from $-\infty$ to $+\infty$ the field configuration changes, so that the boundary conditions at $-\infty$ and $+\infty$ are different, rather as in Figs. 10.1 and 10.3 for the sine–Gordon kink. In that case, of course, the relevant co-ordinate is a spatial one. An obvious way to interpret the present solution is as an evolution in *time*, rather than space. This suggests redrawing the boundary S^3 as in Fig. 10.11. I and II are the hypersurfaces $x_4 \to \infty$ and $x_4 \to -\infty$ and III is the hypercylindrical surface joining them. Then

$$q = \frac{1}{24\pi^2}\left[\int\int_{\text{I-II}} d^3\sigma\varepsilon_{4ijk}\,\text{Tr}(\bar{A}_i\bar{A}_j\bar{A}_k)\right.$$
$$\left. + \int_{-\infty}^{\infty} dx_4 \int_{\text{III}} d^2\sigma_i\varepsilon_{i\nu\kappa\lambda}\,\text{Tr}(\bar{A}_\nu\bar{A}_\kappa\bar{A}_\lambda)\right] \tag{10.107}$$

where $\bar{A}_\mu = S^{-1}(\partial_\mu S) = igA_\mu$.

Now, as remarked above, A_μ is not a pure gauge over the whole volume. The required expression for A_μ is

$$A_\mu = \frac{\tau^2}{\tau^2 + \lambda^2}\left(\frac{-i}{g}\right)(\partial_\mu S)S^{-1} \tag{10.108}$$

where λ is a constant.[26] As $x_4 \to \pm\infty$ this tends to the pure gauge form (10.98), but in the 'interior' of the 4-volume V^4 is such that $F_{\mu\nu} \neq 0$, as required. This expression for A_μ is a solution of the field equations, and is the one which should be used in the expression for q above. However, q is gauge invariant so it is convenient to choose a gauge in which $A_4' = 0$ so that the integral over the 'cylinder' III in (10.107) vanishes (since the condition for a non-vanishing integral is that one of the indices ν, κ, λ should be 4).

Fig. 10.11. The instanton boundary. I is $x_4 \to \infty$, II is $x_4 \to -\infty$.

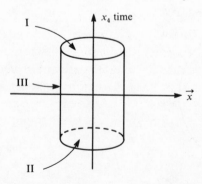

Such a gauge transformation is[‡]

$$A'_\mu = U A_\mu U^{-1} - i(\partial_\mu U) U^{-1} \qquad (10.109)$$

when

$$\left.\begin{array}{l} U = \exp\left[\dfrac{i\mathbf{x}\cdot\boldsymbol{\sigma}}{(\tau^2 + \lambda^2)^{\frac{1}{2}}}\theta\right], \\[4mm] \theta = \tan^{-1}\left[\dfrac{x_4}{(\tau^2 + \lambda^2)^{\frac{1}{2}}} - \dfrac{\pi}{2}\right] \end{array}\right\} \qquad (10.110)$$

and A_μ is equal to $\tau^2(\tau^2 + \lambda^2)^{-1}$ times the expression (10.98) (with (10.99)). It is seen that

$$A'_4 = 0$$

so that q reduces to the difference between two integrals, on the surfaces $x_4 \to -\infty$ and $x_4 \to \infty$. A'_i is a complicated expression which in the cases $x_4 \to \pm\infty$ reduces to

$$\left.\begin{array}{ll} x_4 \to \infty, & A'_i \to i(g_n)^{-1}(\partial_i g_n), \\[2mm] x_4 \to -\infty, & A'_i \to i(g_{n-1})^{-1}(\partial_i g_{n-1}), \end{array}\right\} \qquad (10.111)$$

with

$$g_n = (g_1)^n, \quad g_1 = \exp\left[-i\pi\dfrac{\mathbf{x}\cdot\boldsymbol{\sigma}}{(\tau^2 + \lambda^2)^{\frac{1}{2}}}\right] \qquad (10.112)$$

g_n is clearly an element of the group $SU(2)$, but g_n and g_m $(n \neq m)$ are *not homotopic*. In particular g_1 and $g_0 = 1$ are not homotopic; that is, it is not possible to find a function $g(g_1, a)$ with a a continuous variable between 0 and 1 such that $g(g_1, 1) = g_1$ and $g(g_1, 0) = 1$ (see §3.4). *The instanton therefore describes a solution of the gauge-field equations in which, as x_4 evolves from $-\infty$ to $+\infty$, a vacuum (belonging to homotopy class $n-1$) evolves into another vacuum (belonging to homotopy class n) and the Pontryagin index is*

$$q = n - (n-1) = 1.$$

In between these vacua is a region when the field tensor $F_{\mu\nu}$ is non-vanishing, and therefore there is *positive field energy*. The Yang–Mills vacuum is therefore infinitely degenerate, consisting of an infinite number of homotopically non-equivalent vacua. The instanton solution represents a transition from one vacuum class to another. Physics enters the scene when we ask what is the *amplitude* for this transition. Classically, of course, it is zero, since there is an energy hump in between two vacua. But because of

---

[‡] This is taken from A. Chakrabarti, talks given at Rencontre de Rabat, May 1978 (unpublished). See also Jackiw & Rebbi.[(28)]

quantum mechanics there is a *barrier penetration factor*. We now consider this.

Quantum tunnelling, θ-vacua and symmetry breaking

What we shall argue is that the barrier penetration amplitude is

$$e^{-S_E}, \quad S_E = \text{Euclidean action}. \tag{10.113}$$

To see this consider the problem of the motion of a single particle through a 1-dimensional potential well, in the quasi-classical (WKB) approximation. If $V > E$ the process is classically ($h = 0$) forbidden, but the actual tunnelling amplitude is

$$\exp\left\{ -\frac{1}{\hbar} \int_a^b [2m(V - E)]^{\frac{1}{2}} \, dx \right\} \equiv \exp\left(-\frac{1}{\hbar} S_E \right) \tag{10.114}$$

where S_E is *defined* by the integral above. We shall proceed to show that S_E is, in fact, the action for imaginary times. For consider the case where $E > V$, and the transition is classically allowed. In this case the wave function oscillates, and the number of oscillations is given by

$$\frac{1}{\hbar} \int_a^b p \, dx = \frac{1}{\hbar} \int_a^b [2m(E - V)]^{\frac{1}{2}} \, dx. \tag{10.115}$$

On the other hand,

$$\int p \, dx = \int p \dot{x} \, dt = \int (H + L) \, dt = \int (E + L) \, dt.$$

If the total energy is normalised to zero, then

$$\int p \, dx = \int L \, dt = S$$

which is the total action for the motion from a to b. Now the only difference between (10.114) and (10.115) is that the sign of $E - V$ is reversed. However, the sign of V in the equation of motion

$$m\ddot{x} = \frac{\partial V}{\partial x}$$

is reversed if we replace t by it. Hence S_E, defined in (10.114), is the action for imaginary times. In the case of field theory, this becomes the action in Euclidean space. The tunnelling amplitude, then, is given by (10.113).

What is the action for our instanton? It is easily calculated from the inequality

$$\text{Tr}(F_{\mu\nu} - \tilde{F}_{\mu\nu})^2 \geqslant 0. \tag{10.116}$$

Noting that

$$\varepsilon_{\mu\nu\rho\sigma} \varepsilon_{\mu\nu\kappa\lambda} = 2(\delta_{\rho\kappa}\delta_{\sigma\lambda} - \delta_{\rho\lambda}\delta_{\sigma\kappa})$$

it follows immediately that

$$\tilde{F}_{\mu\nu}\tilde{F}_{\mu\nu} = F_{\mu\nu}F_{\mu\nu}$$

so that (10.116) yields

$$\operatorname{Tr} F_{\mu\nu}F_{\mu\nu} \geqslant \operatorname{Tr} \tilde{F}_{\mu\nu}F_{\mu\nu}. \tag{10.117}$$

The solution (10.108), however, possesses the property of *self-duality* (see (10.89):

$$F_{\mu\nu} = \tilde{F}_{\mu\nu}. \tag{10.118}$$

(This is a crucial property of instantons, and many treatments use it as a starting point. In view of the Bianchi identity (10.96) self-duality guarantees that the field equations (10.95) are satisfied.) It follows that (10.117) becomes an equality for instantons. Noting that the action (in Euclidean space) is

$$
\begin{aligned}
S &= -\frac{1}{4}\int F^a_{\mu\nu}F^a_{\mu\nu}\,\mathrm{d}^4x \\
&= -\frac{1}{2}\int \operatorname{Tr} F_{\mu\nu}F_{\mu\nu}\,\mathrm{d}^4x,
\end{aligned}
\tag{10.119}
$$

the equality (10.117) together with (10.104) yields

$$S = -\frac{8\pi^2}{g^2}q = -\frac{8\pi^2}{g^2} \tag{10.120}$$

since $q = 1$. Hence the tunnelling amplitude between the pure vacuum and the gauge rotated vacuum is of the order

$$\mathrm{e}^{-8\pi^2/g^2}. \tag{10.121}$$

Now that we have established that in a Yang–Mills quantum theory the vacuum is infinitely degenerate, with non-zero transition amplitudes between the gauge rotated vacua belonging to different homotopy classes, it follows that the true ground state of Hilbert space may be written

$$|\mathrm{vac}\rangle_\theta = \sum_{n=-\infty}^{\infty} \mathrm{e}^{in\theta}|\mathrm{vac}\rangle_n \tag{10.122}$$

where n is an integer labelling the homotopy class. It is characterised by a particular value of θ, and the coefficients $\mathrm{e}^{in\theta}$ ensure the invariance (up to a phase) of $|\mathrm{vac}\rangle_\theta$ under gauge transformations g_1 (see (10.112)). They have the effect

$$|\mathrm{vac}\rangle_n \overset{g_1}{\to} |\mathrm{vac}\rangle_{n+1} \tag{10.123}$$

and hence

$$|\mathrm{vac}\rangle_\theta \overset{g_1}{\to} \mathrm{e}^{-i\theta}|\mathrm{vac}\rangle_\theta. \tag{10.124}$$

Gauge transformations of the type g_1 (or g_n), which change the homotopy class, are sometimes called 'large' gauge transformations. 'Small' gauge transformations are those continuously deformable to the identity (for example, infinitesimal ones), which do not change the homotopy class.

Vacua of the type (10.117) are known as θ-*vacua*, and they have several important consequences in particle physics. If $\theta \neq 0$ the vacuum state is complex, and time reversal invariance is violated. From the CPT theorem, it then follows that CP invariance is violated. Further, since under parity $g_1 \rightarrow (g_1)^{-1}$, unless $\theta = 0, P$ is also violated. The observed scale of T violation in physics requires $\theta < 10^{-5}$.[‡] A satisfactory explanation of why θ is so small yet not zero is yet to be found.

Finally, 't Hooft has drawn attention to a remarkable consequence of the existence of instantons when fermions are also present. Consider a theory with N massless quarks, where N is a 'flavour' index. It has a chiral symmetry $SU(N)_L \otimes SU(N)_R \otimes U(1)$. The axial current J_μ^5, however, has an anomaly (cf. equation (9.265))

$$\partial_\mu J_\mu^5 = \frac{Ng^2}{16\pi^2} F_{\mu\nu}^a \tilde{F}_{\mu\nu}^a.$$

Comparing with (10.104), however, gives

$$\partial_\mu J_\mu^5 = 2Nq$$

so that in the field of an instanton with $q = 1$ there is a violation of axial charge Q^5 by

$$\Delta Q^5 = 2N.$$

This results in decays such as

$$p + n \rightarrow e^+ + \bar{\nu}_\mu \quad \text{or} \quad \mu^+ + \bar{\nu}_e$$

which violate baryon and lepton number (which are not gauge symmetries). The probability of these decays is, however,

$$
\begin{aligned}
e^{-16\pi^2/g^2} = e^{-16\pi^2/e^2 \sin^{-2}\theta_W} \\
= e^{-4\pi \times 137 \times \sin^2\theta_W} \\
= e^{-602.6} = 10^{-262}
\end{aligned}
$$

if $\sin^2 \theta_W \approx 0.35$. This gives a deuteron lifetime of the order of 10^{225} s $\approx 10^{218}$ yr. Such an enormously large number is typical of the results of instanton calculations. It would be interesting if some of the large numbers in physics owed their origin to considerations of this type.

The methods of the present chapter are in essence *non-perturbative*; firstly, because a perturbation around a pure vacuum will never produce an

---

‡ F. Wilczek, *Physical Review Letters*, **40**, 279 (1978).

excitation above a vacuum belonging to a *different homotopy class*; and secondly, because the semi-classical approximation is also non-perturbative. This has given a large measure of impetus and excitement to topological methods in the last few years, because of the knowledge that areas of physics are being explored which are completely inaccessible to perturbation theory. Some hope is held out, for example, that quark confinement may be explained by these methods. In any case, a new spectre has opened on the world, and non-Abelian gauge theories like electroweak theory, QCD and grand unification (and gravity?) are now seen to have a much richer structure than had hitherto been dreamed of.

Summary
[1]The kink solution to the sine–Gordon equation is exhibited. The stability of the kink is due to the topology of the boundary conditions. [2]It is shown that in two (or more) space dimensions finite energy solitons may only exist if there is also a gauge field. The corresponding solution in 2-dimensional space (or 3-dimensional space with cylindrical symmetry) is a line carrying magnetic flux, identified with the Abrikosov flux line in superconductivity. Such vortex lines exist when the gauge group is $U(1)$, but not when it is $SU(2)$. In the case of $O(3)$, there is only one value for the charge per unit length of the vortex. [3]The magnetic monopole is introduced, and Dirac's quantisation condition derived. Wu and Yang's fibre bundle formulation of the Dirac monopole is briefly outlined. [4]Certain spontaneously broken non-Abelian gauge theories possess solutions with magnetic charge, the so-called 't Hooft–Polyakov monopoles. If the gauge group is G, and the unbroken subgroup is H, the condition that monopoles exist is that $\pi_2(G/H)$ is non-trivial. Hence 't Hooft–Polyakov monopoles do not exist in the Weinberg–Salam model. It is shown how a gauge transformation relates the Dirac and 't Hooft–Polyakov monopoles. [5]The instanton is a topologically non-trivial solution to the pure (not spontaneously broken) gauge-field equations. It describes a configuration with energy localised in time as well as in space. Its topological nature is described. The vacuum is infinitely degenerate, and some physical consequences of this, depending on quantum tunnelling, are outlined.

Guide to further reading

For recent comprehensive reviews of solitons in non-linear theories (excluding gauge theories), see
(1) A. Scott, F. Chu & D. McLaughlin, *Proceedings of the Institute of Electrical and Electronics Engineers*, **61**, 1443 (1973).
(2) G.B. Whitham, *Linear and Nonlinear Waves*, Wiley, 1974.
For short reviews, see

(3) S. Coleman, in A. Zichichi (ed.), *New Phenomena in Subnuclear Physics*, Part A, Plenum Press, 1977.

(4) G.C. Wick, in A. Zichichi (ed.), *Understanding the Fundamental Constituents of Matter*, Plenum Press, 1978.

Early examples of kinks in physics are discussed in

(5) D. Finkelstein & C.W. Misner, *Annals of Physics*, **6**, 230 (1959); D. Finkelstein, *Journal of Mathematical Physics*, **7**, 1218 (1966).

For a recent example of possible application to elementary particles, see

(6) L.D. Faddeev, *Letters in Mathematical Physics*, **1**, 289 (1976).

Quantisation of solitons is discussed in (3) and in

(7) A. Neveu, *Reports on Progress in Physics*, **40**, 709 (1977).

Vortex lines were first shown to exist in gauge theories by

(8) H.B. Nielsen & P. Olesen, *Nuclear Physics*, **B61**, 45 (1973).

For an application to the Weinberg–Salam model see

(9) Y. Nambu, *Nuclear Physics*, **B130**, 505 (1977).

For a review, see

(10) A. Jaffe & C. Taubes, *Vortices and Monopoles*, Birkhäuser Verlag, 1980.

Dirac's paper on magnetic monopoles is

(11) P.A.M. Dirac, *Proceedings of the Royal Society*, **A133**, 60 (1931).

See also

(12) G. Wentzel, *Supplement of the Progress of Theoretical Physics*, **37 & 38**, 163 (1966).

The Wu–Yang formulation of Dirac's theory is

(13) T.T. Wu & C.N. Yang, *Physical Review D***12**, 3845 (1975).

For further developments of the fibre bundle formulation of the Dirac monopole, see

(14) A. Trautman, *International Journal of Theoretical Physics*, **16**, 561 (1977); M. Minami, *Progress of Theoretical Physics*, **62**, 1128 (1979); L.H. Ryder, *Journal of Physics A*, **13**, 437 (1980).

Reviews of the Dirac monopole are to be found in

(15) B. Felsager, *Geometry, Particles and Fields*, chapter 9, Odense University Press, 1981.

(16) S. Coleman 'Monopoles Revisited', in A. Zichichi (ed.), *The Unity of the Fundamental Interactions*, Plenum Press, 1983.

(17) P. Goddard & D.I. Olive, *Reports on Progress in Physics*, **41**, 1357 (1978).

(18) N.S. Craigie, P. Goddard & W. Nahm (eds.), *Monopoles in Quantum Field Theory*, World Scientific, Singapore, 1982.

The original papers on the 't Hooft and Polyakov monopoles are

(19) G. 't Hooft, *Nuclear Physics*, **B79**, 276 (1974).

(20) A.M. Polyakov, *JETP Letters*, **20**, 194 (1974); *Soviet Physics JETP*, **41**, 988 (1976).

The topological origin of these monopoles was pointed out by

(21) J. Arafune, P.G.O. Freund & C.J. Goebel, *Journal of Mathematical Physics*, **16**, 433 (1975).

(22) Yu. S. Tyupkin, V.A. Fateev & A.S. Shvarts, *JETP Letters*, **21**, 42 (1975); M.I. Monastyrskii & A.M. Perelomov, *ibid.*, **21**, 43 (1975).

For reviews, see refs. (3), (10), (16–18) above, and

(23) Y. Nambu, *Physica*, **96A**, 89 (1979).

(24) R. Rajaraman, *Solitons and Instantons*, North-Holland Publishing Co., 1982.

(25) K. Huang, *Quarks, Leptons and Gauge Fields*, World Scientific, Singapore, 1982.

Instantons were discovered by

(26) A.A. Belavin, A.M. Polyakov, A.S. Schwartz & Yu. S. Tyupkin, *Physics Letters*, **59B**, 85 (1975).

Important developments were made by

(27) G. 't Hooft, *Physical Review Letters*, **37**, 8 (1976) and *Physical Review D***12**, 3432 (1978); R. Jackiw & C. Rebbi, *Physical Review Letters*, **37**, 172 (1976); C.G. Callan, Jnr., R.F. Dashen & D.J. Gross, *Physics Letters*, **63B**, 334 (1976).

Mathematical aspects of instantons are explored in

(28) R. Jackiw & C. Rebbi, *Physics Letters*, **67B**, 189 (1977); M.F. Atiyah, N.J. Hitchin, V.G. Drinfeld & Yu. I. Manin, *Physics Letters*, **65A**, 185 (1978); V.G. Drinfeld & Yu. I Manin, *Communications in Mathematical Physics*, **63**, 177 (1978), M.F. Atiyah, N.J. Hitchin & I.M. Singer, *Proceedings of the Royal Society of London A*, **362**, 425 (1978).

General topological aspects of instantons and monopoles are outlined in

(29) J. Nowakowski & A. Trautman, *Journal of Mathematical Physics*, **19**, 1100 (1978); A. Trautman, *Czechoslovak Journal of Physics*, **B29**, 107 (1979).

Reviews of instantons are to be found in

(30) S. Coleman, in A. Zichichi (ed.), *The Whys of Subnuclear Physics*, Plenum Press, 1979.

(31) A.I. Vainshtein, V.I. Zakharov, V.A. Novikov & M.A. Shifman, *Soviet Physics Uspekhi*, **25**, 195 (1982).

(32) R.J. Crewther & B. Schroer, in P. Urban (ed.), *Facts and Prospects in Gauge Theories, (Acta Physica Austriaca Supplementum XIX)*, Springer-Verlag, 1978. See also ref (15), chapter 5, refs. (24) and (25).

A review of the mathematical aspects of instantons is to be found in

(33) V.G. Drinfel'd & Yu. I Manin, *Mathematical Physics Reviews* (Soviet Scientific Reviews, section C), **1**, 27 (1980).

A very readable introductory account of the differential geometry of gauge fields, including the topology of monopoles and instantons, is

(34) T. Eguchi, P.B. Gilkey & A.J. Hanson, *Physics Reports*, **66**, 213 (1980).

Index

438